MICROBIAL DIFFERENTIATION

Other Publications of the
*Society for General Microbiology**
THE JOURNAL OF GENERAL MICROBIOLOGY
THE JOURNAL OF GENERAL VIROLOGY

SYMPOSIA

1 THE NATURE OF THE BACTERIAL SURFACE
2 THE NATURE OF VIRUS MULTIPLICATION
3 ADAPTATION IN MICRO-ORGANISMS
4 AUTOTROPHIC MICRO-ORGANISMS
5 MECHANISMS OF MICROBIAL PATHOGENICITY
6 BACTERIAL ANATOMY
7 MICROBIAL ECOLOGY
8 THE STRATEGY OF CHEMOTHERAPY
9 VIRUS GROWTH AND VARIATION
10 MICROBIAL GENETICS
11 MICROBIAL REACTION TO ENVIRONMENT
12 MICROBIAL CLASSIFICATION
13 SYMBIOTIC ASSOCIATIONS
14 MICROBIAL BEHAVIOUR, 'IN VIVO' AND 'IN VITRO'
15 FUNCTION AND STRUCTURE IN MICRO-ORGANISMS
16 BIOCHEMICAL STUDIES OF ANTIMICROBIAL DRUGS
17 AIRBORNE MICROBES
18 THE MOLECULAR BIOLOGY OF VIRUSES
19 MICROBIAL GROWTH
20 ORGANIZATION AND CONTROL IN PROKARYOTIC AND
   EUKARYOTIC CELLS
21 MICROBES AND BIOLOGICAL PRODUCTIVITY
22 MICROBIAL PATHOGENICITY IN MAN AND ANIMALS

* Published by the Cambridge University Press, except for the first Symposium, which was published by Blackwell's Scientific Publications Limited.

MICROBIAL
DIFFERENTIATION

TWENTY-THIRD SYMPOSIUM OF THE
SOCIETY FOR GENERAL MICROBIOLOGY
HELD AT
IMPERIAL COLLEGE LONDON
APRIL 1973

CAMBRIDGE
Published for the Society for General Microbiology
AT THE UNIVERSITY PRESS
1973

Published by the Syndics of the Cambridge University Press
Bentley House, 200 Euston Road, London NW1 2DB
American Branch: 32 East 57th Street, New York, N.Y.10022

Library of Congress Catalogue Card Number: 72–95405

ISBN: 0 521 20104 7

Printed in Great Britain
at the University Printing House, Cambridge
(Brooke Crutchley, University Printer)

CONTRIBUTORS

ANDERSON, J. G., Department of Applied Microbiology, University of Strathclyde.

ASHWORTH, J. M., Department of Biochemistry, School of Biological Sciences, University of Leicester.

BAKER, J. R., MRC Biochemical Parasitology Unit, The Molteno Institute, University of Cambridge.

BARTNICKI-GARCIA, S., Department of Plant Pathology, University of California, Riverside, California 92502, U.S.A.

BONNER, J. T., Department of Biology, University of Princeton, New Jersey 08540, U.S.A.

BRADLEY, S., Department of Biochemistry, University of Liverpool.

CARR, N. G., Department of Biochemistry, University of Liverpool.

CHATER, K. F., John Innes Institute, Colney Lane, Norwich.

CROSS, G. A. M., MRC Biochemical Parasitology Unit, The Molteno Institute, University of Cambridge.

DONACHIE, W. D., MRC Molecular Genetics Unit, Department of Molecular Biology, University of Edinburgh.

DWORKIN, M., Department of Microbiology, University of Minnesota, Minneapolis, Minnesota 55455, U.S.A.

GARROD, D., Department of Biochemistry, School of Biological Sciences, University of Leicester.

GOODAY, G. W., Department of Biochemistry, University of Aberdeen.

HALVORSON, H. O., Rosenstiel Basic Medical Sciences Research Center, Brandeis University, Waltham, Massachusetts 02154, U.S.A.

HENRY, S. A., Rosenstiel Basic Medical Sciences Research Center, Brandeis University, Waltham, Massachusetts 02154, U.S.A.

HOPWOOD, D. A., John Innes Institute, Colney Lane, Norwich.

JONES, N. C., MRC Molecular Genetics Unit, Department of Molecular Biology, University of Edinburgh.

KEYNAN, A., Department of Microbiological Chemistry, The Hebrew University, Hadassa Medical School, Jerusalem, Israel.

MITCHISON, J. M., Department of Zoology, University of Edinburgh.

NEWTON, B. A., MRC Biochemical Parasitology Unit, The Molteno Institute, University of Cambridge.

SAUER, H. W., Zoologisches Institut der Universität Heidelberg, 69 Heidelberg 1, W. Germany.

SINGH KLAR, A. J., Rosenstiel Basic Medical Sciences Research Center, Brandeis University, Waltham, Massachusetts 02154, U.S.A.

SMITH, J. E., Department of Applied Microbiology, University of Strathclyde.

SZULMAJSTER, J., Laboratoire d'Enzymologie du C.N.R.S., 91-Gif-sur-Yvette, France.

TEATHER, R., MRC Molecular Genetics Unit, Department of Molecular Biology, University of Edinburgh.

TINGLE, M., Rosenstiel Basic Medical Sciences Research Center, Brandeis University, Waltham, Massachusetts 02154, U.S.A.

CONTENTS

EDITORS' PREFACE

We sometimes forget that although cell differentiation is a characteristic of higher organisms it is by no means their exclusive prerogative. In this Symposium we have set out to show, in fact, that microbial organisms can have complex and sophisticated patterns of differentiation and indeed, in some cases, construct multicellular or macroscopic structures which are fully comparable with the 'bodies' of higher organisms. For this reason much interest has recently arisen in the possible use of microbes as 'model systems' for the study of cell differentiation. However, as pointed out by J. T. Bonner in his Introduction, what is important is 'not what one organism tells you about another, but what it tells you about itself'. There can be few, in fact, who do not feel that 'their' organism is much more than a model – whatever their initial motives for choosing to work with that organism might have been. Thus, although many of our contributors (e.g. A. Keynan) have stressed the wider applicability of the work they have surveyed, the particular appeal of the system they study is still very apparent.

Nor must it be forgotten that cell differentiation is far from being restricted to the eukaryotes. Indeed the transformation of a vegetative bacterial cell into a spore discussed by J. Szulmajster and the converse process discussed by A. Keynan involve as extensive and dramatic a change as any seen in the eukaryote world. These two processes are connected with changes in the cellular environment which affect the capacity of the cell to grow and divide. In these, and many other similar examples, there must be a close connection between the controls which regulate the division cycle of the cell discussed by W. D. Donachie and the controls which initiate the differentiation event(s).

Prokaryotes are often thought of as unicellular organisms, and indeed so they often are, but M. Dworkin reminds us that this is by no means always the case and the sociable behaviour of the *Myxobacteriaceae* which he documents shows how complex apparently simple systems can be. However, the most complex example of differentiation amongst the prokaryotes must be either the blue-green algae discussed by N. G. Carr, where the filamentous species possess a variety of differentiated structures and a complex life cycle, or the *Streptomyces* described by K. F. Chater & D. A. Hopwood. In both these cases there is clearly going to be a rapid deployment of the formidable techniques of genetic analysis and as these organisms come to be studied more extensively they will no doubt cause many surprises.

The simplest eukaryotes show differentiation phenomena which are very similar to those shown by the prokaryotes, and H. O. Halvorson in his article on sporulation in yeasts points out how close and yet how different this organism is from a prokaryote.

J. M. Mitchison challenges the old-established proposition that growth and differentiation are antithetic properties of a cellular system and clearly shows that the vegetative cell cycle in yeast encompasses two of the most important characteristics of a differentiating system – morphogenesis and the periodic synthesis which are the manifestations of ordered gene expression.

The filamentous fungi present yet another aspect of differentiation in that in most cases growth is limited to the apical tip area of the hyphae. The recent advances in cytology, physiology and biochemistry of fungal hyphae are discussed by S. Bartnicki-Garcia and it is clear that these studies will permit a more precise formulation of the mechanisms responsible for apical growth and therefore hyphal morphology.

The involvement of hormones in eukaryotic differentiation has been widely considered. G. W. Gooday describes the historical and present-day status of hormones in fungal differentiation. The identification and chemical characterisation of the trisporic acids undoubtedly emphasises the advantages of a multidisciplinary approach to problems of differentiation.

Reading these articles we have been struck by the tremendous amount of work which is going on in this field at the moment and the surprising absence of any widespread industrial application of the knowledge which is being acquired. This one-sided situation cannot last long and undoubtedly the ability to control the form and function of micro-organisms and particularly the mould fungi, must surely lead to a wider involvement of such studies with industrial processes.

We thank our contributors for producing manuscripts which were such a pleasure for us to read and put together and the staff of the Cambridge University Press who made the putting together as painless as possible.

Department of Biochemistry   JOHN M. ASHWORTH
University of Leicester

Department of Applied Microbiology   JOHN E. SMITH
University of Strathclyde

DEVELOPMENT IN LOWER ORGANISMS

J. T. BONNER

Department of Biology, Princeton University
Princeton, New Jersey 08540, *U.S.A.*

For various reasons the development of a human being makes a rather poor 'model system' (to use the current phrase) to study the development of a microbe. The principal difficulty is that *Homo sapiens* hardly lends itself to easy experimentation: in general, the lowlier the organism the more we know about its molecular biology. The main point, of course, is not what one system tells you about another, but what it tells you about itself. And once we know something of a variety of different organisms up and down the evolutionary scale, it is possible to make some interesting statements about the 'comparative anatomy' of development. I mean anatomy at the molecular level and of course most particularly at the level of molecular control.

The fact that we know much more about such mechanisms in microbial systems has tended to give a considerable asymmetry to our 'comparative anatomy'. There has also been some confusion as to what relation certain phenomena in one group mean in terms of another. Let me begin this discussion by indicating a few basic concepts of development and what they mean at the different levels of organization. It will not be a discourse on the definition of words: not an old-fashioned grapple with semantics, but an attempt to describe the phenomena that occur at the different levels in such a way that genuine comparisons can be made. With this background we can then go on to look at the variety of different kinds of lower organisms, and not only compare them, but show how each has certain virtues for the study of particular aspects of development.

This symposium is called 'microbial differentiation'. We all agree that the formation of a fruiting body in a mould or the formation of a heterocyst in a blue-green alga constitutes differentiation. But if we confine ourselves to single cells the problems begin: is spore formation in a bacterium, where the whole cell turns from one state to another, the same thing as heterocyst formation in a multicellular filament? The answer is simply that one has two kinds of differentiation: temporal and spatial. In the former there is a change in time only (e.g. bacterial spore differentiation), while in the latter some cells undergo temporal differentiation and the others not, so that one can have one or more types of cells existing together in one multicellular organism (e.g. heterocyst formation in blue-green algae).

The far more difficult case (as J. M. Mitchison has pointed out to me) is the cell cycle. There are changes going on from one moment to the next, so in a sense this is a simple case of temporal differentiation. But since the cell always returns to its initial state it does not seem to be quite the same thing as, for instance, spore formation. I think it quite wrong to argue that the cell cycle is fundamentally different, but that it nevertheless provides a useful 'model system'. It is quite likely that certain control events take place in the cell cycle which also play a role in other differentiations, but the mechanisms of the cell cycle are especially interesting in their own right. However, there still remains the question of how cell cycle events relate to the more generally accepted examples of differentiation.

Before answering this question, there is one further difficulty in using the concept of differentiation in single cells. It is usually imagined that a single cell is differentiated into parts: for instance, in a eukaryotic cell not only does it have a nucleus with all its structure, but also ribosomes, basal bodies, mitochondria, Golgi apparatus, plastids, and so forth. The question can be put in this way: are cells ever undifferentiated? It is true that some patterns of organelles may come and go, but the majority of the cell parts, for example the chromosomes or the mitochondria, stem directly from identical cell parts by multiplication or replication.

The way around these difficulties is to remember that development is synonymous with reproduction in its most general sense. All organisms have life cycles and each cycle is a direct reproduction, a direct copy of the previous cycle. If the organism is a single cell then the life cycle and the cell cycle may be one and the same. If the organism is multicellular then the life cycle is made up of a series of cell cycles. In this sense differentiation becomes a condition (a molecular composition in time and space) which is simply a copy of that same condition in the previous life cycle. It can be a metaphase plate, a bacterial spore, a heterocyst-studded filament in a blue-green alga, or the central nervous system of a man. What we seek, then, is to understand how protoplasm is controlled so that these structures can recur with such cyclic precision.

From the point of view of experimental analysis we must look for three different kinds of control mechanisms: (1) the control of the production of substances, (2) the control of the time of appearance of certain substances, and (3) the control of the spacing of the substances, that is, their localization. Again our existing knowledge is quite lopsided. We know much more about the control of substance production than we do about the control of timing or substance localization. In fact, the whole exposing of the methods of transcriptional control, from repressors to sigma factors, has been one of the most successful and important advances of microbiology. The continued and increasing interest in translational controls and their importance in development is a matter of special concern at the moment, not only to the microbiologist, but also to those interested in higher organism development. Finally, and here is the greatest unexplored area of all, there are clearly a large number of secondary controls, such as feed-back inhibitions and hormone-mediated reactions. Often these link back to the genome at some point in their complex series of enzyme-controlled steps so that the secondary reactions are never totally divorced from the primary activity of the genome.

Unfortunately, timing mechanisms are surrounded by much greater mystery. In prokaryotes there is reason to believe that the life span of the *messenger* RNA may play an important role, but this is less likely to be the case in eukaryotes where, in some instances, it is believed that the messengers are stable and persist for long periods of time. Enzymes are catalysts and will therefore control the rate of reactions; so clearly this is a factor, especially when one has a series of enzyme-controlled reactions in a cascade. This is important in those cases where one reaction depends upon the product of a previous reaction; where there is a set sequence of events. No doubt many examples of such timing mechanisms will be given in the pages that follow, but one last point should be stressed. Much work has been done in recent years on circadian rhythms or biological clocks, which are known to exist in many single-celled organisms. There is, however, no evidence so far that these rhythms directly control developmental processes. Rather they seem to be a mechanism whereby the organism can time its stages of development with environmental changes such as night and day, or spring and fall (by responding to increasing or decreasing day lengths). But I feel sure we still have much to learn about the role of biological clocks.

The problems of the localization of substances in space and its control is one of the most pressing ones in modern developmental biology. Here we are not only concerned with the forces that distribute substances within cells, but also the distribution of substances in multicellular systems. Besides the localized production of a substance, one can also move substances by diffusion and even by cell locomotion (morphogenetic movement). There is, in fact, a considerable variety of ways in which localization can occur, and in many instances where we have some understanding of the control mechanisms, certain key substances (e.g. hormones, or what are more generally referred to as morphogens) play an essential role in the pattern formation. It is not possible, in this brief introduction, to illustrate all the different kinds of localization, and what evidence we have for mechanism. Let it be said that there is little reason to doubt that ultimately the control of localization may be ascribed to the DNA of the genome, although the chain of events between the initial protein synthesis and the ultimate spacing may be exceedingly tortuous and indirect.

A few examples of differentiation in microbes may be helpful in illustrating the framework I have set forth in the above discussion. To consider microbes or lower organisms as an entity is, of course, quite illogical. With certain exceptions (especially the lack of development of a nervous system) these lower forms have all the main features of higher animals and higher plants. Their great virtue is that because of their relative simplicity they make available for experimental analysis some aspect that may be quite hidden and unavailable in a complex higher organism. One of the most striking examples of this comes from *Acetabularia*, a green alga with an anatomy so singular that it has permitted Hämmerling and those who have followed him to discover all sorts of important developmental information concerning the role of the nucleus and the cytoplasm.

The control of the production of substances has been studied (as we have already said) with exceptional success in single-celled organisms. This is not only true of *E. coli* but also other prokaryotes, and eukaryotes, such as yeast and other fungi, slime moulds of all sorts, protozoa, and algae. The initial step is to describe the appearance of new substances in the life cycle, enzyme synthesis being a particularly desirable kind of measure. In this way one can make a catalogue of the substances produced and even a time sequence for their production. This raises two problems: what controls the production of each substance, and what does the substance do? The former problem is usually attacked by the use of protein synthesis inhibitors to find the moment of translation and RNA synthesis inhibitors to find the moment of transcription. Furthermore one wants to discover what factors, internal or external, will affect the period of translation and transcription. The substance may be an enzyme and one wants to know its substrate and its product, and what controls the activity of the enzyme; are there feed-back controls? The product of the reaction may now be important in a subsequent reaction, so in this way one must piece together a series of controlled events.

The timing of these sequences soon becomes a matter of central concern. What starts one event or stops another? All aspects of the cell cycle revolve around this problem: what sequences are rigid and is it possible to dissociate one sequence from another? What controls the sequence: is it chromosomal events, or substrate-product feed-back events, or both? The very same questions may be asked for spore formation in bacteria, or fruiting body differentiation in fungi, myxo-mycetes and cellular slime moulds.

Most single cells do not provide easy material for the all-important developmental problem of localization, although there are some notable exceptions such as the formation of cortical patterns in Protozoa, or the beginning of polarity of the eggs of the brown alga, *Fucus*. Some of the best examples come from multicellular lower organisms and I would like to mention four cases which have certain basic similarities.

One example is found among the soil fungi. The first indication of spatial arrangements may be seen as a mycelium spreads through the soil (or an agar plate) for it is obvious that the advancing tips of the hyphae repel one another so that there is an optimal spacing of the feeding tips. This is apparently effected by a chemical which is given off by the growing hyphae that inhibits the growth of neighbouring hyphae and thereby causes their spatial distribution. The beauty of this system is that one has action at a distance in two dimensions and therefore it is admirably suited for experimentation. But besides this hormone-controlled spacing mechanism, there are a number of other such hormone effects. During fruiting in certain ascomycetes or basidio-mycetes where there is a compound fruiting body (e.g. a mushroom) there can be specific sites where the hyphae are attracted to each other rather than repelled. The same phenomenon is seen in the hyphal anastomoses that produces heterokaryons. Finally when the mushroom itself forms there must be an internal system of gradients or patterns of substances which is responsible for the shaping of the fruiting body. In this case the morphogens need not act at a distance, but directly from cell to cell. It not only means that large molecules might be involved, but also more complex variations of simple diffusion mechanisms are possible. The gradients can be established by a mutual stimulation of morphogen production or suppression in cell–cell contacts of actively metabolizing cells. There is even evidence, in mushrooms, of a hormone produced in the primordial gill region which stimulates and controls the elongation of the stalk. So the pattern of the vegetative hyphae and the pattern of the fruiting body are controlled by morphogens which are themselves localized and in turn cause further localization of substances, in this instance by stimulating or inhibiting growth movements. The question of how the morphogen is asymmetrically distributed in the first place can also be answered: its pattern was either passed directly from the previous generation, or due to some instability phenomenon; as Turing showed, an even distribution of a morphogen can break up into a non-uniform pattern. Finally, the fruiting bodies themselves may be spaced in relation to one another. This is again action at a distance by some morphogen: one fruiting body may inhibit the formation of others in its immediate vicinity. This then is a spacing between multicellular organisms: it is a localization, a pattern on the population or social level. To say that micro-organisms have many of the developmental capabilities of higher animals and plants is no exaggeration.

A very similar example of localization may be found in the cellular slime moulds. The feeding amoebae are separate and repel one another, but once the food is gone they form central collection points and the cells stream towards these points. Not only is the non-random distribution of these centres caused by an inhibitory morphogen, but the orientation of the cells towards the centres is caused by the diffusion gradient of a chemotactic agent (acrasin). Once the cell mass is formed there is a redistribution or sorting out of the cells with the result that there is a strictly proportional relation between the number of anterior cells which will form stalk cells and the number of posterior cells which will form spores. It is suspected that this difference is achieved by the unequal distribution of key substances, but the facts are still wanting. Finally the fruiting bodies, as they rise into the air, give off a volatile substance that affects the orientation. If two fruiting bodies rise close to one another they lean away from each other as they rise; if a single fruiting body is found in a confined place it will, by means of this volatile morphogen, orient itself so as to be precisely in the middle of the cavity or cleft. Again this shows a communication between multicellular individuals at a distance and it accounts for the fact that the fruiting bodies always stand at right angles from the substratum regardless of the orientation with respect to gravity, for they are too small to be affected by gravity.

Some of the earlier workers at the turn of the century pointed out the remarkable similarities between the formation of fruiting bodies in the multinucleate or plasmodial myxomycetes, the uninucleate, amoeboid cellular slime moulds which we have just described, and the prokaryotic myxobacteria. In the latter one has aggregation of bacterial rods by chemotaxis and in some of the more elaborate forms there is a differentiation of multicellular cysts. To these three one can add the soil fungi with their spreading mycelial network. One assumes that in each of these cases there must have been strong selection pressure for these small fruiting bodies; they must have a form which ensures effective spore dispersal.

The fact that the cell building block can consist of a bacterial cell, a uninucleate amoeba, a syncytial plasmodium, or a rigid-walled hypha, and yet achieve basically similar kinds of evenly dispersed, small fruiting bodies, is indeed a remarkable fact. By having such an array of similar structures built in totally different ways, one has an ideal opportunity to study their comparative molecular anatomy. This can be done not only on their systems for the localization of substances during development, but on their systems of producing and timing the production of substances during development. If one now adds all the other diverse bacteria, fungi, algae and protozoa, the raw material available, among lower forms, for significant studies on development seems almost unlimited.

Note: Detailed examination of the points covered in this brief introduction, including references, will be found in a forthcoming book now in preparation.

THE BACTERIAL CELL CYCLE

W. D. DONACHIE, N. C. JONES AND R. TEATHER

MRC Molecular Genetics Unit, Department of Molecular Biology, University of Edinburgh

INTRODUCTION

The growth and development of all higher animals and plants involves the growth and division of their component cells. This growth and division may be accompanied by progressive cell differentiation as the initial cell line develops into the various tissues that comprise the whole organism. Nevertheless, since the growth and division of single cells is a common denominator in all such developmental sequences, it is appropriate to include studies of the cell cycle itself in any general consideration of differentiation.

In addition to the fact that the cell cycle is an obligatory component of most embryological development, it is also true that the growth of a single cell, from its creation at the division of its parent through the steps required to enable it to divide in its turn, is a process of differentiation. The newly formed cell differs from a cell about to divide not only in size but, as we shall see, in its composition, and this composition changes qualitatively in a fixed sequence throughout the cell cycle.

In general the process of cellular differentiation can be studied more easily in bacteria than in higher cells. *Escherichia coli* is probably the best understood organism in terms of its molecular biology and this makes the investigation of the cell cycle easier than in organisms where the nature of fundamental molecular processes, such as the regulation of gene activity, is still ill understood. We believe therefore that the investigation of the cell cycle in *E. coli* and other bacteria should be able to proceed much more rapidly than in other organisms. Although this process may prove to differ in some ways from that of higher cells (as it obviously must when the relative simplicity of the bacterial cell is considered) it will probably, like earlier work on the molecular biology of bacteria, prove illuminating and helpful to the investigation of eukaryotic systems.

THE DNA REPLICATION CYCLE

The nature of chromosome replication in Escherichia coli

The genome of *E. coli* consists of a single closed circle of DNA. This circle, about 1200 μm in circumference, has been the object of intensive study, both genetical and biochemical. Nevertheless, probably no more than 10 % has so far been identified with specific genetic functions (see Taylor, 1970) and the biochemical mechanism of its replication is still not understood (see Gross, 1972, for review). Even less is known about the spatial and temporal organisation of this enormous molecule within a cell which is itself only about 2 μm in length.

On one level, however, the control of the replication of this genome can be fully described. This is in terms of the way its replication is integrated into the bacterial cell cycle.

Replication begins on the bacterial chromosome at a fixed site (the 'origin') and proceeds in both directions from that point until the two replication forks reach a point (the 'terminus') which appears to be located on the circle approximately opposite to the origin (Masters & Broda, 1971; Bird, Louarn, Martuscelli & Caro, 1972; McKenna & Masters, in press). This process is shown in Fig. 1.

The time required to duplicate the genome is therefore limited by the rate at which nucleotides can be added to the growing daughter strands at the two replication forks. As Cairns first showed, this process is slow relative to the rate of growth of the bacterial cell itself (Cairns, 1963). In fact the complete sequential duplication of this molecule, from origin to terminus, takes approximately 40 min (at 37°) under a wide variety of growth conditions, although the length of time required to double the rest of the components of the cells can range from 60 min to as little as 20 min under these same conditions (Clark & Maaløe, 1967; Helmstetter, 1967, 1968, 1969; Cooper & Helmstetter, 1968).

Control of the timing of initiation

This slow rate of movement of the replication forks around the chromosome therefore results in the curious situation that the time taken to duplicate a chromosome may be twice as long as the cell cycle itself, in conditions where cells are growing rapidly (in a rich growth medium). Nevertheless, such fast-growing cells are able to divide every 20 min and each daughter cell receives a complete genome. This is achieved by a simple control system which appears to be identical in the two distantly related species in which it has been studied, *E. coli* (a gram-negative organism) and *Bacillus subtilis* (a gram-positive organism). This system

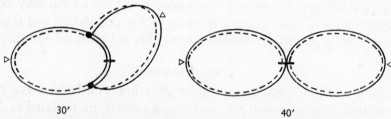

Fig. 1. Diagrammatic representation of the replication of the genome of *E. coli*. DNA replication begins at a fixed site (the origin) on the 'chromosome' (actually a topologically circular double helix of DNA, here drawn as two parallel lines representing the two single strands of the helix). The black dots represent the positions of the replication complexes. Newly synthesised strands of DNA are shown as dotted lines. Replication proceeds at a constant rate until the two replication forks meet at the opposite side from the origin. This point, the terminus, is shown as a bar. The whole process takes about 40 min (in most growth media at 37 °C).

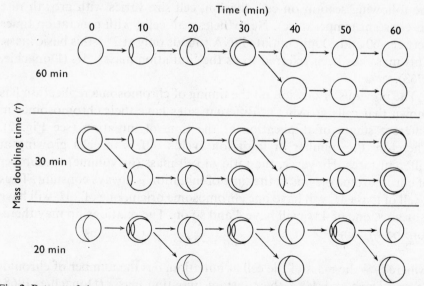

Fig. 2. Pattern of chromosome replication over a 60 min period in cells growing at different rates. The double helix is represented this time as a single line. In every case, each pair of replication forks takes 40 min to travel around the chromosome from the origin to the terminus. A new round of replication is initiated at intervals equal to the mass doubling time of the cells, whether or not the previous round has been completed.

was first described in *B. subtilis* (Yoshikawa, O'Sullivan & Sueoka, 1964; Oishi, Yoshikawa & Sueoka, 1964) and later shown to apply also to *E. coli* (Cooper & Helmstetter, 1968; Masters, 1970).

At different growth rates of the cell (with the exception of very slow growth rates) the rate of addition of nucleotides to the replication forks remains the same but new rounds of chromosome replication are initiated at every doubling in cell mass. Thus in fast-growing cells (mass doubling time 20 min) a new 'round' of replication is initiated at the origin every 20 min, despite the fact that previously initiated replication forks have not yet reached the terminus. In this way, the total amount of DNA in the population will be doubled at the same frequency as total cell mass itself. Each chromosome in such a population will therefore have several sets of replication forks. This is illustrated for *E. coli* in Fig. 2.

Initiation mass

Since a new round of chromosome replication is initiated at every mass doubling, at any growth rate, successive rounds are initiated at $2^n . M_i$ (where n is any integer and M_i is the mass of the cell at the first initiation event). This introduces the question of the value of M_i under different growth conditions. Surprisingly, it turns out that this value is constant (at least to a first approximation) over a wide variety of growth conditions. This value appears to be equal to the mass of a newly divided cell growing with a generation time (τ) of 60 min. As will be explained in the following section on cell division, cell size varies with growth rate (at constant temperature). Nevertheless, all cells, with generation times between 60 and 20 min initiate DNA rounds only at $2^n \times$ this basic mass. This mass has been referred to as the 'initiation mass', M_i (Donachie, 1968).

The previous discussion on the timing of chromosome replication has shown that cells growing at different rates have their chromosomes in different stages of duplication at the time of initiation (see Fig. 2). Therefore the number of initiation events differs in cells growing at different rates. However, the ratio of cell mass (or volume) to number of chromosome origins at the time of initiation is always constant. Thus a cell of mass M_i will have one chromosome origin, one of $2M_i$ will have 2 origins, one of $4M_i$ will have 4 and so on. The relationship may therefore be written:

$$m/o = M_i$$

(where m is the mass of the cell at initiation, o is the number of chromosome origins and M_i is the constant initiation mass) (Donachie, 1968; Donachie & Masters, 1969).

Exceptions

The simple rules for the control of DNA replication in *E. coli* do not appear to hold under conditions of very slow growth in poor media (Cooper & Helmstetter, 1968). At doubling times greater than about 60 min, the time from initiation to termination (C) lengthens (to $2/3\tau$) and initiation may take place at a cell mass less than M_i (Helmstetter, personal communication).

In certain DNA mutants also, initiation may take place at cell masses different from $2^n . M_i$ (Worcel, 1970).

These exceptions need occasion no surprise; it is the constancy of the relationship $m/o = M_i$ under so many growth conditions which is surprising. The explanation of this relationship is not known at this time but it would seem unlikely that cell mass *per se* is the signal for initiation. Two main kinds of mechanism have been suggested. One is that a replication complex is built up, perhaps at a specific location on the cell surface, at a rate proportional to overall growth rate and that this is 'mature' when a certain amount of cell growth has taken place. Initiation then takes place by the combination of the chromosome origin with the replication complex (Donachie, 1968; Helmstetter, Cooper, Pierucci & Revelas, 1968; Bleecken, 1971). Two new replication complexes would then start to be made at the same time as the first complex matured. An alternative model has been suggested by Pritchard, Barth & Collins (1969) who suggest that an inhibitor of initiation could be made over a short interval immediately following the beginning of chromosome replication and that further increase in cell volume would be required to dilute this inhibitor below a critical threshold. However, no-one yet knows how initiation is controlled.

CONTROL OF CELL DIVISION

The timing of cell division

The process of cell division must also be related to this process of chromosome replication, if each newly formed cell is to receive a copy of the genome. The mechanism by which this is achieved is still unknown but the timing of division can be accurately predicted under most growth conditions since cell division normally takes place about 20 min after the completion of each round of chromosome replication (Cooper & Helmstetter, 1968). It is clear therefore that cell division does not take place at any particular cell size and, in fact, one consequence of this system of regulation is that cell size

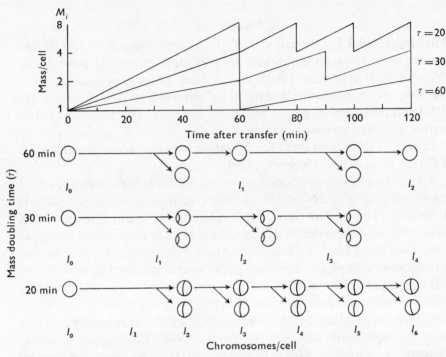

Fig. 3. The timing of cell division and of rounds of chromosome replication. In this hypothetical experiment three single cells, each of mass $1 \cdot M_i$, are inoculated into three different media, in which the mass doubling times are 60, 30 and 20 min respectively. In each case the first division takes place 60 min after inoculation into the new medium. In consequence, cell size at this first division is a function of the growth rate in the particular medium. (In fact, $\log (\text{mass/cell}) = k/\tau$.) Each successive division takes place 60 min after each successive doubling of the initial unit cell mass (M_i). After the first division, therefore, the interval between successive divisions is equal to the mass doubling time.

At 0 min, each cell has a single unreplicated chromosome and starts replicating it immediately. Successive rounds of chromosome replication are initiated at each successive doubling of the unit cell mass (M_i). $I_0, \ldots I_n$, represent the times of successive initiations in the three cell lines.

must increase with increase in growth rate. The way in which this comes about is illustrated in Fig. 3. In this figure, a cell which has been growing with a generation time of 60 min is transferred, immediately after division (i.e. at mass M_i), to one of three different growth media. The mass doubling times in these media are 60 min, 30 min and 20 min respectively. (These different growth rates may be attained by adding various extra nutrients to the basic salt glycerol growth medium.) In cells growing with a 60 min doubling time, initiation of chromosome replication takes place immediately after division. To construct Fig. 3 (which is diagrammatic and approximate), it is assumed that growth rates change immediately on transfer to the new medium, that initiation takes place immediately on transfer to the new medium, that initiation takes place at every mass doubling thereafter and that cell division takes place 20 min after every termination. It can be seen that, as a consequence, cell size at division (or average cell size) is four times greater when the doubling time is 20 min than when it is 60 min. Such a fast-growing cell also has two chromosomes, each half-replicated, immediately after division.

Coupling between DNA replication and cell division

Division normally follows the completion of each round of chromosome replication. It has also been shown in E. coli that completion of a round of replication is a necessary prerequisite for division (Clark, 1968; Helmstetter & Pierucci, 1968). Thus, if termination of a round of replication is prevented (by specific inhibition of DNA synthesis) then the cell division, which would normally follow 20 min after that termination, is also prevented. Each division event is connected in this way to the termination of a specific round of chromosome replication and not to DNA synthesis per se. This was shown by inhibiting DNA synthesis in fast-growing cells where DNA synthesis is continuous (see above). If a termination has taken place less than 20 min before the inhibition of DNA synthesis, then one division can take place even though subsequent rounds of replication have been blocked.

It is clear therefore that the processes leading to each specific cell division become uncoupled from the process of DNA replication at the completion of the duplication of a specific chromosome. Alternatively, the termination of each round of chromosome replication produces exactly enough of some 'division substance' for one division only.

Evidence has been obtained in this laboratory that the transcription of a specific gene (or genes) is required if cell division is to be completed and that this transcription normally takes place at, or immediately after, termination of each chromosome round (Jones & Donachie, in preparation). This was demonstrated in the following way.

An asynchronous population of cells of E. coli was first synchronised for chromosome replication by inhibiting protein synthesis for a period of time sufficient to allow completion of all rounds of replication. Further DNA synthesis was then prevented (by thymine deprivation or the addition of nalidixic acid) and protein synthesis allowed to resume. The mass of the population was allowed to double in the absence of DNA synthesis. (This period of protein synthesis is also required for subsequent cell division as will be discussed in the following section.) At this time all cells have reached or surpassed the required initiation mass (see previous discussion) and chromosome replication is therefore initiated in every cell when the block to DNA synthesis is removed. If protein synthesis is also allowed to continue, then cell division takes place, synchronously, almost immediately after the completion of this round of chromosome replication (Donachie, Hobbs & Masters, 1968; Donachie, 1969). However, if further protein synthesis is prevented during this period at any time up to the time of termination, then cell division is prevented. Experiments involving the addition and removal of various inhibitors of protein or RNA synthesis for various periods of time during this synchronous round of chromosome replication show clearly that cell division can take place only if there is a short period (5 min or less) of transcription and protein synthesis subsequent to termination of the round of chromosome replication.

It is interesting to note that, under these conditions, there is only a very short gap between termination of chromosome replication and cell division. It is clear that the normal delay of 20 min is not obligatory.

The nature of the 'termination protein' is not known at present (although preliminary labelling experiments suggest that it may be identified as one of the protein components of the cell envelope). One possibility is that it has something to do with a change in state of the cell membrane at termination which is required for cell division to be completed. For example, this might involve the dissociation of the chromosome from the old replication complex and the subsequent conversion of this complex into a septum (Donachie, 1969; Donachie & Begg, 1970). It is clear however that this event is required to allow the final assembly of preformed membrane proteins and mucopeptide precursors into the septum.

Another interesting question raised by the synthesis of the termination protein is that of the mechanism of its induction. If its synthesis is truly dependent on the replication of the chromosome terminus, then this may represent a novel control mechanism for gene transcription.

The division clock

As we have seen, cell division normally takes place about 60 min after each doubling in number of initiation mass equivalents (i.e. at $2^n . M_i + 60$ min). The interesting fact is that this time interval appears to be constant at different cellular growth rates (with the exception once again of very slowly growing cells). Since the first two-thirds of this interval runs concurrently with chromosome replication and since termination is a necessary event for cell division, it has been suggested that the chromosome replication cycle could provide the clock for division

(Clark, 1968). This seems unlikely for a number of reasons. For example, this model still leaves unexplained the constancy of the normal interval (20 min) between termination and cell division, which would therefore have to contain a second 'clock' process which was also independent of growth rate. Also, in the experiments on the role of the termination protein, outlined above, it was shown that division could follow almost immediately after termination if division had previously been prevented by inhibition of DNA synthesis. In this case it is evident that most of these processes which normally take place between termination and division had taken place *before* termination, i.e. the division clock had been running in the absence of chromosome replication. In addition to such evidence, certain mutants of *E. coli* exist in which cell division can continue after a block to DNA synthesis. In these mutants, since chromosome duplication has been prevented, normal sized but DNA-less cells are produced (Inouye, 1969; Hirota, Ryter & Jacob, 1968). Thus the timing of division is normal in these mutants although the chromosome replication cycle has been prevented. Even more clearly, wild-type *Bacillus subtilis* are able to continue to divide and give rise to normal sized but anucleate cells when DNA synthesis has been specifically blocked (Donachie, Martin & Begg, 1971).

There is therefore probably a clock for division which is separate from that for DNA replication. The nature of the clock is unknown but experiments on the necessity of prior protein synthesis for division (Pierucci & Helmstetter, 1969) strongly suggest that a constant period of protein synthesis of about 40 min is required before division can take place. That this is part of a time 'clock' is shown by the fact that this required period is constant at different growth rates.

This required period of protein synthesis is then followed by an interval of about 20 min during which further protein synthesis (except for the termination protein) is not required. This period therefore presumably represents some process involving the assembly or modification of preformed proteins into some structure needed for cell division. The final step in this process would then require the participation of the termination protein (which, in the normal cell cycle, would have been produced about 20 min before).

This scheme for the cell cycle in bacteria is outlined in Fig. 4.

The localisation of division

The division of bacteria is almost always in such a position as to give two daughter cells of equal volume. In the best studied species (*Streptococcus faecalis*, *Bacillus subtilis* and *Escherichia coli*), the cells are rods or

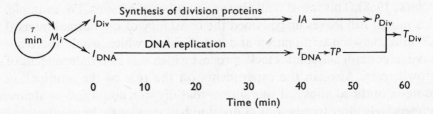

Fig. 4. Model of the cell cycle in *Escherichia coli*. Doubling of the initiation mass (M_i) takes place every mass doubling time (τ min). At each doubling two processes are initiated approximately simultaneously. These are the initiation of DNA replication (I_{DNA}) and the initiation of the sequence of events leading to division (I_{Div}). Termination of chromosome replication (T_{DNA}) at 40 min induces the synthesis of termination protein (TP). The first 40 min of the division sequence involves protein synthesis which is then followed by the initiation of assembly (IA). (N.B. Assembly could actually start earlier.) After 15–20 min more the cell has reached a stage (P_{Div}) where interaction between some septum 'primordium' and termination protein leads to cell division (T_{Div}).

Our picture of the timing of the main events in the cycle is therefore of a periodic event which occurs at intervals equal to the mass doubling time of the cell and at multiples (2^n) of a constant cell mass (M_i). This event triggers two parallel but separate sequences of events which take *constant* periods of time to complete, largely independent of the rate of cell growth. One of these processes is chromosome replication and the synthesis of the termination protein, and requires 40–45 min to complete. The other is a sequence of protein synthesis, followed by another process which may be assembly of some septum precursor. This sequence requires nearly 60 min and, at the end of it, there is an interaction between the septum precursor(s) and the termination protein to give the final septum and cell division. This last event takes only a few minutes.

spheres which divide successively in the same plane (relative to preceding division planes). Other bacteria, like many eukaryotes at early cleavage stages, divide successively in different planes, often at right angles to one another. Such species (e.g. *Micrococcus radiodurans*, *Gaffkya*, *Sarcina*, etc.) divide and grow to give rise to ordered arrays of cells in either two or three dimensions. So far, work has concentrated on the simplest case of cells in which the plane of successive divisions does not vary.

Cell division is a process which involves the invagination of the outer layers, or 'envelope', of the bacterial cell. In *E. coli* (and in other gram-negative organisms), this envelope is complex in structure but, to a first approximation, may be considered to consist of three main layers (for review, see Rogers, 1970). These are the inner membrane (lipoprotein), outside this is the rigid 'sacculus' (consisting of mucopeptide cross-linked polymers) and outside this again the 'outer membrane' (lipoprotein and lipopolysaccharide). This is a highly simplified picture of the cell envelope but one which is adequate for our present discussion. The envelopes of gram-positive cells (such as *Streptococcus* and *Bacillus*) are somewhat simpler and may be pictured as an inner cell membrane enclosed in a thick 'wall' composed mainly of mucopeptide and teichoic acids. In all cases, however, cell division involves the co-ordinate growth of these layers at a unique site in the cell envelope. The present section is concerned with the question of how the location of this site is determined.

Two main groups of hypotheses have been considered. The first is that the site is determined by its distance from the cell poles. When the poles are far enough apart, a site for cell division somehow forms between them. (Note that this does not imply that division will take place at the same time. The site is only a 'potential division site' until the division clock, discussed above, determines the actual moment of division.) Such a process would presumably imply the existence of some gradient of chemical activity or structural deformation extending out from each pole. A gradient model of this kind would be feasible, no matter how the envelope was synthesised (i.e. by intercalation of new material at many sites or by growth in one or a few zones).

The second type of hypothesis is that there are only a few growth sites in the cell envelope (in one or more of the layers) and that the potential division site is located either at such sites or at the consequent junctions between new and old cell envelope areas.

Various techniques have been employed to determine the mode of growth of the cell envelope, with differing results in different species. Cole (1965 for review) and others have used immunofluorescent labelling of specific surface antigens to follow the distribution of pre-existing envelope material during cell growth and division. Electron microscopy has been used, in conjunction with various techniques for marking the pre-existing cell surface, by others (see Ryter, 1968; Higgins & Shockman, 1971 for reviews). Radioactive labelling of various envelope components (e.g. lipids, proteins and mucopeptide) has been used extensively. Finally other techniques such as the assay of the distribution of various surface antigens (Autissier, Jaffe & Kepes, 1971) and the measurement of the location of autolytic sites during growth (Donachie & Begg, 1970) have been used in some cases.

This work has so far produced a clear picture of envelope growth only for *S. faecalis* (see Higgins & Shockman, 1971). In this nearly spherical cell, growth takes place by addition of new wall material to a central ring which represents the location of the next septum. Autolytic enzymes cleave this new material, which then peels apart to form new wall for the two growing halves of the cell. When cell volume has doubled, the addition of new material at the central ring seems to accelerate (or the rate of cleavage decreases) so that the ring grows inwards to form a septum. Autolytic cleavage then splits the septum in two to separate the daughter cells. At the same time, new wall growth begins in each

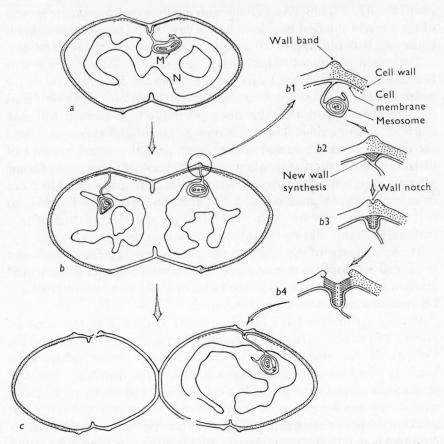

Fig. 5. Growth of the cell envelope during the cell cycle of *Streptococcus faecalis* (taken from Higgins & Shockman, 1971). The model proposes that linear wall elongation is a unitary process which results from wall synthetic activity at the leading edges of the nascent cross wall. The diplococcus in *a* is in the process of growing new wall at its cross wall and segregating its nuclear material to the two nascent daughter cocci. In rapidly growing exponential phase cultures before completion of the central cross wall, new sites of wall elongation are established at the equators of each of the daughter cells at the junction of old, polar wall (stippled) and new equatorial wall beneath a band of wall material that encircles the equator (*b*). Beneath each band, a mesosome is formed while the nucleoids separate and the mesosome at the central site is lost. The mesosome appears to be attached to the plasma membrane by a thin membranous stalk (*b1*). Invagination of the septal membrane appears to be accompanied by centripetal cross wall penetration (*b2*). A notch is then formed at the base of the nascent cross wall which creates two new wall bands (*b3*). Wall elongation at the base of the cross wall pushes newly made wall outward. At the base of the cross wall, the new wall peels apart into peripheral wall, pushing the wall bands apart (*b4*). When sufficient new wall is made so that the wall bands are pushed to a subequatorial position (e.g. from *c* to *a* to *b*) a new cross wall cycle is initiated. Meanwhile the initial cross wall centripetally penetrates into the cell, dividing it into two daughter cocci. At all times the body of the mesosome appears to be associated with the nucleoid. Doubling of the number of mesosomes seems to precede completion of the cross wall by a significant interval. Nucleoid shapes and the position of mesosomes are based on projections of reconstructions of serially sectioned cells. Reprinted from *Critical Reviews in Microbiology*, **1**, 1, 1971 by The Chemical Rubber Company. © 1971 The Chemical Rubber Company. With permission.

daughter cell at the junction between the wall formed in the cell cycle which has just finished and that which was formed in previous cycles. These new sites are again in the centre of each of the new cells (Fig. 5).

The work on the more extensively studied species, *B. subtilis* and *E. coli*, has unfortunately not yet produced such a clear picture. However in *B. subtilis*, Ryter (1971) has obtained evidence that preformed flagella (in a mutant temperature-sensitive for flagellar synthesis) are distributed amongst daughter cells in a way which strongly indicates one or two localised regions of envelope growth per cell. In *E. coli* evidence for the existence of a small number of localised regions of growth has come from studies on the distribution of permease molecules (located in the inner membrane) amongst progeny cells (Autissier *et al.* 1971). Although there are estimated to be several hundred permease molecules per cell, these were found to be all distributed into only 50 % of progeny cells after two or three generations (the exact number of generations depending on the growth of the cells and hence on cell size). The number of cells containing permease thereafter remained constant. Other evidence for a localised site of mucopeptide synthesis, located in the centre of cells, has recently been obtained in pulse-labelling experiments using [^{3}H]DAP (diaminopimelic acid) by autoradiography of the isolated sacculi (Schwarz, Ryter & Hirota, personal communication). Evidence for conserved segregation of large areas of the outer membrane during cell division has been obtained by Leal & Marcovich (1971), who followed the segregation of receptor sites for phage T_6. Donachie & Begg (1970) have measured the location of a site of autolytic activity in the envelope during growth of single cells and concluded, from the constancy of the distance between this site and one of the cell poles during cell elongation, that growth of the envelope must necessarily be taking place at a localised site and in a polarised manner (see below).

Such observations suggest that growth of some or all of the envelope layers takes place at only a few sites in these organisms, as well as in *Streptococcus*. However, other experiments have produced conflicting results. Thus, immunofluorescence labelling of the *E. coli* envelope (Beachey & Cole, 1966), radioactive labelling of the mucopeptide (van Tubergen & Setlow, 1961; Lin, Hirota & Jacob, 1971) and of the envelope lipids or proteins (Lin *et al.* 1971) has shown only a random dispersal of labelled material amongst progeny cells during successive cell divisions. Such observations have been interpreted in terms of diffuse intercalation of new envelope components at numerous sites all over the cell surface. However, none of these experiments excludes the possibility that there is extensive turnover of envelope components, if such components are used preferentially in further envelope synthesis. Indeed, the recent experiments of Ryter, Schwarz & Hirota have shown that localised incorporation of DAP is only demonstrable in short pulses. If this newly incorporated material is followed by a chase of un-labelled DAP, label becomes distributed over the whole cell surface by some process which is as yet not understood. This additional observation explains why the earlier experiments on DAP labelling showed no localisation of mucopeptide synthesis. This example clearly shows the difficulties involved in labelling a highly dynamic structure such as the cell envelope.

The problem of the exact mode of growth of the cell surface is therefore not yet solved but the balance of the evidence is, at the moment, in favour of the existence of a few localised sites of envelope growth in the envelope of these bacteria so far studied. This does not mean to say that there are not also other processes going on which involve the translocation of some envelope components from one location to another. However such a form of growth does imply the existence of junctions between new and older materials at precise locations in the cell surface. Such junctions could provide the spatial information required for the localisation of cell division. (See Higgins & Shockman, 1971 for more extensive discussion.)

Bacteria may contain more than a single site of potential division, depending on the size of the cell. This has been shown in *E. coli* in which division has been blocked specifically (by low concentrations of penicillin) while cell growth was allowed to continue. The number and location of the potential division sites was then determined, for cells of different lengths, by releasing the inhibition and measuring the dividing cells. In this way it was found that the number of potential division sites increases at each doubling of a basic cell volume. This basic volume was found to be equivalent to that of a cell of mass equal to the initiation mass (M_i) for DNA synthesis (Donachie & Begg, 1970 and unpublished). Therefore at the same cell sizes ($2^n \cdot M_i$) at which new rounds of chromosome replication and the division clock are initiated, new sites for cell division are formed.

Such sets of observations have been combined into a 'unit cell' model for bacterial growth (Donachie & Begg, 1970) in which it is assumed that a periodic event in the cell envelope (occurring at $2^n \cdot M_i$) results in a doubling of each pre-existing envelope growth site, giving rise to two new membrane replication complexes at which new rounds of chromosome replication can be initiated and also resulting in the conversion of the pre-existing growth site into a site of potential cell division (which

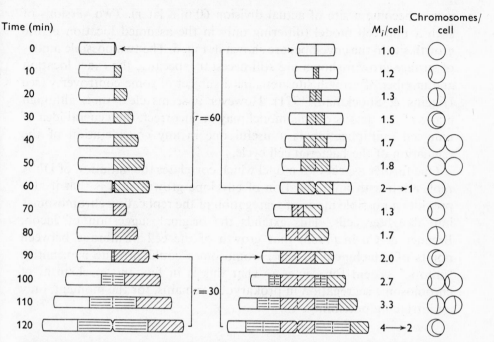

Fig. 6. Growth of the cell envelope in *Escherichia coli* according to two variants of the unit cell model.

At 0 min a unit cell (mass/cell = M_i) is inoculated into a medium where the mass doubling time (τ) is 60 min. After division at 60 min, one of the two daughter cells is inoculated into a richer medium where $\tau = 30$ min. The first division in this new medium takes place 60 min after this, by which time the new daughter cells are each 2 unit cell equivalents in mass. (For simplicity, cell mass is drawn as proportional to cell length. In fact cell diameter also increases on transfer to richer medium so that the relative proportions of cells with different growth rates are somewhat different from those shown. The relative cell masses are however as shown.)

The steady-state cell cycles in the two media are shown by large arrows.

Vertical solid lines across the cells show the edges of the growing zones in the cell envelope. The number of these zones doubles whenever the number of unit cell equivalents doubles.

The rate of increase in cell volume (length in this figure) is assumed to be proportional to the number of growth zones and also to τ. The rate per zone therefore depends on the growth medium but the rate of volume increase always doubles at each doubling in unit cell equivalents.

Cell division takes place 60 min after each doubling in number of unit cell equivalents, at a position which corresponds to the position of the edge of the growing zone 60 min earlier. This position (the future site of cell division) is shown throughout as an interrupted vertical line.

The location of each successive cycle of growth is shown by a different direction of cross-hatching.

The left-hand sequence represents cell growth according to a model in which a unit cell has a single growth zone located at one pole (Donachie & Begg, 1970). The other sequence represents the possibility that each unit cell has a central growth zone (Autissier, Jaffe & Kepes, 1972). Both models provide a possible way in which the location of the site of cell division could be determined by localised regions of growth of the cell envelope.

The number of chromosomes per cell, together with their state of replication, is also shown, on the assumption that initiation takes place at every doubling in number of unit cell equivalents and that each round of replication takes 40 min to complete.

would become a site of actual division 60 min later). Two versions of such a unit cell model (differing only in the assumed location of the growth site in the envelope) are shown in Fig. 6. The two possible modes of surface growth shown are still necessary, because the exact location and number of growth sites remains a subject of some controversy (see Higgins & Shockman, 1971). However it seems clear that, although many of the details of this model may be incorrect, the general idea of unit cell multiplication is a useful one in any consideration of the regulation of the bacterial cell cycle.

One further aspect of a model which correlates the initiation of DNA replication with multiplication of envelope growth sites is that it also provides a possible mode of segregation of the replicating chromosomes into daughter cells. This extends the original suggestion of Jacob, Brenner & Cuzin (1963) that growth of the cell membrane between points of attachment of the chromosomes could provide a primitive 'mitotic' system for bacteria. That this is in fact the mechanism of chromosome segregation in prokaryotes remains, for the moment, only an intriguing possibility.

ENZYME SYNTHESIS DURING THE CELL CYCLE

The bacterial cell cycle is often described only in terms of DNA replication and cell division, but it is now clear that the overall composition of the cell also changes in respect of many enzymes not obviously directly involved in either of these processes.

One possible source of this variation is an obvious consequence of the sequential replication of the genome, in that the potential to synthesise any enzyme ought to double when its structural gene is replicated. This has been shown to be the case by experiments in which the induced or derepressed rates of synthesis of several enzymes have been measured at various points in the DNA replication cycle. The times at which these rates double have been shown to correspond to the times of replication of the corresponding structural genes (Masters & Pardee, 1965; Donachie & Masters, 1966; Pato & Glaser, 1968; Helmstetter, 1968). If the numbers of genes were the only factors determining the rate of enzyme synthesis then one would expect that every enzyme would be synthesised continuously throughout the cell cycle. In this simplest case the rate of synthesis should be linear with a doubling in rate at the time of gene duplication. Although this pattern has been seen in several cases, the synthesis of many other enzymes has been shown to follow other patterns.

Two basically different systems have been investigated: synchronous vegetative cultures of *E. coli*, *B. subtilis* or *Rhodopseudomonas spheroides*, and synchronous spore outgrowth in the genus *Bacillus*. Three methods are commonly used to prepare synchronous vegetative cultures: the dilution of stationary phase cultures into fresh medium (Yanagita & Kaneko, 1961; Masters, Kuempel & Pardee, 1964; Cutler & Evans, 1966), the membrane elution technique of Helmstetter & Cummings (1963, 1964), and size selection using sucrose gradient centrifugation (Mitchison & Vincent, 1965). The latter two methods have the advantage of minimising metabolic disturbances which can seriously affect the synthesis of many enzymes, while the starvation and dilution method can produce an almost perfect synchrony that will persist for many generations. A list of the enzymes studied, together with their mode of synthesis, can be found in a recent book by Mitchison (1971, p. 176).

As has been mentioned above, the synthesis of many enzymes in synchronous cultures does not follow the simple pattern of constant synthesis with a doubling in rate at the time of gene duplication. These enzymes are synthesised in a brief burst during the cell cycle, so that the amount of enzyme in the cell is constant for part of the cycle, doubles abruptly, and then is constant for the rest of the cycle. In a synchronous population of cells the total amount of such an enzyme rises in a series of steps, with one step in each cell cycle. In all cases which have been investigated it has been found that such periodic enzyme synthesis takes place under conditions where a regulatory feed-back system is operating. Such conditions, which have been called 'autogenous' because the rate of enzyme synthesis is then determined in part by the rate at which the enzyme produces its own end-product repressor (Masters & Pardee, 1965), are the usual ones under which most biosynthetic enzymes are produced in growing cells.

The fact that such periodic synthesis is not seen under conditions of constant repression or derepression has given rise to the idea that this periodic synthesis is the consequence of oscillations in the normal feed-back regulatory system (Masters, Kuempel & Pardee, 1964; Kuempel, Masters & Pardee, 1965; Masters & Donachie, 1966). Indeed, computer studies have shown that such oscillations are an intrinsic property of simple feed-back systems, if suitable constants are assumed (Pardee, 1966; Goodwin, 1966). Such a model for periodic enzyme synthesis has been supported by the observations that addition of inducer can stimulate synthesis of such enzymes and addition of end product can abolish synthesis at all times in the cell cycle (see Donachie & Masters, 1969 for review). Also, the timing of the periods of synthesis in the cell cycle can be displaced for specific enzymes by a single pulse of repressor (Masters & Donachie, 1966).

The observed periodicity of autogenous enzyme synthesis is the same as that of the cell cycle, but this is not a necessary attribute of oscillating feed-back systems. It is therefore necessary to explain this periodicity by some other mechanism. Goodwin (1966) has proposed that the regular periodic duplication of individual genes could provide an entrainment mechanism which would make the periodicity of enzyme synthesis equal to that of gene duplication.

This model would also predict that, since the replication of the bacterial genome is sequential, the order of synthesis of various bio-synthetic enzymes under autogenous conditions could be the same as the order of replication of the corresponding structural genes. Just such a correspondence was reported by Masters & Pardee (1965) for three enzymes and their genes in *B. subtilis*. A similar correlation has recently been reported for five enzymes in germinating spores of *B. subtilis* (Kennet & Sueoka, 1971). However it seems clear that, in this latter example at least, the order of enzyme synthesis cannot be determined by entrainment mechanisms of the kind envisaged by Goodwin, since gene duplication in germinating spores does not begin until *after* the first sequence of enzyme synthesis has been completed. The situation in germinating spores is different from that in vegetative cells in that the synthesis of at least some enzymes cannot be induced at all times but only at times corresponding to their periods of 'spontaneous' synthesis (Steinberg & Halvorson, 1968a, b). The possibility therefore exists that, in this situation, the genome becomes available for transcription in an ordered fashion so that individual genes are transcribed in sequence. There are, however, other less dramatic explanations possible for such variations in inducibility of enzymes, including fluctuations in catabolite levels during spore germination. Such classical explanations must be excluded before it is necessary to hypothesise that some novel mechanism of gene transcription is taking place during spore outgrowth.

Outgrowth is well characterised in terms of morphological changes, RNA, DNA and protein synthesis (see Hansen, Spiegelman & Halvorson, 1970; Keynan, this Symposium p. 97). The dormant spore contains little or no functional mRNA though it does contain RNA polymerase (Sakakibara, Saito & Ikeda, 1965). Overall protein synthesis is continuous during outgrowth but only a very few kinds of protein synthesised in the early stages (Kobayashi *et al.* 1965) and the proteins being synthesised change with time (Torriani, Garride & Silberstein, 1969). It has also been shown that the species of mRNA being synthesised change with time (Hansen, Spiegelman & Halvorson, 1970). Steinberg & Halvorson (1968*a*, *b*) studied the timing of synthesis of certain enzymes during outgrowth. They found not only that synthesis of these is normally restricted to a certain period during outgrowth but, that in the case of the two inducible enzymes studied, they were inducible only during this restricted period.

It is known that the RNA polymerase of *B. subtilis* is altered in the β subunit during the process of sporulation and differs in template specificity from the polymerase of the vegetative cell (Losick & Sonenshein, 1969; Sonenshein, Losick & Shorenstein, 1970). Presumably the RNA polymerase found in the spore, during early outgrowth at least, is of this kind but it is easily imaginable that a third form of RNA polymerase might be responsible for the ordered transcription found during outgrowth (Kennet & Sueoka, 1971). However such a mechanism cannot explain the ordered enzyme synthesis found during the vegetative cycle in *B. subtilis* (Masters & Pardee, 1965), since, as we have said, under these conditions it has been shown that any enzyme can be induced or repressed at any time during the cell cycle. Therefore if there is a sequential reading mechanism responsible for the order of enzyme synthesis in the vegetative cell cycle it must coexist with normal feedback systems controlling transcription of these genes. It is difficult to envisage the operation of such a dual system and even more difficult to envisage the necessity of such systems for the temporal control of the cell cycle. For both of these reasons it is best to admit that the regulation of enzyme synthesis in the cell cycle is not fully understood.

It is clear that while a great deal of work remains to be done in describing variations in enzyme levels during the cell cycle, those enzymes which have been studied up till now have little or no direct role in either chromosome replication or cell division. Unfortunately there are as yet no suitable methods of assaying enzymes directly involved in DNA replication or cell division.

At least some autolytic enzymes have been shown to become active at a particular time in the cycle and to act only at the position of the presumptive site of septum synthesis (Schwarz, Asmus & Frank, 1969; Donachie & Begg, 1970; Hoffmann, Messer & Schwarz, in preparation). However the role of these enzymes in cell division is not definitely established.

There is at present therefore no information about the spatial or temporal control of the enzymes directly concerned with the major events of the cell cycle.

CAULOBACTER

The bacterium *Caulobacter* provides a very convenient system by which the control of cellular differentiation can be investigated, though its potential as a source of information is only beginning to be realised. The cell cycle is unique in that the vegetative cycle consists of two cell types, a motile 'swarmer' cell and a non-motile 'stalked' cell. Synchronous populations are readily obtained by selecting newly formed motile cells (Newton, 1972) and the system is susceptible to analysis by the established techniques of bacterial genetics.

The life cycle of *Caulobacter crescentus* has been described elsewhere (Poindexter, 1964; Shapiro, Agabian-Keshishian & Bendis, 1971) and will be only briefly summarised here (see Fig. 7). Only the stalked cell can divide, giving rise to two very different daughter cells; another stalked cell and a motile 'swarmer' cell. The swarmer cell has a single polar flagellum and a holdfast (a 'sticky' region) near the base of the flagellum. After a brief period of motility the swarmer cell loses its flagellum and forms a stalk at the site of the holdfast. The stalk is a continuation of the lipopolysaccharide and mucopeptide layers of the cell wall and is associated with a complex intracytoplasmic membrane structure at its base (Schmidt & Stanier, 1966). This stalked cell can now begin those processes of the cell cycle which are common to the two sister cells, including DNA replication, the synthesis of many pili, and the synthesis of the flagellum, holdfast and basal membrane structure at the 'undifferentiated' pole. The sister stalked cell begins these processes immediately; thus the time of division of the two sister cells differs by the time required for the morphogenesis of the swarmer cell to the stalked form. Hence, unlike the bacteria most usually studied, the vegetative cycle in *Caulobacter* shows a number of morphological differentiation events occurring in a fixed temporal sequence.

Newton (1972) has recently investigated the role of transcription in the control of this cycle using the antibiotic rifampicin, which specifically inhibits the initiation of transcription (Wehrli, Nüesch, Knüsel & Staehelin, 1968). His results are summarised in Fig. 7. Each step in development could be inhibited by the addition of rifampicin up to a critical point in the cell cycle. Addition after this time had no effect. For example, division and stalk formation could be inhibited up to 10 min before the event, loss of motility up to 40 min before the event, and DNA synthesis up to the normal time of initiation [similar observations on a requirement for RNA synthesis for the initiation of DNA synthesis have recently been made for the bacteriophage M_{13} by Brutlag, Shekman

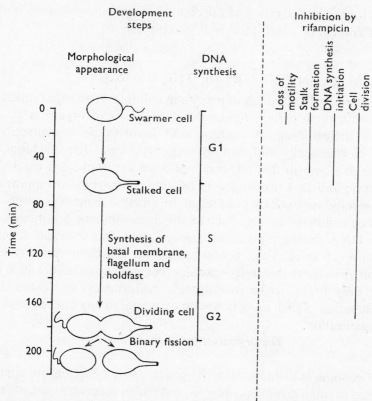

Fig. 7. The cell cycle of *Caulobacter crescentus* (taken from Newton, 1972).

& Kornberg (1971) and for *E. coli* by Lark (1972)]. It is interesting to note that while DNA synthesis begins immediately in the stalked cell the swarmer shows a considerable lag. This differential control of DNA synthesis in the two progeny cells would appear to rule out a model for the control of initiation in *Caulobacter* which involved the dilution of a cytoplasmic repressor during cell growth (as proposed for *E. coli* by Pritchard, Barth & Collins, 1969).

These studies show that development in *Caulobacter* is controlled, at least in part, by differential gene expression at the level of transcription. However the wide variation in timing between the transcriptional event and the execution of the step suggests that other levels of control must also operate.

The potentially most important feature of *Caulobacter* in studies of the bacterial cell cycle is that, in contrast to other bacteria, the cycle includes a number of readily visible morphogenetic events with a clearly defined spatial location within the cell. Thus *Caulobacter* provides a good system for the investigation of the control of spatially oriented events, a type of control of which little is at present known.

MYSTERIES OF THE ORGANISM

The bacterial cell, although in many ways simpler than eukaryotic cells, shares with them certain fundamental problems of organisation. For example the cell cycle, as we have seen, involves the co-ordination of many biochemically different growth processes. The bacterial cell behaves as a unit in that all its components are duplicated together in each cycle and that it changes its relative composition in appropriate ways in response to changes in its environment. Particular components of this co-ordinated system, such as the three dimensional organisation of the DNA during replication, transcription and segregation or the interactions between these processes and the cell membrane (as in cell division), present in themselves problems of great complexity which have as yet only begun to be investigated. Nevertheless we feel that the understanding of such processes is fundamental to any understanding of cell organisation.

The organisation of the genome

Gene order

The genome is a linear array of genes, some of which are spatially arranged in small groups as operons, but which otherwise presents every indication of being randomly ordered. Thus genes affecting similar processes may be scattered around the whole map. Nevertheless, as we have already discussed, this spatial order may be reflected in the temporal order of synthesis of several enzymes during the cell cycle. Such an ordered sequence of change in levels of different enzymes may well play a role in the ordering of other necessary events in the cell cycle. The possibility therefore exists that the developmental sequence of the cell cycle may be determined in part by the linear sequences of genes on the chromosome. It should be noted that a fixed sequence of synthesis of different enzymes is also a characteristic property of the eukaryotic cycle (see Mitchison, 1971), as well as of developmental sequences in general.

Another way in which the location of certain genes on the chromosome may be important in the regulation of cell function is as a consequence of the change in relative proportions of different genes at different growth rates. At fast growth rates there are more replication forks per chromosome than at low growth rates (see above and Fig. 2) and, consequently, relatively more copies of genes close to the origin than of those close to the terminus. This change in relative proportions of most genes will presumably be compensated to some extent by the operation of the normal feed-back systems of end-product inhibition and repression. However, if there are genes in which no such regulatory system operated, this change in relative proportions will presumably result in a change in the relative proportions of their gene products in the cell (Sueoka *et al.* 1970; Van Dijk-Salkinoja & Planta, 1971; Hughes, 1971). Thus the genes coding for ribosomal proteins and ribosomal RNA are close to the origin in both *B. subtilis* (Oishi & Sueoka, 1965; Smith *et al.* 1969) and *E. coli* (see Taylor, 1970). If, as seems likely, these genes are 'constitutive' under all cell growth conditions, then the relative proportion of ribosomes should increase with increasing growth rate. Just such an increase is observed (Maaløe & Kjeldgaard, 1966). Interestingly, not only is an increased concentration of ribosomes within the cell a necessary consequence of increased growth rate (because of this change in gene proportions) but it is in turn also itself a prerequisite for such an increase in growth rate. This is because the rate of polypeptide elongation per ribosome is approximately constant at different growth rates and therefore any increase in rate of protein synthesis requires an increased number of ribosomes per cell (Lacroute & Stent, 1968). This example shows clearly the mutual interdependence of different processes in the cell. To decide whether this proposed mechanism for the regulation of ribosome synthesis is correct will now require the experimental translocation of ribosomal genes relative to the chromosome origin.

Replication

The process of DNA replication has always represented a conceptual problem because of the obvious mechanical difficulties involved in replicating a double helix of such length (1100–1400 μm). The problem has been discussed in some detail by Watson & Crick (1953), Delbrück & Stent (1957) and Cairns & Davern (1968). Two models that were proposed many years ago, breakage-and-reunion (Delbrück & Stent, 1957) and various forms of active unwinding of the helix (Levinthal & Crane, 1956; Cairns, 1963; Maaløe & Kjeldgaard, 1966) have still not been tested.

Sequential replication of DNA requires that daughter strands of opposite polarity be synthesised concurrently; i.e. in the 5′-to-3′ and in the 3′-to-5′ direction. The DNA polymerases characterised to date, from all sources, will proceed only in the 5′-to-3′ direction in *in vitro* systems. There have been many models suggested to explain this discrepancy, though the possibility remains very strong that we have not yet isolated and characterised correctly the enzyme(s) actually responsible for normal DNA replication. Sugino & Okazaki (1972) have succeeded in showing that T_4 phage DNA synthesis proceeds only in the 5'-to-3' direction.

Most models for DNA synthesis assume that synthesis can in fact only proceed in the 5'-to-3' direction and suggest some form of discontinuous synthesis of one or both daughter strands (for a recent review see Gross, 1972). Those models all suggest that one implication of a single direction of synthesis is an asymmetry in the mode of replication of the two daughter strands, and there are a few experimental observations that suggest that a basic asymmetry exists in some phage systems (Inman & Schnös, 1971). The evidence in bacterial systems is not so clear but a consensus appears to be emerging that one of the two daughter strands is synthesised continuously, the other discontinuously (see Gross, 1972).

Physical organisation of the DNA in the cell

The bacterial genome consisting of a single DNA molecule nearly 1000 times the length of the cell itself is condensed into a 'nuclear' region that is of the order of 0.5 μm in diameter. Maaløe & Kjeldgaard (1966) have pointed out that in such a structure the average distance between adjacent daughter strands of DNA must be no more than 30 Å. While most electron micrographs show an apparently disordered arrangement some show an approximately ordered arrangement. If it is difficult to imagine the replication of an extended DNA molecule in solution it is much more so in a dense and presumably ordered structure. The rate of movement of a replication fork is about 14 μm/min (40000 base pairs/min), and throughout this process the nuclear material retains its condensed structure. It is obvious that this must involve continual breaking and remaking of bonds of some sort between all parts of the structure. In speculating about the nature of the condensed state it is interesting to note that when DNA replication is specifically inhibited the dense packing is quickly lost and loosely packed DNA fibrils are formed throughout the cell (Donachie, Martin & Begg, 1971; unpublished observations). Fong (1967) suggested that the compact structure could be the result of supercoiling induced by the rewinding of the duplex during replication, which would be consistent with the observation that such condensed structures disappear when replication stops. However, Stonington & Pettijohn (1971) succeeded in isolating the genome of *E. coli* as a folded DNA–RNA–protein complex. They found that the structure was dependent on RNA for the maintenance of its integrity, in that ribonuclease treatment caused rapid unfolding. This suggests that supercoiling is not the sole reason for the condensed structure of bacterial DNA in the cell.

The extent of our ignorance about DNA replication, even in bacteria, is emphasised when it is realised that not only is the DNA in constant motion within the nucleoid relative to the replication fork, but that simultaneously the two growing daughter strands are being segregated (again by an unknown mechanism) in a process that eventually results in the formation of two separate nuclei.

Maaløe & Kjeldgaard (1966) have also pointed out that if the spacing between DNA helices is regular then it is not possible for RNA polymerase or ribosomes to enter the nuclear region. This would suggest that only genes which were at the surface of the nuclear region could be transcribed. Nevertheless many experiments have shown that any gene can be transcribed at all times in the cell cycle if it is specifically induced or derepressed. Therefore every gene must be available for transcription at least in every few minutes throughout the cell cycle. Whether this reflects yet another dynamic process of the condensed DNA or whether it indicates that, for example, all operators are exposed at the surface, is unknown. In this respect it is interesting to note that the diameter of the bacterial nucleus is of the order of the length of the average gene. Therefore it is possible for every gene to have at least one end at the surface of the nuclear region. The process of transcription could then involve the attachment of the polymerase leading to the local unfolding of the genome.

This brings one to the problem of recognition of specific regions of the DNA by proteins involved in DNA replication, transcription, modification, and restriction. In principle such recognition could be for specific linear sequences of bases, but another attractive possibility is that recognition sequences may cause specific alterations in the secondary structure of the DNA. Brom (1971) has shown that AT-rich regions may have an altered helical structure and it is known that large AT-rich regions do exist in *E. coli* and *B. subtilis* (Yamagishi & Takahashi, 1971). The binding sites for many proteins such as micrococcal nuclease, the *lac* repressor, and RNA polymerase are AT-rich and this altered secondary structure may play a role in recognition. It is also possible that localised base pairing within a single strand might play a role.

To truly appreciate the possible roles of the secondary or tertiary structure of DNA (or RNA) in the cell cycle we must turn to recent work on the small RNA phages such as Qβ and f2. In these phages the RNA serves as a template for both transcription and translation. It also contains recognition and protein binding regions and extensive regions of secondary structure that contribute to the regulation of transcription. Only three proteins are coded for by the phage genome (reviewed by Kozak & Nathans, 1972). The present understanding of the role played by secondary structure in these phages depends on the extensive nucleotide sequence data that are now available. The translational control depends on the maintenance of a secondary structure that prevents ribosomes from reaching the appropriate binding sites, and at least two of the three proteins specified by the phage genome interact with the genome to alter this secondary structure as the developmental cycle progresses. The process of replication of the RNA also plays a role. Although this extreme genetic economy in regulation is probably not necessary to the bacterial cell it seems possible that regulatory mechanisms involving the secondary or tertiary structure of DNA and RNA may play a role in the regulation of transcription and translation in prokaryotes. Kennell (1968) has shown that a high proportion of the *E. coli* chromosome is transcribed only very rarely or not at all. Genes transcribed at very low frequencies could be dependent on disruption of secondary structure, such as that occurring during replication, for their transcription.

Transcription and translation also present mechanical problems related to those presented by DNA replication. The electron micrographs by Miller and his colleagues (1970) show that RNA polymerase molecules follow one another closely on active operons and that ribosomes attach to and begin to translate the growing mRNA chains immediately. It is very unlikely that this large complex of RNA polymerase, nascent mRNA, ribosomes, and nascent protein molecules could be rotating around the DNA molecule during transcription, so we are once again faced with the problem of either spinning the entire DNA molecule during transcription (in both directions depending on which strand is transcribed for a given gene) or introducing single stranded breaks in the DNA molecule.

The organisation of the cell envelope

One of the earliest ideas about the organisation of cells (before the discovery of genes) was that it was in large part determined by the cell wall (Schleiden & Schwann, 1838). This idea fell out of favour with the discovery of the importance of chromosomal genes coding for various elements of the cell but there remains evidence that physical structure serves to some extent as a template for further growth and thus may determine the pattern of that growth (Tartar, 1961; Sonneborn, 1963).

The shape of the bacterial cell, whether a rod or a sphere, is determined by the rigid mucopeptide layer of the cell envelope. This is clear because cells in which the mucopeptide has been removed take up a spherical shape in liquid while the isolated mucopeptide retains the shape of the intact cell (Schwarz, Asmus & Frank, 1969). However it is clear that the mucopeptide itself does not determine the shape of the progeny cells since partial or complete removal of this layer does not change the morphology of progeny cells derived from the altered cell (Landman, Ryter & Fréhel, 1968). Also, point mutations that cause alterations in cell morphology are known (Adler, quoted in Taylor, 1970; Alstyne & Simon, 1971), suggesting some form of direct genetic control of cell shape. There is as yet no evidence that physical structures in bacteria can serve as a template for subsequent generations. However the structure of the cell surface must be extremely important not only in determining cell shape but also in determining the site of cell division and presumably the site of DNA replication and segregation. The attachment of DNA to the cell membrane in bacteria is now well documented but the original suggestion of Jacob, Brenner & Cuzin (1963) that the DNA is attached to a precise point on the surface whose location is important for the proper segregation of DNA daughter strands remains at this time no more than an intriguing possibility. Present methods of electron microscopy do not seem to be adequate to show any precise attachment point.

The one event which is clearly localised on the cell envelope and thus represents a local discontinuity in the structure is septum formation. The way in which the localisation of this site might be determined has already been discussed but it is clear that we do not know whether the site arises *in situ* in a previously homogeneous envelope or whether the discontinuity that will give rise to the site of cell division is present throughout the cell cycle (as was suggested, for example, by the work of Donachie & Begg, 1970, and Schwarz *et al.* personal communication). In the latter case the possibility exists that these discontinuities arise one from another, in a process akin to the duplication of an organelle (for example, the centriole) in eukaryotic cells, during cell growth. The way in which this localised site appears should be an object for fruitful further study, representing as it does a simple example of morphogenesis in the cell cycle.

The events that occur at the site at the time of cell division are complex and apparently involve the synthesis of structures which are basically identical to the rest of the cell wall but oriented differently with respect to the cell axis. It is not known whether they also involve the synthesis of novel elements, such as contractible fibrils, as they do in eukaryotes. There is evidence for some structural differences in the cell wall and septum. Fan *et al.* (1972*a, b*) have shown that the cell ends are more resistant to autolytic degradation than the sides, but it is not at present possible to analyse local variations in structure of so complex an aggregate of macromolecules in any precise way.

Jacob *et al.* (1963), as mentioned above, suggested that the growth of the cell surface in localised zones could serve as a segregation mechanism for sister DNA molecules attached at opposite ends of the growing zone. We have seen that the required kind of zonal growth does take place in bacteria but that the localised attachment of DNA has not been demonstrated. If such attachment occurs it seems unlikely that it can be a permanent one since Ryter & Jacob (1966) have shown that particular labelled DNA strands segregate at random at each division. (However Eberle & Lark (1966) have presented evidence that is consistent with a more permanent attachment.) The recent proof that replication in *E. coli* is bidirectional (Masters & Broda, 1971; Bird, Louarn, Martuscelli & Caro, 1972; Hohlfeld & Vielmetter, personal communication) makes the model of Jacob *et al.* more attractive. The earlier assumption that the origin and terminus of replication were adjacent meant that segregation could not begin as soon as DNA replication was initiated, as now appears to be possible.

'Clocks'

The duration of the cell cycle is determined in part by the genome and in part by the environment. In bacteria the length of the cycle can be very easily controlled by changing the composition of the growth medium and this has been, to a large extent, responsible for the discovery of the various regulatory mechanisms controlling the cell cycle which we have outlined in this article. Analogous control systems may be discovered in eukaryotic cells when the length of the cell cycle comes similarly under experimental control. Meanwhile the work with bacteria has revealed several novel aspects of the cell cycle, including the existence of several clock-like systems which are responsible for the timing of various events.

The most obvious of these systems is the process of DNA replication which, as we have seen, plays a central role in the cell cycle. Under a large number of environmental conditions (at constant temperature) in which the growth rate of the cell can vary between about 1 and 3 doublings per hour, each pair of replication forks takes a constant time to travel from the origin to the terminus of the chromosome (see Fig. 2).

The constancy of the rate of addition of nucleotides to the growing DNA chains is presumably a consequence of the existence, at each replication fork, of a single replication complex or small number of enzyme molecules which are saturated with substrate over the range of substrate concentrations found in cells at these growth rates. The general explanation for the constancy of the rate of DNA replication is therefore probably a trivial one. However the fact that the sequential duplication of the genome is so prolonged, relative to the cell cycle, implies that the proportions of different genes change both during the cell cycle and in cells growing at different rates. We have asked the question whether there has been, as a consequence, evolutionary selection of gene order so that appropriate ratios of different genes will occur both at different stages in the cell cycle and under different environmental conditions. There is presumptive evidence to believe that the latter may be true for the genes concerned with ribosome synthesis but experimental evidence is still lacking to decide whether the order of genes is of any major importance to the cell. The other timed event in which the duplication of the genome plays a role is in the induction of the synthesis of a specific protein or proteins which appears to take place when each pair of replication forks reaches the chromosome terminus. However this protein does not seem to be required until much later in the cycle and therefore it is not clear that the exact duration of rounds of DNA replication is important in the timing of division.

The possible reasons why each DNA round takes a constant period of time in the cell cycle are obvious but it is not obvious that this process is used as a clock to determine the timing of other events in the cell cycle. The second clock-like process which we have discussed, the so-called 'division clock' (Fig. 4), is much more mysterious in its molecular basis but its consequences in the cell cycle are much more obvious. The molecular processes involved are still largely unknown but they include a fixed period of protein synthesis and another fixed period which might, for example, involve the assembly of a structure from preformed elements and, in total, these processes require about one hour to complete. The peculiarities of these preparations for division are that they take approximately the same time at different growth rates and that, in fast growing cells with generation times of less than one hour, such division preparations must have run for one hour before each division. This implies that several such processes, each at different stages, must be running in parallel within the same cell at any instant. This, in turn, strongly suggests the spatial separation within the cell of the preparations for individual divisions. The simplest model is one in which assembly of precursors for each septum takes place at a site which represents the future location of that septum. The growth of each septum would therefore begin one hour before its final completion. (One difficulty in the way of this model is that no sign of such a septum precursor is usually seen in conventional electron micrographs of thin sections. However this may well be an artefact of the fixation procedure; Steed & Murray, 1966.) If the clock process is indeed the addition of newly synthesised (and perhaps unstable) elements to a division primordium, then the same formal explanation for the constant time from initiation to completion of the septum can be given as in the case of DNA replication. Addition of precursors could take place at a fixed number of sites, or be catalysed by a fixed number of enzyme molecules, arranged in a ring around the cell circumference (see Fig. 6).

Both DNA replication and the sequence of events leading to division are initiated at about the same time in the cell cycle (Fig. 6) and the possibility therefore exists that they are in fact triggered by the same event. In the unit cell model, we have assumed that this event is the appearance of new surface sites which serve both as attachment sites for new rounds of DNA replication and as the eventual sites of cell division after these rounds have been completed. The problem then arises as to how these sites themselves arise. All one knows, at the moment, is that the initiation events take place at every doubling of a unit mass (M_i), which suggests that the process leading to the formation of new sites takes place at a rate proportional to the rate of overall cell growth (Fig. 6). This process is therefore not a clock in the same sense as those just discussed, which both proceed at constant rates at different cell growth rates. Instead of taking place at regular intervals of time, this master-process takes place at regular increments of mass. It is a 'mass clock', not a 'time clock'. This system is also a replicating system in the sense that two new rounds of DNA replication and two new division sites are generated for each pre-existing site at each doubling of the unit mass. The possible molecular basis for such a multiplicative process of what are presumably structures in the cell envelope remains completely unknown. That just such a duplication of surface rings can occur is however clearly seen in the cell cycle of S. faecalis (Fig. 5). It would be extremely interesting to know whether new DNA rounds are initiated in this organism at the same time as the first appearance of the pair of new surface rings. Unfortunately there are as yet no studies of this kind in this organism.

One other event that takes place at the same time as the initiation of DNA replication is a doubling in the rate of cell growth (Ward &

Glaser, 1971). This supports the original suggestion (Donachie & Begg, 1970) that the newly formed surface rings are also sites at which all net growth of the cell surface takes place. The 'clock' for the initiation of events in the cell cycle is therefore probably also partly responsible for the regulation of overall growth rate.

This discussion suggests that a great deal of the temporal organisation of the bacterial cell cycle is the consequence of the spatial differentiation of its surface layers. The simple view of the bacterial cell as a membrane enclosing a number of independently operating biochemical systems, each specified by the genome, has therefore been replaced by a more complex picture in which the organisation and localisation of certain complex enzyme systems within the cell envelope is of key importance in the regulation and integration of the major events in the cell cycle. Every component in the cell must be specified by the genome but the spatial arrangement of at least some of these components may prove not to be determined by the genome but rather by the pre-existing arrangement of the same components in parent cells. It will not however be known whether this is true until it is possible to attempt to reconstitute a viable cell from its constituent molecules. If such surface organisation is in this sense 'inherited' in bacteria it is clear that the organisation must be primarily in the membrane layers rather than in the rigid mucopeptide layer, since most of the mucopeptide can be removed without impairing the ability of the resultant protoplast eventually to regain its normal shape on reconstituting the missing layer.

CONCLUSIONS

This article has emphasised the vegetative cell cycle in one or two bacterial species but other aspects of differentiation in prokaryotes have been ignored. We have therefore not mentioned the more complex developmental systems of the Myxobacteria and the *Streptomyces* or the process of sporulation in *Bacillus* which are dealt with elsewhere in this Symposium.

We have attempted to survey briefly the main features of the bacterial cell cycle, not only to describe those aspects which are well understood but also to emphasise those processes which are very little understood. We have not hesitated to speculate and to construct models of the main processes in the cell cycle. There is no need to defend the value of models in organising thinking about particular problems but it must be emphasised that the models in this essay are based on a relatively limited amount of information. The study of the bacterial cell cycle, as an integrated process, is only just beginning and we cannot expect models made at this stage to be more than tentative. However it seems already clear that the study of the cell cycle as a whole is revealing the existence of novel molecular control systems.

REFERENCES

ALSTYNE, D. V. & SIMON, M. (1971). Division mutants of *Bacillus subtilis*: isolation and PBS1 transduction of division-specific markers. *Journal of Bacteriology*, **108**, 1366–79.

AUTISSIER, F., JAFFE, A. & KEPES, A. (1971). Segregation of galactoside permease, a membrane marker during growth and cell division in *Escherichia coli. Molecular and General Genetics*, **112**, 275–88.

BEACHEY, E. H. & COLE, R. M. (1966). Cell wall replication in *Escherichia coli*, studied by immunofluorescence and immunoelectron microscopy. *Journal of Bacteriology*, **92**, 1245–51.

BIRD, R. E., LOUARN, J., MARTUSCELLI, J. & CARO, L. (1972). Origin and sequence of chromosome replication in *Escherichia coli. Journal of Molecular Biology*, **70**, 549–66.

BLEECKEN, S. (1971). 'Replisome'-controlled initiation of DNA replication. *Journal of Theoretical Biology*, **32**, 81–92.

BROM, S. (1971). Secondary structure of DNA depends on base composition. *Nature, New Biology, London*, **232**, 174–6.

BRUTLAG, D., SHEKMAN, R. & KORNBERG, A. (1971). A possible role for RNA polymerase in the initiation of M13 DNA synthesis. *Proceedings of the National Academy of Sciences, U.S.A.* **68**, 2826–9.

CAIRNS, J. (1963). The chromosome of *Escherichia coli. Cold Spring Harbor Symposia on Quantitative Biology*, **28**, 43–5.

CAIRNS, J. & DAVERN, C. I. (1968). The mechanics of DNA replication in bacteria. In *Replication and Recombination of Genetic Material*, ed. W. J. Peacock & R. D. Brock, pp. 53–60. Canberra: Australian Academy of Science.

CLARK, D. J. (1968). The regulation of DNA replication and cell division in *Escherichia coli* B/r. *Cold Spring Harbor Symposia on Quantitative Biology*, **33**, 823–38.

CLARK, D. J. & MAALØE, O. (1967). DNA replication and the division cycle in *E. coli. Journal of Molecular Biology*, **23**, 99–112.

COLE, R. M. (1965). Symposium on the fine structure and replication of bacteria and their parts. III. Bacterial cell-wall replication followed by immunofluorescence. *Bacteriological Reviews*, **29**, 326–44.

COOPER, S. & HELMSTETTER, C. E. (1968). Chromosome replication and the division cycle of *Escherichia coli* B/r. *Journal of Molecular Biology*, **31**, 519–40.

CUTLER, R. G. & EVANS, J. E. (1966). Synchronization of bacteria by a stationary-phase method. *Journal of Bacteriology*, **91**, 469–76.

DELBRÜCK, M. & STENT, G. S. (1957). On the mechanism of DNA replication. In *The Chemical Basis of Heredity*, ed. W. B. McElroy & B. Glass, pp. 699–736. Baltimore: John Hopkins Press.

DONACHIE, W. D. (1968). Relationship between cell size and time of initiation of DNA replication. *Nature, London*, **219**, 1077–9.

DONACHIE, W. D. (1969). Control of cell division in *Escherichia coli*: Experiments with thymine starvation. *Journal of Bacteriology*, **100**, 260–8.

DONACHIE, W. D. & BEGG, K. J. (1970). Growth of the bacterial cell. *Nature, London*, **227**, 1220–4.

DONACHIE, W. D., HOBBS, D. G. & MASTERS, M. (1968). Chromosome replication and cell division in *Escherichia coli* 15T⁻ after growth in the absence of DNA synthesis. *Nature, London* **219**, 1079–80.

DONACHIE, W. D., MARTIN, D. & BEGG, K. J. (1971). Independence of cell division and DNA replication in *Bacillus subtilis*. *Nature, New Biology, London*, **231**, 274–6.

DONACHIE, W. D. & MASTERS, M. (1966). Evidence for polarity of chromosome replication in F⁻ strains of *Escherichia coli. Genetical Research, Cambridge*, **8**, 119–24.

DONACHIE, W. D. & MASTERS, M. (1969). Temporal control of gene expression in bacteria. In *The Cell Cycle*, ed. G. M. Padilla, G. L. Whitson & J. L. Cameron, pp. 37–76. New York: Academic Press.

EBERLE, H. & LARK, K. G. (1966). Chromosome segregation in *Bacillus subtilis. Journal of Molecular Biology*, **22**, 183–6.

FAN, D. P., BECKMANN, M. M. & CUNNINGHAM, W. P. (1972a). Ultrastructural studies on a mutant of *Bacillus subtilis* whose growth is inhibited due to insufficient autolysin production. *Journal of Bacteriology*, **109**, 1247–57.

FAN, D. P., PELVIT, M. C. & CUNNINGHAM, W. P. (1972b). Structural difference between walls from ends and sides of the rod-shaped bacterium *Bacillus subtilis. Journal of Bacteriology*, **109**, 1266–72.

FONG, P. (1967). Packing of the DNA molecule. *Journal of Theoretical Biology*, **15**, 230–5.

GOODWIN, B. C. (1966). An entrainment model for timed enzyme synthesis in bacteria. *Nature, London*, **209**, 479–81.

GROSS, J. D. (1972). DNA replication in Bacteria. *Current Topics in Microbiology and Immunology*, **57**, 39–74.

HANSEN, J. N., SPIEGELMAN, G. & HALVORSON, H. O. (1970). Bacterial spore outgrowth: its regulation. *Science, New York*, **168**, 1291–8.

HELMSTETTER, C. & CUMMINGS, D. (1963). Bacterial synchronization by the selection of cells at division. *Proceedings of the National Academy of Sciences, U.S.A.* **50**, 767–74.

HELMSTETTER, C. & CUMMINGS, D. (1964). An improved method for the selection of bacterial cells at division. *Biochimica et Biophysica Acta*, **82**, 608–10.

HELMSTETTER, C. E. (1967). Rate of DNA synthesis during the division cycle of *Escherichia coli* B/r. *Journal of Molecular Biology*, **24**, 417–27.

HELMSTETTER, C. E. (1968). Origin and sequence of chromosome replication in *Escherichia coli* B/r. *Journal of Bacteriology*, **95**, 1634–41.

HELMSTETTER, C. E. (1969). Sequence of bacterial reproduction. *Annual Reviews of Microbiology*, **23**, 223–38.

HELMSTETTER, C. E., COOPER, O., PIERUCCI, O. & REVELAS, E. (1968). The bacterial life sequence. *Cold Spring Harbor Symposia on Quantitative Biology*, **33**, 809–22.

HELMSTETTER, C. E. & PIERUCCI, O. (1968). Cell division during inhibition of DNA synthesis in *Escherichia coli. Journal of Bacteriology*, **95**, 1627–33.

HIGGINS, M. L. & SHOCKMAN, G. D. (1971). *CRC Critical Reviews in Microbiology*, **1**, 29.

HIROTA, Y., RYTER, A. & JACOB, F. (1968). Thermosensitive mutants of *Escherichia coli* affected in the processes of DNA synthesis and cellular division. *Cold Spring Harbor Symposia on Quantitative Biology*, **33**, 677–93.

HUGHES, S. (1971). Variations in enzyme production as a function of the physiological state of bacteria. M.Sc. thesis, University of Edinburgh.

42   W. D. DONACHIE, N. C. JONES AND R. TEATHER

INMAN, R. B. & SCHNÖS, M. (1971). Structure of branch points in replicating DNA; presence of single-stranded connections in λDNA branch point. *Journal of Molecular Biology*, **52**, 319–25.

INOUYE, M. (1969). Unlinking of cell division from DNA replication in a temperature-sensitive DNA synthesis mutant of *Escherichia coli*. *Journal of Bacteriology*, **99**, 842–50.

JACOB, F., BRENNER, S. & CUZIN, F. (1963). On the regulation of DNA replication in bacteria. *Cold Spring Harbor Symposia on Quantitative Biology*, **28**, 329–48.

KENNELL, D. (1968). Titration of the gene sites on DNA by DNA–RNA hybridization. II. The *Escherichia coli* chromosome. *Journal of Molecular Biology*, **34**, 85–103.

KENNET, R. H. & SUEOKA, N. (1971). Gene expression during outgrowth of *Bacillus subtilis* spores. The relationship between gene order on the chromosome and temporal sequence of enzyme synthesis. *Journal of Molecular Biology*, **60**, 31–44.

KOBAYASHI, Y., STEINBERG, W., HIGA, H., HALVORSON, H. O. & LEVINTHAL, C. (1965). In *Spores*, vol. III, ed. L. L. Campbell & H. O. Halvorson, pp. 200–12. Ann Arbor, Michigan: American Society for Microbiology.

KOZAK, M. & NATHANS, D. (1972). Translation of the genome of a RNA phage. *Bacteriological Reviews*, **36**, 109–34.

KUEMPEL, P. L., MASTERS, M. & PARDEE, A. B. (1965). Bursts of enzyme synthesis in the bacterial duplication cycle. *Biochemical and Biophysical Research Communications*, **18**, 858–67.

LACROUTE, F. & STENT, G. S. (1968). Peptide chain growth of β-galactosidase in *Escherichia coli*. *Journal of Molecular Biology*, **35**, 165–73.

LANDMAN, O., RYTER, A. & FRÉHEL, C. (1968). Gelatin-induced reversion of protoplasts of *Bacillus subtilis* to the bacillary form: electron-microscopic and physical study. *Journal of Bacteriology*, **96**, 2154–70.

LARK, K. G. (1972). Evidence for the direct involvement of RNA in the initiation of DNA replication in *Escherichia coli* 15T⁻. *Journal of Molecular Biology*, **64**, 47–60.

LEAL, J. & MARCOVICH, H. (1971). Segregation of T6 phage receptors during cell division in *Escherichia coli* K12. *Annales de l'Institut Pasteur*, **120**, 467–74.

LEVINTHAL, C. & CRANE, H. R. (1956). On the unwinding of DNA. *Proceedings of the National Academy of Sciences, U.S.A.* **42**, 436–8.

LIN, E. C. C., HIROTA, Y. & JACOB, F. (1971). On the process of cellular division in *Escherichia coli*. VI. Use of a methocel-autoradiographic method for the study of cellular division in *Escherichia coli*. *Journal of Bacteriology*, **108**, 375–85.

LOSICK, R. & SONENSHEIN, A. L. (1969). Change in the template specificity of RNA polymerase during sporulation of *Bacillus subtilis*. *Nature, London*, **224**, 35–7.

MAALØE, O. & KJELDGAARD, N. O. (1966). In *The Control of Macromolecular Synthesis*, pp. 188–97. New York & Amsterdam: W. A. Benjamin.

MASTERS, M. (1970). Origin and direction of replication of the chromosome in *Escherichia coli* B/r. *Proceedings of the National Academy of Sciences, U.S.A.* **65**, 601–8.

MASTERS, M. & BRODA, P. (1971). Evidence for the bidirectional replication of the *E. coli* chromosome. *Nature, New Biology, London*, **232**, 137–40.

MASTERS, M. & DONACHIE, W. D. (1966). Repression and the control of cyclic enzyme synthesis in *Bacillus subtilis*. *Nature, London*, **209**, 476–9.

MASTERS, M., KUEMPEL, P. L. & PARDEE, A. B. (1964). Enzyme synthesis in synchronous cultures of bacteria. *Biochemical and Biophysical Research Communications*, **15**, 38–42.

MASTERS, M. & PARDEE, A. B. (1965). Sequence of enzyme synthesis and gene replication during the cell cycle of *Bacillus subtilis*. *Proceedings of the National Academy of Sciences, U.S.A.* **54**, 64–70.

MILLER, O. L., HAMKALO, B. A. & THOMAS, C. A. (1970). The visualization of bacterial genes in action. *Science, New York*, **169**, 392–5.

MITCHISON, J. M. (1971). In *The Biology of the Cell Cycle*, pp. 159–80. London: Cambridge University Press.

MITCHISON, J. M. & VINCENT, W. S. (1965). Preparation of synchronous cell cultures by sedimentation. *Nature, London*, **205**, 987–9.

NEWTON, A. (1972). Role of transcription in the temporal control of development in *Caulobacter crescentus*. *Proceedings of the National Academy of Sciences, U.S.A.* **69**, 447–54.

OISHI, M. & SUEOKA, N. (1965). Location of genetic loci of ribosomal RNA on *Bacillus subtilis* chromosome. *Proceedings of the National Academy of Sciences, U.S.A.* **54**, 483–91.

OISHI, M., YOSHIKAWA, H. & SUEOKA, N. (1964). Synchronous and dichotomous replications of the *Bacillus subtilis* chromosome during spore germination. *Nature, London*, **204**, 1069–73.

PARDEE, A. B. (1966). In *Metabolic Control Colloquium of the Johnson Research Foundation*, p. 239. New York: Academic Press.

PATO, M. L. & GLASER, D. A. (1968). The origin and direction of replication of the chromosome of *Escherichia coli* B/r. *Proceedings of the National Academy of Sciences, U.S.A.* **60**, 1268–74.

PIERUCCI, O. & HELMSTETTER, C. E. (1969). Chromosome replication, protein synthesis and cell division in *Escherichia coli*. *Federation Proceedings*, **28**, 1755–60.

POINDEXTER, J. S. (1964). Biological properties and classification of the *Caulobacter* group. *Bacteriological Reviews*, **28**, 231–95.

PRITCHARD, R. H., BARTH, P. T. & COLLINS, J. (1969). Control of DNA synthesis in bacteria. In *Microbial Growth, Symposium of Society of General Microbiology*, **19**, 263–97.

ROGERS, H. J. (1970). Bacterial growth and the cell envelope. *Bacteriological Reviews*, **34**, 194–214.

RYTER, A. (1968). Association of the nucleus and the membrane of bacteria: a morphological study. *Bacteriological Reviews*, **32**, 39–54.

RYTER, A. (1971). Flagella distribution and a study on the growth of the cytoplasmic membrane in *Bacillus subtilis*. *Annales de l'Institut Pasteur*, **121**, 271–88.

RYTER, A. & JACOB, F. (1966). Ségrégation des noyaux chez *Bacillus subtilis* au cours de la germination de spores. *Comptes Rendus de l'Académie des Sciences, Paris*, **263**, 1176.

SAKAKIBARA, Y., SAITO, H. & IKEDA, Y. (1965). Incorporation of radioactive amino acids and bases into nucleic acid and protein fractions of germinating spores of *Bacillus subtilis*. *Journal of General and Applied Microbiology*, **11**, 243–54.

SCHLEIDEN & SCHWANN (1838). Quoted in *The Encyclopaedia Britannica*, 10th edit. (1911).

SCHMIDT, J. M. & STANIER, R. Y. (1966). The development of cellular stalks in bacteria. *Journal of Cell Biology*, **28**, 423–36.

SCHWARZ, U., ASMUS, A. & FRANK, H. (1969). Autolytic enzymes and cell division of *Escherichia coli*. *Journal of Molecular Biology*, **41**, 419–29.

SHAPIRO, L., AGABIAN-KESHISHIAN, N. & BENDIS, L. (1971). Bacterial differentiation. *Science, New York*, **173**, 884–92.

SMITH, L., GOLDTHWAITE, C. & DUBNAU, D. (1969). The genetics of ribosomes in *Bacillus subtilis*. *Cold Spring Harbor Symposia on Quantitative Biology*, **34**, 85–9.

SONNEBORN, J. M. (1963). Does preformed cell structure play an essential role in cell heredity? In *The Nature of Biological Diversity*, ed. J. M. Allen, pp. 165–221. New York: McGraw-Hill.

SONENSHEIN, A. L., LOSICK, R. & SHORENSTEIN, R. G. (1970). Structural alteration of RNA polymerase during sporulation. *Nature, London*, **227**, 910–13.

STEED, P. & MURRAY, R. (1966). The cell wall and cell division of gram-negative bacteria. *Canadian Journal of Microbiology*, **12**, 263–70.

STEINBERG, W. & HALVORSON, H. O. (1968a). Timing of enzyme synthesis during outgrowth of spores of *Bacillus cereus*. I. Ordered enzyme synthesis. *Journal of Bacteriology*, **95**, 469–70.

STEINBERG, W. & HALVORSON, H. O. (1968b). Timing of enzyme synthesis during outgrowth of spores of *Bacillus cereus*. II. Relationship between ordered enzyme synthesis and DNA replication. *Journal of Bacteriology*, **95**, 479–89.

STONINGTON, G. & PETTIJOHN, D. (1971). The folded genome of *E. coli* isolated in a protein–DNA–RNA complex. *Proceedings of the National Academy of Sciences, U.S.A.* **68**, 6–9.

SUEOKA, N., ARMSTRONG, R., HARFORD, N., KENNET, R. & O'SULLIVAN, A. (1970). *Proceedings of the 1st International Symposium on the Genetics of Industrial Micro-organisms*. Prague.

SUGINO, A. & OKAZAKI, R. (1972). Mechanism of DNA chain growth. *Journal of Molecular Biology*, **64**, 61–85.

TARTAR, V. (1961). *The Biology of Stentor*. New York: Pergamon Press.

TAYLOR, A. L. (1970). Current linkage map of *Escherichia coli*. *Bacteriological Reviews*, **34**, 155–75.

TORRIANI, A., GARRIDE, L. & SILBERSTEIN, Z. (1969). In *Spores*, vol. IV, ed. L. L. Campbell, p. 247. Ann Arbor, Michigan: American Society for Microbiology.

VAN DIJK-SALKINOJA, M. & PLANTA, R. (1971). Rate of ribosome production in *Bacillus licheniformis*. *Journal of Bacteriology*, **105**, 20–7.

VAN TUBERGEN, R. P. & SETLOW, R. B. (1961). Quantitative radioautographic studies on exponentially growing cultures of *Escherichia coli*. *Biophysical Journal*, **1**, 589–625.

WARD, C. B. & GLASER, D. A. (1971). Correlation between rate of cell growth and rate of DNA synthesis in *Escherichia coli* B/r. *Proceedings of the National Academy of Sciences, U.S.A.* **68**, 1061–4.

WATSON, J. D. & CRICK, F. H. C. (1953). The structure of DNA. *Cold Spring Harbor Symposia on Quantitative Biology*, **18**, 123–31.

WEHRLI, W., NÜESCH, J., KNÜSEL, F. & STAEHELIN, M. (1968). Action of rifamycins on RNA polymerase. *Biochimica et Biophysica Acta*, **157**, 215–17.

WORCEL, A. (1970). Induction of chromosome re-initiations in a thermosensitive DNA mutant of *Escherichia coli*. *Journal of Molecular Biology*, **52**, 371–86.

YAMAGISHI, H. & TAKAHASHI, L. (1971). Heterogeneity in nucleotide composition of *Bacillus subtilis* DNA. *Journal of Molecular Biology*, **57**, 369–71.

YANAGITA, J. & KANEKO, K. (1961). Synchronization of bacterial cultures by preincubation of cells at high densities. *Plant and Cell Physiology*, **2**, 443–9.

YOSHIKAWA, H., O'SULLIVAN, A. & SUEOKA, N. (1964). Sequential replication of the *Bacillus subtilis* chromosome. III. Regulation of initiation. *Proceedings of the National Academy of Sciences, U.S.A.* **52**, 973–80.

INITIATION OF BACTERIAL SPOROGENESIS

J. SZULMAJSTER

Laboratoire d'Enzymologie du C.N.R.S., 91-Gif-sur-Yvette, France

'Que de mots pour si peu de choses...' (FLAUBERT)

Initiation to what?

The use of the term 'initiation' nowadays has become generalised to signify the introduction of a person to a science, an art or to a profession. Heretofore, it has designated those ceremonies by means of which one could be introduced to a knowledge of certain 'mysteries'.

Ethnologists are inclined to distinguish three types of initiation: tribal initiation, which transforms adolescents into adults; religious initiation, which permits one access to secret societies or closed brotherhoods, and magic initiation, which permits one to abandon the normal human condition and to have access to supernatural powers. Tribal, religious or magic initiation seems to be a process designed to achieve psychologically, the passage of man from a supposed inferior to a superior state.

The initiation we will be concerned with in this paper is restricted to primitive bacterial populations which have the hereditary property to undergo metamorphosis under proper conditions. As a result, the active rod-shaped bacterium enters a quiescent state (the dormancy state) in the form of a spore. It is not certain at present if this type of initiation implies the achievement, either by the bacterium or by the investigator, of the passage from an inferior to a superior state of contemplation.

What is the problem?

At the end of exponential growth, when an essential nutrient becomes exhausted in the medium, and the growth rate decreases, spore-forming bacteria undergo progressive morphological and structural changes (accompanied by modification in their enzymatic equipment) which culminate in the formation of an endospore. Microbial sporulation involves the intracellular metamorphosis of a fragile vegetative bacterial cell into a highly resistant, dormant spore structure. This conversion is regulated in time and space by a set of controls operating upon the succession of the biochemical events. In principle, these controls might be exerted at all levels, i.e. to regulate the transcription of DNA into RNA, the translation of RNA into proteins and the activity of the proteins as enzymes.

Bacterial sporogenesis thus represents a unique biological problem whose regulation is intermediate in complexity between the regulation of the production and activity of bacterial enzymes on the one hand and the differentiation of specialised tissues in more complex organisms on the other hand. It is not surprising therefore that, during the last ten years, this problem has become attractive to molecular biologists as a subject for the study of cell differentiation.

The whole process of sporulation involves continuous morphological changes which have been conventionally timed from the end of exponential growth designated as T_0 and hourly periods after T_0 as T_1, T_2, T_n etc. Six morphologically well recognised steps have been established by electron microscope studies, first on *Bacillus cereus* by Young & Fitz-James (1959*a–c*) and on *Bacillus subtilis*, by Ryter (1965). Similar morphological stages have been observed in other spore-forming bacilli by Ohye & Murell (1962), Ellar & Lundgren (1966), Bayen, Frehel, Ryter & Sebald (1967). The designation of these developmental stages serves now as a basis for following the chronology of genetic and biochemical sporulation events. There is no need here to describe once more the anatomy of these stages, which can be found in addition to the above cited references, in many books and reviews concerning bacterial sporogenesis (Robinow, 1960; Gould & Hurst, 1969; Murell, 1967; Balassa, 1971).

The biochemical events associated with the formation of a mature spore are under the control of genes or operon-like groups of genes. Genetic studies from different laboratories (Schaeffer, Ionesco, Ryter & Balassa, 1963; Schaeffer, 1969; Spizizen, Reilly & Dahl, 1963; Spizizen, 1965; Takahashi, 1965*a–c*; Hoch & Spizizen, 1969; Balassa, 1969; Balassa & Yamamoto, 1970; Balassa, 1971) have led to the conclusion that the spore genes are randomly distributed along the *Bacillus subtilis* chromosome. It is generally believed that the spore genome is repressed during rapid vegetative growth and derepressed at the beginning of the stationary growth phase, although the exact nature of this repression or derepression remains obscure. Obviously, the development of a spore must provide a mechanism for the functional transition from the total genetic potentiality of the vegetative cell to a selective activation of specific spore genes and simultaneously to selective repression of certain vegetative genes. To understand the regulation of such a mechanism it is fundamental to know how the overall program of the sporulation process is initiated, i.e. if there exists a primary locus for initiation and what is its biochemical expression. It is also important to know at what level initiation is controlled.

It is interesting to notice here from an historical point of view that Doran (1922) had already distinguished two classes of phenomena which influence dormancy and he termed them the 'internal' and 'external' factors. Although the 'internal' factors were at this time poorly understood and defined as the expression of 'vitality', the 'external' factors were designated more precisely and they included temperature, oxygen, nutrients and toxic substances. Obviously these elements are related to what are called now, in the broader sense, environmental stimuli or external signals which trigger the entire process of sporulation. The nature of these signals that shift the cell from growth-associated activities to morphogenetic ones still remains unknown. However, the idea that the levels of various catabolites in the medium may act as determining factors in the initiation process will be discussed in the light of the recent findings concerning the mechanism of catabolite repression of enzyme induction.

INITIATION AND COMMITMENT

Perhaps the use of the two terms 'initiation' and 'commitment' needs a few comments, because their meaning is somewhat confusing in the literature. These two expressions have been utilised frequently as if they were synonymous. In fact, a bacterial cell can initiate the sporulation process without being committed to the formation of a mature spore.

In the previous section we have already explained what we intend by the term 'initiation' and this notion will be further developed later in this paper.

Let us now consider the problem of commitment. The term 'commitment' is generally applied to differentiating biological systems and is defined as the '*point de non retour*', i.e. the time at which the biochemical and morphological events associated with the developmental system are definitely channelled towards the differentiated form and cannot be undone. In the case of sporulating bacteria 'commitment' is defined as the time at which the cells become irreversibly engaged in the sporulation process, even after their transfer to a fresh growth medium. This definition stems from the work of Hardwick & Foster (1952), in whose laboratory the concept of 'endotrophic sporulation' was born and where the notion of 'irreversibility' was developed. These authors have shown that vegetative cells of various species of *Bacillus*, when removed from the synthetic growth medium at a time before the point of maximum growth and the onset of sporulation, and transferred into distilled water or phosphate buffer will sporulate within a two hour period after ten hours incubation. As a result of their experiments, Hardwick &

Foster (1952) advanced the hypothesis that spore development is strictly an endogenous process, termed 'endotrophic', occurring independently of exogenous nutrients. A more detailed examination of the events in endotrophic sporulation was made possible through the observation, by the same authors, of the glucose effect. Normally about 90 % of the cells sporulate when suspended in water. However, when glucose was added within the first 5 h after the cells were transferred into distilled water, subsequent sporulation was completely suppressed. Glucose added to the cells after 6, 7 or 8 h was progressively less effective and when added after 8 h was totally ineffective in suppressing sporulation. It was concluded that the addition of glucose up to the sixth hour causes the reversal of sporulation events, while these events are irreversible when glucose is added after the sixth hour.

Major objections to the concept of 'endotrophic' sporulation have since been formulated by a great number of investigators (Powell & Hunter, 1953; Murell, 1955; Black & Gerhardt, 1963). It seems that in Hardwick & Foster's experiments lysis of a number of transferred cells converted the water to a dilute nutrient medium, which supported the sporulation of surviving cells. Young & Fitz-James (1959 b) raised a more general objection to the notion of sporulation as an endogenous process. They assume that the levels of calcium and cysteine (Vinter, 1959), greater in the spores than those found in the mother cells, could not be derived entirely from intracellular components.

The problem of commitment is periodically being investigated in many laboratories with different species of spore-forming bacteria and the results obtained rarely lead to the same conclusion, even when referred to the same organism. Thus in Bacillus subtilis a very early stage of commitment has been shown by the glucose effect (Freese et al. 1969; Freese, Klopat & Galliers, 1970). The presence of glucose in a growth medium during exponential phase suppressed sporulation, but when added at any time after T_0, it had no effect. Fréhel & Ryter (1969) working also with Bacillus subtilis have shown that if sporulating cells were diluted into fresh medium before stage II, sporulation was blocked. Balassa (1971) has observed such inhibition even until stage IV, depending on the degree of dilution of the fresh medium.

On the other hand, according to Sterlini & Mandelstam (1969), who examined the biochemistry of commitment in Bacillus subtilis, there is no single point of commitment for the whole sporulation process. Instead, the cells, after initiation in a deficient medium, become successively committed to one sporulation event after another. Thus, fairly early on, they are committed to producing alkaline phosphatase but not to refractility, whereas, somewhat later, they are committed to re-fractility but not to producing dipicolinate or heat-resistant spores.

Sporulating cells of *Bacillus cereus* reach a stage of commitment at the completion of the forespore and the beginning of cortical synthesis. Prior to this, a culture of cells at stage II or III will revert to vegetative growth upon the addition of fresh medium (Fitz-James & Young, 1969). In this case the spore septum appears to be converted to a transverse septum with the deposition of wall material between the membrane layers. Thus two cells are obtained, one long, which starts to divide first, and one short, which starts to divide later. Similar results were obtained by Lundgren, Karp & Lang (1969) who reported that stage IV cells of *Bacillus cereus*, separated from the sporangium and resuspended in fresh medium are viable and able to resume growth without forming a spore.

Faced with the wide differences in the time of commitment, even with respect to the same micro-organism, Orrego, Laporte, Martin, Arnaud & Szulmajster (unpublished results, 1972) have recently reinvestigated the mystery of this phenomenon in *Bacillus subtilis*, Marburg strain using partially synchronised cultures with respect to sporulation process (for different methods of synchronisation of spore formation see the review of Slepecky, 1969).

The synchronisation was achieved by two transfers of cells which had just reached the stationary phase on rich nutrient broth medium to the same fresh medium at a cell concentration that would not allow more than 2 or 3 cell divisions at an exponential rate. The synchronisation of the sporulating culture was followed by examination in the electron microscope, of the different morphological stages as a function of time after T_0 (Fig. 1). It appeared that under these conditions of growth at T_0, 85–95 % of the bacterial population reached stage 0–I, recognisable by the formation of the axial filament from the two nuclei of the vegetative cell. It was seen further that it takes about 90–100 min for the majority of these cells to reach sporulation stage II, which is charac-terised by the formation of spore septum. As sporulation proceeded, there seemed to be a scattering of all the stages. However, at T_5–T_6, when the cells reached stage V–VI (appearance of refractility) there was a rapid increase in the frequency of the last stages. These observations lead to the conclusion that, under our conditions of growth, cells are synchronised in the sense that at the end of exponential phase almost all of them had passed the stage 0 boundary.

When samples of the synchronised culture taken at different times during sporulation were transferred to fresh medium and examined for

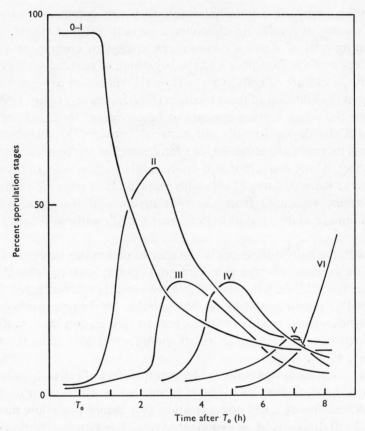

Fig. 1. Distribution of morphological sporulation stages in a partially synchronised culture of *Bacillus subtilis* (see text). Starting from T_0 (end of exponential growth), samples were taken at hourly intervals for electron microscope examinations.

spore formation, the following observations were made: when the transfer took place at any time before T_2 and the cells were examined for thermoresistance at T_5, T_6, T_7 and T_9, it was observed that compared to the control (original culture) there was a 90 % decrease in the number of heat-resistant cells in the bacterial population. This means that almost all the cells engaged in the sporulation process reversed and resumed growth rapidly. This was reflected also by an increase in the optical density of the culture. Examination of the transferred cells in the electron microscope revealed that the spore septum had been converted into a transverse septum characteristic of cell division. However, when the transfer took place at T_2 or later, almost the same number of thermoresistant bacteria was found in the transferred medium as in the original culture. Electron microscope photographs of samples taken at different times after the transfer have shown the appearance of subsequent sporulation stages in the same proportions as in the control culture. These results are rather similar to those obtained by Fréhel & Ryter (1969) with *Bacillus subtilis*, and Millet with *Bacillus megaterium* (personal communication).

It is interesting to note that sporulating cells transferred to a fresh medium at T_5 or at T_7 start to germinate intracellularly about 30 min after their transfer (Plate 1).

In conclusion, it should be said that commitment to sporulation in *Bacillus subtilis* and probably in other spore-forming bacteria is an extremely complex event. The determination of the exact time that a cell becomes irreversibly engaged in spore formation would be facilitated if one could follow the fate of a single cell during the process. Unfortunately, all the synchronisation techniques now available yield bacterial populations in which synchronisation is incomplete and is further dispelled during the various sporulation stages.

INITIATION AND TRICARBOXYLIC ACID ENZYMES

Since several investigators have attributed a role to the tricarboxylic acid (TCA) cycle enzymes in the initiation of spore development (Fortnagel & Freese 1968; Balassa, 1971), it is necessary to examine here this assertion in the light of the available data with respect to the time at which the induction of these enzymes takes place.

It was first demonstrated by Nakata & Halvorson (1960) that when vegetative cells of *Bacillus cereus* were grown in a yeast extract-minerals medium containing glucose as the main carbon source, there was a sharp drop in the pH of the medium when glucose was exhausted. The pH minimum occurred at about T_1 and results mainly from the accumulation of acetic and pyruvic acids. It was further shown that the utilisation of these acids and the rise of the pH coincide with the induction of the TCA cycle enzymes (Hanson, Srinivasan & Halvorson, 1963a, b). These investigators came to the conclusion that the enzymes of the TCA cycle are repressed by glucose during vegetative growth and that utilisation *via* the TCA cycle enzymes was necessary for sporulation to occur.

Hanson, Blicharska & Szulmajster (1964) and Szulmajster & Hanson (1965) provided further evidence for the involvement of the TCA cycle enzymes in sporulation using *Bacillus subtilis* mutants deficient in aconitase, a key enzyme in the citric acid cycle. These mutants are unable to synthesise glutamate and have lost the capacity to sporulate even when the required amino acid was supplied to the medium at the beginning of growth. Transformation of such a phenotypically aconitase-less and asporogenic mutant (Sp⁻, glut⁻) by the DNA of the wild type strain leads to the recovery of both enzyme activity and sporulation. The same authors have also shown that the aconitase-less mutants are genotypically Sp⁺ and that their inability to sporulate may be the result of a physiological disturbance in the events occurring during the sporulation process. The block in the TCA cycle appeared to be the cause of this disturbance. The experiments indicate that in a sporogenic strain, a functional TCA cycle is indispensable for the expression of some of the spore genes. Freese *et al.* (1969) suggested that, in the citric acid cycle mutants, sporulation is prevented because not enough ATP is available to allow sporulation to take place, although in some of these mutants the level of ATP was artificially maintained and nevertheless sporulation did not occur.

Whatever the function of the TCA cycle in sporulation might be, it must be emphasised that the changes in the metabolic pattern observed, i.e. acetate utilisation and induction of the TCA cycle enzymes, occur generally in sporulating cells of *Bacillus cereus* or *Bacillus subtilis* at stage II–III, characterised by the formation of the spore septum (Hanson *et al.* 1963*a*, *b*; Szulmajster & Hanson, 1965; Murell, 1967). Obviously, the initial biochemical event of spore development is situated some time before this stage as it will be seen later on. Therefore, it certainly would be more realistic to assign to the TCA cycle an important role in the expression, not of genes involved in initiation of the sporulation program, but of those controlling some relatively late events.

BIOCHEMICAL EVENTS OCCURRING AFTER PHAGE INFECTION AND DURING BACTERIAL SPORE DEVELOPMENT

The analogy of biochemical events involved in phage multiplication after infection of the host cell and those associated with the formation of a mature spore inside the mother-cell has not escaped the attention of many investigators working on these or similar biological problems (Luria, 1960; Travers, 1971). It is advisable therefore to give a brief evaluation of the progress accomplished in the last few years in elucidating the mechanism of transcription in the phage developmental system. Attention will be focused mainly on recent studies of the role of the RNA polymerase during phage infection and of the nature of the various initiation specificities which reside in its own structure. These investigations are now guiding the studies on the mechanism of initiation and subsequent control of sporulation process.

Regulation of bacteriophage transcription

Bacteriophage-infected cells provide a useful model for studying the mechanisms of regulation which govern sequential gene expression at the transcription level in procaryotic developmental systems.

For the purpose of this paper only a brief summary of the present knowledge of the transcription in T_4-infected cells will be given as an example. The reader will find more detailed information of the various positive and negative controls operating in phage-infected cells in recent excellent reviews (Szybalski et al. 1969; Calendar, 1970; Travers, 1971; Losick, 1972 in press).

The formation of different mRNA species during early and late stages of infection was first observed with T-even phages (Khesin & Shemyakin, 1962; Kano-Sueoka & Spiegelman, 1962) and it was concluded that viral protein synthesis was regulated by 'switching on and off' various genes or groups of genes of the phage chromosome which determine the transcription of different RNA sets. The evidence for this conclusion derives from extensive in vivo and in vitro studies. Three well-defined genetic regions have been localised on the T_4 DNA. These three classes of genes, which are regulated independently, were designated according to their order of expression as immediate–early, delayed–early, and late. [It is now well established that most of the bacterial RNA polymerase studies can be separated into two major components: the core enzyme containing α and $\beta\beta'$ subunits associated reversibly to a further subunit, σ (sigma), which ensures accurate and efficient initiation of RNA synthesis at a specific promotor site of the DNA template (Burgess, 1969; Burgess, Travers, Dunn & Bautz, 1969; Bautz, Bautz & Dunn, 1969). This σ factor then confers on the enzyme the ability to discriminate between genes or group of genes to be transcribed.] The transcription, in vitro, of the immediate–early class was found to be under the control of the Escherichia coli sigma factor (Bautz et al. 1969). Hybridisation-competition experiments, first utilised so elegantly by Khesin & Shemyakin (1962), have shown that σ is also involved in vivo in the expression of T_4 genes immediately after infection. The evidence consisted of experiments in which ^{3}H-labelled T_4 RNA was synthesised in vitro in the presence and absence of sigma and hybridised with T_4 DNA in the presence of unlabelled competitor RNA isolated from Escherichia coli cells harvested at various times after infection with

T_4 phage. The transcription of delayed–early class of RNA seems to be directed by a T_4 sigma-like factor (or by a protein which requires σ for its function) coded for by the phage (Travers, 1969, 1970). This factor is not detected until 5 to 15 min after infection, after which it disappears. In conclusion then, host sigma factor and T_4 sigma-like factor stimulate transcription of T_4 DNA by host core polymerase, each one directing the transcription of unique segments of the T_4 genome.

Both immediate–early and delayed–early RNA hybridise exclusively with the l strand of T_4 DNA (Guha & Szybalski, 1968).

Late RNA synthesis is dependent on the products of genes 55 and 33 and T_4 DNA replication (Bolle, Epstein, Salser & Geiduschek, 1968). It was suggested (Cascino, Riva & Geiduschek, 1970) that the replicating DNA may reflect a particular chemical state of the DNA and that breaks in this DNA may be essential for initiation of late transcription. It was also suggested by Khesin (1970) that synthesis of late mRNA is directed by a phage DNA which differs in some respect from the DNA which invades the cell. Those differences may reside in the secondary structure of the DNA or may be related to its localisation within the cell, for example binding to cell membrane. In a sensitive host cell the synthesis of all classes of T_4 RNA is sensitive to the drug rifamycin throughout the developmental cycle (Haselkorn, Vogel & Brown, 1969). In a mutant host cell which produces a rifamycin-resistant RNA polymerase, the synthesis of T_4 RNA is immune to the drug. Since rifamycin seems to interact with the β subunit (Rabussay & Zillig, 1969) it can be assumed that at least this subunit of the host core polymerase is conserved and is required for the synthesis of all classes of phage RNA.

Alterations of RNA polymerase subunits during T_4 infection

There is evidence that during T_4 infection the *Escherichia coli* host RNA polymerase undergoes structural changes in two consecutive steps: alteration and modification (Schachner, Seifert & Zillig, 1971). Within one minute after infection a very rapid alteration takes place which is independent of protein synthesis. It leads to a reduction of the ratio of template activities of T_4 to calf thymus DNA for the purified *Escherichia coli* RNA polymerase. According to the authors this indicates a reduced σ activity which may be due either to an alteration of sigma factor itself, or to a reduced affinity of σ for an altered core polymerase. The other change, called modification, requires protein synthesis and leads to structural changes of all the subunits (Schachner & Zillig, 1971) and to a complete loss of sigma from the purified enzyme (Bautz & Dunn, 1969).

The fate of the *Escherichia coli* sigma remains obscure. It is possible that Travers' T_4 sigma-like factor, mentioned above, is the product of the alteration of *Escherichia coli* σ. It was further shown that early during infection a modification of the α subunit is due to the covalent addition of 5'-adenylate (Goff & Weber, 1970). The β' and β subunits are apparently also modified between 10 and 15 min after infection as demonstrated by different experimental approaches (Travers, 1970; Seifert, Rabussay & Zillig, 1971).

In addition to these changes, it was found by Stevens (1970, 1972) that early in infection, the core polymerase is associated with newly synthesised polypeptides of molecular weight about 10000 and 22000 daltons. The functions of these phage coded subunits remain unknown.

In conclusion, it can be said that phage development is associated with two major processes: (*a*) alterations and modifications of all the subunits of the pre-existing host RNA polymerase. These changes may play a role in the shutting off of the transcription of the host genes. (*b*) Synthesis of new factors which may direct the sequential expression of the phage genes. The exact biological significance of these changes in the regulation of the transcription during phage development is still unclear.

Regulation of transcription during spore development

Another 'simple' example of intracellular differentiation is bacterial sporogenesis, the subject we are concerned with here. Initiation of spore development implies, in the first place, a switch from the expression of vegetative genes to those involved in sporulation. What do we know about the number and nature of these genes?

Using asporogenous mutants of *Bacillus subtilis*, the studies from different laboratories (Spizizen, 1965; Schaeffer, 1969; Takahashi, 1965*b*, *c*; Hoch & Spizizen, 1969; Hoch, 1971; Balassa, 1971) performed by transformation and transduction have indicated that there are a great number of genetic loci (estimated now to be several hundred) at which a mutation can affect sporulation. These genes are widely scattered along the *Bacillus subtilis* chromosome and some of them seem to be clustered in operon-like units, although no operator typical mutation has ever been demonstrated. Many of these genes are structural genes and code for specific enzymes, but most of them are probably regulatory genes which may be involved in the control of the synthesis and function of the products of these structural genes. This is probably true for all other developmental systems. Most of the asporogenous mutants so far studied can be divided grossly into two classes. To one class belong the early-blocked mutants, i.e. those affected in the first morphological stages of sporulation. These mutants exhibit unidirectional pleiotropic effects, preventing the expression of all the late functions. It can then be said that at least some of the genes affected in the early-blocked mutants are regulatory genes controlling the biochemical events involved in the late process of sporulation.

Mutants belonging to this class have been the subject of extensive genetic studies by Hoch & Spizizen (1969), Schaeffer (1969), Balassa (1971), Takahashi (1965*b*, *c*), and biochemical studies by Szulmajster, Bonamy & Laporte (1970) and Leighton *el al.* (1972) with thermosensitive sporulation mutants. The importance of such mutants for the studies of genetic controls of initiation will be further discussed in a later section.

Studies of the regulation of transcription have been hampered because, unlike phage systems, only a rough correlation exists between the stages of spore development and the appearance of certain biochemical events (Mandelstam, 1969; Szulmajster & Schaeffer, 1961; Szulmajster, 1964; Chasin & Szulmajster, 1969; Bernlohr & Leitzman, 1969; Hanson, Peterson & Yousten, 1970; Kornberg, Spudich, Nelson & Deutscher, 1968). The relation between early events (such as excretion of extracellular enzymes and antibiotics) and sporulation is uncertain, and, moreover, no genes coding for any spore-specific late events (such as synthesis of dipicolinic acid, cortical material (Tipper & Pratt, 1970) and cysteine incorporation (Vinter, 1959) have been identified.

It nevertheless remains true that the transition, at the genetic level, of a vegetative cell to the sporulation state is accompanied by morphological and structural changes, by dramatic shifts in metabolic patterns and by the appearance of new enzymes. All these changes are expressed at the transcriptional level by the synthesis of new classes of mRNA (Doi & Igarashi, 1964; Aronson, 1965*a*, *b*; Yamagishi & Takahashi, 1968), first demonstrated during sporulation by Spotts & Szulmajster (1962) and Balassa (1963).

The stability of the sporulating mRNA has been, for many years, a subject of some conflict. Because the existence of stable mRNA has serious implications as to the extent of translational control in sporulating system, it is worthwhile to reconsider this problem briefly.

Using actinomycin D, an inhibitor of RNA synthesis, it was found that sporulating cells of *Bacillus subtilis* (Szulmajster, Canfield & Blicharska, 1963; Szulmajster & Canfield, 1965; Balassa, 1963, 1966) and *Bacillus cereus* (Fitz-James, 1965; Kobayashi *et al.* 1965) synthesise unstable mRNA species with an half-life of about 2 min, similar to that found for the mRNA in vegetative cells (Levinthal, Fan, Higa & Zimmerman, 1963). Investigations by the same authors on the effect of actinomycin D and chloramphenicol on the formation of mature spores led to the further conclusion that most, if not all, stages of sporulation process, which require protein synthesis, are under the control of short-lived mRNAs. More recently, Leighton & Doi (1971) confirmed these results by using rifamycin (inhibitor of DNA-dependent RNA polymerase) added to *Bacillus subtilis* cells at different times during sporulation.

However, Aronson (1965*b*) working with *Bacillus cereus* and using pulse-labelling and hybridisation techniques, arrived at a different conclusion. According to his experiments, after initiation of sporulation, in addition to the unstable mRNA species, there was formed a persistently stable fraction of membrane-bound spore mRNA which was conserved even in the presence of actinomycin D. Sterlini & Mandelstam (1969), considering that commitment might be due to the formation of stable mRNA, reinvestigated the effect of low doses of actinomycin D (1 μg/ml, instead of the 10 to 100 μg/ml used in the previous described experiments) on three sporulation-dependent characters, synthesis of alkaline phosphatase (earlier enzyme), appearance of refractility and heat resistance (late events). In each case they found that the commitment time (effect of the drug) for each character preceded by about one to two hours the expression of the character. These results would suggest the existence of relatively stable messenger RNA, since the traits studied are expressed two hours after their synthesis.

A critical review of some of the problems concerning mRNA during sporulation has been given recently by Balassa (1971) and I will avoid repetition. I would like nevertheless to emphasise the main difficulty in the interpretation of all the above experiments, including those where the techniques of hybridization and competition were used. They all deal with bulk mRNA resulting from the transcription not only of different classes of spore specific genes, which may have different stabilities, but also of many vegetative genes which are still expressed during sporulation (Aubert & Millet, 1963, 1965; Aronson, 1965*a*; Yamakawa & Doi, 1971; DiCioccio & Strauss, 1971). Relevant to this criticism are the results obtained by Chasin & Szulmajster (1967) concerning the stability of mRNA of two sporulation-specific enzymes involved in the biosynthesis of dipicolinic acid (DPA), the dihydrodipicolinate synthetase and DPA-synthetase. The synthesis of the first enzyme was completely inhibited by actinomycin D immediately after its addition, whatever the time of addition. The synthesis of the second enzyme was also inhibited by the drug but only if added at the very beginning of the synthesis.

Once rapid synthesis started the inhibition was less effective and it took 10–12 min to obtain complete inhibition of the enzyme formation. These results indicate that even the mRNA of a spore specific enzyme is only slightly more stable than the vegetative ones. More studies on similar spore specific enzymes would tell us if the relative greater stability observed by Chasin & Szulmajster is a general characteristic of such enzymes.

In discussing in some detail the problem of stability of sporulating messenger RNA the aim of the author was not to discard the possibility or the importance of a control during sporulation at the translational level (Szulmajster, Kerjan & Maia, 1970) but rather to direct the reader's attention to the predominance of unstable mRNA and therefore to the likelihood that transcriptional control plays a large role in the initiation and subsequent events of the entire spore process.

It should be mentioned that strand selection also occurs during sporulation, and this may be part of a regulatory mechanism (Yamakawa & Doi, 1971; DiCioccio & Strauss, 1971).

During sporulation, as during phage infection, the DNA-dependent RNA polymerase plays a central role in being able to select spore-specific promotor sites on the bacterial DNA. Although the modulation of the RNA polymerase subunit complex during sporulation is less known than that during phage infection, it shares with the phage system a common fact, that the β subunit responsible for rifampicin sensitivity is maintained throughout sporulation process like during the entire period of phage development (Losick & Sonenshein, 1969). Another similarity with the phage system is the modification of one of the β subunits (discussed in the next section), although the physiological significance of this modification for sporulation still remains to be clarified.

INITIATION OF SPORULATION AT THE TRANSCRIPTION LEVEL

A most promising approach to the study of the initiation of bacterial sporogenesis is based on the recent findings of the structure and mode of action of the DNA-dependent RNA polymerase in normal and in phage-infected cells of *Escherichia coli*.

I have already described (see previous section) the major components of the bacterial DNA-dependent RNA polymerase.

A number of other factors have been identified in *Escherichia coli* which are involved in the regulation of the transcription of specific regions of the DNA. These are the ρ (rho) termination factor (Roberts,

1969), the ψ (psi) factor (Travers, Kamen & Schleif, 1970) which binds to RNA polymerase and directs the transcription of genes coding for ribosomal RNA (although the specificity of this factor is now questioned (Haseltine, 1972; Pettijohn, 1972), and the cyclic AMP-receptor protein (CRP) that regulates the transcription of the *lac* and *gal* operons, presumably by binding at the promotor site of the genome and increasing the frequency of initiation (Riggs, Reiness & Zubay, 1970; Zubay, Chambers & Cheong, 1970; Emmer, de Crombrugghe, Pastan & Perlman, 1970). These protein factors can be considered as elements of a positive control over gene expression.

The concept of sigma factor proteins sheds new light upon all attempts to explain this type of control of gene expression and becomes particularly attractive for the interpretation of the mechanism of regulation in developmental systems which involve a sequential gene expression as exemplified by phage-infected bacteria and by the sporulation process. Studies in the last few years relative to these problems have indeed greatly justified some of the hopes and expectations. The experiments reported by Losick and his associates (Losick & Sonenshein, 1969; Losick, Shorenstein & Sonenshein, 1970) give strong support to these ideas as far as the sporulation system is concerned.

These authors were the first to demonstrate that the transition of a vegetative spore-forming vegetative cell to the sporulating state is associated with a specific molecular event which is expressed both *in vivo* and *in vitro*. They observed that when vegetative cells of *Bacillus subtilis* were infected by the virulent phage ϕe, they give rise to a large progeny. However, when sporulating cells were infected by the same phage, the cessation of the synthesis of at least three early phage gene products was observed and it was suggested by the authors that as a consequence, the bacteria no longer support phage multiplication. This strange comportment of the phage was explained by the same workers by means of in-vitro experiments using partially purified RNA polymerase isolated from both exponential phase and sporulating cells of *Bacillus subtilis*. They observed that the vegetative enzyme was able to transcribe either ϕe DNA or poly d-AT while the RNA polymerase from the sporulating cells was only able to transcribe poly d-AT but not the phage DNA. Furthermore, Losick *et al.* (1970) have also found that during sporulation, one of the β subunits of the RNA polymerase was modified and replaced by a polypeptide with a lower molecular weight. The obvious and intellectually most satisfying conclusion of these experiments was that the change in template specificity of the RNA polymerase was responsible for the arrest of the multiplication of the phage ϕe in

Table 1. *Molecular weight of subunits of RNA polymerase from different strains of Bacillus subtilis isolated from vegetative, sporulating cells and from dormant spores*

Origin of core enzyme	Molecular weight of subunits					Template DNA transcribed with veg. σ (stimulation)*	References
	σ	β	β'	α	Unknown polypeptide		
Vegetative cells:							
B. subtilis, 12A (trp⁻, Sp⁻, prot⁻)	55000	146000	146000	43000	96000	T_4	a
B. subtilis, 3610 (wt)	57000	155000	155000	45000	120000	ϕe	b
B. subtilis, 168 (leu⁻, met⁻, thr⁻, su⁻)	56000	150000	150000	43000	—	ϕ29	c
B. subtilis, WB746 (wt)	—	155000	160000	45000	—	—	d
Sporulating cells:							
B. subtilis, 3610 (wt)	60000	110000	155000	45000	120000	No stimulation	b
B. subtilis, WB74 (wt)	—	110000	160000	45000	—	—	d
Spores:							
B. subtilis, 168 (wt)	55000 (inactive)	129000	146000	43000	0	ϕe, T_4, thymus, B. sub. (no stimulation by spore σ)	a

* Stimulation by core enzyme of the transcription of DNA templates in presence of vegetative sigma factor.
a, Maia et al. 1971; b, Losick et al. 1970; c, Avila et al. 1970; d, Leighton et al. 1972.

sporulating cells and that this change was the consequence of the structural modification of the vegetative core RNA polymerase. Sonenshein & Losick (1970) further concluded that the modification was essential for the sporulation process, for a rifampycin-resistant mutant which had lost the capacity to sporulate did not show the β modification in its RNA polymerase.

One step further in these studies was taken by the demonstration by Kerjan & Szulmajster (1969) that a sigma-like factor is associated with the vegetative RNA polymerase of *Bacillus subtilis*. This factor, called σ_s, was able to stimulate the transcription of T_4 DNA by the vegetative RNA polymerase of several other strains of *Bacillus subtilis* (Losick *et al.* 1970; Avila, Hermoso, Vinŭela & Salas, 1970, 1971). These investigators agree that the vegetative *Bacillus subtilis* sigma factor has a molecular weight approximately half that of the *Escherichia coli* sigma factor (Burgess, 1969). Table 1 summarises the molecular weights of subunits of RNA polymerase from different strains of *Bacillus subtilis* and the stimulation by the sigma factor of the transcribing capacity of various template DNAs used in the different laboratories. These results seem to indicate that the base composition of the σ promotor site in the three different phage DNAs (ϕe, T_4 and ϕ29) have elements in common and may even be identical.

The RNA polymerase isolated from dormant spores of *Bacillus subtilis* (Maia, Kerjan & Szulmajster, 1971) differs from the vegetative enzyme, in at least three properties: (*a*) one of the β subunits (146000 daltons) is replaced in spores by a lower molecular weight subunit (129000 daltons); (*b*) the spore sigma protein factor is only partially dissociated from the holoenzyme and (*c*) the spore sigma factor is inactive in stimulating the transcription of T_4 DNA used as template. However, it is interesting to notice that the vegetative sigma factor was able to stimulate considerably the transcription by the spore core enzyme of all the DNAs used as templates and in particular the DNA from *Bacillus subtilis*. These results are somewhat different from those obtained by Losick *et al.* (1970), who observed that the vegetative sigma factor was unable to restore the activity of the sporulating core enzyme with ϕe DNA as template. These authors then concluded that the loss of stimulation capacity of the sporulating sigma factor is due to the modification of the β subunit in the RNA polymerase of the sporulating cells. This discrepancy may be due to the use of different template DNA by the two groups.

An alternative explanation is that the modification at T_6, observed by Losick and his colleagues, is different from the modification observed in the spore, the former being not complemented by vegetative σ while the latter is. This seems to me an attractive hypothesis since the core enzyme from the spore may have to be ready to read 'vegetative' genes during germination. It seems nevertheless that the modified spore core enzyme is functional provided a 'correct' sigma or sigma-like factor is available to allow initiation of RNA synthesis. From this point of view, the conclusion of Losick *et al.* (1970) that the observed modified core RNA polymerase from sporulating cells is responsible for the 'switch off' of the vegetative genome has to be reconsidered. In addition, there are no data available as to the time during sporulation when the β modification takes place (see Szulmajster, 1972). It is not excluded that some other alterations in the core enzyme of early stage sporulation cells might occur which could be responsible for the 'switch off' of some vegetative genes. Such alterations are known to occur immediately after phage infection of *Escherichia coli* (Seifert *et al.* 1971; Schachner, Seifert & Zillig, 1971). Whatever the structural modification of the RNA polymerase might be, it appears that the first sign of initiation of the sporulation process is the change in the template specificity of the RNA polymerase (Losick & Sonenshein, 1969). This observation was assumed by these authors to mean that in a mutant resistant to the drug rifampicin, the alteration of the polymerase is such that it retained the vegetative template specificity, and also prevented the bacteria from sporulating (Sonenshein & Losick, 1970).

It was further reported by Hussey, Losick & Sonenshein (1971) that as a consequence of the change in template specificity, ribosomal RNA synthesis was turned off early during sporulation of *Bacillus subtilis*. This was also confirmed by the in-vitro experiments with purified RNA polymerase from vegetative and sporulating cells of *Bacillus subtilis* (Hussey, Pero, Shorenstein & Losick, 1972). It was important to know if this interesting observation is a general characteristic of sporulating cells, in other words, if this phenomenon is directly related to the sporulation process. Szulmajster & Bonamy (unpublished results, 1972) have re-examined the incorporation of [^{3}H]uracil into ribosomal RNA in a number of sporogenous and asporogenous mutants grown in two different media. The asporogenous mutants used were all blocked at stage 0 of the sporulation process. Sporulation was induced in partially synchronised cultures (see Fig. 1), either by the resuspension method of Sterlini & Mandelstam (1969) or by that of Leighton *et al.* (1972).

In Fig. 2(a, b, c, d) are summarised the results obtained with various sporogenic and asporogenic strains of *Bacillus subtilis* in the two dif-

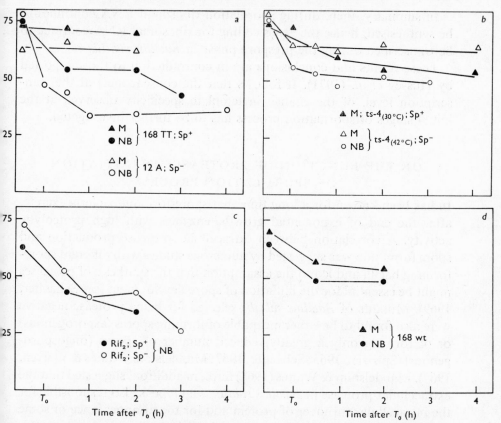

Fig. 2. Incorporation of [³H]uracil into ribosomal subunits of various sporogenic and asporogenic strains of *Bacillus subtilis*. Cells, labelled 20 min with [5-³H]uracil at various times during growth and stationary phase, were mixed with unlabelled vegetative cells and the extracts prepared by lysozyme treatment. After removal of cell debris and dialysis the extracts were sedimented through a linear sucrose gradient (5 to 20 %) in Tris–Mg buffer. The percentage of RNA in ribosomal subunits is the percentage of total radioactivity in the gradient located in the 50S and 30S peaks. M: resuspension medium of Sterlini & Mandelstam (1969). NB: double strength nutrient broth medium (Leighton *et al.* 1972).

ferent resuspension media. The curves show the percentage of incorporation of [³H]uracil into ribosomal subunits as a function of time during growth cycle. The following conclusions can be drawn from these experiments: (i) the ratio of ribosomal to total incorporation of [³H]-uracil decreases by only 25–40 % at the end of exponential growth and this effect seems to be due to the nature of the resuspension medium and not to the Sp⁺ or Sp⁻ phenotype of the strain. (ii) The incorporation of [³H]uracil into ribosomal RNA continues further during the stationary growth phase and is independent of the sporulation capacity of the cells.

In summary, then, during sporulation ribosomal RNA continues to be synthesised, hence the genes coding for ribosomal RNA continue to be transcribed during the prespore phase in *Bacillus subtilis*.

These results and conclusions are in contradiction to those reported by Hussey *et al.* (1971). It follows that the consequence, at the transcription level, of the change in template specificity observed at the initial stages of sporulation process has to be further investigated.

ON THE FUNCTION OF PROTEASES IN INITIATION OF SPORULATION PROCESS

In has been known for a long time that spore-forming bacteria excrete, after the end of exponential growth, enzymes with high proteolytic activity. A correlation between extracellular protease production and spore formation was suggested by numerous studies with different spore-forming bacilli and led to the assumption that the synthesis of proteases might be essential for the initiation of spore development (see Schaeffer, 1969). Mutants of *Bacillus subtilis* selected for being protease-negative were also found to be either incapable of forming spores (asporogenous) or of forming only a greatly reduced number of spores (oligosporogenous) (Spizizen, 1965; Schaeffer, 1967; Mandelstam, Waites & Warren, 1967). Mandelstam & Waites (1968) have, in addition, suggested that the extracellular protease in *Bacillus subtilis* wild type is also responsible for the intracellular turnover of protein and for the disappearance of some protein antigens. Mutants that have lost the ability to produce this protease were unable to sporulate and showed no protein turnover, whereas revertants which have regained the protease activity restored also protein turnover and sporulation capacity.

However, a causal relation between protease production and sporulation has so far not been rigorously established. Mutants of *Bacillus subtilis* were found with reduced extracellular proteolytic activity, which are still able to sporulate (Freese & Fortnagel, 1967; Shoer & Rappaport, 1972). Sporulation in *Bacillus brevis* occurs normally without the production of detectable amounts of extracellular protease (Slapikoff, Spitzer & Vaccaro, 1971). Mutants of *Bacillus cereus* have been found that produce very low levels of extracellular proteases but are still capable of sporulating (Levisohn & Aronson, 1967; Aronson, Angelo & Holt, 1971). Similarly, Millet & Aubert (1964, 1969) have shown that in *Bacillus megaterium* there is no correlation between the extracellular proteolytic activity and sporulation. Finally, a class of thermosensitive stage 0 sporulation mutants of *Bacillus subtilis* have been isolated by

Szulmajster, Bonamy & Laporte (1970) that have lost the capacity to sporulate at 42 °C, but produced apparently the same levels of protease activity as at 30 °C, the temperature at which sporulation takes place at a high rate.

It appears, therefore, that the real physiological function of the proteases is still undetermined. It is possible that in the spore-forming bacilli these proteases have a multiple function. In their extracellular form they could supply nutrients from any proteins present in the environment. Intracellularly, they could play a role in turnover and enable the cell to change its spectrum of proteins and to synthesise, from endogenous sources, spore structural components (Mandelstam, 1971). It was also proposed that the proteases play a role in eliminating repressors of sporulation (Mandelstam & Waites, 1968).

The interest in the problem of the involvement of proteases in sporulation process has been revived by the recent report by Leighton *et al.* (1972). These authors reported the isolation of a temperature-sensitive *Bacillus subtilis* mutant which is affected in its structural gene for the 'serine protease'. This mutant was able to grow normally at 30 °C and at 47 °C but at the latter temperature did not sporulate and did not produce significant amounts of the serine protease. Electron microscope examination indicated that this mutant was morphologically blocked at or prior to stage 0. Temperature shift experiments have shown that serine protease activity is required throughout the sporulation sequence. They came to the conclusion that this proteolytic activity is required for the conversion of the vegetative core RNA polymerase to the sporulation core RNA polymerase. This conclusion was supported by two sets of data: (1) RNA polymerase isolated from the thermosensitive mutant cells grown at 47 °C, contrary to the wild type cells, continues to transcribe ϕe DNA (although at a much lower rate) during the stationary phase of growth and shows no modification in the β subunit of the RNA polymerase (Losick *et al.* 1970). This modification was observed when the cells were grown at 30 °C. (2) Purified vegetative *Bacillus subtilis* RNA polymerase undergoes the β modification *in vitro* when incubated in presence of the serine protease.

It is important at this point to remember that in *Bacillus subtilis* three proteolytic extracellular enzymes have been observed when the cells were grown in nutrient broth (Millet, 1970). Two of the proteases were producing during the early stages of sporulation (from $T_0-T_{4.5}$). One of these is a serine protease and accounts for 80 % of the total proteolytic activity. The other belongs to the metal group of enzymes and accounts for 20 % of the total proteolytic activity. The third enzyme, probably an esterase, showed a high esterolytic activity (85 %) and only low proteolytic activity (15 %).

The data presented by Leighton and his colleagues (1972) strongly suggest that the extracellular serine protease (isolated by the method of Millet, 1970) is necessary for the initiation of the overall sporulation process in *Bacillus subtilis*. According to them, this initiation is, in the first place, expressed *in vivo* by the modification of the core RNA polymerase. They assume then, in agreement with Losick *et al.* (1970), that the β modification of the RNA polymerase is a prerequisite for the 'switch on' of the spore genes. However, the experiments of Leighton *et al.* (1972) and their interpretation need to be re-examined in relation to those obtained by Losick *et al.* (1970) and also in the light of some recent experiments carried out in this laboratory. It has to be emphasised that the available data do not allow one to assert that the change in template specificity observed at the end of exponential growth results from the β modification, considering that the latter has been demonstrated by Losick *et al.* (1970) in the polymerase from cells at T_6. It is quite possible that during sporulation as during phage infection, alterations of the RNA polymerase could take place in all the subunits of the enzyme at different times and are of various kinds.

Whatever the physiological significance of the β modification may be, the question of how this modification of the RNA polymerase takes place *in vivo* remains unknown. There are, of course, always some difficulties in answering such a question, in particular when the results obtained are based on experiments carried out in a test tube. However, by using conditions which are more likely to approach physiological conditions the investigator might get as near as possible to the in-vivo state and there should be some reasonable hope that inside the cell the event under examination takes place in a similar way. Some of the conclusions of Leighton and his colleagues described above are based on in-vitro experiments which, we believe, are far removed from the physiological conditions inside a sporulating cell, and also far removed from what is known about the properties of the serine protease of *Bacillus subtilis* (Millet, 1970). It is doubtful that this extracellular enzyme acts *in vivo* on the DNA-dependent RNA polymerase located in the nuclear fraction of the cell. This scepticism is not gratuitous but is based on the recent results obtained by Millet, Kerjan, Aubert & Szulmajster (1972). The isolation from *Bacillus megaterium* and *Bacillus subtilis* of an intracellular protease (located exclusively in the cytoplasm) was described by Millet (1971). It was shown that this protease has no activity with the substrates normally used for carboxypeptidases A and B. The only substrates so far tested which are hydrolysed by this protease are the insoluble azocoll, the β chain of carboxymethylated insulin and cytochrome c. In addition, the specific activity of this intracellular enzyme is very low during exponential growth and increases sharply before the appearance of refractility in the sporulating bacteria. It seemed to us, therefore, that this protease rather than the extracellular enzyme should be a good candidate for a proteolytic modification of the DNA-dependent RNA polymerase.

Studies were therefore undertaken in this direction with the isolation and complete purification of the intracellular protease from a mutant of *Bacillus megaterium*, defective in the single extracellular protease (termed megateriopeptidase) present in the wild type (Millet *et al.* 1972). By the use of this mutant contamination of the enzyme by the extracellular protease was thus avoided.

When purified *Bacillus subtilis* vegetative RNA polymerase was incubated in the presence of the *Bacillus megaterium* intracellular proteolytic enzyme under suitable conditions, a structural modification was observed which rendered the RNA polymerase similar to that of the dormant spores. Furthermore, the in-vitro modified RNA polymerase retained the capacity to transcribe different DNA templates as does the spore enzyme. Finally, what is more important for the discussion here, is that when highly purified extracellular serine protease from *Bacillus subtilis* was incubated under the same conditions but at a concentration about 100 times lower than the intracellular enzyme, virtually complete degradation of all the subunits of the RNA polymerase was observed. This is also true even when the incubation was carried out for 2 min instead of 30 min and at 4 °C instead of 37 °C. A similar result was obtained when the serine protease was replaced in the incubation mixture by trypsin.

These results, therefore, provide strong evidence for the structural conversion of the vegetative RNA polymerase of *Bacillus subtilis* by an intracellular proteolytic enzyme and this mechanism might well be the physiological means by which the transcriptional enzyme is modified *in vivo*. There are many examples in the literature of proteolytic conversion which concern various enzymatic systems (Klee, 1969; Cassio & Waller, 1971). Such a mechanism is particularly attractive for explaining the changes in the behaviour of different enzymes occurring during sporulation or other developmental processes. A classical example with respect to sporulation is that reported by Sadoff *et al.* (1970) for the conversion of the vegetative fructose 1,6-diphosphate aldolase isolated from *Bacillus cereus* to the spore form enzyme by a protease purified from the same micro-organism. In general, by this type of mechanism, the differentiated cell economises on the synthesis of a great number of macromolecules involved in *de novo* protein synthesis.

Finally, in discussing a possible role of extracellular proteases in sporulation process, one cannot discard the idea that a common step might control the regulation of the production of proteases and that of certain enzymes involved in the synthesis of an early spore specific structural component (Schaeffer, Millet & Aubert, 1965). This hypothesis is substantiated by the experiments of Levisohn & Aronson (1967), showing that protease production is regulated by the level of one or more catabolic or biosynthetic intermediates.

USE OF CONDITIONAL SPORULATION MUTANTS
FOR THE STUDIES OF
INITIATION PROCESS OF SPORE FORMATION

One way of studying the biochemical nature of initiation of the sporulation process is the use, if available, of conditional mutants affected in this function under restrictive and not under permissive conditions. Temperature-sensitive mutations belong to this class, and have already proven to be extremely useful tools in studies of activating enzymes (Neidhardt, 1966), ribosomes (Philips, Schlessinger & Apirion, 1969; Tai, Kessler & Ingraham, 1969), DNA synthesis (Kohiyama, Cousin, Ryter & Jacob, 1966) and other aspects of macromolecular synthesis in prokaryotic organisms. Obviously temperature-sensitive mutants affected at particular stages of spore development would be extremely useful in elucidating the biochemical events and their regulation during bacterial sporogenesis. Surprisingly few studies utilising temperature-sensitive sporulation mutants have been reported (Lundgren & Beskid, 1960; Lundgren & Cooney, 1962; Leighton *et al.* 1972).

In order to gain information on the nature of the primary initiation step and on the number of indispensable functions leading to the formation of a mature spore, we have, during the last years, focused our attention on studies of temperature-sensitive sporulation mutants of *Bacillus subtilis* isolated in my laboratory (Szulmajster *et al.* 1970). These mutants are able to perform all the vegetative functions at 30 °C (permissive temperature) or at 42 °C (restrictive temperature) but are unable to sporulate at the latter temperature.

Genetic mapping of two such mutants (t_s-4 and t_s-B8) by transduction using the phage PBS1 (Takahashi, 1961) have shown that the mutation is localised between *ura* and *cys c* on the *Bacillus subtilis* chromosome

(Anagnostopoulos, Arnaud & Szulmajster, unpublished results, 1971). Electron microscope studies of two of these mutants, t_s-4 and t_s-B8, have shown that they were blocked at stage 0 of spore development (Plate 2).

Physiological studies on the t_s-4 mutant have strengthened the idea that these thermosensitive sporulation mutants are deficient in an initial step of sporulation process at the non-permissive temperature. This was shown by shifting the mutant cultures from the permissive to the non-permissive temperature and vice versa at different times during the growth cycle and then analysing their sporulating capacity. For example, the thermosensitive mutant t_s-4 was grown at 42 °C for different periods of time, shifted to 30 °C and left at this temperature until T_{22} for spore counts. If the shift was done before T_0-T_1 the cells were unable to regain the sporulation capacity. Similar results were obtained if the cells were shifted from 30 °C to 42 °C (Szulmajster et al. 1970). These experiments therefore strongly suggested that the temperature-sensitive event occurs in a very early stage of spore development. Is this event directly related to spore formation? As already stated in this paper, it is now generally believed that the developmental process in a sporulating cell is a continuing, time-ordered sequence of biochemical reactions initiated by an early event (Halvorson, 1965). If so, the t_s-4 mutant, grown at the restrictive temperature, would be expected to be defective in spore-specific enzymes involved in the late process of sporulation. In other words, this mutant should be of pleiotropic nature and behave as some of the SpoA$^-$ mutants (Schaeffer, 1969; Michel & Cami, 1969; Hoch & Spizizen, 1969). This is exactly what we observed. It is now well established that one of the late spore-specific events is the biosynthesis of dipicolinic acid (DPA) (Bach & Gilvarg, 1966; Chasin & Szulmajster, 1967, 1969). DPA is a low molecular-weight component specific to bacterial spores and completely absent in vegetative cells (Powell, 1953). DPA (in the form of its calcium salt) has been shown to be associated with the heat-resistant properties of bacterial spores (Black, Hashimoto & Gerhardt, 1960; Levinson, Hyatt & Moore, 1961). Spores deficient in DPA as a result of mutation have decreased heat-resistance and altered rates of germination (Aronson, Henderson & Tincher, 1967; Fukuda & Gilvarg, 1968; Halvorson & Swanson, 1969). It was shown by Chasin & Szulmajster (1967, 1969) that in Bacillus subtilis the specific activities of the dihydrodipicolinate synthetase and the dipicolinate synthetase increase sharply just before the onset of sporulation and reach a maximum at about T_9, at which time the majority of the cell population becomes thermoresistant. They have also demonstrated that this increase in activity corresponds to a *de novo* synthesis of these two enzymes. This synthesis precedes both the production of DPA and the occurrence of thermal resistance in the culture.

The t_s-4 mutant, when grown at the non-permissive temperature, was devoid of both of these enzymes in the stationary phase of growth, while in the parent strain the normal increase in activity was observed both at 30 °C and 42 °C.

It was important to show that only spore-specific enzymes were affected by the thermosensitive mutation. This was achieved by measuring the activities of alanine dehydrogenase and aconitase, enzymes not directly related to the sporulation process. There was no significant difference in the behaviour of these two enzymes in the cultures grown at the two different temperatures (Szulmajster *et al.* 1970). These results lead to the conclusion that the t_s-4 thermosensitive sporulation mutant is blocked early in the specific sequence of biochemical reactions leading to the formation of the mature spore.

In general, early blocked sporulation mutants exhibit a pleiotropic phenotype which is diagnostic for regulatory elements. It was suggested on the basis of genetic studies (Hoch & Spizizen, 1969) that the early events in sporulation might be under positive control mediated by a regulatory protein whose presence is necessary for the expression of spore structural genes. It is therefore possible that the functional absence of such a regulatory protein, brought about by growing the t_s-4 mutant at the non-permissive temperature, results in a pleiotropic phenotype which, among other things, has lost the ability to synthesise the DPA required for the formation of a mature spore.

Whatever may be the nature of the lesion in the earlier sporulation function in this mutant, it can be assumed that the expression of the affected gene could be controlled either at the transcription or at the translation level or at both.

As noticed above, infection of *Bacillus subtilis* by the virulent phage ϕe at a critical stage during sporulation resulted in failure to produce phage, while a non-sporulating mutant infected at the corresponding stationary phase gave rise to the normal burst size of this phage (Sonenshein & Roscoe, 1969).

Infection of the t_s-4 partially synchronised cultures and the parental strain by the phage ϕe at various times during growth cycle at 42 °C has shown that at a critical stage in the stationary phase the average burst size is higher in the mutant cell than in the cells of the parental strain. No such difference was observed when the cells of these two strains were infected at 30 °C. This behaviour of the phage ϕe is paralleled by changes in activity of the RNA polymerase in the presence of φe DNA or d-AT as templates (Orrego, Kerjan & Szulmajster, unpublished results, 1972). The exact significance of these results is not clear yet. It is possible that some structural changes in the RNA polymerase are responsible for this phage behaviour as in the case studied by Losick & Sonenshein (1969).

AND NOW WHAT? (ANTICONCLUSIONS)

After this tour of the various problems related to sporulation, it is time to return to the starting point, and to ask, once again, how is the sporulation process initiated? Pertinent to this question is the more general problem of how the subsequent events of spore development are controlled, once the process is initiated.

First signal

It is generally agreed that the initial trigger event for sporulation is starvation, which, in turn, supposed that the levels of various catabolites in the environment provide the starting signal for the entire process (Grelet, 1951a, b, 1957). This leads one directly to the hypothesis that during vegetative growth glucose causes inhibition of sporulation in the presence of an assimilable nitrogen source by the mechanism known as catabolite repression (Schaeffer et al. 1965). After cessation of growth, however, when nutrients are exhausted from the medium, this repression is released. For a better understanding of the mechanism of this release one has then assumed that there must be some sort of 'carrier' which transmits the environmental signal to the informational centre inside the cell where the fraction of information to be read in the first place is selected.

The effect of glucose on the repression of sporulation has to be re-examined in the light of the present knowledge of the so-called 'glucose effect' on inducible enzymes. It is now well established that glucose reduces the level of cyclic adenosine 3',5'-monophosphate (cyclic AMP) in the cell (Makman & Sutherland, 1965). Originally a role of 'second messenger' was attributed to cyclic AMP in eukaryotic cells for transmitting the effect of an hormone acting on the adenylate cyclase at the cell membrane. Later on, cyclic AMP was shown to be necessary for the stimulation of inducible enzymes in *Escherichia coli*, for the formation of bacterial flagella, and for a number of other biological systems (see review of Pastan & Perlman, 1970). It was further shown that the stimulation of inducible enzyme synthesis in the presence of cyclic AMP required a specific protein, the cyclic AMP receptor protein (CRP) (de

Crombrugghe *et al.* 1971). Recent studies with purified components of the in-vitro transcription system (RNA polymerase, ρ factor and cyclic AMP receptor protein) have shown that CRP regulates the transcription of *lac* and *gal* operons by binding at the promotor sites of DNA and increasing the frequency of initiation (Eron *et al.* 1971; Nissley *et al.* 1971). These authors have suggested that the formation of a complex between DNA and cyclic AMP binding protein is probably necessary for binding and action of the RNA polymerase.

In summary then, these studies clearly indicate that the release of specific repression of inducible enzymes (negative control) by the inducer is not sufficient for full expression of the operon genes (*lac* and *gal* for example) but requires an additional positive control system, namely, cyclic AMP and CRP.

It can reasonably be assumed that the repression of sporulation by glucose is due to fluctuations of the level of cyclic AMP present in vegetative and sporulating cells. Because the enzyme which synthesises cyclic AMP, the adenylate cyclase, is located at the cell membrane, it was suggested that this enzyme plays a role in the response of the cell to environmental conditions. It is possible that in bacterial sporogenesis it is the adenylate cyclase that plays a role as 'carrier' between the external environment and the corresponding receptive sites in the cell. However, no adenylate cyclase has been found in *Bacillus subtilis* (Ide, 1971) and there are no data available at present as to the levels of cyclic AMP in vegetative and sporulating cells.

State of the DNA

Consider now the two main elements of the transcriptional machinery, RNA polymerase and DNA, and their role in the initiation of the sporulation program.

According to Dawes, Kay & Mandelstam (1971) and Mandelstam, Sterlini & Kay (1971), the time at which the cell is sensitive to initiation of sporulation is limited to the period just after completion of the chromosome. The starvation stimulus therefore must be applied while the DNA is still replicating. Once replication is completed in rich medium, the cells will continue their vegetative cell division. These results are in agreement with the earlier observation by Szulmajster & Canfield (1965) using the ³²P-suicide technique and with the similar results obtained by Ryter & Aubert (1969) using the technique of autoradiography combined with electron microscopy. Both types of experiments have indicated that sporulation starts after the last cell division.

RNA polymerase

The next step is the initiation of transcription of the correct segments of the DNA by the RNA polymerase under the control of a number of factors. It is reasonable to admit, as does Travers (1971), that at the transcriptional level, only the control of initiation specificity can provide the necessary flexibility for most of the transcriptional transitions. This is because, in most morphogenetic processes as in bacterial sporogenesis, bacterial germination and phage infection, the same basic elements of the transcriptional machinery are being utilised by the cell.

The first observable sign of the activation of sporulation genes in *Bacillus subtilis* is a change in the template specificity of the RNA polymerase (Losick & Sonenshein, 1969) (see above). Further evidence that the RNA polymerase is directly involved in the sporulation process is based on the studies of rifampicin-resistant mutants of *Bacillus subtilis* (Sonenshein & Losick, 1970). These studies have revealed a class of mutants that, by virtue of a single mutational event, has gained simultaneously drug resistance *in vivo* and *in vitro*, and lost the ability to sporulate. This mutation also prevents the change in template specificity of the RNA polymerase at a critical stage in growth cycle. A further modification of RNA polymerase which has been observed is that of alteration of a β subunit of the core polymerase (Losick *et al.* 1971). The idea of these authors that this modification of the core RNA polymerase in sporulating cells may be responsible for the loss of the vegetative template specificity cannot be accepted without further evidence relative to the exact time at which the modification takes place (Szulmajster, 1972).

Relevant to the problem of modification of the RNA polymerase is the suggestion of Hussey *et al.* (1971) that as a result of this alteration ribosomal RNA genes 'turn off' during sporulation. This suggestion should be taken with care, for, in our hands, incorporation of [^{3}H]uracil into ribosomal subunits continued during the stationary phase regardless of whether or not the strains of *Bacillus subtilis* were sporogenic or asporogenic.

It is possible that the β modification or some other alterations occur at an earlier stage of spore development and might be responsible for the turning off of some vegetative genes. The primary question, however, which still remains to be answered, is how the transcription of the spore genes is initiated. Based on our experiments (Kerjan & Szulmajster, 1969) I am inclined to believe that the interaction of the vegetative *Bacillus subtilis* sigma-like factor with the modified core enzyme is responsible for the initiation. Should this be the case, then the proteolytic conversion by the intracellular protease of the vegetative RNA polymerase into the sporulating enzyme form described earlier in this paper becomes an essential prerequisite step in the control of initiation mechanism (Millet *et al*. 1972).

Once the transcription is initiated, other important questions can be raised, for example: how is subsequent sequential gene activation controlled; in other words, how is the switch from 'early' to 'late' mRNA synthesis regulated during spore development; are there new sigma factors formed, and if so, the problem is again posed of the regulation of their biosynthesis, again returning one to the initial question.

It is possible that the switching of RNA synthesis from 'early' to the 'late' genes during spore development may be partially caused by change in the affinity of RNA polymerase for the initiator sites of these operons. By analogy to what was suggested for the synthesis of late mRNA in phage-infected cells (Khesin, 1970) it is also possible that gene selection during sporulation may involve, in addition to the changes in affinity by the modified RNA polymerase, a special form of template DNA and a specific location of the DNA within the cells. It was suggested (Travers, 1971) that a specific nuclease might generate a single strand break at a certain point on the DNA template and that such a break might serve as an initiation site for RNA polymerase.

It is clear that further experiments are necessary for the complete elucidation of the mechanism of control of gene expression during bacterial sporogenesis. The discovery of the complexity of the RNA polymerase and of a number of its regulatory factors is encouraging, and provides the necessary versatility to confer the potential for specificity of transcription and modulation of positive and negative control mechanisms in prokaryotic cells.

In this paper we have been concerned mainly with the problem of controls of initiation, and of some of the subsequent events during sporulation, at the level of transcription. This, of course, does not diminish the importance of other types of controls operating in the sporulation process, particularly control at the translational level (see Szulmajster, 1972).

Recent studies by Dube & Rudland (1970) on T_4 phage infected cells can serve as a model for investigation in the spore system. It was shown that after infection a new set of protein initiation factors coded by the phage genome are made, and displace the host cell factors. As a result, the ribosomes of the *Escherichia coli* host cells recognise preferentially the start signals of the T_4 mRNAs. It follows then that the host cell protein machinery is switched off while phage proteins start to be synthesised. This mechanism would be similar to that occurring at the transcriptional level, through the function of sigma factors.

It is quite conceivable that a similar mechanism of control might operate at the translation level during spore development. Substitutions or modifications of the vegetative initiation factors would confer on the 30S ribosomes a new specificity for initiation of spore proteins. There is no doubt that studies of these aspects will be carried out in the near future.

Finally, I would like to add one conclusion to the 'anticonclusions' and that is to say that the subject under review in this paper is in a particularly dynamic state of development. This implies that we have to avoid preconceived ideas and premature generalisations.

The author is grateful to Drs Norman Strauss and Cristian Orrego for critically reading the manuscript and valuable discussions. This work was partially supported by grants from the 'Délégation Générale à la Recherche Scientifique et Technique', the Commissariat à l'Énergie Atomique, France, and the Fondation pour la Recherche Médicale Française.

REFERENCES

ARONSON, A. J. (1965a). Characterization of messenger RNA in sporulating *Bacillus cereus*. *J. Mol. Biol.* **11**, 576.

ARONSON, A. J. (1965b). Membrane-bound messenger RNA and polysomes in sporulating bacteria. *J. Mol. Biol.* **13**, 92.

ARONSON, A. J., ANGELO, N. & HOLT, S. C. (1971). Regulation of extracellular protease production in *Bacillus cereus* T: characterization of mutants producing altered amounts of protease. *J. Bacteriol.* **106**, 1016.

ARONSON, A. J., HENDERSON, E. & TINCHER, A. (1967). Participation of the lysine pathway in dipicolinic acid synthesis in *Bacillus cereus* T. *Biochem. Biophys. Res. Commun.* **26**, 454.

AUBERT, J. P. & MILLET, J. (1963). L'induction de la β-galactosidase au cours de la sporulation chez *Bacillus megaterium*. *C. R. Acad. Sci. Paris*, **256**, 5442.

AUBERT, J. P. & MILLET, J. (1965). Induction de la β-galactosidase chez *Bacillus megaterium* au cours de la sporulation. In *Mécanismes de régulation chez les microorganismes. Coll. Int. Centre Nat. Rech. Sci. France*, **124**, 545.

AVILA, J., HERMOSO, J. M., VINÜELA, E. & SALAS, M. (1970). Subunit composition of *B. subtilis* RNA polymerase. *Nature*, **226**, 1244.

AVILA, J., HERMOSO, J. M., VINÜELA, E. & SALAS, M. (1971). Purification and properties of DNA-dependent RNA polymerase from *Bacillus subtilis* vegetative cell. *Eur. J. Biochem.* **21**, 526.

BACH, M. L. & GILVARG, C. (1966). Biosynthesis of dipicolinic acid in sporulating *Bacillus megaterium*. *J. Biol. Chem.* **241**, 4563.

BALASSA, G. (1963). Renouvellement de l'acide ribonucléique au cours de la sporulation de *Bacillus subtilis*. *Biochim. Biophys. Acta*, **76**, 410.

BALASSA, G. (1966). Synthèse et fonction des ARN messagers au cours de la sporulation de *Bacillus subtilis*. *Ann. Inst. Pasteur*, **110**, 175.

BALASSA, G. (1969). Bacterial genetics of bacterial sporulation. I. Unidirectional pleiotropic interactions. *Molec. Gen. Genetics*, **104**, 73.

BALASSA, G. (1971). The genetic control of spore formation. In *Current Topics in Microbiology and Immunology*, vol. 56, p. 99. Berlin: Springer-Verlag.

BALASSA, G. & YAMAMOTO, T. (1970). Biochemical genetics of bacterial sporulation. III. Correlation between morphological and biochemical properties of sporulation mutants. *Molec. Gen. Genetics*, **108**, 1.

BAUTZ, E. K. F., BAUTZ, F. A. & DUNN, J. J. (1969). *E. coli* σ factor: a positive control element in phage T₄ development. *Nature*, **223**, 1022.

BAUTZ, E. K. F. & DUNN, J. J. (1969). DNA-dependent RNA polymerase from phage T₄-infected *E. coli*: an enzyme missing a factor required for transcription of T₄ DNA. *Biochem. Biophys. Res. Commun.* **34**, 230.

BAYEN, H., FREHEL, C., RYTER, A. & SEBALD, M. (1967). Etude cytologique de la sporulation chez *Clostridium histolyticum*: souche sporogène et mutants de sporulation. *Ann. Inst. Pasteur*, **113**, 163.

BERNLOHR, R. W. & LEITZMAN, C. (1969). Control of sporulation. In *The Bacterial Spore*, ed. G. W. Gould & A. Hurst, p. 183. New York: Academic Press.

BLACK, S. H. & GERHARDT, P. (1963). 'Endotrophic' sporulation. *Ann. New York Acad. Sci.* **102**, 755.

BLACK, S. H., HASHIMOTO, T. & GERHARDT, P. (1960). Calcium reversal of the heat susceptibility and dipicolinate deficiency of spores formed 'endotrophically' in water. *Can. J. Microbiol.* **6**, 213.

BOLLE, A., EPSTEIN, R. H., SALSER, W. & GEIDUSCHEK, E. P. (1968). Transcription during bacteriophage T₄ development: requirement of late messenger synthesis. *J. Mol. Biol.* **33**, 339.

BURGESS, R. R. (1969). Separation and characterization of the subunits of ribonucleic acid polymerase. *J. Biol. Chem.* **244**, 6168.

BURGESS, R. R., TRAVERS, A. A., DUNN, J. J. & BAUTZ, E. K. F. (1969). Factor stimulating transcription by RNA polymerase. *Nature*, **221**, 43.

CALENDAR, R. (1970). The regulation of phage development. *Ann. Rev. Microbiol.* **24**, 241.

CASCINO, A., RIVA, S. & GEIDUSCHEK, E. P. (1970). DNA ligation and the coupling of T₄ late transcription to replication. *Cold Spring Harbor Symp. Quant. Biol.* **35**, 213, ed. L. Silvestri.

CASSIO, D. & WALLER, J. P. (1971). Modification of methionyl-tRNA synthetase by proteolytic cleavage and properties of the trypsin-modified enzyme. *Europ. J. Biochem.* **20**, 283.

CHASIN, L. A. & SZULMAJSTER, J. (1967). Biosynthesis of dipicolinic acid in *Bacillus subtilis*. *Biochem. Biophys. Res. Commun.* **29**, 648.

CHASIN, L. A. & SZULMAJSTER, J. (1969). Enzymes of dipicolinic acid biosynthesis in *Bacillus subtilis*. In *Spores*, vol. IV, ed. L. L. Campbell, p. 133. Bethesda, Md.: Am. Society for Microbiol.

DE CROMBRUGGHE, B., CHEN, B., ANDERSON, W., NISSLEY, P., GOTTESMAN, M. & PASTAN, I. (1971). *Lac* DNA, RNA polymerase and cyclic AMP receptor protein, cyclic AMP, *lac* repressor and inducer are the essential elements for controlled *lac* transcription. *Nature, New Biol.* **231**, 139.

DAWES, I. W., KAY, D. & MANDELSTAM, J. (1971). Determining effect of growth medium on the shape and position of daughter chromosomes and on sporulation in *Bacillus subtilis*. *Nature*, **230**, 567.

DiCioccio, R. A. & Strauss, N. (1971). Annealing studies of transcription in *B. subtilis*. *Biochem. Biophys. Res. Commun.* **45**, 212.

Doi, R. H. & Igarashi, R. T. (1964). Genetic transcription during morphogenesis. *Proc. Nat. Acad. Sci. U.S.A.* **52**, 755.

Doran, W. L. (1922), quoted by Sussman, A. S. & Halvorson, H. O. (1966). In *Spores, their dormancy and germination*, ed. Harper & Row, p. 3.

Dube, S. K. & Rudland, P. S. (1970). Control of translation by T_4 phage: altered binding of disfavoured messengers. *Nature*, **226**, 820.

Ellar, D. J. & Lundgren, D. G. (1966). Fine structure of sporulation in *Bacillus cereus* grown in a chemically defined medium. *J. Bacteriol.* **92**, 1748.

Emmer, M., de Crombrugghe, B., Pastan, I. & Perlman, R. (1970). Cyclic AMP receptor protein of *E. coli*: its role in the synthesis of inducible enzymes. *Proc. Nat. Acad. Sci. U.S.A.* **66**, 480.

Eron, L., Ardilti, R., Zubay, G., Connaway, S. & Beckwith, J. R. (1971). An adenosine 3′:5′-cyclic monophosphate-binding protein that acts on the transcription process. *Proc. Nat. Acad. Sci. U.S.A.* **68**, 215.

Fitz-James, P. (1965). Spore formation in wild type and mutant strains of *B. cereus* and some effects of inhibitors. In *Mécanismes de régulation chez les microorganismes. Coll. Int. Centre Nat. Rech. Sci. France*, **124**, 529.

Fitz-James, P. & Young, I. E. (1969). Morphology of sporulation. In *The Bacterial Spore*, ed. G. W. Gould & A. Hurst, p. 39. New York: Academic Press.

Fortnagel, P. & Freese, E. (1968). Analysis of sporulation mutants. II. Mutants blocked in the citric acid cycle. *J. Bacteriol.* **95**, 1431.

Freese, E. & Fortnagel, P. (1967). Analysis of sporulation mutants. I. Response of uracil incorporation to carbon sources and other mutant properties. *J. Bacteriol.* **94**, 1957.

Freese, E., Fortnagel, P., Schmitt, R., Klopat, W., Chappelle, E. & Picciolo, G. (1969). Biochemical genetics of initial sporulation stages. In *Spores*, vol. IV, ed. L. L. Campbell, p. 82. Bethesda, Md.: Am. Society for Microbiol.

Freese, E., Klopat, W. & Galliers, E. (1970). Commitment to sporulation and induction of glucose–phosphoenolpyruvate–transferase. *Biochim. Biophys. Acta*, **222**, 265.

Fréhel, C. & Ryter, A. (1969). Réversibilité de la sporulation chez *Bacillus subtilis*. *Ann. Inst. Pasteur*, **117**, 297.

Fukuda, A. & Gilvarg, C. (1968). The relationship of dipicolinate and lysine biosynthesis in *Bacillus subtilis*. *J. Biol. Chem.* **243**, 3871.

Goff, C. G. & Weber, K. (1970). A T_4-induced RNA polymerase subunit modification. *Cold Spring Harbor Symp. Quant. Biol.* **35**, 101.

Gould, G. W. & Hurst, A. (eds.) (1969). *The Bacterial Spore*. New York: Academic Press.

Grelet, N. (1951a). Le déterminisme de la sporulation de *Bacillus megaterium*. I. L'effet de l'épuisement de l'aliment carboné en milieu synthétique. *Ann. Inst. Pasteur*, **81**, 430.

Grelet, N. (1951b). Le déterminisme de la sporulation de *Bacillus megaterium*. II. L'effet de la pénurie des constituants minéraux du milieu synthétique. *Ann. Inst. Pasteur*, **82**, 62.

Grelet, N. (1957). Growth limitation and sporulation. *J. Appl. Bacteriol.* **20**, 315.

Guha, A. & Szybalski, W. (1968). Fractionation of the complementary strands of coliphage T_4 DNA based on the asymmetric distribution of the poly U and poly U-G binding sites. *Virology*, **34**, 608.

Halvorson, H. O. (1965). Sequential expression of biochemical events during intracellular differentiation. In *Symp. Soc. Gen. Microbiol.* **15**, 343.

HALVORSON, H. O. & SWANSON, A. (1969). Role of dipicolinic acid in the physiology of bacterial spores. In *Spores*, vol. IV, ed. L. L. Campbell, p. 121. Bethesda, Md.: Am. Society for Microbiol.

HANSON, R. S., BLICHARSKA, J. & SZULMAJSTER, J. (1964). Relationship between tricarboxylic acid cycle enzymes and sporulation. *Biochem. Biophys. Res. Commun.* **17**, 1.

HANSON, R. S., PETERSON, J. A. & YOUSTEN, A. A. (1970). Unique biochemical events in bacterial sporulation. *Ann. Rev. Microbiol.* **24**, 53.

HANSON, R. S., SRINIVASAN, V. R. & HALVORSON, H. O. (1963*a*). Biochemistry of sporulation. I. Metabolism of acetate by vegetative and sporulating cells of *Bacillus cereus. J. Bacteriol.* **85**, 451.

HANSON, R. S., SRINIVASAN, V. R. & HALVORSON, H. O. (1963*b*). Biochemistry of sporulation. II. Enzymatic changes during sporulation of *Bacillus cereus. J. Bacteriol.* **86**, 45.

HARDWICK, W. A. & FOSTER, J. W. (1952). On the nature of sporogenesis in some aerobic bacteria. *J. Gen. Physiol.* **35**, 907.

HASELKORN, R., VOGEL, M. & BROWN, R. D. (1969). Conservation of rifamycin sensitivity of transcription during T_4 development. *Nature*, **221**, 836.

HASELTINE, W. A. (1972). *In vitro* transcription of *Escherichia coli* ribosomal RNA genes. *Nature*, **235**, 329.

HOCH, J. A. (1971). Selection of cells transformed to prototropy for sporulation markers. *J. Bacteriol.* **105**, 1200.

HOCH, J. A. & SPIZIZEN, J. (1969). Genetic control of some early events in sporulation of *Bacillus subtilis*, 168. In *Spores*, vol. III, ed. L. L. Campbell & H. O. Halvorson, p. 112. Ann Arbor, Michigan: Am. Society for Microbiol.

HUSSEY, C., LOSICK, R. & SONENSHEIN, A. L. (1971). Ribosomal RNA synthesis is turned off during sporulation of *Bacillus subtilis. J. Mol. Biol.* **57**, 59.

HUSSEY, C., PERO, J., SHORENSTEIN, R. G. & LOSICK, R. (1972). *In vitro* synthesis of ribosomal RNA by *Bacillus subtilis* RNA polymerase. *Proc. Nat. Acad. Sci. U.S.A.* **60**, 407.

IDE, M. (1971). Adenyl cyclase of bacteria. *Arch. Biochem. Biophys.* **144**, 262.

KANO-SUEOKA, T. & SPIEGELMAN, S. (1962). Evidence for a non random reading of the genome. *Proc. Nat. Acad. Sci. U.S.A.* **48**, 1942.

KERJAN, P. & SZULMAJSTER, J. (1969). DNA-dependent RNA polymerase from vegetative cells and from dormant spores of *B. subtilis*. III. Isolation of a stimulating factor. *FEBS Letters*, **5**, 288.

KHESIN, R. B. (1970). Studies on the RNA synthesis and RNA polymerase in normal and phage infected *E. coli* cells. In *Lepetit Colloquia Biol. Med.* 1. *RNA polymerase and transcription*, p. 167. Amsterdam: North-Holland.

KHESIN, R. B. & SHEMYAKIN, M. F. (1962). Some properties of information RNA's and their complexes with DNA. *Biochimiya*, **27**, 761.

KLEE, C. B. (1969). The proteolytic conversion of polynucleotide phosphorylase to a primer-dependent form. *J. Biol. Chem.* **244**, 2558.

KOBAYASHI, Y., STEINBERG, W., HIGA, A., HALVORSON, H. O. & LEVINTHAL, C. (1965). In *Spores*, vol. III, ed. L. L. Campbell & H. O. Halvorson, p. 200. Ann Arbor, Michigan: Am. Society for Microbiol.

KOHIYAMA, M., COUSIN, D., RYTER, A. & JACOB, F. (1966). Mutants thermosensibles d'*Escherichia coli* K12. *Ann. Inst. Pasteur*, **110**, 30.

KORNBERG, A., SPUDICH, J. A., NELSON, D. L. & DEUTSCHER, M. P. (1968). Origin of proteins in sporulation. *Ann. Rev. Biochem.* **37**, 51.

LEIGHTON, T. J. & DOI, R. H. (1971). The stability of messenger ribonucleic acid during sporulation in *Bacillus subtilis. J. Biol. Chem.* **246**, 3189.

LEIGHTON, T. J., FREESE, P. K., DOI, R. H., WARREN, A. J. & KELLN, R. A. (1972). Initiation of sporulation in *Bacillus subtilis*: requirement for serine protease activity and RNA polymerase modification. In *Spores*, vol. v, ed. H. O. Halvorson, R. Hanson & L. L. Campbell, p. 285. Ann Arbor, Michigan: Am. Society for Microbiol.

LEVINSON, H. S., HYATT, M. T. & MOORE, F. E. (1961). Dependence of the heat resistance of bacterial spores on the calcium:dipicolinic ratio. *Biochem. Biophys. Res. Commun.* **5**, 417.

LEVINTHAL, C., FAN, D. P., HIGA, A. & ZIMMERMAN, R. A. (1963). The decay and protection of messenger RNA in bacteria. *Cold Spring Harbor Symp. Quant. Biol.* **28**, 183.

LEVISOHN, S. & ARONSON, A. I. (1967). Regulation of extracellular protease in *Bacillus cereus*. *J. Bacteriol.* **93**, 1023.

LOSICK, R. (1972). *In vitro* transcription. *Ann. Rev. Biochem.* (In Press.)

LOSICK, R., SHORENSTEIN, R. G. & SONENSHEIN, A. L. (1970). Structural alteration of RNA polymerase during sporulation. *Nature*, **227**, 910.

LOSICK, R. & SONENSHEIN, A. L. (1969). Change in the template specificity of RNA polymerase during sporulation of *Bacillus subtilis*. *Nature*, **224**, 35.

LUNDGREN, D. G. & BESKID, G. (1960). Isolation and investigation of induced asporogenic mutants. *Can. J. Microbiol.* **6**, 135.

LUNDGREN, D. G. & COONEY, J. J. (1962). Chemical analysis of asporogenic mutants of *Bacillus cereus*. *J. Bacteriol.* **83**, 1287.

LUNDGREN, D. G., KARP, D. F. & LANG, D. R. (1969). Structure-function relationship in bacterial sporulation. In *Spores*, vol. iv, ed. L. L. Campbell & H. O. Halvorson, p. 20. Ann Arbor, Michigan: Am. Society for Microbiol.

LURIA, S. (1960). The bacterial protoplasm: composition and organization. In *The Bacteria*, vol. i, ed. I. C. Gonsalus & R. Y. Stanier, p. 1. New York: Academic Press.

MAIA, J. C. C., KERJAN, P. & SZULMAJSTER, J. (1971). DNA-dependent RNA polymerase from vegetative cells and from spores of *Bacillus subtilis*. IV. Subunit composition. *FEBS Letters*, **13**, 269.

MAKMAN, R. S. & SUTHERLAND, E. W. (1965). Adenosine 3′,5′-phosphate in *Escherichia coli*. *J. Biol. Chem.* **240**, 1309.

MANDELSTAM, J. (1969). Regulation of bacterial spore formation. *Symp. Soc. Gen. Microbiol.* **19**, 377.

MANDELSTAM, J. (1971). Recurring patterns during development in primitive organisms. In *Control Mechanisms of Growth and Differentiation. Symp. Soc. Exp. Biol.* **25**, 1.

MANDELSTAM, J., STERLINI, J. M. & KAY, D. (1971). Sporulation in *Bacillus subtilis*. *Biochem. J.* **125**, 635.

MANDELSTAM, J. & WAITES, W. M. (1968). Sporulation in *Bacillus subtilis*. The role of exoprotease. *Biochem. J.* **109**, 793.

MANDELSTAM, J., WAITES, W. M. & WARREN, W. C. (1967). Exoprotease activity and its relationship to some other biochemical events involved in the sporulation of *Bacillus subtilis*. *Abstr. 7th Int. Congr. Biochem., Tokyo Symp.* V-2, **1**, 253.

MICHEL, J. F. & CAMI, B. (1969). Sélection de mutants de *Bacillus subtilis* bloqués au début de la sporulation. III. Nature des mutations sélectionnées. *Ann. Inst. Pasteur*, **116**, 3.

MILLET, J. (1970). Characterization of proteinases excreted by *Bacillus subtilis*, Marburg strain during sporulation. *J. Appl. Bacteriol.* **30**, 207.

MILLET, J. (1971). Caractérisation d'une endopeptidase cytoplasmique chez *Bacillus megaterium* en vie de sporulation. *C. R. Acad. Sci.* **272**, 1806.

MILLET, J. & AUBERT, J. P. (1964). Rôle d'une protease exocellulaire de *Bacillus megaterium. C. R. Acad. Sci.* **259**, 2555.

MILLET, J. & AUBERT, J. P. (1969). Étude de la megateriopeptidase, protease exocellulaire de *Bacillus megaterium*. Biosynthèse et rôle physiologique. *Ann. Inst. Pasteur,* **117**, 461.

MILLET, J., KERJAN, P., AUBERT, J. P. & SZULMAJSTER, J. (1972). Proteolytic conversion *in vitro* of *B. subtilis* vegetative RNA polymerase into the homologous spore enzyme. *FEBS Letters,* in press.

MURELL, W. G. (1955). In *The Bacterial Endospore.* University of Sydney, Sydney, Australia.

MURELL, W. G. (1967). The biochemistry of the bacterial endospore. *Adv. Microbiol. Physiol.* **1**, 133.

NAKATA, H. M. & HALVORSON, H. O. (1960). Biochemical changes occurring during growth and sporulation of *Bacillus cereus. J Bacteriol.* **80**, 801.

NEIDHARDT, F. C. (1966). Role of amino acids activating enzymes in cellular physiology. *Bact. Rev.* **30**, 701.

NISSLEY, S. P., ANDERSON, W. B., GOTTESMAN, M. E., PERLMAN, R. L. & PASTAN, I. (1971). *In vitro* transcription of the *Gal* operon requires cyclic adenosine monophosphate and cyclic adenosine monophosphate receptor protein. *J. Biol. Chem.* **246**, 4671.

OHYE, D. F. & MURELL, W. G. (1962). Formation and structure of spore of *Bacillus coagulans. J. Cell Biol.* **14**, 111.

PASTAN, I. & PERLMAN, R. (1970). Cyclic adenosine monophosphate in bacteria. *Science,* **169**, 339.

PETTIJOHN, D. E. (1972). Ordered and preferential initiation of ribosomal RNA synthesis *in vitro. Nature, New Biol.* **235**, 204.

PHILIPS, S. L., SCHLESSINGER, D. & APIRION, D. (1969). Mutants in *Escherichia coli* ribosomes: a new selection. *Proc. Nat. Acad. Sci. U.S.A.* **62**, 772.

POWELL, J. F. (1953). Isolation of dipicolinic acid (pyridine-2:6-dicarboxylic acid) from spores of *Bacillus megaterium. Biochem. J.* **54**, 210.

POWELL, J. F. & HUNTER, J. R. (1953). Sporulation in distilled water. *J. Gen. Physiol.* **36**, 601.

RABUSSAY, D. & ZILLIG, W. (1969). A rifampicin resistant RNA-polymerase from *E. coli* altered in the β-subunit. *FEBS Letters,* **5**, 104.

RIGGS, A. D., REINESS, G. & ZUBAY, G. (1970). Purification and DNA-binding properties of the catabolite gene activator protein. *Proc. Nat. Acad. Sci. U.S.A.* **68**, 1222.

ROBERTS, J. W. (1969). Termination factor for RNA synthesis. *Nature,* **224**, 1168.

ROBINOW, C. F. (1960). Morphology of bacterial spores, their development and germination. In *The Bacteria,* vol. I, ed. I. C. Gunsalus & R. Y. Stanier, p. 207. New York: Academic Press.

RYTER, A. (1965). Étude morphologique de la sporulation de *Bacillus subtilis. Ann. Inst. Pasteur,* **108**, 40.

RYTER, A. & AUBERT, J. P. (1969). Étude autoradiographique de la synthèse de l'ADN au cours de la sporulation de *Bacillus subtilis. Ann. Inst. Pasteur,* **117**, 601.

SADOFF, H. L., CELIKOL, E. & ENGELBRECHT, H. L. (1970). Conversion of bacterial aldolase from vegetative to spore form by a sporulation-specific protease. *Proc. Nat. Acad. Sci. U.S.A.* **66**, 844.

SCHACHNER, M., SEIFERT, W. & ZILLIG, W. (1971). A correlation of changes in host and T_4 bacteriophage specific RNA synthesis with changes of DNA-dependent RNA polymerase in *Escherichia coli* infected with bacteriophage T_4. *Eur. J. Biochem.* **22**, 520.

PLATE 1

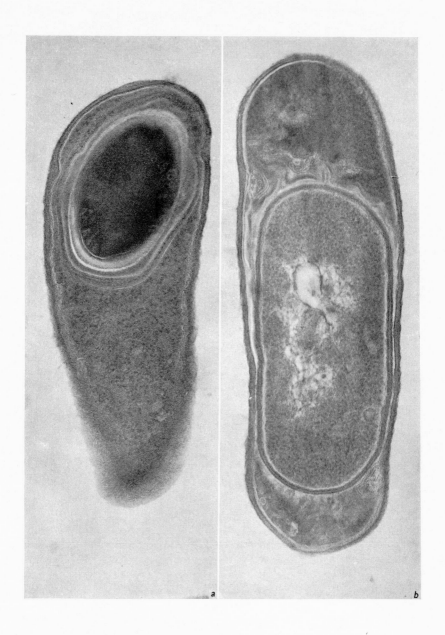

PLATE 2

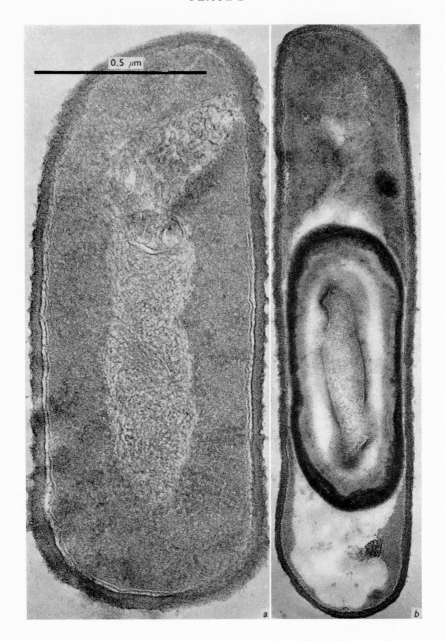

0.5 μm

a

b

SCHACHNER, M. & ZILLIG, W. (1971). Fingerprint maps of tryptic peptides from subunits of *Escherichia coli* and T₄-modified DNA-dependent RNA polymerase. *Eur. J. Biochem.* **22**, 513.

SCHAEFFER, P. (1967). Asporogenous mutants of *Bacillus subtilis*, Marburg. *Folia Microbiol.* **12**, 291.

SCHAEFFER, P. (1969). Sporulation and the production of antibiotics, exoenzymes and endotoxines. *Bact. Rev.* **33**, 48.

SCHAEFFER, P., IONESCO, H., RYTER, A. & BALASSA, G. (1963). La sporulation de *Bacillus subtilis*: étude génétique et physiologique. *Coll. Int. Centre Nat. Rech. Sci., Marseille*, p. 553.

SCHAEFFER, P., MILLET, J. & AUBERT, J. P. (1965). Catabolic repression of bacterial sporulation. *Proc. Nat. Acad. Sci. U.S.A.* **54**, 571.

SEIFERT, W., RABUSSAY, D. & ZILLIG, W. (1971). On the chemical nature of alteration and modification of DNA-dependent RNA polymerase of *E. coli* after T₄ infection. *FEBS Letters*, **16**, 175.

SHOER, R. & RAPPAPORT, H. P. (1972). Analysis of a *Bacillus subtilis* proteinase mutant. *J. Bacteriol.* **109**, 575.

SLAPIKOFF, S., SPITZER, J. L. & VACCARO, D. (1971). Sporulation in *Bacillus brevis*: studies on protease and protein turnover. *J. Bacteriol.* **106**, 739.

SLEPECKY, R. A. (1969). Synchrony and the formation of bacterial spores. In *The Cell Cycle*, ed. G. M. Padilla, G. L. Whitson & I. L. Cameron, p. 77. New York: Academic Press.

SONENSHEIN, A. L. & LOSICK, R. (1970). RNA polymerase mutants blocked in sporulation. *Nature*, **227**, 906.

SONENSHEIN, A. L. & ROSCOE, D. H. (1969). The course of phage φe infection in sporulating cells of *Bacillus subtilis* strain 3610. *Virology*, **30**, 265.

SPIZIZEN, J. (1965). Analysis of asporogenic mutants in *Bacillus subtilis*. In *Spores*, vol. III, ed. L. L. Campbell & H. O. Halvorson, p. 64. Ann Arbor, Michigan: Am. Society for Microbiol.

SPIZIZEN, J., REILLY, B. E. & DAHL, B. (1963). Transformation of genetic traits associated with sporulation in *Bacillus subtilis*. *Proc. 11th Int. Congr. Genet.* **1**, 31.

SPOTTS, C. R. & SZULMAJSTER, J. (1962). Synthèse de l'acide ribonucléique et des protéines chez *Bacillus subtilis* sporogène et asporogène. *Biochim. Biophys. Acta*, **61**, 635.

STERLINI, J. M. & MANDELSTAM, J. (1969). Commitment to sporulation in *Bacillus subtilis* and its relationship to development of actinomycin resistance. *Biochem. J.* **113**, 29.

STEVENS, A. (1970). An isotopic study of DNA-dependent RNA polymerase of *E. coli* following T₄ phage infection. *Biochem. Biophys. Res. Commun.* **41**, 367.

STEVENS, A. (1972). New small polypeptides associated with DNA-dependent RNA polymerase of *Escherichia coli* after infection with bacteriophage T₄. *Proc. Nat. Acad. Sci. U.S.A.* **69**, 603.

SZULMAJSTER, J. (1964). Biochimie de la sporogénèse chez *B. subtilis*. *Bull. Soc. Chim. Biol.* XLVI, 443.

SZULMAJSTER, J. (1972). Concluding remarks on the regulation of macromolecular synthesis during sporulation. In *Spores*, vol. V, ed. H. O. Halvorson, R. Hanson & L. L. Campbell, p. 58. Ann Arbor, Michigan: Am. Society for Microbiol.

SZULMAJSTER, J., BONAMY, C. & LAPORTE, J. (1970). Isolation of a temperature-sensitive sporulation mutant of *Bacillus subtilis*. *J. Bacteriol.* **101**, 1027.

SZULMAJSTER, J. & CANFIELD, R. E. (1965). Changements biochimiques associés à la sporulation de *B. subtilis*. In *Mécanismes de régulation chez les micro-organismes. Coll. Int. Centre Nat. Rech. Sci. France*, **124**, 587.

SZULMAJSTER, J., CANFIELD, R. E. & BLICHARSKA, J. (1963). Action de l'actino-mycine D sur la sporulation de *Bacillus subtilis*. *C. R. Acad. Sci. Paris*, **256**, 2057.

SZULMAJSTER, J. & HANSON, R. S. (1965). Physiological control of sporulation in *Bacillus subtilis*. In *Spores*, vol. III, ed. L. L. Campbell & H. O. Halvorson, p. 162. Ann Arbor, Michigan: Am. Society for Microbiol.

SZULMAJSTER, J., KERJAN, P. & MAIA, J. C. C. (1970). Regulation of sporogenesis in *Bacillus subtilis*. In *1st Int. Symp. Genetics of Ind. Microorg.*, *Prague*, in press.

SZULMAJSTER, J. & SCHAEFFER, P. (1961). Biochemical analysis of sporogen and asporogen mutants of *B. subtilis*. *Biochem. Biophys. Res. Commun.* **6**, 217.

SZYBALSKI, W., BØVRE, K., FIANDT, M., GUHA, A., HRADEENA, Z., KUMAR, S., LOZERON, H. A., MAHER, Sr. W. M., NIJKAMP, H. J. J., SUMMERS, W. C. & TAYLOR, K. (1969). Transcriptional controls in developing bacteriophages. *J. Cell. Physiol.* **74**, 33, Supp. 1.

TAI, P. C., KESSLER, D. P. & INGRAHAM, J. (1969). Cold-sensitive mutations in *Salmonella typhimurium* which affect ribosome synthesis. *J. Bacteriol.* **97**, 1298.

TAKAHASI, I. (1961). Genetic transduction in *Bacillus subtilis*. *Biochem. Biophys. Res. Commun.* **5**, 171.

TAKAHASHI, I. (1965a). Transduction of sporogenesis in *Bacillus subtilis*. *J. Bacteriol.* **89**, 294.

TAKAHASHI, I. (1965b). Mapping of spore markers on the chromosome of *Bacillus subtilis*. In *Spores*, vol. III, ed. L. L. Campbell & H. O. Halvorson, p. 138. Ann Arbor, Michigan: Am. Society for Microbiol.

TAKAHASHI, I. (1965c). Localization of spore markers on the chromosome of *Bacillus subtilis*. *J. Bacteriol.* **89**, 1065.

TIPPER, D. J. & PRATT, I. (1970). Cell wall polymers of *Bacillus sphaericus* 9602. II. Synthesis of the first enzyme unique to cortex synthesis during sporulation. *J. Bacteriol.* **103**, 305.

TRAVERS, A. A. (1969). Bacteriophage σ factor for RNA polymerase. *Nature*, **223**, 1107.

TRAVERS, A. A. (1970). Positive control of transcription by a bacteriophage sigma factor. *Nature*, **225**, 1009.

TRAVERS, A. A. (1971). Control of transcription in bacteria. *Nature, New Biol.* **229**, 69.

TRAVERS, A. A., KAMEN, R. I. & SCHLEIF, R. F. (1970). Factor necessary for ribosomal RNA synthesis. *Nature*, **228**, 748.

VINTER, V. (1959). Differences in cystein content between vegetative cells and spores of *Bacillus cereus* and *Bacillus megaterium*. *Nature*, **183**, 998.

YAMAGISHI, H. & TAKAHASHI, I. (1968). Genetic transcription in asporogenous mutants of *Bacillus subtilis*. *Biochim. Biophys. Acta*, **155**, 150.

YAMAKAWA, T. & DOI, R. H. (1971). Preferential transcription of *Bacillus subtilis* light deoxyribonucleic acid strands during sporulation. *J. Bacteriol.* **106**, 305.

YOUNG, E. I. & FITZ-JAMES, P. C. (1959a). Chemical and morphological studies of bacterial spore formation. I. Formation of spores in *Bacillus cereus*. *J. Biophys. Biochem. Cytol.* **6**, 467.

YOUNG, E. I. & FITZ-JAMES, P. C. (1959b). Chemical and morphological studies of bacterial spore formation. II. Spore and parasporal protein formation in *Bacillus cereus* var. Alesti. *J. Biophys. Biochem. Cytol.* **6**, 483.

YOUNG, E. I. & FITZ-JAMES, P. C. (1959c). Chemical and morphological studies of bacterial spore formation. III. The effect of 8-azaguanine on spore and parasporal protein formation in *Bacillus cereus* var. Alesti. *J. Biophys. Biochem. Cytol.* **6**, 499.

ZUBAY, G., CHAMBERS, D. A. & CHEONG, L. C. (1970). Cell-free studies on the regulation of the *lac* operon. In *The lactose operon*, ed. J. R. Beckwith & D. Zipser, p. 375. Cold Spring Harbor Laboratory, N.Y.

EXPLANATION OF PLATES

PLATE 1

Electron photomicrographs of intracellular germination. Cells at T_5 (*a*) were transferred to a fresh nutrient broth medium and photographed 30 min later (*b*). The spore with its cortex and coats visible in (*a*) underwent rapid morphological changes characteristic of the germinating spore.

PLATE 2

Electron photomicrographs of the t_s-4 mutant grown at 42 °C (*a*) and 30 °C (*b*). Samples were taken from cultures 20 h after the end of exponential growth (T_{20}). (From Szulmajster *et al.* 1970.)

THE TRANSFORMATION OF BACTERIAL ENDOSPORES INTO VEGETATIVE CELLS

A. KEYNAN

*Department of Microbiological Chemistry, The Hebrew University
Hadassah Medical School, Jerusalem, Israel*

INTRODUCTION

One of the main challenges posed by modern biology is the clarification of the mechanisms which lead to cell differentiation. The complexity involved in the investigation of the changes which occur on the molecular level in the embryo of multicellular organisms during cell differentiation has led many investigators to look for simple unicellular systems in which such phenomena occur. This approach has led to an increasing interest in the study of sporulation and germination in spore-forming bacteria.

It is the object of this review to describe the changes which occur during the transformation of bacterial endospores into vegetative cells and to enquire whether this system, and the information derived from it, might be relevant to the understanding of the mechanism of cell differentiation in general. One must stress the fact that bacteria are prokaryotes and ask whether it can be assumed that mechanisms responsible for differentiation in prokaryotes are identical to those responsible for the differentiation which takes place in eukaryotes. As our knowledge of comparative cell biology broadens, we realize more and more that there are crucial differences in the structure and control mechanisms of prokaryotic and eukaryotic cells (for review see Charles & Knight, 1970). The doubts voiced by a number of scientists on whether events which occur in prokaryotic cells are identical or similar to those which occur in eukaryotes, should certainly be taken into account. On the other hand, the cumulative experience of molecular biology during the last decades has shown that, although there are differences between the various control mechanisms active in bacteria and those of higher cells, there is a good chance that whatever we discover in bacteria will be relevant to the understanding of these processes in higher cells. Modern biology is full of examples of this truism. The discovery of a number of metabolic pathways in bacterial cells and their control mechanisms has proved to be of universal, rather than species-specific interest. One also sometimes forgets that it was through the study of bacteria that the genetic code was uncovered.

While summarizing our knowledge on the processes responsible for the transformation of bacterial endospores into vegetative cells, we might ask ourselves whether this specific system possesses characteristics which make it suitable for the investigation of cellular differentiation at the molecular level in general. *Prima facie*, the germinating spore seems to have several features which enable it to be used as a model system. A spore suspension can be induced to germinate synchronously and thus gives a homogeneous system in which one can study events at the molecular level with relative ease. Moreover, to a certain extent, one can even manipulate differentiation events in the outgrowing spore. The process can be conveniently stopped temporarily and reversibly at certain stages, and even the direction of the differentiation can be changed. Thus the outgrowing spore has the potential to form a vegetative cell, but can also be induced to form a new spore or alternatively a cyst-like cell, all this before the first cell division has taken place (Vinter & Slepecky, 1965; Vinter, Chaloupka, Šťastná & Čáslavská, 1972).

If we consider the transformation of the spore into a vegetative cell to be a process of differentiation, we are assuming that the spore is different in its structure and composition from the vegetative cell which it finally forms. It is, therefore, important to begin this review by discussing the differences between spores and vegetative cells.

HOW MUCH DO SPORES DIFFER FROM VEGETATIVE CELLS?

Inherent in the concept of differentiation is the notion that one kind of cell is transformed into another kind, different in structure, function and chemical composition. According to current thinking such transformation occurs through preprogrammed and time-ordered changes in the pattern of synthesis of proteins which are responsible for the composition and structure of a cell. During the process of differentiation changes which occur in this programme of protein synthesis lead to the creation of a new cell type. The patterns of protein synthesis leading to new kinds of cells are thought to derive from different parts of the same genome. Differentiation on the molecular level is therefore considered to be the activation of different parts of the same multi-potential genome (Jacob & Monod, 1963).

The differences in the properties of spores and vegetative cells have been known for a long time. Spores are heat and radiation resistant, ametabolic, refractile and cannot be stained. A dramatic difference between the structure of the spore and the vegetative cell has also been shown by electron microscopy. Spores are surrounded by proteinaceous coats and have a wide 'cortex' beneath the spore coat composed of a mucopeptide. All these structures surround a core which is the spore protoplast. Recently published results (for review see Murrell, Ohye & Gordon, 1969; Sussman & Halvorson, 1966; Tipper & Gauthier, 1972) indicate that the chemical composition of the specific spore structures is also very different. In addition to the differences in macromolecules several low molecular substances are known to be unique to the spore state. One of the best studied of these is dipicolinic acid (DPA), which appears in high concentration in spores.

Very good evidence now exists to prove that the high sulphur-containing spore coat protein is distinct from, and different to, the proteins of the vegetative state. This has been shown by immunological, as well as by chemical methods (Aronson & Horn, 1972; Tipper & Gauthier, 1972).

The cortex which takes up a great deal of the spore space is composed of spore-specific peptidoglycans. Although they resemble some of the vegetative cell wall peptidoglycans, their basic unit is different and there is a difference in the way in which they are assembled in the polymer (Tipper & Gauthier, 1972). Additional proof that the cortex mucopeptides differ from cell wall mucopeptides has come from the fact that a mutant has been isolated in which the cell wall composition is normal but which is incapable of forming a cortex during sporulation. Beneath the cortex surrounding the core is another layer recently designated the germ cell wall. This consists initially of the inner forespore membrane of the sporangium and will eventually, after germination, develop into the cytoplasmic membrane of the vegetative cell. Its composition seems to be similar but not identical to the composition of the cytoplasmic membrane of vegetative cell, and it is surrounded by a lysozyme-sensitive peptidoglycan similar to that described in bacterial cell walls. This layer, unlike the vegetative cell wall, does not contain teichoic acid (for review see Pearce & Fitz-James, 1971; Tipper & Gauthier, 1972).

The picture is not so clear and defined when the composition of the spore core and its enzymes is considered. Some spore-specific enzymes are formed in great quantity during sporulation, for instance L-alanine racemase, alanine dehydrogenase, glucose dehydrogenase, alkaline phosphatase, adenosine deaminase, etc. (Mandelstam, 1969). The identification of spore-specific enzymes is complicated by another factor as was pointed out recently by Mandelstam (1969). The enzymes which appear during spore formation and have been considered sporulation specific, might actually be vegetative enzymes which have been repressed by nutrients in the growth medium and are derepressed when the medium is exhausted and growth stops. They will appear even in mutants blocked in spore formation or can sometimes be induced by changes in the growth medium. Enzymes of the glyoxylic and tricarboxylic acid (TCA) cycles can be cited as examples.

Some of the enzymes which are actively formed during spore formation can also be found in vegetative cells; others which are known to exist in vegetative cells can also be found in spores. Although these enzymes catalyse the same reactions in spores as in vegetative cells, they sometimes differ from vegetative enzymes in several ways. They are usually heat resistant *in vivo* and under certain conditions also *in vitro*. They often differ in their kinetic properties from those of vegetative enzymes. Those which have been studied in detail have a lower molecular weight than enzymes in the vegetative cell, but apparently have a similar primary protein structure. Some spore enzymes have been shown to be formed by a proteolytic degradation of vegetative enzymes. Several of the spore enzymes differ from vegetative enzymes by being particle bound. Both the spore enzymes as well as particle bound vegetative enzymes might be slightly altered by proteolysis, and therefore exhibit different physical characteristics. A quantitative evaluation of the degree to which spore and vegetative enzymes differ is not yet possible, since we do not know how many of the spore enzyme proteins differ in their primary structure from the proteins of vegetative cells (for review see Kornberg, Spudich, Nelson & Deutscher, 1968; Mandelstam, 1969; Sadoff, 1969).

The synthesis of some of these enzymes, glucose dehydrogenase and alkaline phosphatase, does seem to be an integral part of the sporulation process, as indicated by the fact that mutants blocked in the synthesis of these enzymes will either not form spores or do so very poorly (Glenn & Mandelstam, 1971).

The respiratory system of spores is also quite different from that of vegetative cells. Resting spores do not respire but spore extracts contain a soluble FMN-dependent NADH oxidase and the same system can be shown to be active in the germinated spore. Vegetative cells, on the other hand, have a cytochrome-containing particulate enzyme. There is conflicting information on cytochromes in spores. Respiration of germinating spores is more resistant to cyanide and carbon monoxide. Keilin (1958) found very little functional cytochrome in spores and Doi & Halvorson (1961) none at all. On the other hand, Felix & Lundgren (1972) in their study of membrane formation in *Bacillus cereus* during sporulation, reported the existence in spores of cytochrome-containing membranes. It is, of course, possible that cytochrome exists in spores but is not metabolically active after germination.

From the above description it can be concluded that the spore differs markedly from the vegetative cell in structure and in molecular composition. It also exhibits functional differences in its enzymes and respiratory system. It is not clear at this stage to what extent the spore enzymes differ in their primary protein structure from similar enzymes of the vegetative cells.

PROCESSES RESPONSIBLE FOR THE TRANSFORMATION OF SPORES INTO VEGETATIVE CELLS

Having shown that the spore differs in both its morphological structure and chemical composition from the vegetative cell, and contains several spore-specific substances, we may assume that the transformation of the spore into a vegetative cell can be considered a process of cell differentiation. Comparing this process to what is known of other systems of cell differentiation, we would expect it to consist of activating the genome of the resting spore and the consequent initiation of a programme of vegetative protein synthesis.

Actually the early changes which occur during the transformation of spores into vegetative cells do not result from changes in the genome, and are in no way dependent on RNA or protein synthesis. The accumulated evidence of a great number of research workers in this field seems to indicate that three distinct and sequential processes, induced by different external stimuli and different in their nature, are responsible for the transformation of spores into vegetative cells. These have been called activation, germination and outgrowth. They are sequential, and the occurrence of each stage is dependent on the previous process having taken place. They are induced by different external factors, inhibited by different inhibitors and mediated by different kinds of biochemical reactions. The evidence for the existence of these separate sequential steps has recently been extensively reviewed and will therefore be summarized only briefly here (Gould, 1969; Gould & Dring, 1972; Halvorson, Vary & Steinberg, 1966; Keynan, 1972; Keynan & Evenchik, 1969; Lewis, 1969; Strange & Hunter, 1969).

The resting, freshly harvested endospore is reluctant to germinate even under optimal germination conditions, and its germination rate is a function of aging (time of storage after harvesting). Obviously some change(s) which occur in resting spores as a function of time (aging), are responsible for the difference in their response to inducers of germination. Aging can be mimicked by heat activation (60° for one hour), or exposure to low pH or mercaptoethanol. Such conditioning of spores for germination is called activation. Activation is reversible in the sense that the germination rate of the spores, if stored after activation, will decrease as a function of time. Activation is not inhibited by the presence of metabolic inhibitors and there is no evidence that it is metabolism-mediated. It occurs only in the presence of water and is pH-dependent. The rate of activation increases at low pH and cations may inhibit it. Most of these facts are consistent with the hypothesis that activation consists of configurational changes of macromolecules (for review see Keynan & Evenchik, 1969).

When activated spores are exposed to certain substances which are usually species-specific and germination-inducing (e.g. L-alanine, glucose, adenosine), the resistant, dormant spore is transformed into a sensitive metabolically active cell within a short time. This irreversible change has been called germination. Germination is metabolism-mediated and can be inhibited by several metabolic poisons (for review see Vinter, 1970). However, it occurs in the presence of inhibitors of macromolecular synthesis (actinomycin D, chloramphenicol) and is thus not dependent on RNA or protein synthesis (Steinberg, Halvorson, Keynan & Weinberg, 1965). During germination 30 % of the dry weight of a spore (including DPA, calcium and cortical materials) is excreted into the medium. Germination seems to be a process during which several spore-specific substances are broken down, the spore state is irreversibly terminated and metabolism is activated. It does not consist of, nor is it dependent on, the synthesis of new macromolecules. It must therefore be viewed as a process of biochemical degradation responsible for the termination of the cryptobiotic state, and not as a growth or differentiation process. Germination results in a 'germinated spore', an actively metabolizing cell which still has all the typical spore enzymes, is surrounded by the proteinaceous spore coat, and has the shape of the spore, but is not refractile and can easily be stained. If germination occurs in water, without nutrients, the germinated spore will not develop further. But when exposed to nutrients, it will start to swell, elongate and produce new kinds of proteins, form a cell wall and then turn into a typical vegetative cell. This process is known as outgrowth and can be inhibited by inhibitors of macromolecular synthesis. It therefore depends on the synthesis of new macromolecules, particularly RNA and protein, and can be viewed as a process of growth and differentiation.

Thus of these three processes: activation, germination and outgrowth, the first and second can be considered as processes responsible for the termination of the cryptobiotic stage, and the third, outgrowth, as a process of differentiation.

Although the germination reaction, i.e. the reaction leading from a dormant to a germinated spore, is not a 'differentiation' phenomenon *sensu strictu*, as explained above, it might be of some interest to developmental biologists because as we shall see it involves activation of pre-existing enzymes – a phenomenon which might be of importance not only to students of the bacterial spore but also to those interested in developing systems in general.

THE PROCESS OF SPORE GERMINATION

As stated above, germination is the irreversible termination of the cryptobiotic state resulting in a metabolizing cell which is still quite different from the vegetative one. It is now being realized that the process of germination does not consist of a single reaction but is a multi-stage phenomenon involving a chain of possibly different kinds of reactions. There is also good evidence that a single spore might possess a multiplicity of different mechanisms which initiate the breaking of dormancy. As this subject has recently been thoroughly and extensively reviewed, (for review see Gould, 1969; Gould & Dring, 1972; Halvorson, Vary & Steinberg, 1966; Sussman & Halvorson, 1966), only a short summary will now be given emphasizing those aspects of the problem relevant to developmental biology.

The mechanism responsible for germination seems to be unique, and specifically designed for the termination of cryptobiosis, not a part of the metabolic pattern of the vegetative cell. This can be deduced from the fact that the optimal conditions for growth and metabolism are very often different from those for germination, and different inhibitors affect these two different processes. Germination is usually induced by some specific initiators, amino acids, nucleosides or sugars which are specific to different spore species or even strains. In addition to these species-specific substances, germination can also be initiated in a non-specific way by mechanical means, surface active substances and chelating agents, of which calcium dipicolinate has been widely investigated (see Gould & Dring, 1972). Most researchers in this field assume that the germination reaction takes place in at least three stages, a multi-stage enzyme-mediated trigger reaction which in some way activates a lytic enzyme, which in turn breaks down a lot of the cortex material which is finally excreted, thereby 'releasing' the dormant spore protoplast.

Although there is no detailed understanding of the trigger reaction which starts the process of germination, there is good evidence that this process is mediated by metabolic reactions involving several enzymes. There is evidence that germination can be inhibited both by some stereospecific antagonists of initiators and also by several metabolic poisons (Vinter, 1970). This fact is even more convincingly illustrated by the work of Prased, Diesterhaft & Freese (1972), who used mutants altered in their germination requirements. For instance, they found that mutants of *Bacillus subtilis* lacking various of their glycolytic enzymes exhibited changes in their initiation requirements. Further, mutants without L-alanine dehydrogenase (which apparently is the enzyme responsible for the first reaction leading to germination of spores when L-alanine is used as a germinating agent) will not germinate, or only partly so, with L-alanine (Freese, Park & Cashel, 1964). The metabolic nature of the reaction can also be demonstrated by the fact that the initiating substance is degraded during germination.

Although germination is metabolism-mediated, the explanation for it is not necessarily initiation by an enzyme substrate reaction leading to the breakdown of the initiating substance. Initiation of the reaction might also be explained as the result of an allosteric change of an enzyme molecule induced by a small molecule (for review see Halvorson, Vary & Steinberg (1966)). There is good evidence that the initiation of germination comprises a complicated multi-stage reaction, triggered by a specific substance which activates metabolism. The evidence for the assumption that we are dealing with a multi-stage phenomenon, even taking into consideration the trigger reaction, can be shown in the following ways. In *B. cereus*, D-alanine inhibits initiation of germination induced by L-alanine if added before or together with L-alanine, but it has no influence if added shortly afterwards when no changes in the spore can yet be observed. In *Bacillus licheniformis*, the optimal conditions of the initiating trigger reaction are different from those of the reactions which follow (Halmann & Keynan, 1962).

All these facts indicate that the initiation of germination is a reaction which differs from those that follow it. Therefore it will take place even if the initiating substance is removed from the spore suspension shortly after the initiation, as described above. During germination heat sensitivity is lost, calcium, DPA and breakdown substances are excreted and the spore under the phase microscope darkens, becomes stainable and metabolically active. Vary & Halvorson (1964) have shown that in the individual spore all these changes occur within a limited period of time, in some cases as little as 50 sec. Gould & Dring (1972) have lately compiled evidence confirming Levinson & Hyatt's (1966) early observation that the loss of the different spore characters during germination is not simultaneous, but sequential in time. Heat resistance is lost first and phase darkening usually occurs last. A lytic enzyme is apparently activated after the initial trigger reaction during the germination reaction which is responsible for the breakdown of the spore cortex as was first suggested by Powell & Strange (1953). Good evidence for this step was supplied by Gould & Hitchins (1965) who showed that spores pretreated with thioglycolic acid to make them more permeable could be germinated by lysozyme and also by Strange & Dark's sporolytic enzyme (Gould & King, 1969) isolated from spore debris.

DPA also seems to be involved in some way in germination, as shown by the fact that it is very difficult to germinate spores containing very little or no DPA. Furthermore, external addition of DPA will restore the ability to germinate to spores low in DPA (Halvorson & Swanson, 1969; Keynan & Halvorson, 1962; Keynan, Murrell & Halvorson, 1961). Externally added calcium dipicolinate is a non-specific universal inducer of germination (Riemann & Ordal, 1961). DPA has also been shown to release the sporolytic enzyme from the spore debris to which it is bound (Gould & King, 1969). Ions have been proved necessary for the induction of germination. Although their role in the germination reaction is unknown, their importance is highlighted by Fitz-James' (1971) experiment described later (p. 94), in which the influence of calcium and magnesium ions on the state of the spore protoplast could be shown. There is no real proof for any of the existing hypotheses for the explanation of the germination reaction. A synthesis of the most commonly accepted theories would envisage the following sequence of events: a chemical initiator triggers a chain of metabolic reactions, as a result of which an internal release of DPA takes place, an event which can, to some extent, be mimicked by the external addition of calcium and DPA. A release of DPA might activate the sporolytic enzyme, and conceivably chelate several cations which are in some way responsible for the dormant state of the spore protoplast. No conclusive evidence is available to support either these mechanisms or this sequence of events.

ACTIVATION OF DORMANT ENZYMES
DURING GERMINATION

As mentioned above, germination takes place and metabolism starts during germination even under conditions in which macromolecular synthesis is excluded. This fact, as well as the fact that active enzymes can be extracted from dormant spores leads, therefore, to the assumption that metabolism is initiated during germination by activation of pre-existing enzymes and not through their *de novo* synthesis. The reason for their inactivity in the spore state is not known, but various theories have been suggested (for review see Keynan, 1972). This phenomenon of the activation of pre-existing enzymes during developmental processes has been described in many and diverse developing systems in particular during fertilization of the sea urchin egg and development in seeds of higher plants. The mechanism of this process is not known and might vary in different systems. Such enzyme activation has been imitated *in vitro* when enzymically non-active particles were isolated from pea seeds and activated with the help of trypsin *in vitro*.

THE GERMINATED SPORE – HOW MUCH DOES IT DIFFER FROM THE VEGETABLE CELL?

It is clear from the previous description that differentiation events, if defined as consisting of gene activation for a new pattern of RNA and protein synthesis, actually begin after germination is completed, with outgrowth leading to the formation of a vegetative cell. Differentiation thus begins with the germinated spore, which can be regarded as a rudiment of a vegetative cell. The germinated spore will start to meta-bolize, generate energy, form macromolecules and within a time period of 60–100 min produce a cell similar to a vegetative cell. It is, therefore, important to describe the composition and biochemical potential of the cell (the germinated spore) which is the starting point of this process.

Structure of the germinated spore

A great part of the spore cortex is excreted during germination. The germinated spore develops through changes in the spore core or spore protoplast whilst still engulfed in the shell of the spore coats. The fact that protoplasts of the germinated spores developed from the spore cores was proved by Fitz-James (1971) when he isolated stable spore cores through digestion of coatless spores with lysozyme. These spore cores remained stable in the presence of calcium and absence of magnesium. They were semi-refractile and very similar in appearance to the core inside the spore. When magnesium was added in the presence of sucrose, refractility was lost and so-called 'spore protoplasts' formed. They lysed in the absence of sucrose, but in its presence could develop into vegetative cells. The magnesium, thus, induces changes in refractility. Changes in the properties accompanying it have not yet been investigated, but resemble the process of spore germination. During germination the spore core also swells and takes up water and is apparently kept intact by the spore coat. The fact that in Fitz-James' experiments spore protoplasts were only stable in the presence of sucrose indicates that an osmotically fragile cell wall surrounds the core.

Enzyme composition

It is assumed that enzymes in the germinated spore are identical in structure to spore enzymes, although their physical state in the dormant spore probably differs from their state after germination (Church & Halvorson, 1957; Sadoff, 1969; Stewart & Halvorson, 1954). Students of the outgrowth reactions are of course more interested in the formation of enzymes and structures formed during outgrowth which have not been shown to exist in spores. However, much of the work done on spore enzyme composition has been concentrated on spore-specific enzymes, those formed during sporulation and those present in the dormant spore. Less information is available on vegetative enzymes which are missing in spores and are synthesized during outgrowth.

Enzymes present in vegetative cells, but absent in spores include aspartic transcarbamylase and TCA cycle enzymes (Deutscher, Chambon & Kornberg, 1968). The lack of TCA cycle enzymes is demonstrated not only by the absence of a functional TCA cycle in germinated spores (Goldman & Blumenthal, 1960, 1964) but also by the absence of three specific TCA cycle activities thus far examined. As mentioned above, some controversy exists as to the existence and function of cytochrome in spores. Some workers did not find cytochrome in spores, but others have lately reported that it does occur. Even if this is so, good evidence exists to show that the enzyme system which carries electrons to oxygen in germinated spores is different from the cytochrome-containing particle which carries electrons to oxygen in the vegetative cells. In germinated spores the enzymes seem to be a soluble flavoprotein oxidase, stimulated by FMN and DPA and containing diaphorase. The spore electron transport system is more sensitive to atebrine and less sensitive to cyanide (Doi, 1961).

The protein synthesizing system in germinated spores

Extracts of dormant spores do not contain a functional protein synthesizing system, but such a system can be shown to exist in spores immediately after germination, even if germination occurs in the presence of chloramphenicol. Spores contain RNA and ribosomes. Some messenger-like RNA has been shown by Chambon, Deutscher &

Kornberg (1968) to exist in spores but is apparently not functional for protein synthesis, because spores germinated in the presence of actinomycin D will not form protein (Gould & Hitchins, 1965; Idriss & Halvorson, 1969; Steinberg, Halvorson, Keynan & Weinberg, 1965). New mRNA synthesis is therefore needed for differentiation during outgrowth. The inability of spore extracts to synthesize protein is not only due to the lack of messenger – since the addition of mRNA to such extracts does not lead to protein synthesis. The defect in the protein synthesizing mechanism in spores seems to be in the ribosomes. Evidence for such a defect in the ribosomes of dormant spores has been found in *B. cereus* (Kobayashi & Halvorson, 1966, 1968). These ribosomes lacked the ability to bind aminoacyl tRNA.

An analysis of spore ribosomal subunits indicates a number of unusual small subunits (Woese, 1961). Kobayashi (1972) has shown that the small defective subunits are not precursors of vegetative cell ribosomes, but rather their degradation products. By analysing spore ribosomal protein, using polyacrylamide gel electrophoresis, he demonstrated that some vegetative ribosomal protein is missing in spore ribosomes. Spore ribosomes could be activated for protein synthesis (up to 80-fold) through the addition of vegetative ribosomal proteins. Kobayashi therefore suggested that during sporulation some of the ribosomal proteins are detached. The detached ribosomal protein seems to be retained intact in the resting spore, and during germination it apparently binds to the defective subunits, thereby restoring activity to the ribosomes.

Although, as stated before, tRNA is present in spores, its quantity and quality differ from those in vegetative cells. Leucyl, glycyl and histidyl tRNAs are always present in lower concentration in spores than in vegetative cells, and lysyl and leucyl tRNAs were different in spores and vegetative cells (Vold & Minatogawa, 1972). RNA polymerase is considered by some researchers to be responsible for varying gene expression during differentiation. Differences in the structure of RNA polymerase in spores and vegetative cells will be discussed later (p. 110).

To sum up, the germinated spore which turns into the vegetative cell is actually the swollen spore protoplast, surrounded by the spore coat. Although differences in the biochemical functions of this rudimentary cell and the vegetative cell have been reported, there is still no convincing evidence for differences between the primary protein structure of the germinated spore protoplast and that of the vegetative cell. It is certain that several of the vegetative enzymes are missing, that the cell wall is not complete and the electron transport seems to be different.

Therefore one can assume that the synthetic processes taking place during outgrowth are processes of completion of the missing cell components, rather than a drastic change in the composition of most of the proteins of the spore protoplast itself.

A GENERAL OUTLINE OF THE EVENTS WHICH OCCUR DURING OUTGROWTH

The following is a chronological description of sequential events which occur in the germinated spore during outgrowth. Respiration and energy metabolism start simultaneously with germination. The first macromolecular synthesis recorded is the formation of RNA, which seems to begin immediately on germination. Protein synthesis lags behind RNA synthesis, the lag time usually being around 2 min (Torriani & Levinthal, 1967). If germination is carried out in the presence of actinomycin D, the germinated spore will neither produce RNA nor protein, indicating that no preformed functional messenger exists in the spores (for review see Hansen, Spiegelman & Halvorson, 1970). A certain level of RNA synthesis occurs even in the presence of chloramphenicol (Sakakibara, Saito & Ikeda, 1965) when this inhibitor is added during or before germination. If chloramphenicol is added 12 min after initiation a much higher rate of RNA synthesis can be shown to occur, thus indicating that although some transcription occurs in the absence of protein synthesis, its increased rate after germination is dependent on protein synthesis (Cohen & Keynan, 1970).

As previously stated, the spore has non-functional ribosomes, and as a functional protein-synthesizing system can be isolated immediately after germination, even in the presence of chloramphenicol, we must assume, therefore, that during the germination reaction ribosome activation occurs. During the first stages of germination, RNA synthesis can be stopped immediately by adding actinomycin D, which stops protein synthesis within 3–5 min. The time lag between the termination of protein synthesis after addition of actinomycin D indicates the absence of any stable mRNA in this system, at least at this stage. Synthesis of RNA and protein occurs to some extent in the absence of externally added amino acids or nucleosides and is therefore probably based on utilization of existing precursor pools in germinated spores and/or on the turnover of pre-existing macromolecules (Setlow & Kornberg, 1970; Torriani & Levinthal, 1967).

The first morphological change which occurs in the germinated spore during outgrowth is swelling, due to intake of water. Increase in volume

4

by swelling varies in different species but is usually two to three times the volume of the spore before germination. This swelling reaction is the main demarcation between the germinated and the outgrowing spore. Swelling occurs only under appropriate nutrient conditions. If conditions for outgrowth are not satisfactory, the cell remains in the form of a germinated spore and refrains from swelling for a period up to a few days. It will resume development when conditions are suitable. Swelling can also be inhibited by specific inhibitors which do not affect germination and outgrowth. The swollen germinated spore protoplast is still inside the spore coat and starts to develop the vegetative cell wall. Depending on the species, it either emerges from the spore coat, or the coat is absorbed during outgrowth (for review see Strange & Hunter, 1969).

As previously stated, cortical material is excreted into the medium. Part of the cortical material stays between the spore protoplast and the spore coat – and analysis of the fate of radioactively labelled cortex has shown that this material can be incorporated into the newly synthesized cell wall (Vinter, 1965a). Under optimal conditions emergence will occur in 50 to 60 min and the first cell division in 100–150 min. Flagella may be seen in some cases within 60 min. The chromatin bodies of the spore divide into two (*B. cereus*) or four chromatin bodies (*Bacillus megaterium*) during emergence of the cell and after further elongation, cell division occurs (Strange & Hunter, 1969).

Although RNA and protein synthesis can be shown to occur at the beginning of germination, DNA synthesis can be demonstrated only after around 100 min. It takes up to 40 min until a functional cell wall is formed which can stabilize the spore protoplast. Although the initiation of respiration is not dependent on protein synthesis, a protein synthesis dependent increase in respiration occurs 15 min after the start of germination (Steinberg, Halvorson, Keynan & Weinberg, 1965). Respiration is endogenous but its rate increases with the addition of external substrate. The increase in respiration rate follows the various morphological stages of outgrowth, the rate increases during the swelling, emergence and elongation of the outgrowing cell (Strange & Hunter, 1969). Although there is some disagreement as to the function of cytochrome in the spore, its increase in the spore during outgrowth can be measured as can the change in the pathway of electron transport (Felix & Lundgren, 1972a).

As indicated in the above paragraph, various events which occur during outgrowth are sequential and time-ordered. Such sequential time-ordered protein synthesis can be demonstrated during outgrowth when one measures the synthesis of some specific enzymes. These experiments will be reviewed in the following section. As mentioned before, germinated spores have no TCA cycle enzymes. At what exact stage these enzymes appear during outgrowth is a matter of controversy – some reports claim that they appear in cells only after several generations (Goldman & Blumenthal, 1960); on the other hand some evidence of the appearance of a TCA cycle for oxidation of pyruvate has been shown (Halvorson & Church, 1957) after germination. These contradicting results may be partly due to the different media used by different authors for the outgrowing spore (for review and discussion see Sussman & Halvorson, 1966).

Following this general description, the next section will bring some evidence to show that protein synthesis during outgrowth is sequential and time-ordered, that it depends on transcription control and that transcription leads to the sequential appearance of a variety of different mRMA species, in a time-ordered way. Finally, we shall discuss the suggestion for a possible mechanism of such transcription control.

THE SEQUENTIAL SYNTHESIS OF DIFFERENT PROTEINS DURING OUTGROWTH

The appearance of a variety of different proteins as a function of time during outgrowth can be demonstrated either by the isolation of proteins which differ in their physical properties, or by showing the *de novo* production of various enzymes as a function of time after outgrowth. As no evidence for the appearance of stable messenger during outgrowth exists, protein synthesis seems to be transcription controlled and the sequence of appearance of different kinds of protein molecules may therefore reflect the sequence of gene activation during outgrowth. This is, therefore, quite a fruitful approach for the investigation of cell differentiation phenomena during this period. While investigating the sequential 'programme' of protein synthesis, we are studying indirectly the 'programme' of sequential gene activation.

The existence of a time-ordered appearance of proteins was first shown by fluorescence antibody staining of outgrowing spores (Norris & Baillie, 1964) and by pulse labelling proteins with one amino acid during outgrowth, at two different times after germination. Thus a different distribution of labels in chromatograms of proteins extracted at these two time intervals indicated that different proteins are synthesized at different times after outgrowth (Hoyem, Rodenberg, Douthit & Halvorson, 1968). Additional direct evidence was obtained through using the technique of isolating proteins by gel electrophoresis after pulse labelling at different times during outgrowth. Radioautography of such gels indicates that different kinds of proteins are synthesized at different times during outgrowth (Kobayashi, Steinberg, Higa, Halvorson & Levinthal, 1965; Torriani & Levinthal, 1967). Additional evidence for such time-ordered sequence of protein synthesis is also found in the elegant work of Steinberg & Halvorson (1968), showing the sequential appearance of enzymes during outgrowth. They demonstrated that synthesis of α-glucosidase, L-alanine dehydrogenase, alkaline phosphatase and histidase during outgrowth occurs at a certain predictable time and only continues for a definite time. Some of them are inducible enzymes, but they can be induced only when their background levels are synthesized anyway, indicating a specific time at which the genome is responding to inducer. This discontinuous, time-ordered synthesis of different kinds of proteins has been shown to occur also in a variety of other differentiating systems, and is typical of differentiation phenomena.

THE DEPENDENCE OF PROTEIN SYNTHESIS DURING OUTGROWTH ON TRANSCRIPTION

In a variety of developing systems protein synthesis is dependent upon a pre-existing stable messenger. During sporulation evidence for the formation of stable mRNA has also been reported by Aronson (1965). Recently, additional evidence for the existence of stable mRNA molecules during sporulation has been published by Sterlini & Mandelstam (1969). They demonstrated that several sporulation-specific enzymes were synthesized after actinomycin D had been added to sporulating cells for very much longer time periods than the half life of their mRNA, indicating that these enzymes are synthesized by stable mRNA. On the other hand, Chasin & Szulmajster (1969) while investigating the resistance of DPA synthesis to actinomycin D during spore formation, found that this antibiotic does not inhibit this synthesis in low concentration, but it does so in higher concentrations. They suggested, therefore, that actinomycin D in low concentrations does not penetrate into the fore spore and some protein can therefore be synthesized during sporulation in the presence of actinomycin D. The evidence for a stable messenger during sporulation is not yet conclusive (Freese, 1972). The problem of stable mRNA during sporulation is tied up with the problem of translational control. Such control is not necessarily based on stable mRNA – it could also result from changes occurring in the translation mechanism itself; one such example is the modification of the host cell ribosomes occurring during phage infection which affects the translation efficiency of phage messengers. The subject of translational control during sporulation has recently been reviewed by Doi & Leighton (1972).

On the other hand, there is no evidence for the existence of stable mRNA during outgrowth. On the contrary it seems that in this system protein synthesis is transcription-dependent only. When actinomycin D is added to outgrowing spores it stops not only RNA synthesis but also, after a short delay, protein synthesis. The half-life of mRNA has been measured in rapidly growing cells and in outgrowing spores (Levinthal, Keynan & Higa, 1962). The same short half-life of about 4 min could be demonstrated in both cell types (Kobayashi *et al.* 1965). The fact that protein synthesis can be stopped by the addition of actinomycin D at any time during early outgrowth after a short delay, identical in time with the mRNA half-life, indicates that protein synthesis in this system is dependent on continuous RNA synthesis. Rodenberg, Steinberg, Piper, Nickerson, Vary, Epstein & Halvorson (1968) investigated this dependence in detail, examining the rate of protein synthesis and its relationship to the rate of the synthesis of various RNA molecules. They found a constant ratio between the rate of protein synthesis and the rate of labile RNA synthesis and suggested, therefore, that the rate of formation of the former depends on the rate of formation of the latter. No dependence of protein synthesis on the rate of production of new ribosomes could be shown in these experiments.

Additional evidence of continuous transcriptional control of protein synthesis during outgrowth has been demonstrated by Hansen, Spiegelman & Halvorson (1970), who showed that addition of actinomycin D to outgrowing *B. cereus* spores terminated synthesis of each of three different enzymes which appear at regular intervals after outgrowth. The length of time for which the outgrowing spore would support enzyme synthesis after the addition of actinomycin D is equivalent to the life span of mRNA in vegetative cells. All the evidence cited supports the notion that protein synthesis during outgrowth is transcription-dependent. One can prove that these sequential appearances of proteins are not caused by release from pre-existing enzymes since their appearance stops immediately after the addition of chloramphenicol and their *de novo* synthesis during outgrowth can also be shown by radioactive labelling and radioautography experiments (Sakakibara, Saito & Ikeda, 1965).

EVIDENCE FOR SEQUENTIAL mRNA SYNTHESIS

The whole set of enzymes responsible for transcription does exist in the dormant spore. This can be deduced from the fact that transcription starts immediately after spores germinate even if new protein synthesis is inhibited by chloramphenicol. The rate of transcription depends on the composition of the medium and increases during the first few minutes of outgrowth. This increase is inhibited by inhibitors of protein synthesis (Cohen & Keynan, 1970). The exact pattern of RNA synthesis during outgrowth is not known, and reports in the literature are controversial. Armstrong & Sueoka (1968) described a double sequence of RNA synthesis in *B. subtilis* in which mRNA was synthesized only 16 min after initiation of germination, while tRNA and ribosomal RNA were synthesized immediately germination started. In *B. cereus* it was shown that part of the RNA synthesized initially is neither ribosomal RNA nor tRNA (Spiegelman, Dickinson, Idriss, Steinberg, Rodenberg & Halvorson, 1969). There is some controversy in the literature on the comparative rate of tRNA to rRNA synthesis during outgrowth of *B. cereus* spores (for review see Hansen, Spiegelman & Halvorson, 1970; Keynan, 1969).

It has been shown (Doi & Igarashi, 1964; Donnellan, Nags & Levinson, 1965) by DNA/RNA hybridization competition experiments that this RNA fraction, formed during outgrowth and which is degraded a few minutes after addition of actinomycin D, is different from the one formed during sporulation but competes with vegetative mRNA in hybridization experiments and is therefore probably similar to it. In the context of the concept of sequential transcription it is, of course, of special interest to see if different species of mRNA formed during outgrowth as a function of time differ from one another. Hansen, Spiegelman & Halvorson (1970) improved the method of DNA/RNA hybridization competition experiments which enabled them to distinguish predominant (but not minor) species of mRNA which appear during outgrowth. Using this method, and comparing mRNA formed at 8, 17 and 60 min, they could show that different forms of mRNA are formed at different times of outgrowth, thus indicating that there is sequential time-ordered transcription and that, therefore, transcriptional control occurs during outgrowth. These data are in agreement with the notion that certain genes are transcribed early during outgrowth and others later. On the other hand, there is also good evidence that some cistrons, such as the ribosomal RNA cistron (Spiegelman *et al.* 1969), are transcribed continuously during outgrowth.

Summarizing these findings it can be seen that in *B. cereus*, tRNA and rRNA are produced continuously, although possibly at different rates, during outgrowth and that there is a definite time-ordered, sequential change in the formation of species of mRNA as a function of outgrowth time.

DNA SYNTHESIS DURING OUTGROWTH

Net synthesis of new DNA starts very late during outgrowth in *B. cereus* and *B. subtilis*. In synchronously sporulating cells net synthesis of DNA starts suddenly 120–160 min after germination and continues for about 90 min (for review, see also Keynan, 1969; Wake, 1967).

Some incorporation of radioactive precursors into DNA occurs immediately after germination but this has been attributed to repair synthesis (Yoshikawa, 1965). Repair synthesis of DNA is usually not inhibited by the addition of chloramphenicol but replication of DNA is so inhibited (Maaløe & Hanawalt, 1961). The incorporation of radioactive precursors during outgrowth does not occur in the presence of chloramphenicol, therefore Steinberg & Halvorson (1968) suggested that this synthesis is not DNA 'repair' but must be DNA replication. Lately Lammi & Vary (1972), repeating these experiments, suggested that although this synthesis is inhibited by chloramphenicol, it still might be a repair reaction, and that the chloramphenicol prevents this reaction indirectly by preventing the synthesis of enzymes necessary for DNA repair. In order to test this hypothesis, Lammi & Vary have used a pyrimidine analogue (HPVra) which has been shown by Brown (1971) to block replication but not repair. This analogue stopped incorporation of precursors immediately, showing that the incorporation of radioactive precursors into DNA before net synthesis occurs is replication and not repair.

The various sequential macromolecular syntheses occurring during outgrowth do not depend on DNA synthesis or replication, as can be shown by the fact that such time-ordered protein synthesis occurs also in the presence of an inhibitor of DNA synthesis or in a thymidine auxotroph outgrowing in the absence of thymidine (Steinberg & Halvorson, 1968). Even the membrane synthesis involved in elongation and cell division is not necessarily inhibited by inhibitors of DNA synthesis (Dawes & Halvorson, 1972).

THE FORMATION OF CELL MEMBRANES AND
CELL WALL DURING OUTGROWTH

The synthesis of cell membranes has been studied recently by Dawes &
Halvorson (1972). They added labelled glycerol and orthophosphate to
outgrowing spores of *B. cereus* strain T and measured membrane lipid
synthesis. In this organism the membrane contains nearly all the cellular
lipids. Thus the amount of newly formed lipids in cell membranes could
be measured by the radioactivity of the chloroform/methanol extracts of
the trichloroacetic acid precipitate of cells to which radioactive glycerol
and orthophosphate had been added. By this method it was shown that
membrane synthesis is discontinuous during outgrowth; it begins at a
low rate at 10 min, synthesis stops at 50 min and is resumed at about
110 min. It then continues in an approximately linear fashion, until
180 min, and after that increases very suddenly. No turnover of
membrane could be shown to occur during the first 150 min of outgrowth when outgrowing spores with prelabelled membranes were
incubated in the presence of unlabelled glycerol or phosphate, but after
this time interval, turnover reached rates of about 30 % per hour. The
incorporation of the labelled glycerol for the first two hours of outgrowth
is, therefore, not due to turnover but represents a net synthesis of cell
membrane. Chloramphenicol did not affect the rate of glycerol incorporation into membranes for the first 100 min after germination, indicating that the enzymes involved in membrane lipid synthesis pre-exist
in the spore. During the first 100 min, lipid incorporation into the
membrane was not coupled with protein synthesis. The membrane of
the germinated spore apparently could incorporate lipids even without
de novo protein synthesis, but the rate of lipid incorporation into
membrane became sensitive to cholramphenicol after 110 min. During
the period of accelerated lipid incorporation into membranes (i.e. after
150 min), chloramphenicol addition led to an immediate fixation of the
rate of incorporation at the level which existed just before addition of the
antibiotic. The enzymes responsible for the 'late' lipid incorporation
into membranes are therefore beginning to be synthesized only after
about 110 min. The formation of these new enzymes are transcription-
dependent, because their formation could be prevented by rifampicin.
The increased rate of lipid incorporation after 110 min occurs at the
same time as DNA synthesis and division start, and the very rapid
incorporation at 180 min corresponds to septum formation when the
outgrowing spore divides. This suggested a dependence of lipid incorporation into membranes on DNA synthesis, but when the problem

was investigated using an inhibitor of DNA synthesis (nalidixic acid), it became evident that the inhibition of DNA synthesis had no effect on the rate of lipid incorporation into membrane since the timing or rate of incorporation of radioactive glycerol was not affected by nalidixic acid. On the other hand, septum formation was prevented by adding nalidixic acid *before*, but not after, the beginning of net DNA synthesis (Dawes & Halvorson, 1972).

Cell wall formation during outgrowth was investigated by Vinter (1965 b) who measured the incorporation of radioactive diaminopimelic acid (DAP) into a fraction insoluble in hot trichloracetic acid, which comprises the newly synthesized cell wall material. By adding actinomycin D to outgrowing spores at sequential times, he showed that messenger for cell wall formation starts to be formed 10 min after germination, but new cell wall formation itself could be first detected 20 min after germination. The enzymes responsible for cell wall formation are formed 20–40 min after germination begins. This can be shown by studying the ability of chloramphenicol to inhibit cell wall formation, as a function of time. After 40 min cell wall formation continues, even in the presence of chloramphenicol, indicating that during the time interval of 10–40 min, all enzymes needed to maintain cell wall synthesis have been formed. Vinter (1965 b) has also published evidence indicating the formation of some cell wall intermediates, before the actual cell wall is formed. Cell wall formation occurs up to elongation time inside the spore wall; the cells therefore do not lyse when penicillin is added, but when cells later begin to elongate, penicillin-induced lysis occurs, indicating that at this stage most of the spore coat has already been degraded and the newly emerging cell depends for its osmotic protection on the newly formed cell wall.

REGULATION OF PROTEIN SYNTHESIS DURING OUTGROWTH BY NUTRIENTS

Protein synthesis will start in outgrowing spores even in media whose composition does not support growth, and is apparently dependent on the turnover of resting spore proteins. Torriani & Levinthal (1967) have shown that only very few proteins are formed under such conditions. By pulse labelling these cells and studying the bands of newly-formed protein by acrylamide gel electrophoresis and autoradiography, it became evident that only a few different kinds of new proteins are formed. These kinds of protein molecules will continue to be synthesized for a while when deficient medium is used. From this experiment it can

be concluded that germination itself does not induce complete transcription of the genome, it induces a limited transcription, forming only a
few proteins. Some of these early proteins do not appear, or at least not
in prominent bands, during later stages of outgrowth; they seem to be
typical of this particular state of the outgrowing spore. Other bands will
also appear in the same quantity, at later stages, when full outgrowth is
induced by complete growth media.

When such cells, in the state of 'arrested differentiation' are transferred to a full growth medium for a few minutes, then pulse labelled by
radioactive amino acids and their proteins extracted and separated on
acrylamide gel, a continuous radioactive smear with no well-defined
bands can be seen in radioautograms. This indicates that a very large
number of different proteins are being synthesized at very similar rates
under these conditions. This pattern remains constant for some time, but
later the rate of synthesis of various individual proteins changes. This
can be deduced from the appearance of distinct bands when examining
radioautograms of acrylamide gels of such cells. These bands are
identical to those from proteins from actively growing vegetative cells.
The authors, when summarizing these findings, commented: 'a continuous distribution of label into different proteins may indicate uniform
transcription of DNA with no control by repression. Thus if the synthetic process is activated before repressors are made, one could obtain
a period of essentially uncontrolled synthesis by all genes of the cell.'

EXPERIMENTAL CONTROL OF THE DIRECTION OF DIFFERENTIATION EVENTS DURING OUTGROWTH

When germinated spores are suspended in a complete nutrient medium,
they will differentiate into vegetative cells. If the medium is not complete, as described above, differentiation can be interrupted and only
part of the genome is transcribed. The germinated spore will, therefore,
produce only a limited number of proteins. It is now known that we can
not only interrupt, but also change, the direction of the differentiation
of outgrowing spores by a change in the composition of the medium.
The germinated spore normally will produce a vegetative cell, but under
some conditions can also produce a new endospore in its old cell coat,
even before the first cell division takes place and, as has recently been
described, it can also produce an additional cell type resembling a
cyst.

The outgrowing spore is, therefore, multi-potential in its differentiation ability and can produce one of three different types of cell, the

vegetative cell, a new spore or a cyst. The following is a description of conditions affecting the direction of differentiation in the germinated outgrowing spore.

Microcycle sporulation

Vinter & Slepecky (1965) were the first to show that dilution of the complex medium used for outgrowth of germinated spores will induce sporulation even before the first cell division occurred. This kind of sporulation inside a germinated, outgrowing spore, is called 'microcycle sporulation'. Microcycle sporulation can also be induced by suspending germinated spores in glucose-free growth medium, limited in an essential nutritional factor. In such microsporulation media, most of the outgrowing spores of *B. megaterium* or *B. cereus* differentiate directly into spores before undergoing cell division (Holmes & Levinson, 1967; MacKechnie & Hanson, 1968; Hoyem, Rodenberg, Douthit & Halvorson, 1968). In the case of *B. cereus*, phosphate limitation was found to be most effective in inducing microcycle germination (Hanson & MacKechnie, 1969). Other factors which have been found to effect the induction of microcycle sporulation were high concentrations of potassium and ammonia relative to glucose in the medium. Spores produced during microcycle are normal in most respects. They contain DPA, are heat resistant, but cannot themselves undergo another cycle of microsporulation. The spore seems also to be different in its DNA content; it has up to three times more DNA than a normal spore but otherwise microcycle sporulation seems to be similar to normal sporulation with the exception that, in this process, DNA synthesis starts earlier (for review, see Rodenberg, O'Kane, Hackel & Cocklin, 1972). Recently, a new method of induction of microcycle sporulation in germinated spores outgrowing in a full growth medium has been demonstrated by Rodenberg et al. (1972). They demonstrated that spore formation could be induced by exposing outgrowing cells to D-cycloserine or vancomycin. Both antibiotics produce limited lesions in cell walls and when added simultaneously induced more than 80 % of the cells to sporulate. The pattern of macromolecular synthesis during antibiotic-induced microcycle sporulation in growth media was similar to that which occurs during microcycle sporulation induced by a deficient medium. This phenomenon of inducing sporulation in a full growth medium by damaging the cell wall, indicates a feed-back from the cell wall forming system to the system activating the part of the genome responsible for sporulation.

The time during outgrowth at which spore formation can be induced has been studied carefully. Holmes & Levinson (1967) have shown that during outgrowth and before microcycle sporulation can be induced, the ability to oxidize acetate has to be developed since fluoroacetate-inhibited microcycle sporulation and this inhibition could be overcome by citrate or other TCA cycle acids. Aconitase activity was absent after germination but developed during outgrowth. Holmes & Levinson concluded, therefore, that the TCA cycle is essential for spore formation. The optimal time for induction of microcycle sporulation using dilution of growth medium was investigated by Vinter & Slepecky (1965) in cells outgrowing in a full medium. Spores immediately after germination, or when beginning to swell, could not be induced to sporulate again but immediately after swelling, under conditions which enable spores to develop up to a stage of elongation, microcycle sporulation was induced. On the other hand, the cells lysed when the medium was diluted after first cell division had occurred. The optimal time of induction of microcycle sporulation corresponded to the time of DNA replication during outgrowth. This agrees with the finding of Dawes, Kay & Mandelstam (1971), who showed that in synchronized, rapidly growing vegetative cells, the ability to sporulate was a function of the time in the division cycle of the vegetative cells and limited to the period just before completion of the chromosome (see also Dworkin, Higgins, Glenn & Mandelstam, 1972).

Differentiation of outgrowing spores into cyst-like cells

Vinter (1972) has recently shown that the outgrowing spore also has the ability to form yet another kind of hypometabolic form – a cyst-like cell. Vegetative bacilli can form under certain conditions cyst-like cells (Vinter, Chaloupka, Šťastná & Čáslavská, 1972). This was shown by studying the reactions of an asporogenic mutant of *B. megaterium* starved for nitrogen. Under these conditions an increase in the cell wall component murein occurred, as measured by incorporation of diamino-pimelic acid. This increase in mucopeptide can also be shown morphologically by electronmicrographs of starved cells. These cells show a marked thickening of the cell wall, and have been shown to have a much smaller ability to incorporate radioactive amino acids and they are considered by Vinter and his co-workers to be in a hypometabolic state. These investigators also studied the behaviour of germinated spores of *B. cereus* and *B. megaterium* exposed to various periods of starvation, followed by 'shift up' to complete medium. In these experiments Vinter was able to demonstrate that yet another kind of cell can be formed by the germinated spore. It is a slightly enlarged, semi- or highly refractile and non-stainable cell, with a marked thickening in its cell wall. The respiratory activity of these cells was considerably lower than that of outgrowing cells and they could not readily form colonies on solid media. Their protein turnover decreased by about 8 % per hour after induction of starvation. They were, therefore, clearly in a hypometabolic state. Vinter *et al.* (1972) described several methods through which they could induce the formation of cyst-like forms in germinated spores. The simplest way was starvation of germinated spores. The degree of hypo-metabolism induced by starvation depends on the starvation time, as shown by the decrease in rate of incorporation of amino acids. This hypometabolic state could not be terminated by the addition of nutrient medium. Even an hour of starvation can induce this state which will last for up to one hour after addition of complete medium. Another method by which a more permanent cyst-like cell was induced from a germinated spore, was by a 'shift up' by addition of peptone and glucose after starvation. The nature of the reaction of the starved germinated spore to such a shift up depended on the glucose to peptone ratio.

An analysis of these experiments revealed that low pH was responsible for the induction of cysts. These cysts are thick-walled refractile cells which have a low rate of respiration and a slow rate of incorporation of amino acids. Cyst formation was protein-synthesis-dependent, since it could be prevented by addition of chloramphenicol. These cysts differ physiologically and morphologically from either vegetative cells or spores, but the structural or molecular differences have not yet been elucidated. These dormant cysts will develop into normal cells when suspended for a long enough time in rich media.

THE MECHANISM OF REGULATION OF TRANSCRIPTION DURING OUTGROWTH

To summarize, all available evidence seems to indicate that the process of outgrowth consists of a sequential time-ordered and pre-programmed synthesis of a variety of different proteins. It also seems that during out-growth this programme of protein synthesis is directly transcription-controlled. The understanding of the specific mechanism regulating the time-ordered and sequential transcription in the outgrowing spore might be relevant to the understanding of the process of differentiation in general. Until a few years ago, theories of transcription control, during cell differentiation, modelled themselves on the concept of derepression similar to the mechanisms responsible for enzyme induction in bacteria (Jacob & Monod, 1963).

The evidence of the last few years seems to indicate that at least in some stages of spore formation transcription is not negatively controlled by the removal of the repressor from the genome, but positively controlled by changes occurring in the DNA-dependent RNA polymerase, the enzyme responsible for transcription. These changes have been measured as alterations in the template specificity of the enzyme, and have in some instances been correlated with structural changes in the enzyme itself. The existence of such a positive control based on changes in structure of RNA polymerase has been demonstrated by Burgess, Travers, Dunn & Bautz (1969) during phage infection of *E. coli* cells. They demonstrated that changes in template specificity of RNA polymerase after phage infection are responsible for the cessation of transcription of the *E. coli* genome and the induction of transcription of the phage genome. They assumed that the DNA template specificity was directed by a detachable segment of the enzyme called sigma factor, and that during phage infection a new sigma factor is produced which directs the transcription of the phage genes (Travers, 1969, 1970). More evidence has lately been accumulated which indicates that transcription during the process of sporulation of Bacilli might also be positively controlled.

This evidence is based on the study of changes which occur in the DNA-dependent RNA polymerase during spore formation in *B. subtilis*. Losick & Sonenshein (1969) have used DNA from the phage ϕe (a virulent phage of *B. subtilis*) in their study of changes occurring in the polymerase of sporulating cells. Although this phage is a virulent one and infects sporulating cells, it does not develop in them. The phage genome is preserved during the spore stage and will lead to phage production and lysis after germination at the start of outgrowth. When the phage is injected into spore-forming cells, Yehle & Doi (1967) could not find any phage mRNA, indicating that phage development is already inhibited at the level of transcription of the phage genome.

When Losick & Sonenshein compared polymerases isolated from sporulating and vegetative cells, they found a difference in their template specificities. Both enzymes could transcribe poly dAT, but only the enzyme from vegetative cells and not the one from sporulating cells, could transcribe the DNA from the phage ϕe. When the structure of the vegetative and sporulating enzymes was compared, it was found that one of the β-subunits of its RNA polymerase was modified and this modified enzyme will not transcribe phage DNA even when sigma factor is added. Therefore it is apparently the core of the enzyme which is modified (Losick, Shorenstein & Sonenshein, 1970). RNA polymerase is known to be inhibited by rifampicin. Rifampicin-resistant *B. subtilis* cells are sometimes asporogenic and do not show any modification in the

β-subunit of the polymerase after cessation of growth (Sonenshein & Losick, 1970). These investigations all seem to indicate that spore formation is correlated with changes in RNA polymerase. Polymerase isolated during spore formation lacks sigma factor, and exhibits a change in the β-unit of its core. A mechanism responsible for such a change has lately been suggested by Leighton, Freese, Doi, Warren & Kelln (1972). These researchers isolated a mutant of *B. subtilis*, asporogenic at 47 °C, but sporogenic at 30 °C. Analysis of this mutant showed that it produces a serine protease which was labile at 46 °C. When serine protease was not functional at 46 °C, neither alteration in RNA polymerase nor sporulation occurred. Synthesis of proteases are probably essential for the initiation of spore development (for review see Schaeffer, 1969; Mandelstam, 1971; Szulmajster, this Symposium). Serine protease has been previously shown to modify vegetative fructose 1,6-diphosphate aldolase into the spore-specific enzyme (Sadoff, Celikol & Engelbrecht, 1970). Leighton *et al.* (1972) suggested, therefore, that it is the serine protease which transfers the vegetative RNA polymerase into the sporulating polymerase. They suggested that the same enzyme might also be responsible for other intracellular proteolytic activities, creating molecules responsible for spore formation, thus suggesting that this protease is the enzyme which 'triggers' sporulation.

The evaluation of the various experiments with isolated polymerase *in vitro* and their relevance to events *in vivo* are complicated by the fact that nearly all *in vitro* experiments are based on synthetic or phage templates, which are not necessarily similar in their function to the physiologically significant templates *in vivo*. In addition, the conditions under which these experiments are carried out influence the rate of polymerase activity to a large degree. In particular the differences in ionic strength used in experiments by various researchers make comparison of their work difficult. These difficulties have recently been extensively reviewed (see Burgess, 1971) and an extensive review of the role of RNA polymerase in spore formation, as well as alternative explanations for the findings of Losick & Sonenshein are given by Szulmajster in this Symposium.

If this change in polymerase is indeed the mechanism of bacterial differentiation and responsible for the different patterns of protein synthesis during sporulation, a similar mechanism could be expected to act at the transcriptional level during the changes leading towards the formation of a vegetative cell from a germinated spore. A key factor for the investigation of this theory is the elucidation of the exact structure of the DNA-dependent RNA polymerase in the resting spore and its alterations during outgrowth. A comparative investigation of the enzyme from resting spores and vegetative cells in *B. cereus* has been published lately by Cohen, Silberstein & Mazor (1972). They found that the vegetative RNA polymerase holoenzyme can use poly dAT as a template for *in vitro* transcription as well as the DNA of the phage CP51, a virulent phage infecting *B. cereus*. On the other hand, spore holoenzyme, while showing good ability for transcription of dAT, showed only little ability for transcription of CP51 DNA, which was even further reduced by subjecting it to phosphocellulose chromatography, which is known to separate the sigma factor from the core of the polymerase. This small activity of spore enzyme to CP51 DNA and the removal of this activity by phosphocellulose chromatography suggested the existence of sigma factor in resting spores. In order to test this, fractions resulting from the phosphocellulose chromatography of RNA polymerase from vegetative cells and spores have been added to the core of spore and vegetative cell enzymes, demonstrating that sigma exists in both vegetative and spore cells. The spore fraction always has lower activity than the vegetative cell fraction when CP51 DNA is used as a template. In order to see if this is due to less binding of the sigma factor to the spore enzyme or to less sigma factor, the following experiment was carried out. The crude spore enzyme extracts prepared from spores of rifampicin-sensitive strains was added to vegetative core enzyme resistant to rifampicin and assayed for its ability to transcribe CP51 DNA (which is an assay for sigma factor). This assay for sigma was carried out in the presence of rifampicin, which binds the β-unit of non-resistant enzyme and inactivates it. In this experiment it was shown that spore enzymes, which were purified through DEAE columns added to vegetative enzyme core in the presence of rifampicin, showed sigma activity, thus demonstrating the existence of sigma factor in spores. When DNA-cellulose columns were used in the purification (which bind the *B. cereus* spore polymerase but not *B. subtilis* sporulation polymerase) it could be shown that the vegetative enzymes purified in this column contains sigma factor, whereas in the case of spore enzymes, the sigma was not retained in the column, indicating that sigma factor is apparently not bound, or only very weakly bound, to the core of the spore enzyme. Sigma has not been found to occur in *B. subtilis* spores (Maia, Kerjan & Szulmajster, 1971).

Cohen *et al.* (1972) have also investigated the changes occurring in RNA polymerase during germination. Changes in enzyme function could be shown in heat-shocked spores, induced to germinate with L-alanine and adenosine as soon as 7 min of germination. When polymerase was isolated 7 min after germination, unlike spore polymerase, it did transcribe CP51 DNA. This rapid change in the template specificity of the polymerase occurs under conditions which do not permit outgrowth, and is one of the first events which can be detected during germination. Such a rapid change in function of another polymerase isolated from *E. coli* has been described. Schachner, Seifert & Zillig (1971) demonstrated that within 1 min after T_4 phage infection of *E. coli*, and in the absence of protein synthesis, an alteration of the enzyme resulting also in a reduced affinity to sigma factor occurred. It is not yet clear if the changes occurring in RNA polymerase during germination of *B. cereus* spores are protein synthesis dependent, but the novel synthesis of peptides is necessary at the beginning of outgrowth for the normal rate of transcription (Cohen & Keynan, 1970).

THE TRANSFORMATION OF SPORES INTO VEGETATIVE CELLS – A SYSTEM FOR THE STUDY OF DIFFERENTIATION

At the conclusion of this review, it might be appropriate to attempt an evaluation of the research potential of the spore to vegetative cell transformation system since the transformation of dormant spores into vegetative cells has been proposed as a model system for the study of cell differentiation (for review see Hansen, Spiegelman & Halvorson, 1970; Keynan, 1969; Halvorson, Vary & Steinberg, 1966) and appears to have all the advantages of a unicellular, synchronized, rapidly-differentiating cell suspension.

As we learn more about the system, it becomes increasingly evident that we cannot assume that all the mechanisms involved in the transformation of bacterial endospores into vegetative cells are identical with those responsible for differentiation in higher organisms. However, it is clear that there are some features of this system unique enough to justify considering it as one of the tools which can help us in the elucidation of the control mechanisms responsible for biological development or differentiation. The following paragraphs attempt to assess, on the one hand, the features which are specific for this system and therefore of no general biological significance, and on the other hand, the advantages, if any, of this system as a model for the study of cell differentiation in general. Before pointing out some of the possible advantages of the system, it is important to understand that there are at least three distinct and different sequential stages responsible for the transformation of spores into vegetative cells, activation, germination and outgrowth. As described above, these processes are fundamentally different from each other both on account of the different type of change they induce in the spore, and because of the biochemical mechanism responsible for them. If we do not keep this fact in mind, the spore system might be a source of confusion rather than clarification for the developmental biologist.

The first two of these processes – activation and germination – are apparently processes which terminate the dormant state (Keynan & Halvorson, 1965) and only the third process – outgrowth – is a developmental process *sensu stricto*. During outgrowth, *de novo* macromolecular synthesis occurs and new enzyme systems and structures are formed. If developmental biologists are interested in differentiation in terms of sequential gene activation, it is this stage which is of most interest, and we will therefore discuss its research potential in some detail. However, before discussing outgrowth, one should perhaps point out that another stage occurring during the spore to vegetative cell transformation – germination – might also be of general biological interest. During germination sudden activation of pre-existing enzymes occurs. The mechanism of the germination process is not yet fully understood. In many other developing systems such sudden activation of pre-existing enzyme systems has been described and investigated. As an example, we might cite the sudden increase in respiration of germinating plant seeds (Mayer & Poljakoff-Mayber, 1963) and in the sea urchin egg immediately after fertilization (Monroy, 1965). In both these cases this sudden increase of respiration occurs in the absence of protein synthesis – therefore the activation of pre-existing enzymes must be involved. In both of these systems not only is respiration started, but a rapid increase in the rate of protein synthesis occurs, and as it occurs also in the presence of actinomycin D, it cannot be transcription-mediated, but must be in some way connected with enzyme activation.

In the germination of spores very similar phenomena occur – respiration increases rapidly and a protein-synthesizing system is activated. Therefore, spore germination can serve as a simple system for the investigation of enzyme activation. There are indications of some similarities in these processes in different organisms at the molecular level. The increase in the rate of protein synthesis in germinating spores and in fertilized sea urchin eggs are in both cases dependent on ribosomal activation. The mechanism for ribosomal activation might well be different in sea urchin eggs in which a ribosome inhibitor, which is released during fertilization, has been described (Felicetti, Gambino, Metafora & Monroy, 1971). In bacterial spores the ribosomes seem to lack an essential protein which is apparently present but separated from the ribosome in the resting spore, and becomes attached to it during germination. More detailed studies will be necessary to elucidate the similarities and differences of these processes. It seems that the nature of enzyme dormancy and the mechanisms of its termination, may be of far more universal significance than hitherto realized. If this is so, spore germination might be a useful system for the study of these phenomena (for review see Keynan, 1972).

Returning to the main subject of our discussion – the evaluation of the research potential of the outgrowing endospore – we have firstly to ask ourselves whether spore outgrowth is similar to cell differentiation. The starting point of this process is not the dormant spore but the germinated spore, which is quite different in composition and structure. It is the spore protoplast in the germinated spore which undergoes differentiation. As our knowledge of spore protoplast protein composition increases, as a result of detailed studies of specific spore enzymes and their similarity to vegetative cell enzymes, it appears that differences between these two cells – the germinated spore protoplast and the vegetative cell – are less than previously assumed (Kornberg, Spudich, Nelson & Deutscher, 1968; Sadoff, 1969). On the other hand, there are without a doubt differences between these two cells which can be clearly demonstrated. They are the lack of a cell wall in the germinated spore, and also the lack of some metabolic pathways, including the vegetative cell respiratory system. During outgrowth, these absent structures and enzyme pathways are synthesized according to a time-ordered sequential pattern. Although we cannot quantify the changes occurring during outgrowth, they seem definite enough to justify the assumption that this is a legitimate system for the investigation of differentiation phenomena.

It appears that the outgrowing spore offers two advantages for the student of differentiation. The sequential time-ordered synthesis of protein in the system seems to be transcription-dependent only – this system therefore is less complicated for the study of differentiation than most others in which translational control has been shown to occur. In the outgrowing spore it should therefore be easier to study directly gene activation as a function of differentiation. The second advantage is our ability to manipulate this system. We can interrupt differentiation and we can determine its direction experimentally – either allowing the germinated spore to form a vegetative cell, or inducing in it microcycle sporulation, or alternatively, we can induce it to form cysts.

The elucidation of the mechanisms involved in the change of direction of differentiation is of course the challenge which this system poses. A review (Szulmajster, 1972) in this publication deals specifically with spore formation and it would be redundant to deal with the induction of spore formation during outgrowth here. However, it might be worth mentioning that some work has been done relevant to the understanding of the mechanisms of spore formation – specifically during microcycle sporulation. The universal signal for the induction of spore formation appears to be starvation which is a somewhat undefined signal. The work of Rodenberg *et al.* (1972), discussed in this review, indicates that in the outgrowing spores, microcycle sporulation can be induced by the anti-biotics vancomycin and D-cycloserine which induce microcycle sporu-lation in the presence of a growth-supporting medium in cells which are clearly not starving. These observations are in line with Sadoff's hypothesis (1972) that the antibiotic peptides excreted during germina-tion might be considered as a kind of hormonal substance responsible for cell differentiation in Bacilli. According to Sadoff (1969) these sporulation peptides are formed by proteolytic enzymes which have also been shown to change vegetative RNA polymerase into sporulation-specific polymerase (Leighton *et al.* 1972). It would be interesting to follow the sequence of these events on the molecular level in outgrowing spores in which sporulation has been induced by antibiotics.

Outgrowth, including sequential time-ordered transcription-depend-ant translation, occurs in the absence of DNA synthesis. On the other hand, there is evidence that the ability of the cell to respond to inducers of sporulation is in some way connected with the cell division or the DNA division cycle as shown first by Dawes, Kay & Mandelstam (1971). Vinter & Slepecky (1965), investigating microcycle sporulation, noticed that only at certain definite stages during outgrowth, correlated with stages of DNA replication, could sporulation be induced. These obser-vations draw our attention to the possibility that some of the controls of differentiation are connected with DNA replication itself.

The developmental programme leading to microcycle sporulation seems to be similar but not exactly identical to that leading to sporula-tion in vegetative cells. The facts that early DNA synthesis occurs during microcycle sporulation and that more DNA is synthesized might be well worth investigating and might lead to a better understanding of the 'switch' inducing sporulation. The cyst-like cell, or the different kinds of cyst-like forms discovered by Vinter *et al.* (1972) and described in this review have not yet been investigated on the molecular level and we do not yet know enough about the process of their formation. The investiga-tion of this process and its similarity or dissimilarity to that of spore formation might be of great interest, because here again we are influencing the activation of what is possibly another part of the genome and which leads to the formation of yet another different cell.

SUMMARY

This review describes our concepts of the processes responsible for the transformation of dormant bacterial endospores into vegetative cells and the mechanisms by which this transformation takes place. Three sequential processes, activation, germination and outgrowth, are responsible for transforming the dormant, resistant, non-metabolizing and non-stainable spore into a heat- and radiation-sensitive, metabolizing and stainable cell.

Activation, which is reversible and not mediated by metabolism, may consist of changes in the tertiary structure of spore macromolecules.

Germination is the irreversible loss of spore properties and is metabolism-mediated. It leads to the excretion of a great deal of the spore cortex. Its mechanism is not understood but it is probably a multi-stage reaction, initiated by an external chemical factor that induces a 'trigger reaction', which activates a sporolytic enzyme, which in turn breaks down macromolecules in the cortex. The breakdown products are excreted together with dipicolinic acid and calcium.

Activation and germination are degradative reactions responsible for the termination of the spore state. The protoplast of the germinated spore starts to take up water and nutrients and begins the process of outgrowth.

Outgrowth is a true growth process consisting of macromolecular synthesis and culminating in the formation of a vegetative cell. During germination, enzymes dormant in the resting spore are activated. These enzymes are responsible for respiration and protein synthesis.

RNA synthesis starts a few minutes before protein synthesis, and evidence is given that protein synthesis is transcription-controlled. At a given time only parts of the cell genome are transcribed, resulting in the appearance of different kinds of messenger RNA as a function of time. Net DNA synthesis starts about one hour after germination. Outgrowth can be interrupted at an early stage by withholding nutrients. Under these conditions only a few kinds of proteins are formed.

The outgrowing spore is multi-potential in its ability to differentiate. It can form a vegetative cell, but under some defined conditions forms a new spore, even before it divides. The outgrowing spore can also be induced under certain definite conditions to form a low-metabolizing cyst-like cell.

Transcription control during outgrowth has not been investigated in detail, but is assumed to be positively controlled by changes in the RNA polymerase which is present in the dormant spore. The research potential of this system as a model for cell differentiation is discussed.

I express my deep gratitude to Brigit Wolman for her most valuable assistance in preparing this manuscript, and to Deborah Rabinowitz for her help in typing and editing.

I gratefully acknowledge Amikam Cohen's help in discussing part of the manuscript, and I would like to thank Jekisiel Szulmajster, Ernst Freese, Joel Mandelstam and Vladimir Vinter for letting me see their unpublished manuscripts.

Part of the work reported on in this review was supported by The Israel Academy of Sciences and Humanities.

REFERENCES

ARMSTRONG, R. L. & SUEOKA, N. (1968). Phase transitions in ribonucleic acid synthesis during germination of *B. subtilis* spores. *Proc. Nat. Acad. Sci. U.S.A.* **59**, 153–60.

ARONSON, A. J. (1965). Characterization of messenger ribonucleic acid in sporulating *B. cereus. J. Mol. Biol.* **11**, 576–88.

ARONSON, A. I. & HORN, D. (1972). Characterization of the spore coat protein of *B. cereus* T. In *Spores*, vol. v, ed. H. O. Halvorson, R. Hanson & L. L. Campbell, p. 19. Washington, D.C.: Am. Society for Microbiol.

BROWN, N. C. (1972). Inhibition of bacterial DNA replication by 6-(*p*-hydroxyphenylazo)-uracil: differential effect on repair and semi-conservative synthesis in *B. subtilis. J. Mol. Biol.* **59**, 1–16.

BURGESS, R. R. (1971). RNA polymerase. *Ann. Rev. Biochem.* **40**, 711–40.

BURGESS, R. R., TRAVERS, A. A., DUNN, J. J. & BAUTZ, E. K. F. (1969). Factor stimulating transcription by RNA polymerase. *Nature*, **221**, 43–6.

CHAMBON, P., DEUTSCHER, M. P. & KORNBERG, A. (1968). Biochemical studies of bacterial sporulation and germination. X. Ribosomes and nucleic acids of vegetative cells and spores of *B. megaterium. J. Biol. Chem.* **243**, 5110–16.

CHARLES, H. P. & KNIGHT, B. C. J. G. (eds.) (1970). Organization and control in prokaryotic and eukaryotic cells. In *Organization and Control in Prokaryotic and Eukaryotic Cells. Soc. Gen. Microbiol.* **20**. London: Cambridge University Press.

CHASIN, L. A. & SZULMAJSTER, J. (1969). Enzymes of dipicolinic acid biosynthesis in *B. subtilis*. In *Spores*, vol. iv, ed. L. L. Campbell, p. 133. Bethesda, Maryland: Am. Society for Microbiol.

CHURCH, B. D. & HALVORSON, H. (1957). Intermediate metabolism of aerobic spores. I. The activation of glucose oxidation in spores of *B. cereus* var. *terminalis. J. Bacteriol.* **73**, 470–6.

COHEN, A. & KEYNAN, A. (1970). Synthesis of a factor stimulating transcription in outgrowing *B. cereus* spores. *Biochem. Biophys. Res. Commun.* **38**, 744–9.

COHEN, A., SILBERSTEIN, Z. & MAZOR, Z. (1972). Ribonucleic acid polymerase from vegetative cells and spores of *B. cereus*. In *Spores*, vol. v, ed. H. O. Halvorson, R. Hanson & L. L. Campbell, p. 247. Washington, D.C.: Am. Society for Microbiol.

DAWES, I. W. & HALVORSON, H. O. (1972). Membrane synthesis during outgrowth of bacterial spores. In *Spores*, vol. v, ed. H. O. Halvorson, R. Hanson & L. L. Campbell, p. 449. Washington, D.C.: Am. Society for Microbiol.

DAWES, I. W., KAY, D. & MANDELSTAM, J. (1971). Determining effect of growth medium on the shape and position of daughter chromosomes and on sporulation in *B. subtilis. Nature*, **230**, 567–9.

DEUTSCHER, M. P., CHAMBON, P. & KORNBERG, A. (1968). Biochemical studies of bacterial sporulation and germination. XI. Protein-synthesising systems from vegetative cells and spores of *B. megaterium. J. Biol. Chem.* **243**, 5117–25.

DOI, R. H. (1961). Control of metabolic activity and its relation to breaking the dormant state. In *Spores*, vol. II, ed. H. O. Halvorson, p. 237. Minneapolis: Burgess Publishing Company.

DOI, R. H. & HALVORSON, H. O. (1961). Comparison of electron transport systems in vegetative cells and spores of *B. cereus. J. Bacteriol.* **81**, 51–8.

DOI, R. H. & IGARASHI, R. T. (1964). Genetic transcription during morphogenesis. *Proc. Nat. Acad. Sci. U.S.A.* **52**, 755–62.

DOI, R. H. & LEIGHTON, T. J. (1972). Regulation during initiation and subsequent stages of bacterial sporulation. In *Spores*, vol. V, ed. H. O. Halvorson, R. Hanson & L. L. Campbell, p. 225. Washington, D.C.: Am. Society for Microbiol.

DONNELLAN, J. E., NAGS, E. H. & LEVINSON, H. S. (1965). Nucleic acid synthesis during germination and postgerminative development. In *Spores*, vol. III, ed. L. L. Campbell & H. O. Halvorson, pp. 152–61. Ann Arbor, Michigan: Am. Society for Microbiol.

DWORKIN, M., HIGGINS, J., GLENN, A. & MANDELSTAM, J. (1972). Synchronization of the growth of *B. subtilis* and its effect on sporulation. In *Spores*, vol. V, ed. H. O. Halvorson, R. Hanson & L. L. Campbell, p. 233. Washington, D.C.: Am. Society for Microbiol.

FELICETTI, L., GAMBINO, R., METAFORA, S. & MONROY, A. (1971). Control mechanisms of protein synthesis in the unfertilized egg and early development of sea urchins. In *Control Mechanisms of Growth and Differentiation. Symp. Soc. Exp. Biol.* **25**, 183–96.

FELIX, J. & LUNDGREN, D. G. (1972a). Characteristics of the electron transport system of *B. cereus* during growth and sporulation. In *Abstr. Ann. Meet. Amer. Soc. Microbiol.*, p. 56.

FELIX, J. & LUNDGREN, D. G. (1972b). Some membrane characteristics of *B. cereus* during growth and sporulation. In *Spores*, vol. V, ed. H. O. Halvorson, R. Hanson & L. L. Campbell, p. 35. Washington, D.C.: Am. Society for Microbiol.

FITZ-JAMES, P. C. (1971). Formation of protoplasts from resting spores. *J. Bacteriol.* **105**, 1119–36.

FREESE, E. (1972). Sporulation of bacilli, a model of cellular differentiation. In *Current Topics in Developmental Biology*, vol. 7, ed. A. A. Moscona & A. Monroy. New York: Academic Press. (In Press.)

FREESE, E., PARK, S. W. & CASHEL, M. (1964). The developmental significance of alanine dehydrogenase in *B. subtilis. Proc. Nat. Acad. Sci. U.S.A.* **51**, 1164–72.

GLENN, A. R. & MANDELSTAM, J. (1971). Sporulation in *B. subtilis* 168. Comparison of alkaline phosphatase from sporulating and vegetative cells. *Biochem. J.* **123**, 129–38.

GOLDMAN, M. & BLUMENTHAL, H. J. (1960). Pathways of glucose catabolism in intact heat-activated spores of *B. cereus. Biochem. Biophys. Res. Commun.* **3**, 164–8.

GOLDMAN, M. & BLUMENTHAL, H. J. (1964). Pathways of glucose catabolism in *B. cereus. J. Bacteriol.* **87**, 377–86.

GOULD, G. W. (1969). Germination. In *The Bacterial Spore*, ed. G. W. Gould & A. Hurst, pp. 397–444. London: Academic Press.

GOULD, G. W. & DRING, G. W. (1972). Biochemical mechanisms of spore germination. In *Spores*, vol. V, ed. H. O. Halvorson, R. Hanson & L. L. Campbell, p. 401. Washington, D.C.: Am. Society for Microbiol.

GOULD, G. W. & HITCHINS, A. D. (1965). Germination of spores with Strange and Dark's spore lytic enzyme. In *Spores*, vol. III, ed. L. L. Campbell & H. O. Halvorson, pp. 213–21. Ann Arbor, Michigan: Am. Society for Microbiol.

GOULD, G. W. & KING, W. L. (1969). Action and properties of spore germination enzymes. In *Spores*, vol. IV, ed. L. L. Campbell, pp. 276–86. Bethesda, Maryland: Am. Society for Microbiol.

HALMANN, M. & KEYNAN, A. (1962). Stages in germination of spores of *B. licheniformis*. *J. Bacteriol.* **84**, 1187–93.

HALVORSON, H. & CHURCH, B. D. (1957). Intermediate metabolism of aerobic spores. II. The relationship between oxidative metabolism and germination. *J. Appl. Bacteriol.* **20**, 359–72.

HALVORSON, H. O. & SWANSON, A. (1969). Role of dipicolinic acid in the physiology of bacterial spores. In *Spores*, vol. IV, ed. L. L. Campbell, pp. 121–32. Bethesda, Maryland: Am. Society for Microbiol.

HALVORSON, H. O., VARY, J. C. & STEINBERG, W. (1966). Developmental changes during the formation and breaking of the dormant state in bacteria. *Ann. Rev. Microbiol.* **20**, 169–88.

HANSEN, J. N., SPIEGELMAN, G. & HALVORSON, H. O. (1970). Bacterial spore outgrowth: its regulation. *Science*, **168**, 1291–8.

HANSON, R. S. & MACKECHNIE, I. (1969). Regulation of sporulation and the entry of carbon into the tricarboxylic acid cycle. In *Spores*, vol. IV, ed. L. L. Campbell, pp. 196–211. Bethesda, Maryland: Am. Society for Microbiol.

HOLMES, P. K. & LEVINSON, H. S. (1967). Metabolic requirements for microcycle sporogenesis of *B. megaterium. J. Bacteriol.* **94**, 434–40.

HOYEM, T., RODENBERG, S., DOUTHIT, H. A. & HALVORSON, H. O. (1968). Changes in the pattern of proteins synthesized during outgrowth and microcycle in *B. cereus* T. *Arch. Biochem. Biophys.* **125**, 964–74.

IDRISS, J. M. & HALVORSON, H. O. (1969). The nature of ribosomes of spores of *B. cereus* T and *B. megaterium. Arch. Biochem. Biophys.* **133**, 442–53.

JACOB, F. & MONOD, J. (1963). In *Cytodifferentiation and Macromolecular Synthesis*, ed. M. Locke, pp. 30–64. New York: Academic Press.

KEILIN, D. (1958). The problem of anabiosis: history and current concept. *Proc. Roy. Soc.* B, **150**, 150–91.

KEYNAN, A. (1969). The outgrowing bacterial endospore as a system for the study of cellular differentiation. In *Current topics in Developmental Biology*, vol. 4, ed. A. A. Moscona & A. Monroy, pp. 1–36. New York: Academic Press.

KEYNAN, A. (1972). Cryptobiosis: a review of the mechanisms of the ametabolic state in bacterial spores. In *Spores*, vol. V, ed. H. O. Halvorson, R. Hanson & L. L. Campbell, p. 355. Washington, D.C.: Am. Society for Microbiol.

KEYNAN, A. & EVENCHIK, Z. (1969). Activation. In *The Bacterial Spore*, ed. G. W. Gould & A. Hurst, pp. 359–96. London: Academic Press.

KEYNAN, A. & HALVORSON, H. O. (1962). Calcium dipicolinic acid-induced domination of *B. cereus* spores. *J. Bacteriol.* **83**, 100–5.

KEYNAN, A. & HALVORSON, H. (1965). Transformation of a dormant spore into a vegetative cell. In *Spores*, vol. III, ed. L. L. Campbell & H. O. Halvorson, pp. 174–9. Ann Arbor, Michigan: Am. Society for Microbiol.

KEYNAN, A., MURRELL, W. G. & HALVORSON, H. O. (1961). Dipicolinic acid content, heat-activation and the concept of dormancy in the bacterial endospore. *Nature*, **192**, 1211.

KOBAYASHI, Y. (1972). Activation of dormant spore ribosomes during germination. II. Existence of defective ribosomal subunits in dormant spore ribosomes. In *Spores*, vol. V, ed. H. O. Halvorson, R. Hanson & L. L. Campbell, p. 269. Washington, D.C.: Am. Society for Microbiol.

KOBAYASHI, Y. & HALVORSON, H. O. (1966). Incorporation of amino acids into protein in a cell-free system from *B. cereus*. *Biochim. Biophys. Acta*, **119**, 160–70.

KOBAYASHI, Y. & HALVORSON, H. O. (1968). Evidence for a defective protein synthesizing system in dormant spores of *B. cereus*. *Arch. Biochem. Biophys.* **123**, 622–32.

KOBAYASHI, Y., STEINBERG, W., HIGA, A., HALVORSON, H. O. & LEVINTHAL, C. (1965). Sequential synthesis of macromolecules during outgrowth of bacterial spores. In *Spores*, vol. III, ed. L. L. Campbell & H. O. Halvorson, p. 200. Ann Arbor, Michigan: Am. Society for Microbiol.

KORNBERG, A., SPUDICH, J. A., NELSON, D. L. & DEUTSCHER, N. P. (1968). Origin of proteins in sporulation. *Ann. Rev. Biochem.* **37**, 51–78.

LAMMI, C. J. & VARY, J. C. (1972). Deoxyribonucleic acid synthesis during outgrowth of *B. megaterium* QM B1551 spores. In *Spores*, vol. V, ed. H. O. Halvorson, R. Hanson & L. L. Campbell, p. 277. Washington, D.C.: Am. Society for Microbiol.

LEIGHTON, T. J., FREESE, P. K., DOI, R. H., WARREN, R. A. J. & KELLN, R. A. (1972). Initiation of sporulation in *B. subtilis*: requirement for serine protease activity and RNA polymerase modification. In *Spores*, vol. V, ed. H. O. Halvorson, R. Hanson & L. L. Campbell, p. 238. Washington, D.C.: Am. Society for Microbiol.

LEVINSON, H. S. & HYATT, M. T. (1966). Sequence of events during *B. megaterium* spore germination. *J. Bacteriol.* **91**, 1811–18.

LEVINTHAL, C., KEYNAN, A. & HIGA, A. (1962). Messenger RNA turnover and protein synthesis in *B. subtilis* inhibited by actinomycin D. *Proc. Nat. Acad. Sci. U.S.A.* **48**, 1631–8.

LEWIS, J. C. (1969). Dormancy. In *The Bacterial Spore*, ed. G. W. Gould & A. Hurst, pp. 301–58. London: Academic Press.

LOSICK, R., SHORENSTEIN, R. G. & SONENSHEIN, A. L. (1970). Structural alteration of RNA polymerase during sporulation. *Nature*, **227**, 910–13.

LOSICK, R. & SONENSHEIN, A. L. (1969). Change in the template specificity of RNA polymerase during sporulation of *B. subtilis*. *Nature*, **224**, 35–7.

MAALØE, O. & HANAWALT, P. C. (1961). Thymine deficiency and the normal DNA replication cycle. *J. Mol. Biol.* **3**, 144–55.

MACKECHNIE, I. & HANSON, R. S. (1968). Microcycle sporogenesis of *B. cereus* in a chemically defined medium. *J. Bacteriol.* **95**, 355–9.

MAIA, J. C. C., KERJAN, P. & SZULMAJSTER, J. (1971). DNA-dependent RNA polymerase from vegetative cells and from spores of *B. subtilis*. IV. Subunit composition. *FEBS Letters*, **13**, 269–74.

MANDELSTAM, J. (1969). Regulation of bacterial spore formation. In *Microbial Growth, Symp. Soc. Gen. Microbiol.* **19**, 377–402.

MANDELSTAM, J. (1971). Recurring patterns during development in primitive organisms. In *Control Mechanisms of Growth and Differentiation. Symp. Soc. Exp. Biol.* **25**, 1–26.

MAYER, A. M. & POLJAKOFF-MAYBER, A. (1963). *Germination of Seeds*. Monograph on plant physiology. Oxford: Pergamon Press.

MONROY, A. (1965). *Chemistry and Physiology of Fertilization*. New York: Holt, Rinehart & Winston.

MURRELL, W. G., OHYE, D. F. & GORDON, R. A. (1969). Cytological and chemical structure of the spore. In *Spores*, vol. IV, ed. L. L. Campbell, pp. 1–19. Bethesda, Maryland: Am. Society for Microbiol.

NORRIS, J. R. & BAILLIE, A. (1964). Immunological specificities of spores and vegetative cell catalases of *B. cereus*. *J. Bacteriol.* **88**, 264–5.

PEARCE, S. M. & FITZ-JAMES, P. C. (1971). Sporulation of a cortexless mutant of a variant of *B. cereus*. *J. Bacteriol.* **105**, 339–48.

POWELL, J. F. & STRANGE, R. E. (1953). Biochemical changes occurring during the germination of bacterial spores. *Biochem. J.* **54**, 205–9.

PRASAD, C., DIESTERHAFT, M. & FREESE, E. (1972). Initiation of spore germination in glycolytic mutants of *B. subtilis*. *J. Bacteriol.* **110**, 321–8.

RIEMANN, H. & ORDAL, Z. J. (1961). Germination of bacterial endospores with Ca and dipicolinic acid. *Science*, **133**, 1703–4.

RODENBERG, S. D., O'KANE, D. J., HACKEL, R. A. & COCKLIN, E. (1972). Factors regulating cellular development in *B. cereus* T spores. In *Spores*, vol. v, ed. H. O. Halvorson, R. Hanson & L. L. Campbell, p. 197. Washington, D.C.: Am. Society for Microbiol.

RODENBERG, S., STEINBERG, W., PIPER, J., NICKERSON, K., VARY, J., EPSTEIN, R. & HALVORSON, H. O. (1968). Relationship between protein and ribonucleic acid synthesis during outgrowth of spores of *B. cereus*. *J. Bacteriol.* **96**, 492–500.

SADOFF, H. L. (1969). Spore enzymes. In *The Bacterial Spore*, ed. G. W. Gould & A. Hurst, pp. 275–99. London: Academic Press.

SADOFF, H. L. (1972). Sporulation antibiotics of *Bacillus* species. In *Spores*, vol. v, ed. H. O. Halvorson, R. Hanson & L. L. Campbell, p. 157. Washington, D.C.: Am. Society for Microbiol.

SADOFF, H. L., CELIKOL, E. & ENGELBRECHT, H. L. (1970). Conversion of bacterial aldolase from vegetative to spore form by a sporulation-specific protease. *Proc. Nat. Acad. Sci. U.S.A.* **66**, 844–9.

SAKAKIBARA, Y., SAITO, H. & IKEDA, Y. (1965). Incorporation of radioactive amino acids and bases into nucleic acid and protein fractions of germinating spores of *B. subtilis*. *J. Gen. Appl. Microbiol.* **11**, 243–54.

SCHACHNER, M., SEIFERT, W. & ZILLIG, W. (1971). A correlation of changes in host and T_4 bacteriophage specific RNA syntheses with changes of DNA-dependent RNA polymerase in *Escherichia coli* infected with bacteriophage T_4. *Eur. J. Biochem.* **22**, 520–8.

SCHAEFFER, P. (1969). Sporulation and the production of antibiotics, exoenzymes and endotoxins. *Bacteriol. Rev.* **33**, 48–71.

SETLOW, P. & KORNBERG, A. (1970). Biochemical studies of bacterial sporulation and germination. XXII. Nucleotide metabolism during spore germination. *J. Biol. Chem.* **245**, 3645–52.

SONENSHEIN, A. L. & LOSICK, R. (1970). RNA polymerase mutants blocked in sporulation. *Nature*, **227**, 906–9.

SPIEGELMAN, G., DICKINSON, E., IDRISS, J., STEINBERG, W., RODENBERG, S. & HALVORSON, H. O. (1969). Classes of ribonucleic acid and protein synthesized during outgrowth of spores of *B. cereus*. In *Spores*, vol. IV, ed. L. L. Campbell, pp. 235–46. Bethesda, Maryland: Am. Society for Microbiol.

STEINBERG, W. & HALVORSON, H. O. (1968). Timing of enzyme synthesis during outgrowth of spores of *B. cereus*. *J. Bacteriol.* **95**, 469–78.

STEINBERG, W., HALVORSON, H. O., KEYNAN, A. & WEINBERG, E. (1965). Timing of protein synthesis during germination and outgrowth of spores of *B. cereus* strain T. *Nature*, **208**, 710–11.

STERLINI, J. M. & MANDELSTAM, J. (1969). Commitment to sporulation in *B. subtilis* and its relationship to development of actinomycin resistance. *Biochem. J.* **113**, 29–37.

STEWART, B. T. & HALVORSON, H. O. (1954). Studies on the spores of aerobic bacteria. II. The properties of an extracted heat-stable enzyme. *Arch. Biochem. Biophys.* **49**, 168–78.

STRANGE, R. E. & HUNTER, J. R. (1969). Outgrowth and the synthesis of macro-molecules. In *The Bacterial Spore*, ed. G. W. Gould & A. Hurst, pp. 445–83. London: Academic Press.

SUSSMAN, A. O. & HALVORSON, H. O. (1966). *Spores, their dormancy and germination.* New York: Harper.

SZULMAJSTER, J. (1972). *Initiation of bacterial sporogenesis.* (In press.)

TIPPER, D. J. & GAUTHIER, J. J. (1972). Structure of the bacterial endospore. In *Spores*, vol. v, ed. H. O. Halvorson, R. Hanson & L. L. Campbell, p. 3. Washington, D.C.: Am. Society for Microbiol.

TORRIANI, A. & LEVINTHAL, C. (1967). Ordered synthesis of proteins during out-growth of spores of *B. cereus. J. Bacteriol.* **94**, 176–83.

TRAVERS, A. A. (1969). Bacteriophage σ factor for RNA polymerase. *Nature*, **223**, 1107–10.

TRAVERS, A. A. (1970). Positive control of transcription by a bacteriophage sigma factor. *Nature*, **225**, 1009–12.

VARY, J. C. & HALVORSON, H. O. (1964). Kinetics of germination of *Bacillus* spores. *J. Bacteriol.* **89**, 1340–7.

VINTER, V. (1965*a*). Spores of microorganisms. XVII. The fate of pre-existing diaminopimelic acid-containing structures during germination and post-germinative development of bacterial spores. *Folia Microbiol.* **10**, 280–7.

VINTER, V. (1965*b*). Spores of microorganisms. XVIII. The synthesis of new cell wall during postgerminative development of bacterial spores and its alteration by antibiotics. *Folia Microbiol.* **10**, 288–98.

VINTER, V. (1970). Germination and outgrowth: effect of inhibitors. *J. Appl. Microbiol.* **33**, 50.

VINTER, V., CHALOUPKA, J., ŠŤASTNÁ, J. & ČÁSLAVSKÁ, J. (1972). Possibilities of cellular differentiation of bacilli into different hypometabolic forms. In *Spores*, vol. v, ed. H. O. Halvorson, R. Hanson & L. L. Campbell, p. 390. Washington, D.C.: Am. Society for Microbiol.

VINTER, V. & SLEPECKY, R. A. (1965). Direct transition of outgrowing bacterial spores to new sporangia without intermediate cell division. *J. Bacteriol.* **90**, 803–7.

VOLD, B. S. MINATOGAWA, S. (1972). Characterization of changes in transfer ribonucleic acids during sporulation in *B. subtilis.* In *Spores*, vol. v, ed. H. O. Halvorson, R. Hanson & L. L. Campbell, p. 254. Washington, D.C.: Am. Society for Microbiol.

YEHLE, C. O. & DOI, R. H. (1967). Differential expression of bacteriophage ge-nomes in vegetative and sporulating cells of *B. subtilis. J. Virol.* **1**, 935–47.

YOSHIKAWA, H. (1965). DNA synthesis during germination of *B. subtilis* spores. *Proc. Nat. Acad. Sci. U.S.A.* **53**, 1476–83.

WAKE, R. G. (1967). A study of the possible extent of synthesis of repair DNA during germination of *B. subtilis* spores. *J. Mol. Biol.* **25**, 217–34.

WOESE, C. R. (1961). Unusual ribosome particles occurring during spore germina-tion. *J. Bacteriol.* **82**, 695–701.

CELL–CELL INTERACTIONS IN THE MYXOBACTERIA

M. DWORKIN

Department of Microbiology, University of Minnesota,
Minneapolis, Minnesota 55455

INTRODUCTION

The commonplace wisdom of bacteriology contains the implicit notion that while some bacteria manifest a rudimentary multicellular organization (e.g. the mycelia of the actinomycetes, the moving colony of *Bacillus circulans*, the fruiting body of myxobacteria), they are, as a group, inherently unicellular and behave essentially independently of each other. On the other hand, it has become clear to us that the myxobacteria not only manifest a surprising variety of organized cell–cell interactions, but that corporate behavior pervades the entire cycle of development and perhaps even growth.

One purpose of this essay is to describe these myxobacterial interactions and to suggest a biological function for them. Another is to illustrate that these bacteria offer a unique prokaryotic model system for studying cell interactions, hitherto investigated only among the eukaryotes.

BACKGROUND

Communication between individual cells in a population (or between tissues) is an essential biological function, necessary for growth and development both of micro-organisms as well as higher forms of life. The biological significance of such cell–cell interactions can be emphasized by classifying them on the basis of five very general functions which they serve: 1. To assemble cells which are dispersed. 2. To relay a signal from specialized receptor cells to other cells which cannot respond directly to the initial stimulus; 3. To regulate population density; 4. To form multicellular structures; 5. To maintain and repair the multicellular *status quo*. It is my purpose in this introduction briefly to discuss these five categories of cell–cell interactions, partly in order to emphasize how incompletely we understand the molecular basis for the interactions, but also to point out the functional analogy between the cell–cell interactions in a prokaryotic system and those in more highly developed systems.

1. *Assembly*

Cells which are initially separated during certain stages of growth and development frequently assemble at focal points for specific purposes, usually when entering a new developmental stage. This is a well known phenomenon, carefully documented for embryogenesis. It is also an important part of the mobilization of the inflammatory response. Although we know that cells migrate, frequently over relatively large distances, to specific sites in the body, we know very little about the stimuli which cause the migration or about the forces which guide the migrating cells. Presumably diffusible, chemotactic substances are involved when cells migrate to the ultimate site of tissue assembly, but these have not yet been demonstrated in the higher systems. Migrations are also seen with microbial systems; for example when it is necessary to assemble a population of cells which has become widely dispersed, as in the case of microbes which have been food gathering. This occurs both in the case of the cellular slime molds (Konijn, Van de Meene, Bonner & Barkley, 1967) and in the myxobacteria (for review, see Dworkin, 1965) which must aggregate in order to begin development. In both cases, diffusible chemotactic aggregates have been demonstrated.

2. *Relay signals between specialized cells*

In a cell population which has become functionally differentiated, one cell type may perceive a stimulus and notify other cells in the population, which then respond to the stimulus. Here again, the ideal mechanism seems to be one involving the transfer, via the extracellular space, of a diffusible molecule. Examples of this type of behavior are classical hormonal effects, and the macrophage–lymphocyte system for antibody synthesis. In the latter case the macrophage perceives an antigenic stimulus and transmits an RNA-antigen complex to the lymphocyte which actually synthesizes the antibody (Abdou & Richter, 1970).

3. *Mechanisms for regulating population density*

There are numerous examples of the production by cells of auto-inhibitors or stimulators of growth. For example, Rubin (1966) has shown that dense populations of chick cells produce a substance that, when added to isolated cells, stimulates their growth rate and enhances their colony forming ability. Eagle & Piez (1962) have shown that HeLa cells at less than 100 cells/ml require a variety of metabolites to be added to the medium; at higher cell densities the requirements are eliminated. In a similar fashion, self-regulation of growth by the production of auto-inhibitors is well documented in a variety of systems including cell cultures (Eagle, 1965) and fungi (Lingappa, Lingappa & Bell, 1972). Self-inhibition in bacteria is also well known; however, the casual inhibition by toxic end products does not have the same teleonomic implications as the inhibition found among multicellular organisms. The role of cell–cell contact is self-inhibition and stimulation of growth is more difficult to analyse. Stoker & Sussman (1965) working with fibroblast cells, have shown that while short-range feeder effects are most effective while the cells are in contact, a diffusible material is involved which will cross large intercellular distances. The role of cell–cell contact in cell growth inhibition is well known and often referred to. There has accumulated, however, a considerable amount of persuasive evidence that cell contact inhibition of growth may also involve a diffusible substance. In many cases, it is possible to obtain multi-layered growth by using large volumes of medium or by perfusing fresh medium over the culture. Thus, Rubin (1966) could obtain growth of chick fibroblasts equivalent to nine monolayers of cells by frequent changes of medium. Recently Ceccarini & Eagle (1971) have shown that the so-called contact inhibition observed in a variety of mammalian cells is a result of artefactual variations in the pH of the medium. The inhibition could be substantially reduced by proper pH stabilization. The role of actual contact inhibition of growth, not involving the passage of an inhibitor molecule, is thus unresolved.

4. *The role of cell interactions during the formation of multicellular structures*

The formation of an organized multicellular structure obviously requires communication and co-operation among cells. A rich variety of such interactions have been described. These include contact inhibition of movement (Abercrombie & Heaysman, 1954), orientation and arrangement in confluent cultures (Stoker & Macpherson, 1961), adhesion (Coman, 1960), changes in cell function (Konigsberg, 1963), and reaggregation (Moscona & Moscona, 1952). In addition, the formation of plasmodia, pseudoplasmodia and fruiting bodies by the slime molds (Bonner, 1963) and of fruiting bodies by the myxobacteria (Kühlwein & Reichenbach, 1968) clearly indicate that cell interactions occur both among eukaryotic and prokaryotic microbes. It seems difficult to explain these co-operative activities without invoking direct cell–cell contacts. The nature of one of these couples has been elegantly examined by Loewenstein (1968) whose work clearly indicates the existence of tight, intercellular junctions. Recently Gilula, Reeves &

Steinbach (1972) have associated gap junctions in Chinese hamster cells with both ionic coupling and the passage of metabolites involved in hypoxanthine metabolism.

Not only would we like to learn more about the nature of the stimuli which guide cells and tissues during the construction of multicellular structures but we are also interested in some of the early, biochemical responses of cells which have received such stimuli. Sussman and his co-workers have investigated the relationship between cell–cell interactions and the expression of specific genes during development of the cellular slime mold, *Dictyostelium discoideum*. Loomis & Sussman (1966) concluded that aggregation of the amoebae results in the transcription of the gene coding for UDP-galactose:polysaccharide transferase, an enzyme required for the synthesis of a mucopolysaccharide found specifically in the spore capsules. More recently, Newell, Franke & Sussman (1972) presented data which indicated that cell–cell interaction is required not only for transcription of certain genes, but continued cell–cell contact is necessary in order to translate certain mRNA molecules. The slime mold system clearly is providing some clues concerning the responses of cells at the molecular level to intercellular interactions.

5. *Maintenance and repair of the multicellular status quo*

It seems intuitively reasonable that the processes involved in the homeostatic maintenance of cell organization include cell interactions. Cells normally apposed to each other in developed tissues would manifest contact inhibition of movement and contact or short range inhibition of growth. A disruption of this normal association, as in a wound, would release the inhibition until the normal association was reestablished. It has been frequently demonstrated (Abercrombie, Heaysman & Karthauser, 1957; Temin & Rubin, 1958; Vogt & Dulbecco, 1960; Stoker & Macpherson, 1961; Stoker, 1964; Barski & Belehradek, 1965; Jamakosmanovic & Loewenstein, 1968) that malignant cells lose this ability to manifest contact inhibition of movement, contact arrangement, monolayer formation or electrical coupling.

THE MYXOBACTERIA

It is unfortunate that studies of the prokaryotic cell, which have facilitated the dramatic successes in biochemistry and molecular biology, have not found a role in the investigations of cell–cell interactions. The myxobacteria seem the most likely candidates for such studies. These

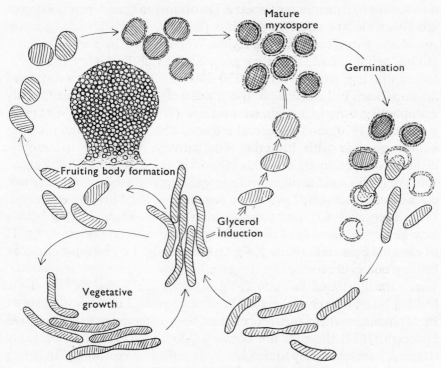

Fig. 1. The life cycle of *Myxococcus xanthus*.

are gram-negative bacteria which are distinguished from other bacteria by two properties – one unusual and the other unique. The unusual property is the nature of their motility. The cells glide across the surface of a solid substrate, e.g. glass or agar. They possess no flagella nor any visible organelles of locomotion. Their movement does not involve pseudopod formation, inching, wiggling, side winding nor any other visible perturbation of the cell. The myxobacteria share this property with some of the blue-green algae and with other gliding bacteria such as *Cytophaga* and *Sporocytophaga*, *Flexibacter*, *Beggiatoa* and others. Their unique property is the ability to go through a life cycle involving both cellular and colonial morphogenesis.

Description of life cycle

The life cycle of *Myxococcus xanthus* is illustrated in Fig. 1. *M. xanthus* is the organism with which much of the recent work on the myxobacteria has been done. Vegetative rods are typical gram-negative cells in terms of their cellular structure and organization. Their cell wall contains peptidoglycan (White, Dworkin & Tipper, 1968), their nucleus is unbounded by a membrane (Voelz & Dworkin, 1962), and their ribosomes are 70 s with 50 s and 30 s subunits (Juengst, 1972). These vegetative rods have two alternative life-styles. When placed in a medium containing peptone and salts, they grow and divide by binary transverse fission with a generation time of 210–270 min. They grow in a dispersed fashion, reach cell densities of 6–8 g wet weight/l, can be cloned, easily disrupted, grown in a defined medium (Dworkin, 1962; Witkin & Rosenberg, 1970), and in general are amenable to most of the manipulations possible with bacteria. Alternatively, when the medium is depleted of specific amino acids (Dworkin, 1963; Hemphill & Zahler, 1968) cells on a solid medium will aggregate (presumably in response to a chemotactic stimulus; Lev, 1954; Jennings, 1961; McVittie & Zahler, 1962; Fluegel, 1963), and form fruiting bodies. These fruiting bodies may be simple mounds of cells resting on the substrate (Plate 1, fig. 1) or elevated by a stalk (Plate 1, fig. 2; Plate 2, fig. 1). The most complex fruiting bodies are formed by the genera *Chondromyces* and *Stigmatella*. These are illustrated in Plate 2, fig. 2 and Plate 3, fig. 1. Within the fruiting body, the vegetative rods become converted to the resting stage. In *M. xanthus* where this process has been extensively examined (see Dworkin, 1972) the cells are converted to round, resistant, optically refractile, metabolically quiescent cells called myxospores. In some myxobacteria, such as *Chondromyces* and *Stigmatella*, the cells are contained in specialized structures called cysts. The relative contribution of the cyst or the individual cells within the cyst toward maintaining the resting state is not clear. Under the appropriate physical and nutritional conditions the cells or cysts germinate, giving rise to the metabolically active, swarming vegetative rods. In the case of *Myxococcus* and other genera this involves the conversion of the individual resting cell. In those organisms where the cells are contained within a cyst, the cyst ruptures, releasing the cells (Plate 3, fig. 2).

A considerable improvement in our ability to obtain biochemical and molecular correlates of the developmental process came with the development of a technique for inducing vegetative rods, growing exponentially in complex medium, to convert rapidly, synchronously and essentially completely to myxospores (Dworkin & Gibson, 1964; Dworkin & Sadler 1966; Sadler & Dworkin, 1966). If 0.5 M glycerol or isopropanol, erythritol, ethylene glycol or one of a variety of other hydroxylated compounds was added to the cells, they converted to myxospores in about 100 min. While the morphological conversion was complete at this time, a variety of biochemical changes continued to take place for at least an additional 4 h. The myxospores thus formed will germinate and, on the basis of a variety of parameters, are identical to those formed within the fruiting body during the normal developmental cycle. Germination occurred either when the cells were placed in the normal growth medium or at a high cell density ($> 5 \times 10^9$/ml) in distilled water. Under the latter conditions, phase darkening occurred at 60–90 min and elongation to rods at about 120 min.

Myxobacterial interactions

A considerable amount of information is now available regarding the basic properties of the myxobacteria, and the stage is thus set for examining some of the more subtle aspects of their development. The remainder of this article will describe the variety of cell interactions which have been reported for the myxobacteria. I shall also speculate as to the function and mechanism of these various interactions. The absence of hard data bearing on most of these problems has at least the virtue of forcing one to keep an open mind.

1. Swarming

One of the most striking manifestations of the co-ordinated behavior of the myxobacteria is the characteristic swarm. Microscopic and time-lapse cinematographic studies (Kühlwein & Reichenbach, 1968) show clearly that, as the cells glide over the substrate they remain closely apposed to each other, continually resisting the tendency to become dispersed. Isolated cells at the edge of the swarm move much more rapidly than the cells within the swarm. One sees them making frequent brief sorties out from the edge, only to return to the main body of the swarm. The swarming behavior of two myxobacteria is illustrated in Plate 4, figs. 1 and 2. Fig. 1 is of the swarm edge of *Polyangium velatum* showing the tongues of cells frequently seen in myxobacterial swarms. Fig. 2 is of the swarm edge of a *Sorangium* species showing the advancing, elevated ridge characteristic of this genus.

The behavior of the pack of cells released from a freshly-germinated cyst of *Chondromyces apiculatus* (Plate 3, fig. 2) is most instructive. These cells will migrate as a group, retaining their cohesiveness over relatively long periods of time. If the swarm comes within a few millimeters of another pack of cells they migrate towards each other, finally fusing (Kühlwein & Reichenbach, 1968).

The question as to the forces which restrict or regulate the motility of the individual cells is an interesting one. In some cases the swarm migrates out from a point (presumably the germinated fruiting body), moving in all directions and forming an expanding circle. The area of this expanding circle will increase as a function of the square of the radius. In other words in order to maintain a constant (and presumably optimal) cell density the distance moved by the cells must increase as the square root of the cell number. Since the cells are dividing as they swarm, it may be the division rate which thus regulates the rate of outward movement of the swarm. It would therefore be interesting simply to try to relate rate of swarm expansion and generation time. If such a relationship does indeed exist it might reflect a fascinating regulatory couple between growth rate and rate of movement. The model can be extended to include the induction of fruiting body formation. It has been amply demonstrated that induction of fruiting is triggered by the reduction in the concentration of certain specific amino acids in the medium (Dworkin, 1963; Hemphill & Zahler, 1968). If, at the time of nutrient depletion, division ceases but motility continues at a low rate, the increasing separation between the non-growing cells may trigger the chemotactic stimulus, leading to aggregation and fruiting. It is indeed common knowledge among workers who have examined the conditions for fruiting body formation, that there is an optimal cell density necessary for rapid fruiting.

Reichenbach (1965) has described an intriguing behavior pattern among some of the myxobacteria. Swarming cells, when examined with time-lapse photomicrography showed what he called 'rhythmic oscillation'. These are waves of cells, moving outward in approximately concentric circles. Rhythmic oscillation in *Podangium lichenicolum* is illustrated in Plate 5. The waves are periodically extinguished only to be overrun by the next wave. The period of the oscillation has not been determined, nor have the waves been subjected to any analysis. It is possible that they are reflecting the movement of synchronized segments of the swarming population. When swarming cells are in the act of division, motility ceases (Kühlwein & Reichenbach, 1968); in addition, DNA synthesis ceases, while the synthesis of RNA and protein is substantially reduced (Zusman, Gottlieb & Rosenberg, 1971). It is possible that the temporary extinction of the moving wave reflects this loss of motility during division.

One characteristic of the myxobacteria is their production of slime, which surrounds the cells. The function of the slime is not known. It is possible, however, that it may play a role in mediating the various cell interactions during swarming and fruiting. Unfortunately, there is, as yet, no evidence for interactions in the myxobacteria *via* direct cell–cell contact. The relatively small size of the cells precludes microelectrode experiments which have been used to demonstrate cell junctions in higher cells. Perhaps electron microscopy of thin-sections of swarming cells might reveal areas of organized contact or junctions.

There has recently been a considerable amount of activity relating levels of cyclic AMP (cAMP) or dibutyryl cAMP, either to motility (Johnson, Morgan & Pastan, 1972), growth (Otten, Johnson & Pastan, 1971) or contact inhibition (Sheppard, 1971) of mammalian cells. Sheppard has, in fact, been able to reverse the loss of contact inhibition in transformed mouse cells by adding dibutyryl cAMP and theophylline, an inhibitor of cAMP phosphodiesterase. He has suggested that contact inhibition is the result of activation, by cell–cell contact, of membrane-bound adenyl cyclase which converts ATP to the growth inhibitor cAMP (Sheppard, 1971). It would be interesting if the relationship in the myxobacteria between cell–cell contact and swarming or motility also involved the alteration of the activity of membrane-bound nucleotide cyclases.

2. *Motility*

Another approach toward the problem of co-ordinated activity has emerged as a result of the isolation by Burchard (1970) of an intriguing motility mutant of *M. xanthus*. This mutant, designated SM, can move only when apposed to other cells; isolated, single cells are non-motile.

In contrast, wild type cells move either as single cells or as part of a swarm. It is not clear what nutritional or environmental factors regulate this choice; it is obvious, however, that in the SM mutant, this regulatory mechanism has been altered or lost. This is the first cell interaction mutant isolated.

3. *Phototaxis*

Aschner & Chorin-Kirsh (1970) have recently described the phototactic behavior of a myxobacterium which seems to be a *Nannocystis*. Swarms of cells migrate away from a higher light intensity. Individual cells, however, do not manifest this behavior but remain indifferent to the light.

4. *Chemotaxis, aggregation and fruiting body formation*

The ability of myxobacterial cells to assemble at focal points (presumably in response to a chemotactic stimulus) and then to construct a macroscopic, elaborately differentiated structure obviously includes a variety of cell–cell interactions. It is not clear to what extent the streaming of cells into the aggregation centers involves the same kinds of interactions as among cells that are swarming. In the former case, the cells often move in streams, radiating inward toward the centers of aggregation. Their movement is presumably oriented by the production of a chemotactic substance, although there is no evidence available concerning its nature. The recent demonstrations by Bonner's group that cyclic AMP (Konijn *et al.* 1967) and folic acid (Pan, Hall & Bonner, 1972) are specific attractants for *Dictyostelium* and *Poly-sphondylium* has revived interest in the problem. The general questions one would wish to answer with regard to the myxobacteria are: (1) What is the attractant? (2) How is its production and excretion regulated? (3) What is the nature of the sensory mechanism which perceives it? (4) How is its perception translated into oriented movement? The detailed and systematic investigations of Adler (1969) on chemotaxis of *Escherichia coli* may answer some of these questions. The role of chemotaxis, however, in the context of a developing system may hold some additional surprises.

Our understanding of the actual processes involved in the construction of the fruiting body is limited. There have been two investigations of the fine structure of myxobacterial fruiting bodies. Voelz & Reichenbach (1969) examined *Stigmatella aurantiaca* and McCurdy (1969) examined *Chondromyces crocatus*. In both cases, the stalk appeared to be acellular and tubular. The diameter of the tubules was approximately that of the vegetative cells. McCurdy (1969) has suggested that the cells literally travel through the tubules during the morphogenetic process. Voelz & Reichenbach (1969), however, have taken a longitudinal section through a stalk which shows quite clearly that the tubules are discontinuous and occasionally contain the hulls of lysed cells. It thus seems unlikely that the cells actively travel through the stalk.

The structural matrix of the fruiting body seems to be comprised of slime. A simple explanation for the morphogenesis of the stalk could be as follows. Aggregated cells produce slime in a random, disoriented fashion. That slime which is exposed to the air dries and hardens, providing a barrier to further expansion in an upward or sideways direction. The slime is thus excreted downwards, lifting the cell mass above the surface. The tubules may simply be a result of lysed cells which were trapped in the slime. No such simple, physical explanation comes to mind however for the construction of the cysts themselves. Time lapse photomicrography of *Chondromyces* (Kühlwein & Reichenbach, 1968) shows that the cysts arise from the cell mass after the stalk has reached its maximal length. The nature of the timing of cyst emergence or the mechanics of their formation remain completely unclear.

5. Regulation of growth and DNA synthesis during germination and outgrowth

On the basis of analysis of grain count distribution in cells of *M. xanthus* labelled with tritiated thymidine, Rosenberg, Katarski & Gottlieb (1967) postulated that *M. xanthus*, growing exponentially in a rich medium, possessed two nucleoids replicating sequentially. They suggested further that during myxospore formation, cycles of DNA replication were completed but none were initiated. Thus, most myxospores should contain either three or four chromosomes per cell. Zusman & Rosenberg (1968) did a similar analysis of DNA synthesis in germinating and outgrowing cells of *M. xanthus*. Their data indicated that 75 % of the myxospores contained three chromosomes and 25 % contained four. Following germination and outgrowth the cells exhibited a peculiar pattern of synchronous division and DNA synthesis. The data could be rationalized by postulating that cell division and DNA synthesis occurred initially only among that segment of the myxospores which contained three chromosomes. These cells underwent a series of cell divisions and DNA synthesis until each cell contained the requisite two chromosomes. During this time, the portion of the population (25 %) containing four chromosomes withheld DNA synthesis and cell division. This resulted eventually in a population of cells which had become homogeneous with respect to DNA content and which then proceeded to divide synchronously. One interpretation of Rosenberg's model suggests a unique cell interaction at the level of DNA synthesis and regulation of cell division.

6. The germination signal

Myxospores of *M. xanthus* will germinate in 40 to 60 min when placed in the complex growth medium. They will also germinate in distilled water, but only if one of two conditions is met. Either a specific germinant must be present, or the cell density must exceed about 5×10^9 cells/ml. We have been intrigued both by the significance of this phenomenon and by the mechanism. A requirement for high cell density is unusual for germination of resting cells; usually the reverse is true. If, however, one views myxospore germination in the larger context of the entire developmental cycle the requirement for high cell density seems consistent. Thus far, all the interactional phenomena described for the myxobacteria result in the cells either resisting dispersion or coming together after having been dispersed. The requirement for high cell density for germination in distilled water may simply guarantee that under conditions whereby the cells will have to swarm to maximize nutrient collection, large numbers must germinate simultaneously. The questions we are asking then about this phenomenon are: (1) What is the auto-signal which both reflects cell density and triggers germination? (2) How is its production and excretion regulated? (3) How is the signal perceived? (4) How is this perception translated into germination?

Our data thus far indicate that when thoroughly washed myxospores were placed in distilled water the cells excreted inorganic phosphate and uracil. The concentration of phosphate excreted by 10^{10} cells/ml reached about 1 mM. Cells at a concentration of $< 10^9$/ml would not germinate unless placed in 1–10 mM phosphate or in the supernatant solution of cells which had germinated at a high cell density. The cellular source of the inorganic phosphate was neither polyphosphate nor RNA, two of the major sources of phosphate in the cell. The excretion of phosphate and uracil did however parallel a drop in CMP present in the acid-extractable material in the cell (Dworkin, unpublished results).

Rosenberg (personal communication) has shown that the level of alkaline phosphatase (AP) in germinating cells is far higher than that found in myxospores. We have further examined the kinetics of appearance and disappearance of AP in cells going through the cycle of myxospore formation and germination. When vegetative cells were induced to form myxospores by the glycerol induction technique (Dworkin & Gibson, 1964) morphological conversion was complete at 120 min. At about 180 min there was an abrupt tenfold increase in AP activity which then levelled off. When these cells were germinated in distilled water there was an additional immediate increase of AP activity. Thus germinating cells contained 50–100 times the specific AP activity found in vegetative rods. This level remained high throughout germination and outgrowth but dropped abruptly as the cells began to divide. It seems a reasonable possibility that the rise in AP activity was responsible for the release of phosphate from CMP in the intracellular nucleotide pool and is thus the enzymatic trigger for release of the germination signal. If this is the case, myxospores which have not yet synthesized AP should be unable to germinate in distilled water. When myxospores were removed from the induction medium at 150 min (30 min before the burst of AP activity) and placed at high cell density in distilled water, these cells lagged in their germination until they duplicated the second burst of AP activity seen in normal germinating cells (Dworkin, unpublished results). It is pertinent that Wilkins (1972) has just reported that induction of alkaline phosphatase in *E. coli* is not directly regulated by orthophosphate concentration; rather, the data point clearly to nucleotides as direct regulators.

We have previously examined RNA synthesis during myxospore induction (Ramsey & Dworkin, 1970) and found a sharp increase in the synthesis of 4–5 S RNA at about 3 h after induction. This coincided quite closely with the increase in AP activity seen during induction. The process of delayed induction of enzymatic activity is characteristic of many developing systems and leads naturally to questions as to the nature of the clock. The interval between induction of myxospore formation and the first burst of AP activity (180 min) coincides fairly well with the generation time of vegetative cells (210 min). Thus DNA polymerase is a reasonable candidate for the clock, with termination of a round of replication as the signal. However, Sadler & Dworkin (1966) showed that net DNA synthesis ceased about 80 min after induction started. Therefore, for DNA polymerase to be involved one must also invoke DNA turnover. Watson (1970) has pointed out the possibility that DNA transcription could act as a timing device in developing cells. Since there is sufficient DNA in *Escherichia coli* to permit 33 h of continuous transcription, 1 % of the genome would suffice for timing a three-hour event.

It seems appropriate here to quote Filner (1969) who, speaking at a symposium on 'Communication in Development', wistfully asked the rhetorical question 'Wouldn't it be nice to have systems in which a developmental event is initiated by a metabolite, particularly one whose biochemistry is well known? Then, with a little bit of luck, perhaps the site of initiation and the chain of biochemical steps leading to the developmental event would be found in known biochemistry?'

CONCLUSIONS AND SPECULATIONS

I wish to propose a theory as to the biological function of the myxo-bacterial life cycle. Such a theory has the merit not only of placing the life cycle in a broader biological context but also of implying relationships which are not otherwise obvious.

Myxobacteria, as a group, are capable of the extracellular hydrolysis of a wide variety of insoluble macromolecules. These include cellulose, protein, starch, and peptidoglycan. They can also use these polymers for growth. This property seems consistent with their ability to move over the surface of a solid substrate. It seems reasonable that generation of optimal concentrations of the lower molecular weight subunits of these polymers will depend on a certain optimal density of cells excreting the hydrolases. In other words, a wolf-pack effect. Thus the basic principle of my thesis is that the function of the myxobacterial life cycle is to maintain at all times the presence or the potential of an optimal density of swarming cells.

It has been known for some time that one can, in general, regulate fruiting body formation by limiting the concentration of nutrients in the medium (Vahle, 1909; Oetker, 1953; Kühlwein, 1950). In *M. xanthus* one can induce fruiting body formation by specifically reducing the concentration of phenylalanine and tryptophan (Dworkin, 1963) or under somewhat different nutritional conditions, any of the required amino acids (Hemphill & Zahler, 1968).

Thus a particular feature of the thesis is that when the swarming cells perceive that a specific diffusible component or components of the macromolecule which they have been hydrolysing has dropped below a certain level they cease their outward-oriented swarming and move inward toward aggregation centers. This may reflect the fact that, at that point, they become able to perceive the presence of a chemotactic aggregant which has been continually produced but ignored. Stated in more mechanistic terms, the reduction in the level of a small molecule may derepress the synthesis or release the inhibition of an enzyme (system) responsible for perceiving the attractant. Alternatively, of course, it may be the production of the aggregant which is regulated rather than the perception.

Aggregation of the vegetative rods is followed by the construction of the fruiting body. Within this fruiting body the individual vegetative cells lose their gliding motility and convert to metabolically quiescent resting cells. In some cases these are contained within cysts and it is not clear whether the cyst or the cell itself is the unit resting structure of the organism. Having achieved the aggregation, it is not obvious what the additional or peculiar function the fruiting structure itself confers. The specific features it adds to the organism are: (1) In those fruiting bodies with a stalk, the cell mass is lifted above the substrate. (2) In the so-called higher myxobacteria (e.g. *Chondromyces* and *Stigmatella*) the cells are contained in cysts. In a sense, these may be prepackaged swarms containing an optimal number of cells. If so, it might be possible to correlate the number of cells in a cyst with the optimal number of cells for constituting a migrating swarm capable of growing on food bacteria. (3) Cells in the mature fruiting body are separated from the external environment either by a cyst wall or by a slime layer. It is interesting that *Azotobacter* cysts have a far greater resistance to desiccation than the free myxospores of *M. xanthus*, to which they are morphologically similar. It may be pertinent that, while the latter form fruiting bodies, the former do not.

The germination process provides another interesting dimension, for the two interactional phenomena that have been described can both be interpreted in terms of the above hypothesis. The model proposed by Rosenberg's group (Zusman & Rosenberg, 1968) to explain the observed pattern of DNA synthesis and growth following germination, proposes that those myxospores which contain a full complement of DNA withhold their growth and DNA synthesis until the remainder of the population completes its required patterns of biosynthesis. The function of such an extraordinary intercellular interaction could be to ensure that, following germination, a homogeneous population can begin swarming in a synchronous fashion. While the germinated cells have indeed been shown to be synchronous with respect to growth it would be most interesting if such synchrony could also be demonstrated in the actual swarm.

I have already discussed the details and what seems a reasonable interpretation of the auto-signal for germination. That is, that under conditions of limited nutritional availability, myxospores will not germinate unless there are sufficient numbers and density of cells to constitute a swarm.

I shall close by pointing out (but not apologetically) that the tone of this article has been both mechanistic and teleological. For I believe (as do others, e.g. Ayala, 1968) that no biological explanation is complete unless it includes both approaches. In that sense, the examination of cell interactions in the myxobacteria offers not only a technically excellent system for obtaining mechanistic insights but a sufficiently circumscribed biological context to allow teleological speculation.

I am grateful to Dr David White for many helpful comments and criticisms and to Dr Hans Reichenbach for kindly supplying Plates 1–5.

The author is a Career Development Awardee of the National Institutes of Health, U.S. Public Health Service.

REFERENCES

ABDOU, N. I. & RICHTER, M. (1970). Role of the bone marrow in the immune response. *Adv. Immunol.* **12**, 201–70.

ABERCROMBIE, M. & HEAYSMAN, J. E. M. (1954). Social behavior of cells in tissue culture. II. Monolayering of fibroblasts. *Exp. Cell Res.* **6**, 293–306.

ABERCROMBIE, M., HEAYSMAN, J. E. M. & KARTHAUSER, H. M. (1957). Social behavior of cells in tissue culture. III. Mutual influence of sarcoma cells and fibroblasts. *Exp. Cell Res.* **13**, 276–91.

ADLER, J. (1969). Chemoreceptors in bacteria. *Science*, **166**, 1588–97.

ASCHNER, M. & CHORIN-KIRSH, I. (1970). Light-oriented locomotion in certain *Myxobacter* species. *Archiv f. Mikrobiol.* **74**, 308–14.

AYALA, F. J. (1968). Biology as an autonomous science. *Amer. Scien.* **56**, 207–21.

BARSKI, G. & BELEHRADEK, K., JR. (1965). Étude microcinématographique du mécanisme d'invasion cancéreuse en cultures de tissu normal associé aux cellules malignes. *Exp. Cell Res.* **37**, 464–80.

BONNER, J. T. (1963). *Morphogenesis: An Essay on Development*, pp. 173–86. New York: Atheneum.

BURCHARD, R. P. (1970). Gliding motility mutants of *Myxococcus xanthus*. *J. Bacteriol.* **104**, 940–7.

CECCARINI, C. & EAGLE, H. (1971). pH as a determinant of cellular growth and contact inhibition. *Proc. Nat. Acad. Sci. U.S.A.* **68**, 229–33.

COMAN, D. R. (1960). Reduction in cellular adhesiveness upon contact with a carcinogen. *Cancer Res.* **20** (2), 1202–4.

DWORKIN, M. (1962). Nutritional requirements for vegetative growth of *Myxococcus xanthus*. *J. Bacteriol.* **84**, 250–7.

DWORKIN, M. (1963). Nutritional regulation of morphogenesis in *Myxococcus xanthus*. *J. Bacteriol.* **86**, 67–72.

DWORKIN, M. (1965). Biology of the Myxobacteria. *Ann. Rev. Microbiol.* **20**, 75–106.

DWORKIN, M. (1972). The Myxobacteria: new directions in studies of procaryotic development. *Critical Reviews in Microbiology*, **2**, no. 1, 435–52.

DWORKIN, M. & GIBSON, S. M. (1964). A system for studying microbial morphogenesis: rapid formation of microcysts in *Myxococcus xanthus*. *Science*, **146**, 243–4.

DWORKIN, M. & SADLER, W. (1966). Induction of cellular morphogenesis in *Myxococcus xanthus*. I. General description. *J. Bacteriol.* **91**, 1516–19.

EAGLE, H. (1965). Metabolic controls in cultured mammalian cells. *Science*, **148** (1), 42–51.

EAGLE, H. & PIEZ, K. (1962). The population-dependent requirement by cultured mammalian cells for metabolites which they can synthesize. *J. Exp. Med.* **116**, 29–43.

FILNER, P. (1969). Control of nutrient assimilation. A growth-regulating mechanism in cultured plant cells. In *Communication in Development. Develop. Biol. Suppl.* 3, ed. A. Lang, pp. 206–26. New York and London: Academic Press.

FLUEGEL, W. (1963). Fruiting chemotaxis in *Myxococcus fulvus* (Myxobacteria). *Minn. Acad. Sci. Proc.* **32**, 120–3.

GILULA, N. B., REEVES, O. R. & STEINBACH, A. (1972). Metabolic coupling, ionic coupling and cell contacts. *Nature*, **235**, 262–5.

HEMPHILL, H. E. & ZAHLER, S. A. (1968). Nutritional induction and suppression of fruiting in *Myxococcus xanthus*. *J. Bacteriol.* **95**, 1018–23.

JAMAKOSMANOVIC, A. & LOEWENSTEIN, W. (1968). Intercellular communication and tissue growth. III. Thyroid cancer. *J. Cell Biol.* **38**, 556–61.

JENNINGS, J. (1961). Association of a steroid and a pigment with a diffusible fruiting factor in *Myxococcus virescens*. *Nature*, **190**, 190.

JOHNSON, G. D., MORGAN, W. D. & PASTAN, I. (1972). Regulation of cell motility by cyclic AMP. *Nature*, **235**, 54–6.

JUENGST, F., JR. (1972). RNA and protein synthesis during germination of *Myxococcus xanthus* myxospores. Ph.D. Thesis, University of Minnesota.

KONIGSBERG, I. R. (1963). Clonal analysis of myogenesis: its relevance to the general problem of the stability of cell-type in cultured animal cells. *Science*, **140**, 1273–84.

KONIJN, T. M., VAN DE MEENE, J. G. C., BONNER, J. T. & BARKLEY, D. S. (1967). The acrasin activity of adenosine-3′,5′-cyclic phosphate. *Proc. Nat. Acad. Sci. U.S.A.* **58**, 1152–4.

KÜHLWEIN, H. (1950). Beiträge zur Biologie und Entwicklungsgeschichte der Myxobakterien. *Archiv f. Mikrobiol.* **14**, 678–704.

KÜHLWEIN, H. & REICHENBACH, H. (1968). Schwarmentwicklung und Morphogenese bei Myxobakterien. Archangium, Myxococcus, Chondrococcus, Chondromyces. Wissenschaftlichen Film C 893/1965. Institut für Wissenschaftlichen Film. Göttingen, W. Germany.

LEV, M. (1954). Demonstration of a diffusible fruiting factor in Myxobacteria. *Nature*, **173**, 501.

LINGAPPA, B. T., LINGAPPA, Y. & BELL, E. (1972). A self-regulator of protein synthesis in the conidia of *Glomerulla cingulata*. *Abstracts of the Annual Meeting of the Am. Soc. Microbiol.* G-59, p. 40.

LOEWENSTEIN, W. R. (1968). Communication through cell junctions. Implications in growth control and differentiation In *The Emergence of Order in Developing Systems, Develop. Biol. suppl.* **2**, ed. M. Locke, pp. 151–83. New York and London: Academic Press.

LOOMIS, W. F., JR & SUSSMAN, M. (1966). Commitment to the synthesis of a specific enzyme during cellular slime mold development. *J. Molec. Biol.* **22**, 401–4.

McCURDY, H. D. JR (1969). Light and electron microscope studies on the fruiting bodies of *Chondromyces crocatus*. *Archiv. f. Mikrobiol.* **65**, 380–90.

McVITTIE, A. & ZAHLER, S. A. (1962). Chemotaxis in *Myxococcus*. *Nature*, **194**, 1299–1300.

MOSCONA, A. & MOSCONA, H. (1952). The dissociation and aggregation of cells from organ rudiments of the early chick embryo. *J. Anat.* **86**, 287–300.

NEWELL, P. C., FRANKE, J. & SUSSMAN, M. (1972). Regulation of four functionally related enzymes during shifts in the developmental program of *Dictyostelium discoideum*. *J. Molec. Biol.* **63**, 373–82.

OETKER, H. (1953). Untersuchungen über die Ernährung einiger Myxobakterien. *Archiv f. Mikrobiol.* **19**, 206–46.

OTTEN, J., JOHNSON, G. S. & PASTAN, I. (1971). Cyclic AMP levels in fibroblasts: relationship to growth rate and contact inhibition of growth. *Biochem. Biophys. Res. Comm.* **44**, 1192–8.

PAN, P. C., HALL, E. M. & BONNER, J. T. (1972). Effect of the second chemotactic substance-folic acid and related compounds on cellular slime molds. *Abstracts of the Annual Meeting of the Am. Soc. Microbiol.* G-243, p. 71.

RAMSEY, W. S. & DWORKIN, M. (1970). Stable messenger ribonucleic acid and germination of *Myxococcus xanthus* microcysts. *J. Bacteriol.* **101**, 531–40.

REICHENBACH, H. (1965). Rhythmische Vorgänge bei der Schwarmentfaltung von Myxobakterien. *Ber. der Deutsch. Bot. Gesellsch.* **78**, 102–5.

ROSENBERG, E., KATARSKI, M. & GOTTLIEB, P. (1967). Deoxyribonucleic acid synthesis during exponential growth and microcyst formation in *Myxococcus xanthus*. *J. Bacteriol.* **93**, 1402–8.

RUBIN, H. (1966) The inhibition of chick embryo cell growth by medium obtained from cultures of Rous sarcoma cells. *Exp. Cell Res.* **41**, 149–61.

SADLER, W. & DWORKIN, M. (1966). Induction of cellular morphogenesis in *Myxococcus xanthus*. II. Macromolecular synthesis and mechanism of inducer action. *J. Bacteriol.* **91**, 1520–5.

SHEPPARD, J. R. (1971). Restoration of contact-inhibited growth to transformed cells by dibutyryl adenosine 3′:5′-cyclic monophosphate. *Proc. Nat. Acad. Sci. U.S.A.* **68**, 1316–30.

STOKER, M. & MACPHERSON, I. (1961). Studies on transformation of hamster cells by polyoma virus in vitro. *Virology*, **14**, 359–70.

STOKER, M. (1964). Regulation of growth and orientation in hamster cells transformed by polyoma virus. *Virology*, **24**, 165–74.

STOKER, M. & SUSSMAN, M. (1965). Studies on the action of feeder layers in cell culture. *Exp. Cell Res.* **38**, 645–53.

TEMIN, H. M. & RUBIN, H. (1958). Characteristics of an assay for Rous sarcoma virus and Rous sarcoma cells in tissue culture. *Virology*, **6**, 669–88.

VAHLE, C. (1909). Vergleichende Untersuchungen über die *Myxobakteriazeen* und *Bakteriazeen*, sowie die *Rhodobakteriazeen* und *Spirilazeen*. *Zentral. f. Bakteriol.* **25**, 178–226.

VOELZ, H. & DWORKIN, M. (1962). Fine structure of *Myxococcus xanthus* during morphogenesis. *J. Bacteriol.* **84**, 943–52.

VOELZ, H. & REICHENBACH, H. (1969). Fine structure of fruiting bodies of *Stigmatella aurantiaca* (Myxobacterales). *J. Bacteriol.* **99**, 856–66.

VOGT, M. & DULBECCO, R. (1960). Virus–cell interaction with a tumor-producing virus. *Proc. Nat. Acad. Sci. U.S.A.* **46**, 365–70.

WATSON, J. D. (1970). *Molecular Biology of the Gene*, 2nd edit., pp. 528–9. New York: W. A. Benjamin.

WHITE, D., DWORKIN, M. & TIPPER, D. (1968). Peptidoglycan of *Myxococcus xanthus*. Structure and relation to morphogenesis. *J. Bacteriol.* **95**, 2186–97.

WILKINS, A. (1972). Physiological factors in the regulation of alkaline phosphatase synthesis in *Escherichia coli*. *J. Bacteriol.* **110**, 616–23.

WITKIN, S. S. & ROSENBERG, E. (1970). Induction of morphogenesis by methionine starvation in *Myxococcus xanthus*: polyamine control. *J. Bacteriol.* **103**, 641–9.

ZUSMAN, D. & ROSENBERG, E. (1968). Deoxyribonucleic acid synthesis during microcyst germination in *Myxococcus xanthus*. *J. Bacteriol.* **96**, 981–6.

ZUSMAN, D., GOTTLIEB, P. & ROSENBERG, E. (1971). Division cycle of *Myxococcus xanthus*. III. Kinetics of cell growth and protein synthesis. *J. Bacteriol.* **105**, 811–19.

EXPLANATION OF PLATES

PLATE 1

Fig. 1. Fruiting body of *Myxococcus fulvus* perched on a bit of cellulose. Magnification approx. 300×.

Fig. 2. Fruiting body of *Myxococcus stipitatus*. Magnification approx. 480×.

PLATE 2

Fig. 1. Fruiting body of *Podangium* spp. Magnification approx. 200×.

Fig. 2. Fruiting body of *Chondromyces apiculatus*. Magnification approx. 500×.

PLATE 3

Fig. 1. Fruiting body of *Stigmatella aurantiaca*. Magnification approx. 650×.

Fig. 2. Germinated cyst of *Chondromyces apiculatus*. Magnification approx. 425×.

PLATE 4

Fig. 1. Swarm edge of *Polyangium velatum*. Magnification 240×.

Fig. 2. Swarm edge of *Sorangium* spp. Magnification 26×.

PLATE 5

Oscillating waves in *Podangium lichenicolum*. Magnification about 40×.

PLATE 1

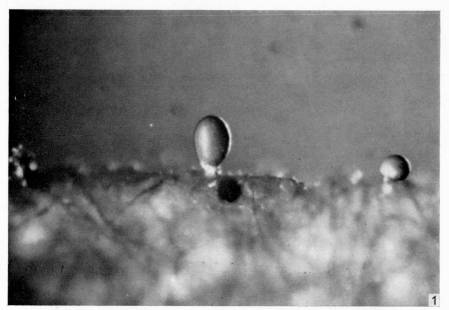

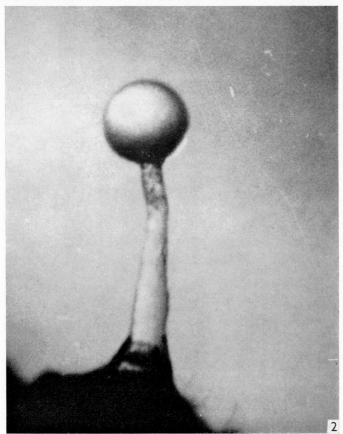

PLATE 2

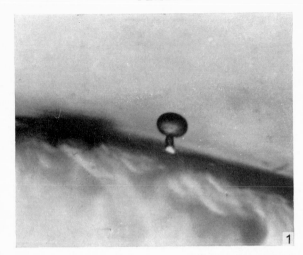

PLATE 3

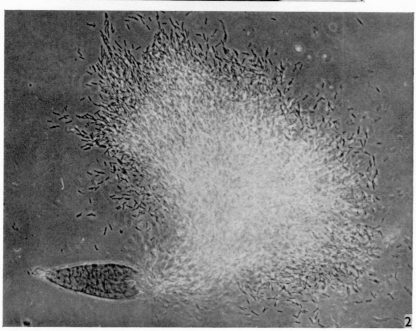

PLATE 4

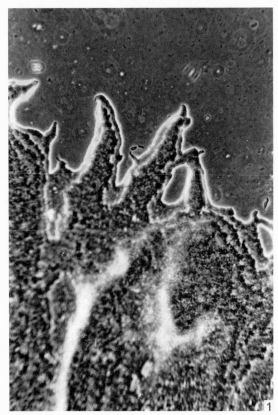

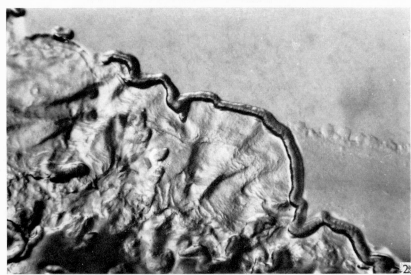

PLATE 5

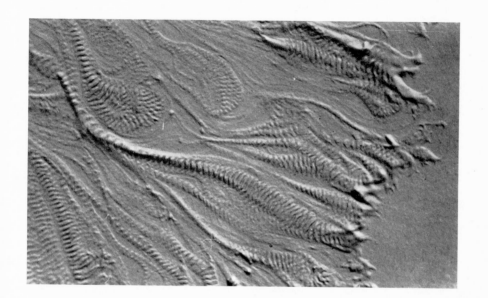

DIFFERENTIATION IN ACTINOMYCETES

K. F. CHATER AND D. A. HOPWOOD

John Innes Institute, Colney Lane, Norwich NOR 70F

INTRODUCTION

All but the simplest actinomycetes have a complex colony structure based on multinucleate, branching mycelia, with differentiation of the colony into regions playing vegetative and reproductive roles. This complex multicellular morphology led early microbiologists to believe that the actinomycetes were either fungi or something intermediate between bacteria and fungi. However, it is now clear that they are typically prokaryotic in ultrastructure (e.g. Hopwood & Glauert, 1960), in cell wall structure (Cummins & Harris, 1958; Pollock *et al.* 1972), in the possession of phages (Welsch, 1969) and in features of their genetic exchange systems (Hopwood, 1967; Sermonti, 1969; Hopwood & Wright, 1972). The variety of differentiation within the actinomycetes is indicated by their classification, which is largely on morphological grounds. For example, the mycelium of the thermophile *Thermoactinomyces vulgaris* produces aerial branches bearing heat-resistant true endospores arranged singly on protrusions from the hyphae (Cross, Walker & Gould, 1968; Dorokhova, Agre, Kalakutskii & Krassilnikov, 1968); in *Dermatophilus*, no aerial growth occurs, motile spores giving rise to hyphae that subdivide both longitudinally and transversely into compartments that eventually become spores (Gordon & Edwards, 1963); in *Streptomyces*, chains of spores are formed following subdivision of the long cells of the aerial branches (Plates 1, 2 and 3); and subdivision of long cells to give spore compartments also occurs in *Actinoplanes*, in which the sporogenous parts of the hyphae are enclosed within sporangia (Lechevalier, Lechevalier & Holbert, 1966).

Among other prokaryotes the range of organisms showing differentiation is rather limited. Most research has been directed at the formation and germination of the spores of bacilli (e.g. Schaeffer, 1969), while less extensive studies have been made of fruiting body formation and the rod/microcyst transition in myxobacteria (Dworkin, 1966; Rosenberg, Katarski & Gottlieb, 1967), cyst formation in *Azotobacter vinelandii* (Hitchins & Sadoff, 1970), rod/sphere metamorphosis in *Arthrobacter* (Ensign & Wolfe, 1964; Krulwich, Ensign, Tipper & Strominger, 1967*a*, *b*), the swarmer/stalked cell cycle in *Caulobacter crescentus* (Poindexter, 1964; Shapiro, Agabian-Keshishian & Bendis, 1971;

Newton, 1972) and heterocyst and spore-formation in the blue-gree alga *Anabaena* (Wolk, 1966, 1967; Wilcox, 1970; Lang & Fay, 1971). Three of these topics will be discussed by other contributors to this Symposium.

We wish to describe two experimental systems within the actinomycetes. Studies on one of these, the spore architecture of *Thermoactinomyces vulgaris*, have barely begun, and are not directly related to the general problems of colony differentiation; we shall briefly describe these studies first, and then devote the rest of the review to studies of differentiation in *Streptomyces coelicolor*.

SPORE GEOMETRY OF *THERMOACTINOMYCES VULGARIS*

Electron microscopic studies of endospore formation in *Thermoactinomyces vulgaris* (Cross, Walker & Gould, 1968; Dorokhova, Agre, Kalakutskii & Krassilnikov, 1968, 1970; Cross, Davies & Walker, 1971; McVittie, Wildermuth & Hopwood, 1972) suggest overall similarity to the analogous process in *Bacillus subtilis*. *B. subtilis* has many advantages over *T. vulgaris* for studying endospore formation, so a similar study in *T. vulgaris* would seem prodigal. One unique feature of the *T. vulgaris* spore is, however, the comparative regularity of its surface geometry. McVittie *et al.* (1972) showed that the spores are polyhedral with 12 pentagonal and about 12 hexagonal faces, each face being slightly concave. This geometry is a function of the outer spore coat, which is composed of parallel arrays of long fibrous striations spaced at about 5 μm. Several layers of these are found, each having a different orientation of the fibres.

An electron micrograph of endospores of a second thermophilic, endospore-producing actinomycete, *Actinobifida dichotomica*, was presented by Williams (1970), and from his micrograph it would appear that the spores are polyhedral but have many fewer faces than do those of *Thermoactinomyces vulgaris*, perhaps having just 12 pentagonal faces (McVittie *et al.* 1972). A detailed comparison of the fine structure and biochemistry of the outer spore coats of *A. dichotomica* and *T. vulgaris* would possibly provide useful information about the control of such spore architecture.

The regulation of the development of regular structures at this level of complexity (the spores are about 1 μm in diameter) is a largely unexplored phenomenon, and one which we hope to analyse through the isolation of mutants with altered spore shape. These studies should be helped by the recent development of a genetic transformation system in *Thermoactinomyces vulgaris* (Hopwood & Wright, 1972).

DIFFERENTIATION IN *STREPTOMYCES COELICOLOR*

Our studies of strain A3(2) of this organism (*S. violaceoruber* according to Kutzner & Waksman, 1959) have been mainly concerned with the initiation of aerial mycelium production and its subsequent metamorphosis into spores. One great advantage of using *S. coelicolor* is its amenability to genetic manipulation (Hopwood, 1967), facilitated by the uninucleate haploid spores which permit efficient cloning of mutants and recombinants. Analysis of recombinants has shown that more than 100 markers mapped to date are all located on a single circular linkage map (Hopwood & Chater, 1973), with the exception of characteristics controlled by a putative plasmid (SCP1) concerned with fertility (Vivian, 1971; see also below).

The life cycle and cytology of Streptomyces coelicolor

The following descriptions apply entirely to the growth of colonies of *S. coelicolor* on chemically defined agar media at 30 °C. After spore germination, which takes at least 4 h, a much-branched network of substrate hyphae develops in the agar and on its surface, giving a colony of about 1 mm diameter after 48 h. At this stage the colony is somewhat shiny and 'bald', but soon becomes white and hairy as aerial mycelium develops. The whiteness rapidly turns to the characteristic grey colour of fully developed colonies, the greyness being a property of mature spores.

Wildermuth (1970a) examined thin sections of colonies at various stages of development. He found that the substrate mycelium was a loose network of hyphae of homogeneous appearance, whereas the young aerial mycelium was more closely packed and the cells often contained vacuoles. In fully sporulating colonies the aerial mycelium was conspicuously heterogeneous: the surface of the colony consisted of mature spores and spore chains mixed with intact hyphal cells, while further from the surface and from the edge of the colony most hyphal cells were disintegrating (Fig. 1). Just below the colony surface many germinated spores were observed, indicating that secondary rounds of growth occur.

A possibly significant distinction between substrate and aerial hyphae is the occurrence of a superficial fibrous sheath covering the outer surface of aerial hyphae and spores (Hopwood & Glauert, 1961a, b; Wildermuth, Wehrli & Horne, 1971). The sheath is composed of paired rodlets arranged in a basketwork pattern on a delicate membrane (Plate 3, fig. 3). *S. coelicolor* spores have smooth silhouettes, but the spores of many other streptomycetes possess warty, spiny or hairy

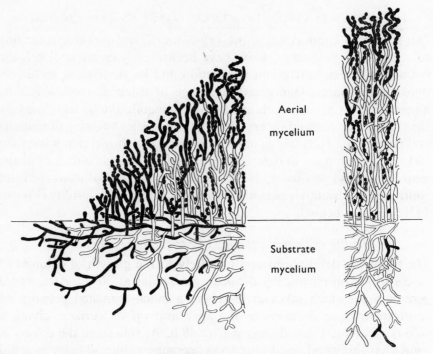

Fig. 1. Idealised diagram of a vertical section through the centre of a *Streptomyces coelicolor* colony at the climax of sporulation. Black represents intact cells and white disintegrating or completely lysed cells (from Wildermuth, 1970 *a*).

protrusions, which in some cases seem to be formed from material corresponding to the rodlets observed in *S. coelicolor* (Rancourt & Lechevalier, 1964; Wildermuth, 1970*b*, 1972*a*, 1972*b*).

Sporulation in streptomycetes is quite different from that of spore-forming bacilli, and has been the subject of several ultrastructural studies (Glauert & Hopwood, 1961; Rancourt & Lechevalier, 1964; Bradley & Ritzi, 1968; Wildermuth & Hopwood, 1970; Williams & Sharples, 1970). Wildermuth & Hopwood (1970) described four stages in spore formation in *S. coelicolor*. In stage 1 long cells of the aerial mycelium become coiled; in stage 2, sporulation septa are synchronously formed at regular intervals (of 1–2 μm; Hopwood, Wildermuth & Palmer, 1970) within such cells, each by the ingrowth of a double annulus continuous with the cell membrane and wall (Plate 2, figs. 1, 2). (This process is morphologically different from the formation of cross-walls in substrate and aerial hyphae unconnected with sporulation; Wildermuth & Hopwood, 1970; Williams & Sharples, 1970.) As sporulation septation is completed, stage 3, the laying down of the thick spore wall, begins (Plate 3, figs. 1, 2). In stage 4, which merges with stage 3, the cylindrical spore compartments become ellipsoidal, with the disintegration of remnants of old cell wall external to the spore wall, and the final stages in spore maturation occur. Ultimately, the spores in a chain are only joined at a very small interface and by the remnants of the fibrous sheath. Mesosomes are seen particularly frequently in stages 2 and 3, often in association with sporulation septa (Plate 2, fig. 2) (Glauert & Hopwood, 1960; Wildermuth & Hopwood, 1970; Wildermuth, 1971). There is some suggestion that spore maturation continues even after chains have broken up (A. McVittie, personal communication). Spores of streptomycetes are not very resistant to heat and contain no dipicolinic acid, a substance associated with heat resistant endospores of bacilli and *Thermoactinomyces* (Church & Halvorson, 1959; Cross *et al.* 1968). As we have seen, they are not endospores but relatively unspecialised compartments of the aerial hyphae.

The specific physiological requirements of each step in morphogenesis in streptomycetes remain virtually unknown and unstudied; even spore germination, which should be easily amenable to physiological experimentation, does not seem to have been examined. The other processes are of course much less easy to study because they do not take place in liquid media. On the other hand they are among the most striking differentiation phenomena in the bacteria and, as such, they are more likely to involve processes with general morphogenetic relevance.

Morphogenetic mutants of Streptomyces coelicolor

The sequential changes in the appearance of *S. coelicolor* colonies described earlier facilitate the isolation of mutants defective in particular morphogenetic processes. Mutants unable to initiate aerial mycelium formation remain 'bald' on prolonged incubation, hence the gene designation *bld* (Chater, 1970), while mutants blocked in spore formation form white aerial mycelium that does not turn grey on prolonged incubation, hence the gene designation *whi* (Hopwood *et al.* 1970). Both kinds of mutant are easily maintained by sequential subculture and by lyophilisation: indeed Hopwood & Ferguson (1969) used a strain carrying the type *bldA* mutation *bldAS48*, then termed S48, as a test of the efficiency of their lyophilisation procedure for streptomycetes. In order to avoid mutants whose morphological abnormality results from some metabolic disorder not specific for aerial growth, only mutants growing at the normal rate have been used in these studies.

Bld mutants

Bld mutants have been isolated after UV or *N*-methyl-*N'*-nitro-*N*-nitrosoguanidine treatment (Delić, Hopwood & Friend, 1970) and occurred at frequencies comparable to those for normal chromosomal markers. Okanishi, Ohta & Umezawa (1970) obtained a high frequency of aerial mycelium-less mutants of *Streptomyces venezuelae* and *S. kasugaensis* after heat treatment (and by acridine treatment of *S. venezuelae*). They attributed this to elimination of a hypothetical plasmid involved in aerial mycelium formation. However, in preliminary experiments in our laboratory designed to test the effect of acridine dyes on the putative fertility plasmid SCP1, no increased frequency of morphologically abnormal colonies of *S. coelicolor* was observed (A. Vivian & M. E. Townsend, personal communication). The idea that plasmid-borne genes are not involved in aerial mycelium formation in *S. coelicolor* receives support from genetic mapping studies with *bld* mutants. All *bld* mutations tested have a location on the linkage map, though only three have been accurately mapped. These three are all situated between the *cysA* and *nicA* loci (e.g. *bldAS48* in Table 2 of Hopwood, 1967). Other *bld* mutations have been located in the vicinity of *argA*, *strA* and the 9 o'clock region (Fig. 2) (D. A. Hopwood, unpublished; Chater, 1970). None has behaved in crosses as if unlinked to chromosomal markers, so we may conclude that *bld* mutants obtained by routine mutagenic procedures are not usually, if ever, due to plasmid elimination.

Studies on *bld* mutants have not yet proceeded beyond this preliminary crude mapping stage, though we expect soon to embark on a more thorough study. It is, however, perhaps worth mentioning that aerial mycelium formation is rather sensitive to various unresearched environmental and genetic background factors, and that some *bld* mutations have been difficult to map because of the ambiguous morphology of a small proportion of the recombinants.

Whi mutants

A much larger number of white colony (*whi*) mutants has been studied. Hopwood *et al.* (1970) isolated about 100 *whi* mutants and after preliminary examination in the light microscope they went on to study three of these in more detail. These mutations, *whi-6*, *-13*, and *-46*, all affect the formation of sporulation septa such that in *whi-6* and *-13* the spacing of the septa was aberrant and in *whi-46* the septa were absent. Mapping studies placed *whi-13* near to the *his* gene cluster at 12 o'clock on the

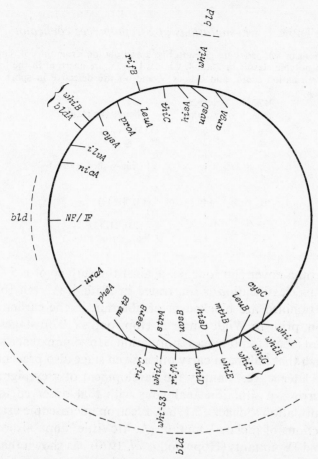

Fig. 2. Map locations of morphogenetic and rifampicin-resistance mutations on the *Streptomyces coelicolor* linkage map. Reference markers (Hopwood, 1967) are given inside the circle together with the location of the fertility determinant, NF/IF (Vivian & Hopwood, 1970). The locations of mutations giving white colonies (*whi*, Hopwood *et al.* 1970; Chater, 1972), 'bald' colonies (*bld*, Chater, 1970; Hopwood & Chater, unpublished), and rifampicin resistance (*rif*, Chater, unpublished) are given outside the circle. Table 1 gives further information on *whi* mutations.

linkage map, and *whi-6* and *-46* very close together between *leuB* and the *cysC, D* markers in the 5 o'clock region. The presence of normal vegetative crosswalls in *whi-46* was genetic confirmation of the conclusion from thin sections that sporulation septa are distinct from vegetative crosswalls.

More than 50 of the *whi* mutants of Hopwood *et al.* (1970) have now been examined in more detail (Chater, 1972; A. McVittie, unpublished) and a rough phenotypic classification has been suggested for ease of reference, based on the microscopic appearance of the material

Table 1. *whi mutations of Streptomyces coelicolor*

Map locations of *whi* genes are given in Fig. 2 and photomicrographs of representatives of the six phenotypic classes in Plate 4. Classes I to IV are aberrant in the formation or spacing of sporulation septa, and classes V and VI are defective in spore maturation (Chater, 1972).

Whi gene	Number of mutants	Phenotypic class
A	6	II (5), IV (1)
B	2	II
C	1	I
D	1	VI (thin-walled spores)
E	2	VI
F	1	V
G	14	I (13), III (1)
H	7	III
I	15	IV (13), III (2)
-53	1	Oligosporogenous

deposited on a cover slip touched against the surface of a 5–7 day old colony. This classification is illustrated in Plate 4. All but four of the mutants examined are blocked at or aberrant in the earlier stages of sporulation, prior to Wildermuth & Hopwood's (1970) stages 3 and 4. This suggests that streptomyces sporulation is to some extent a sequential process such that failure to carry out a given stage also prevents at least one of the later stages, namely the development of grey pigmentation, from occurring, a situation analogous with that observed in *Bacillus subtilis* sporulation (Schaeffer, 1969). Electron microscopic examination of thin sections of a class I mutant (A. McVittie, unpublished) and of class III and IV mutants (Hopwood *et al.* 1970) has shown that the wall thickening found in spores at stage 3 and 4 rarely or never takes place, again demonstrating sequential gene action.

One might argue that this effect could be the result of multiple mutation; however, all the *whi* mutations mapped by Hopwood *et al.* (1970) and Chater (1972) segregated as if they were single point mutations having a characteristic chromosomal location; moreover, with one exception (see below), microscopic examination of white recombinants always showed a phenotype essentially similar to that of the *whi* parent.

To date eight distinct *whi* genes have been identified by mapping, most of them in the lower section of the linkage map (Fig. 2). In general, mutations at each location possess the same phenotype, the two most outstanding exceptions to this being: *whi-13*, which has class IV phenotype yet is very closely linked to five mutations at the *whiA* locus giving the class II phenotype; and *whi-99*, the only mutation giving class V phenotype, which is very closely linked to a group of mutations (*whiG*) giving class I phenotype. The latter phenotypic difference was so great that *whi-99* was considered to be the unique representative of a gene designated *whiF*: however, recent observations suggest that most white recombinant colonies in crosses of the original *whi-99* strain with *whi*[+] strains have the class I phenotype, so that a reinvestigation of the nature of the defect in this strain will be necessary. It is an interesting possibility that the apparently simple class V phenotype of *whi-99* in fact arises by interaction between a class I mutation in *whiG* and a mutation in a second, as yet unrecognised, gene.

The mapping and phenotypic studies of *whi* mutants done so far are summarised in Fig. 2 and Table 1. Five of the six loci concerned with early stages of sporulation (prior to stage 3) are represented by several mutants, suggesting that not many early *whi* genes remain to be discovered (unless each *whi* locus presently recognised comprises more than one gene). If we exclude *whiF* (for the reasons given above) only two loci concerned with spore maturation, *whiD* and *E*, have been identified, having 1 and 2 mutational representatives respectively; hence it is probable that more maturation loci await discovery, and we are at present studying a number of spore-forming mutants to clarify this situation, most of which have been newly isolated by Dr E. Lawson. The possibility should be borne in mind that not all sporulation mutations will give rise to white colonies; however, if a strictly sequential process is centrally involved in sporulation, white colony colour should result from any mutation affecting the central sequence of events.

In *Bacillus subtilis* a large class of sporulation mutants is termed 'oligosporogenous', such mutants producing a reduced number of spores, which are apparently normal (Schaeffer, 1969). The *Streptomyces coelicolor* mutant *whi-53* produces spores having wild-type morphology in thin sections, but the ratio of sporulating to non-sporulating aerial hyphae observed by light or electron microscopy is lower than usual, and the spore chains appear to be shorter (A. McVittie, personal communication). This mutation is located on the linkage map somewhere near the *whiC* gene (Fig. 2). Such oligosporogenous mutants appear to be rare in *S. coelicolor*, but the possibility exists that their colonies are often too grey to be clearly recognised as *whi* mutants.

Complementation tests with whi mutants

Two situations are available in *Streptomyces coelicolor* in which partially diploid cells occur and in which complementation tests are therefore, in principle, feasible. The first, the zygote, is only transient and is found entirely in the substrate mycelium. It is thus separated in both time and distance from aerial mycelium growth and is therefore unsuited to tests on sporulation mutants. The second kind of partially diploid cell, the heteroclone cell, is more permanent though still markedly unstable, and has been used successfully in complementation tests with auxotrophic and UV sensitive mutants (Hopwood, Sermonti & Spada-Sermonti, 1963; Harold & Hopwood, 1970, 1972). A detailed discussion of heteroclones is given elsewhere (Hopwood, 1967) so that their nature will not be discussed here. Because heteroclone colonies are unstable, irregular and small, it is difficult to use their appearance as a criterion in complementation tests of morphological markers; but it may be possible to utilise heteroclones in a less direct way. Since each heteroclone arises from a partially diploid plating unit, the morphology of this plating unit might be used as an indicator of complementation. The problem then becomes the recognition of these partially diploid plating units. One solution may be to exploit the difference in retention of spores and hyphal fragments by filters. If two *whi* mutations of, say, class I phenotype complement, then the appropriate partially diploid genotype should be found exclusively in spores of wild-type morphology. The only other spores formed in crosses involving closely linked *whi* mutations would be rare true *whi+* haploid recombinants, which are easily recognised. All other genotypes found in the progeny would be expected to reside in class I fragments that would be preferentially removed by filtration. Thus in principle the effect of filtration on the formation of appropriate heteroclones could be harnessed as a means of complementation testing of *whi* mutants of phenotypic classes I–IV, in which spores are not found. Such experiments require the construction of many new recombinant strains, and this work is now in progress.

Molecular genetical approaches to the regulation of morphogenesis in Streptomyces coelicolor

The use of an interaction between fertility types

S. *coelicolor* has recently been found to possess a rather complex fertility system involving the three fertility types NF (Normal Fertility; Hopwood, Harold, Vivian & Ferguson, 1969); IF (Initial Fertility; Vivian & Hopwood, 1970); and UF (Ultra-Fertility; Hopwood *et al.* 1969). The characteristics of these fertility types are reviewed in detail elsewhere (Hopwood, 1972; Hopwood & Chater, 1973; Hopwood, Chater, Dowding & Vivian, 1973); the important feature in the context of this review is the inhibition of aerial mycelium formation in UF strains by IF and NF strains (Vivian, 1971). This has been studied mainly with IF cultures, to which most of what follows applies. If an IF culture is grown in close proximity to a UF culture on agar, the parts of the UF mycelium closest to the IF fail to sporulate, a zone of inhibition of several millimetres often being observed. Thus IF colonies produce a diffusible substance affecting this stage of UF morphogenesis. Production of the substance is correlated with possession of the plasmid SCP1, the determinant that converts UF strains into IF (Vivian, 1971).

This circumstance opens up a novel approach to the study of aerial mycelium formation, centring on the isolation of the diffusible substance and the determination of its properties and site of action, with the obvious possibility of isolating UF mutants that produce aerial mycelium in the presence of the inhibitor.

Investigation of possible transcriptional control

In several prokaryotic systems where sequential changes occur in the spectrum of genes expressed, central control is through the template specificity of RNA polymerase. In the development of T4 phage in *Escherichia coli*, for example, the promoters of the pre-early phage genes are recognised by the host's enzyme, this specificity being mediated through the host sigma subunit (which plays an important catalytic role in the interaction between promoters and RNA polymerase). As a result of pre-early gene expression, a new phage-specified sigma factor replaces the host factor and causes a change in promoter specificity of the enzyme such that promoters of the 'delayed early' genes are specifically recognised (Travers, 1970). Further changes in promoter specificity of the enzyme that occur during the phage cycle are accompanied by, and presumably result from, modifications of all three subunit species of the apoenzyme (Seifert, Rabussay & Zillig, 1971). A second example of such transcriptional control is in the sporulation of *Bacillus subtilis*, where the RNA polymerases isolated from vegetative and sporulating cells have different template specificities (Losick & Sonenshein, 1969) and behave differently when subjected to sodium dodecyl sulphate poly-acrilamide gel electrophoresis: in particular, no sigma factor is detectable in the enzyme from sporulating cells, and the β-subunit of the vegetative enzyme is cleaved by a sporulation-specific serine protease to yield an altered subunit in the sporulation enzyme (Losick, Shorenstein & Sonenshein, 1970; Leighton, Freese, Doi, Warren & Kelln, 1971).

Such observations have led us to speculate that analogous mechanisms may operate in differentiation processes in *S. coelicolor* and we are therefore initiating an investigation into the RNA polymerase of this organism, making use of the antibiotic rifampicin. All rifampicin-resistant mutants of *Escherichia coli* possess alterations of the β-subunit of RNA polymerase (Mindlin, Ilyina, Voeykova & Velkov, 1972; Heil & Zillig, 1970), and similar mutants of many other bacteria have been obtained. Of particular interest were the observations of Sonenshein & Losick (1970) and Doi, Brown, Rodgers & Hsu (1970) that among the rifampicin-resistant mutants of *Bacillus subtilis*, many were sporulation-defective, the pleiotropic phenotype resulting from single-site mutations affecting RNA polymerase (Sonenshein & Losick, 1970). We have isolated a number of rifampicin-resistant mutants of *Streptomyces coelicolor*. A few of these possessed morphological abnormalities, but separation of resistance and the morphological abnormality has always been obtained in crosses, indicating multiple mutations. *S. coelicolor rif* mutants fall into at least three groups on the basis of genetic mapping (Chater, unpublished; Fig. 2), and we do not yet know which, if any, of these groups possess altered RNA polymerase. (A sharp reduction in the rate of uptake of ^{3}H-uridine into the TCA-insoluble material of growing wild-type cells occurs within 2–3 min of adding rifampicin ($> 5\ \mu g/ml$), and in the resistant mutants incorporation is much less affected by the antibiotic (Chater, unpublished), indicating that the wild-type RNA polymerase of *S. coelicolor* is sensitive to the antibiotic, as in other prokaryotes.) Representative mutants of all three classes have been used to test crudely the possibility that different RNA polymerases occur in substrate and aerial mycelium. Spores of each of the three mutants and the wild-type were separately added to pieces of cellophane lying on agar medium lacking rifampicin, and after 24 h incubation water was added to prevent aerial hyphae from developing. After a further 48 h incubation the cellophane was transferred to dry plates with and without added rifampicin, and incubation was continued for a further 72 h. Rifampicin inhibited aerial mycelium formation only in the wild-type strain. Assuming that at least one of the mutants had rifampicin-resistant RNA polymerase, it thus appears probable that rifampicin-resistance or sensitivity is conserved throughout the development of substrate and aerial mycelium, indicating that at least one subunit of the RNA polymerase is used in both growth phases (Chater, unpublished).

Studies on the role of transcription in control of sporulation in *Bacillus subtilis* have been helped by the use of the DNA of a phage (ϕe) as an indicator of vegetative transcription; this DNA is transcribed by the vegetative, but not by the sporulation, RNA polymerase (Losick & Sonenshein, 1969). The recent isolation of a number of *S. coelicolor* phages and the development of techniques for their propagation and purification (Lomovskaya, Emeljanova & Alikhanian, 1971; Lomovskaya, Mkrtumian, Gostimskaya & Danilenko, 1972; J. E. Dowding, manuscript in preparation) opens up the possibility of using actinophage DNA as an indicator of vegetative transcription.

Studies on transcription in *Streptomyces griseus* have been made by Szeszák, Szabó & Sümegi (1970), who obtained evidence suggesting that a native DNA fraction contained a factor(s) inhibiting its use as a template for RNA synthesis by endogenous RNA polymerase or by the exogenous enzyme from *Escherichia coli* at low ionic strength. The significance of this observation in the context of morphogenesis is undetermined.

CONCLUSION

The morphological complexity of actinomycetes is at the same time the reason for studies of the kind described in this paper and the greatest barrier to success in them. In particular, the failure of most actinomycetes to undergo their full differentiation in liquid medium makes physiological studies of the process extremely difficult. However, in *Streptomyces coelicolor* we may be able to make fairly homogeneous preparations of various stages of growth and thus to characterise the static features of each. Growth in liquid medium prevents aerial mycelium formation, and allows the definition of aspects of the vegetative phase of growth; while the use of *whi* mutants may facilitate the fractionation of aerial mycelium at various stages of development. Separation of substrate and aerial mycelium in aqueous 2-phase polymer systems, as has been done with spores and vegetative cells of *Bacillus* spp. (Sacks & Alderton, 1961) may also be possible, particularly in view of the characteristic possession of an additional superficial layer, the fibrous sheath, by aerial mycelium.

We wish particularly to thank Dr Anne McVittie for providing Plate 2 and Plate 3, figs. 1 and 2, together with various unpublished results, and for a critical reading of the manuscript. We are grateful to Dr Hansrudolf Wildermuth for providing Plate 3, fig. 3 and to Miss Judith Humphries for assistance with much of the new experimental work described in this review.

REFERENCES

BRADLEY, S. G. & RITZI, D. (1968). Composition and ultrastructure of *Streptomyces venezuelae. Journal of Bacteriology*, **95**, 2358–64.

CHATER, K. F. (1970). Genetic control of morphogenesis in *Streptomyces coelicolor. First International Symposium, Genetics of Industrial Microorganisms, Prague.* Abstract book, 38–9.

CHATER, K. F. (1972). A morphological and genetic mapping study of white colony mutants of *Streptomyces coelicolor*. *Journal of General Microbiology*, **72**, 9–28.

CHURCH, B. D. & HALVORSON, H. O. (1959). Dependence of the heat resistance of bacterial endospores on their dipicolinic acid content. *Nature, London*, **183**, 124–5.

CROSS, T., DAVIES, F. L. & WALKER, P. D. (1971). *Thermoactinomyces vulgaris*. I. Fine structure of the developing endospores. In *Spore Research* – 1971, ed. A. N. Barker, G. W. Gould & J. Wolf, pp. 175–80. London: Academic Press.

CROSS, T., WALKER, P. D. & GOULD, G. W. (1968). Thermophilic actinomycetes producing resistant endospores. *Nature, London*, **220**, 352–4.

CUMMINS, C. S. & HARRIS, H. (1958). Studies on the cell wall composition and taxonomy of Actinomycetales and related groups. *Journal of General Microbiology*, **18**, 173–89.

DELIĆ, V., HOPWOOD, D. A. & FRIEND, E. J. (1970). Mutagenesis by *N*-methyl-*N'*-nitro-*N*-nitrosoguanidine in *Streptomyces coelicolor*. *Mutation Research*, **9**, 167–82.

DOI, R. H., BROWN, L. R., RODGERS, D. & HSU, Y. (1970). *B. subtilis* mutant altered in spore morphology and in RNA polymerase activity. *Proceedings of the National Academy of Sciences, U.S.A.* **66**, 404–10.

DOROKHOVA, L. A., AGRE, N. S., KALAKUTSKII, L. V. & KRASSILNIKOV, N. A. (1968). Fine structure of spores in a thermophilic actinomycete, *Micromonospora vulgaris*. *Journal of General and Applied Microbiology*, **14**, 295–303.

DOROKHOVA, L. A., AGRE, N. S., KALAKUTSKII, L. V. & KRASSILNIKOV, N. A. (1970). Electron microscopic study of spore formation in *Micromonospora vulgaris*. *Microbiology* (a translation of *Mikrobiologiya*), **39**, 589–93.

DWORKIN, M. (1966). Biology of the myxobacteria. *Annual Reviews of Microbiology*, **20**, 75–106.

ENSIGN, J. C. & WOLFE, R. S. (1964). Nutritional control of morphogenesis in *Arthrobacter crystallopoietes*. *Journal of Bacteriology*, **87**, 925–32.

GLAUERT, A. M. & HOPWOOD, D. A. (1960). The fine structure of *Streptomyces violaceoruber* I. The cytoplasmic membrane system. *Journal of Biophysical and Biochemical Cytology*, **7**, 479–88.

GLAUERT, A. M. & HOPWOOD, D. A. (1961). The fine structure of *Streptomyces violaceoruber* (*S. coelicolor*) III. The walls of the mycelium and spores. *Journal of Biophysical and Biochemical Cytology*, **10**, 505–16.

GORDON, M. A. & EDWARDS, M. R. (1963). Micromorphology of *Dermatophilus congolensis*. *Journal of Bacteriology*, **86**, 1101–15.

HAROLD, R. J. & HOPWOOD, D. A. (1970). Ultraviolet-sensitive mutants of *Streptomyces coelicolor* II. Genetics. *Mutation Research*, **10**, 439–48.

HAROLD, R. J. & HOPWOOD, D. A. (1972). A rapid method for complementation testing of ultraviolet-sensitive (*uvs*) mutants of *Streptomyces coelicolor*. *Mutation Research* **16**, 27–34.

HEIL, A. & ZILLIG, W. (1970). Reconstitution of bacterial DNA-dependent RNA polymerase from isolated subunits as a tool for the elucidation of the role of the subunits in transcription. *Federation of European Biochemical Societies Letters*, **11**, 165–8.

HITCHINS, V. M. & SADOFF, H. L. (1970). Morphogenesis of cysts in *Aztobacter vinelandii*. *Journal of Bacteriology*, **104**, 492–8.

HOPWOOD, D. A. (1967). Genetic analysis and genome structure in *Streptomyces coelicolor*. *Bacteriological Reviews*, **31**, 373–403.

HOPWOOD, D. A. (1972). Genetics of the Actinomycetales. In *Actinomycetales: Characteristics and Practical Importance*, ed. G. Sykes, pp. 9–31. London: Academic Press.

HOPWOOD, D. A. & CHATER, K. F. (1973). *Streptomyces coelicolor*. In *Handbook of Genetics*, ed. R. C. King. New York: Van Nostrand Reinhold (in press).

HOPWOOD, D. A., CHATER, K. F., DOWDING, J. & VIVIAN, A. (1973). Advances in *Streptomyces coelicolor* genetics. *Bacteriological Reviews* (in press).

HOPWOOD, D. A. & FERGUSON, H. M. (1969). A rapid method for lyophilizing *Streptomyces* cultures. *Journal of Applied Bacteriology*, **32**, 434–6.

HOPWOOD, D. A. & GLAUERT, A. M. (1960). The fine structure of *Streptomyces coelicolor*. II. The nuclear material. *Journal of Biophysical and Biochemical Cytology*, **8**, 267–78.

HOPWOOD, D. A. & GLAUERT, A. M. (1961a). Electron microscope observations on the surface structures of *Streptomyces violaceoruber*. *Journal of General Microbiology*, **26**, 325–30.

HOPWOOD, D. A. & GLAUERT, A. M. (1961b). Studi di microscopia elettronica sulle strutture di superficie dello *Streptomyces coelicolor* (*S. violaceoruber*). *Annali di Microbiologia ed Enzimologica*, **11**, 173–9.

HOPWOOD, D. A., HAROLD, R. J., VIVIAN, A. & FERGUSON, H. M. (1969). A new kind of fertility variant in *Streptomyces coelicolor*. *Genetics*, **62**, 461–77.

HOPWOOD, D. A., SERMONTI, G. & SPADA-SERMONTI, I. (1963). Heterozygous clones in *Streptomyces coelicolor*. *Journal of General Microbiology*, **30**, 249–60.

HOPWOOD, D. A., WILDERMUTH, H. & PALMER, H. M. (1970). Mutants of *Streptomyces coelicolor* defective in sporulation. *Journal of General Microbiology*, **61**, 397–408.

HOPWOOD, D. A. & WRIGHT, H. M. (1972). Transformation in *Thermoactinomyces vulgaris*. *Journal of General Microbiology*, **71**, 383–98.

KRULWICH, T. A., ENSIGN, J. C., TIPPER, D. J. & STROMINGER, J. L. (1967a). Sphere-rod morphogenesis in *Arthrobacter crystallopoietes*. I. Cell wall composition and polysaccharides of the peptidoglycan. *Journal of Bacteriology*, **94**, 734–40.

KRULWICH, T. A., ENSIGN, J. C., TIPPER, D. J. & STROMINGER, J. L. (1967b). Sphere-rod morphogenesis in *Arthrobacter crystallopoietes*. II. Peptides of the cell wall peptidoglycan. *Journal of Bacteriology*, **94**, 741–50.

KUTZNER, H. J. & WAKSMAN, S. A. (1959). *Streptomyces coelicolor* Müller and *Streptomyces violaceoruber* Waksman and Curtis, two distinctly different organisms. *Journal of Bacteriology*, **78**, 528–38.

LANG, N. J. & FAY, P. (1971). The heterocysts of blue-green algae. II. Details of ultrastructure. *Proceedings of the Royal Society of London*, B, **178**, 193–203.

LECHEVALIER, H. A., LECHEVALIER, M. P. & HOLBERT, P. E. (1966). Electron microscopic observation of the sporangial structure of strains of *Actinoplanaceae*. *Journal of Bacteriology*, **92**, 1228–35.

LEIGHTON, T. J., FREESE, P. K., DOI, R. H., WARREN, R. A. J. & KELLN, R. A. (1971). Initiation of sporulation in *Bacillus subtilis*: requirement for serine protease activity and ribonucleic acid polymerase modification. In *Spores*, vol. v, ed. H. O. Halvorson, R. Hanson & L. L. Campbell, pp. 238–46. Washington, D.C.: American Society for Microbiology.

LOMOVSKAYA, N. D., EMELJANOVA, L. K. & ALIKHANIAN, S. I. (1971). The genetic location of prophage on the chromosome of *Streptomyces coelicolor*. *Genetics*, **68**, 341–7.

LOMOVSKAYA, N. D., MKRTUMIAN, N. M., GOSTIMSKAYA, N. L. & DANILENKO, V. N. (1972). Characterisation of temperate actinophage ϕC31 isolated from *Streptomyces coelicolor* A3(2). *Journal of Virology*, **9**, 258–62.

LOSICK, R. & SONENSHEIN, A. L. (1969). Change in the template specificity of RNA polymerase during sporulation of *Bacillus subtilis*. *Nature, London*, **224**, 35–7.

LOSICK, R., SHORENSTEIN, R. G. & SONENSHEIN, A. L. (1970). Structural alteration of RNA polymerase during sporulation. *Nature, London*, **227**, 910–13.

MCVITTIE, A. M., WILDERMUTH, H. & HOPWOOD, D. A. (1972). Fine structure and surface topography of *Thermoactinomyces vulgaris* endospores. *Journal of General Microbiology*, **71**, in press.

MINDLIN, S. Z., ILYINA, T. S., VOEYKOVA, T. A. & VELKOV, V. V. (1972). Genetical analysis of rifampicin resistant mutants of *E. coli* K12. *Molecular and General Genetics*, **115**, 115–21.

NEWTON, A. (1972). Role of transcription in the temporal control of development in *Caulobacter crescentus*. *Proceedings of the National Academy of Sciences, U.S.A.* **69**, 447–51.

OKANISHI, M., OHTA, T. & UMEZAWA, H. (1970). Possible control of formation of aerial mycelium and antibiotic production in *Streptomyces* by episomic factors. *Journal of Antibiotics*, **23**, 45–7.

POINDEXTER, J. S. (1964). Biological properties and classification of the *Caulobacter* group. *Bacteriological Reviews*, **28**, 231–95.

POLLOCK, J. J., GHUYSEN, J. M., LINDER, R., SALTON, M. R. J., PERKINS, H. R., NIETO, M., LEYH-BOUILLE, M., FRERE, J. M. & JOHNSON, K. (1972). Transpeptidase activity of *Streptomyces* D-alanyl-D-carboxypeptidases. *Proceedings of the National Academy of Sciences, U.S.A.* **69**, 662–6.

RANCOURT, M. W. & LECHEVALIER, H. A. (1964). Electron microscopic study of the formation of spiny conidia in species of *Streptomyces*. *Canadian Journal of Microbiology*, **10**, 311–16.

ROSENBERG, E., KATARSKI, M. & GOTTLIEB, P. (1967). Deoxyribonucleic acid synthesis during exponential growth and microcyst formation in *Myxococcus xanthus*. *Journal of Bacteriology*, **93**, 1402–8.

SACKS, L. E. & ALDERTON, G. (1961). Behaviour of bacterial spores in aqueous polymer two-phase systems. *Journal of Bacteriology*, **82**, 331–41.

SCHAEFFER, P. (1969). Sporulation and the production of antibiotics, exoenzymes and endotoxins. *Bacteriological Reviews*, **33**, 48–71.

SEIFERT, W., RABUSSAY, D. & ZILLIG, W. (1971). On the chemical nature of alteration and modification of DNA-dependent RNA polymerase of *E. coli* after T4 infection. *Federation of European Biochemical Societies Letters*, **16**, 175–9.

SERMONTI, G. (1969). *Genetics of Antibiotic-producing Micro-organisms*. London: Wiley-Interscience.

SHAPIRO, L., AGABIAN-KESHISHIAN, N. & BENDIS, I. (1971). Bacterial differentiation. *Science, New York*, **123**, 884–92.

SONENSHEIN, A. L. & LOSICK, R. (1970). RNA polymerase mutants blocked in sporulation. *Nature, London*, **227**, 906–9.

SZESZÁK, F., SZABÓ, G. & SÜMEGI, J. (1970). RNA synthesis on native DNA complexes isolated from *Streptomyces griseus* and *Escherichia coli*. *Archiv für Mikrobiologie*, **73**, 368–78.

TRAVERS, A. A. (1970). Positive control of transcription by a bacteriophage sigma factor. *Nature, London*, **225**, 1009–12.

VIVIAN, A. (1971). Genetic control of fertility in *Streptomyces coelicolor* A3(2): plasmid involvement in the interconversion of UF and IF strains. *Journal of General Microbiology*, **69**, 353–64.

VIVIAN, A. & HOPWOOD, D. A. (1970). Genetic control of fertility in *Streptomyces coelicolor* A3(2): the IF fertility type. *Journal of General Microbiology*, **64**, 101–17.

WELSCH, M. (1969). Biology of Actinophages. In *Genetics and Breeding of Streptomyces*, ed. G. Sermonti & M. Alaćević, pp. 4–62. Zagreb: Yugoslav Academy of Sciences and Arts.

WILCOX, M. (1970). One-dimensional pattern found in blue-green algae. *Nature, London*, **228**, 686–7.

WILDERMUTH, H. (1970a). Development and organization of the aerial mycelium in *Streptomyces coelicolor*. *Journal of General Microbiology*, **60**, 43–50.

WILDERMUTH, H. (1970b). Surface structure of Streptomycete spores as revealed by negative staining and freeze-etching. *Journal of Bacteriology*, **101**, 318–22.

WILDERMUTH, H. (1971). The fine structure of mesosomes and plasma membrane in *Streptomyces coelicolor*. *Journal of General Microbiology*, **68**, 53–63.

WILDERMUTH, H. (1972a). The surface structure of spores and aerial hyphae in *Streptomyces viridochromogenes*. *Archiv für Mikrobiologie*, **81**, 309–20.

WILDERMUTH, H. (1972b). Morphological surface characteristics of *Streptomyces glaucescens* and *S. acrimycini*, two streptomycetes with 'hairy' spores. *Archiv für Mikrobiologie*, **81**, 321–32.

WILDERMUTH, H. & HOPWOOD, D. A. (1970). Septation during sporulation in *Streptomyces coelicolor*. *Journal of General Microbiology*, **60**, 51–9.

WILDERMUTH, H., WEHRLI, E. & HORNE, R. W. (1971). The surface structure of spores and aerial mycelium in *Streptomyces coelicolor*. *Journal of Ultrastructural Research*, **35**, 168–80.

WILLIAMS, S. T. (1970). Further investigations of Actinomycetes by scanning electron microscopy. *Journal of General Microbiology*, **62**, 67–73.

WILLIAMS, S. T. & SHARPLES, G. P. (1970). A comparative study of spore formation in two *Streptomyces* species. *Microbios*, **5**, 17–26.

WOLK, C. P. (1966). Evidence of a role of heterocysts in the sporulation of a blue-green alga. *American Journal of Botany*, **53**, 260–2.

WOLK, C. P. (1967). Physiological basis of the pattern of vegetative growth of a blue-green alga. *Proceedings of the National Academy of Sciences, U.S.A.* **157**, 1246–51.

EXPLANATION OF PLATES

PLATE 1

Phase-contrast photomicrograph of *Streptomyces coelicolor* grown against a coverslip. Note widely spaced vegetative cross walls (CW), rod-shaped immature spores (IS) and helical chain of mature spores (MS).

PLATE 2

Electron micrographs of thin sections of *Streptomyces coelicolor* aerial mycelium.

Fig. 1. Early stage of sporulation septation. Note synchronous development of septa.

Fig. 2. Late stages of sporulation septation. Note double edge of ingrowing annulus (AA); early stage of rounding off of spores (arrows); and mesosomes (M).

PLATE 3

Electron micrographs of *Streptomyces coelicolor*.

Fig. 1. Thin section of late stage in spore formation; at this stage spores are connected only at a small interface and by the fibrous sheath, the ends having rounded off. Wall thickening has also begun.

Fig. 2. Thin section of mature spores. Note thick wall (W) and fibrous sheath (FS).

Fig. 3. Part of the surface of a frozen-etched spore, showing the double rodlet pattern of the fibrous sheath.

PLATE 4

Phase-contrast photomicrographs of aerial mycelium of *whi* mutants of
Streptomyces coelicolor.

Fig. 1. Class I mutant. No curling or fragmentation.

Fig. 2. Class II mutant. Tight helical curling but no fragmentation.

Fig. 3. Class III mutant. Wavy hyphae with little fragmentation.

Fig. 4. Class IV mutant. Formation of curled fragments.

Fig. 5. Class V mutant. Spores rod-shaped.

Fig. 6. Class VI mutant. Ellipsoidal spores. (In this example, *whi*D16, the spores are heterogeneous and often contain bodies showing Brownian motion; this mutant makes thin-walled spores: A. McVittie, personal communication.)

PLATE 1

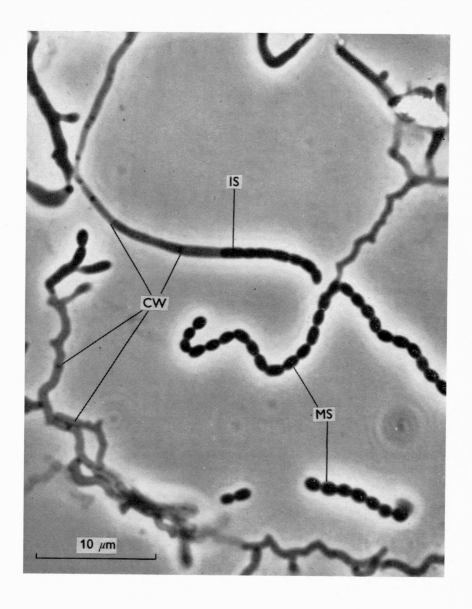

PLATE 2

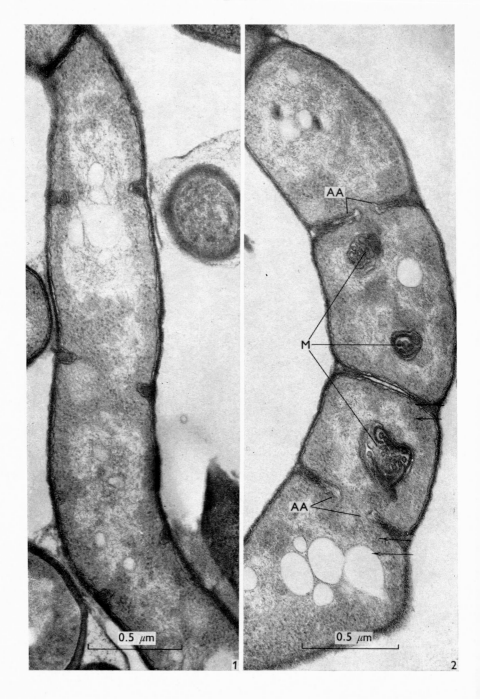

PLATE 3

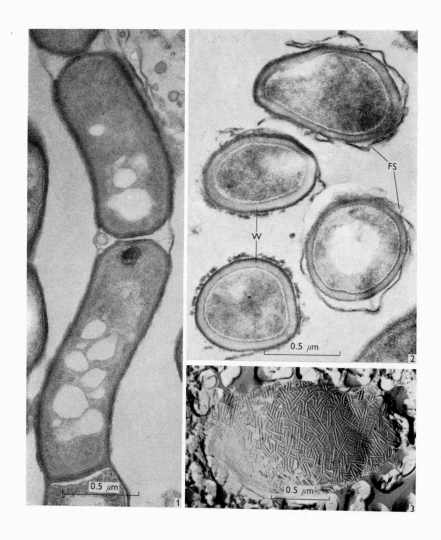

PLATE 4

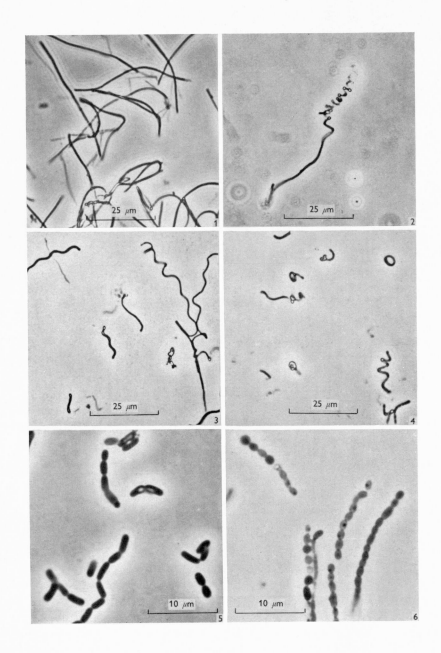

ASPECTS OF DEVELOPMENT IN
BLUE-GREEN ALGAE

N. G. CARR AND S. BRADLEY

Department of Biochemistry, University of Liverpool,
Liverpool L69 3BX

INTRODUCTION

Historical considerations have led to the division of prokaryotes into bacteria and blue-green algae, a demarcation that is increasingly difficult to uphold. The possession of an oxygen-evolving photosynthesis employing water as the reductant does not separate blue-green algae from photosynthetic bacteria in any way more fundamentally than, say, the Athiorhodaceae are separated from Pseudomonads. All other features of blue-green algal biology have counterparts in bacterial species and there exist several groups of micro-organisms, such as Flexibacteria, that may be considered either as bacteria or colourless blue-green algae. The recent recognition that certain species of gliding filamentous prokaryotes, morphologically similar to Flexibacteria, possess bacteriochlorophyll, and presumably carry out bacterial-type photosynthesis, adds particular confusion to the distinction between bacteria and blue-green algae (Pierson & Castenholz, 1971). The most striking metabolic feature of blue-green algae is their considerably restricted range of nutrition and in this they do contrast sharply with the remainder of the prokaryotes. All blue-green algae are phototrophs and, with the exception of certain vitamin requirements by some marine species, can use carbon dioxide as sole source of cell material. This phototrophic nutrition is in some species facultative, and growth in the dark at the expense of sugar has been well described. In some *Nostoc* sp. this dark heterotrophic growth is associated with an alteration in the development cycle of the organism, which will be discussed later in greater detail.

The filamentous species of blue-green algae possess a variety of differentiated structures and a complexity of life cycle quite without parallel in bacteria. The existence, in the oldest strata containing remains of living organisms, of fossil remains of blue-green algae and their apparent possession of at least one form of differentiated cell (see Fay, 1972) is in accord with the conclusion of Fritsch (1945) that blue-green algae are an extremely ancient group and that the diversity of filamentous types represents a plurality of evolutionary development

6

from a unicellular, coccoid form. Some species of the order Chamae-
siphonales range from truly unicellular forms (*C. morphus*) through cells
which are polarised with the formation of an exospore at one end and a
pointed attachment at the other (*C. fuscus*), the attachment being held in
a sheath-like pseudovagina, to long series of cells with a sheath, with
formation of endospores and clear development of a basal attachment
site such as *Stichosiphon regularis* (Fritsch, 1945).

The richness of blue-green algal morphological diversity can, at the
present time, be appreciated in the literature only from line-drawings in
taxonomic floras (Plate 1); there is little published light-photomicro-
graphic record and even less fine-structural analysis by electron
microscopy. The classic descriptions of Geitler (1932) and Fritsch (1945)
are within the framework of an examination of the Higher Algae and as
such do not emphasise the degree of morphological variation described.
Fritsch (1945) accepts the 'relatively low stage of differentiation' of
blue-green algae as part of the feature of a persistent archaic group of
organisms. In the more modern context of blue-green algae having a
prokaryotic structure and the descriptions of nuclear material corresponding closely to that of bacteria (Edelman *et al.* 1967; see Carr &
Craig, 1970) the extent and variation of blue-green algal development
becomes more striking. In certain species there are several different cell
types which are evidently developed from a common genome which, as
far as have been examined in the blue-green algae, do not appear to
have associated histones (Leak, 1967; Makino & Tsuzuki, 1971).

Perhaps the most interesting blue-green algae in this context are the
Gloeotrichia sp. which are illustrated in Plate 1, taken from Bourrelly
(1970). Four different cell types may be discerned in field material,
although the reliable axenic culture of these species in the laboratory has
not been achieved. A basal heterocyst is always present upon which
develops a filament enclosed within a gelatinous sheath. A characteristic feature is that the vegetative cells of the filament, at the end
opposite to the heterocyst, elongate and become narrower to form a
thin hair. At certain stages of development the vegetative cells immediately adjacent to the heterocyst enlarge, fill with granular material and
turn into one or more akinetes, or spore cells. In certain cases the basal
heterocyst may be double. A colony of *Gloeotrichia* sp. or of other
members of the Rivulariaceae consists of member filaments with their
heterocysts at the centre and the whip-like hair on the extremity. Not all
members, such as *Rivularia biasolettiana*, possess akinetes (Plate 2,
fig. 1) and in some species the akinetes are not adjacent to heterocysts
but at a point distant from them in the filament. Much of the experi-

mental work on heterocyst and akinete formation has been done with *Anabaena cylindrica* or *Anabaena flos-aquae* and considerable caution will have to be exercised in extrapolating to other species. A connection between heterocyst and akinete formation clearly exists, in that these specialised cells are not found in a random relationship to each other in any filament, but they may be either maximally separated or in immediate juxtaposition (Plate 1).

THE DEVELOPMENTAL CYCLE OF NOSTOCEAE

The two major differentiated cell-types found in blue-green algae, the akinete and the heterocyst, will be discussed separately below. Early observation had described the occurrence of akinetes and heterocysts in filamentous blue-green algae and the relationship of the latter to a pattern of cell development of Nostoceae in which the colony form, and the arrangement of cells within the filament, showed considerable variation. It was recognised that spores (akinetes) could germinate to long, fila-mentous chains of cells, but that another means of propagation existed, resulting from the transient production of hormogonia. These are short, motile, undifferentiated rows of cells quite different in appearance from the longer, differentiated filaments; a field specimen is shown in Plate 2, fig. 2. Hormogonia, after ceasing to be motile, would develop into pockets of larger cells not arranged in an obviously filamentous manner. The relation-ship of these different forms of growth was complicated by uncertainties as to the number of species involved. The subsequent elucidation of this developmental sequence has been described in the species *Nostoc muscorum* by Lazaroff & Vishniac (1961, 1962) and latterly in *Nostoc commune* by Robinson & Miller (1970). Detailed accounts of current views with some fascinating descriptions of the pioneering nineteenth-century observations have been presented by Lazaroff (1970, 1972).

Some species of blue-green algae can produce, by breakage of the terminal portion of longer filaments, the short, motile hormogonia which have a role in the development sequence and, probably, in the distribution of the organism. In the developmental cycle of *Nostoc muscorum*, when the hormogonia have ceased movement, the terminal cells of the short filaments develop into heterocysts while the inter-calary cells elongate and commence division in a plane parallel to the axis of the filament. This proceeds until each intercalary cell has given rise to a cluster of new cells surrounded by a sheath. This stage of development has been termed aseriate and when *Nostoc muscorum* grows in the dark at the expense of sucrose this type of morphology may be permanently maintained. The next stage in the developmental sequence of *N. muscorum* requires light and it involves the organisation of the clumps of cells (within the containing sheath) to adopt filament form and the subsequent formation of intercalary heterocysts at intervals along the filament. Lazaroff (1972) shows in detail that this process is not just a consequence of photosynthesis and that the spectral quality of the light required and the small quantity necessary indicate a specific photo-induction. It was concluded that red light initiated the differentiation of filamentous form and that allophycocyanin was the probable photoreceptor. When a broad spectrum of light was employed the photo-induction by red light was reversed, apparently by the green component.

After the development of a filament, including heterocysts, breakage of the enclosing sheath results in the release of filamentous forms. These may then, in course of time, break at the point of heterocyst–vegetative cell contact to yield free heterocysts and motile, gliding hormogonia. The hormogonia may swarm together to produce large whorls of moving spiral aggregates which, in turn, release single hormogonia which can start the developmental cycle once more. Alternatively, hormogonia may proceed directly to the development of terminal heterocysts and the aseriate stage without prior aggregation.

The interrelationships in *Nostoc muscorum*, of morphological form and comparatively complex development cycle, cannot at this stage be expressed biochemically. The influence of environmental factors, such as the role of an appropriate sugar in maintaining the aseriate stage through dark growth, and the red-light trigger in the initiation of filamentous morphology, suggests that a biochemical approach may be used to explore this developmental sequence more fully. The extent to which other species of blue-green algae possess a defined development cycle is not clear, but certainly many filamentous species possess akinetes, heterocysts and hormogonia, both collectively or separately.

THE AKINETE

Of the several different types of propagation bodies described in blue-green algae (Fritsch, 1945), the akinete is the most readily identified and has attracted most attention. These cells are spherical or cylindrical and may have a granulated appearance; the outer coat is thicker than that of vegetative cells and is sometimes coloured yellow or brown (Plate 2, fig. 3). Early literature contains descriptions of the resistance of akinetes to desiccation and temperature extremes and of the ability to germinate after long periods (see Fritsch, 1945) and in this they differ from endospores which appear to be only propagating agents and do not have resistant properties.

The ultrastructural characteristics of akinetes from *Anabaena cylindrica* have been described by Wildon & Mercer (1963), the most apparent features being large granules, together with a new thicker outer wall. These workers regarded the akinete of this species as an enlarged vegetative cell with modifications, whereas they suggested that the heterocyst should be thought of as a distinct body formed by differentiation of a vegetative cell. In *Cylindrospermum* the heterocyst is found at the end of a filament and the akinete develops next to it. The fine structural changes of akinete formation in this organism are similar to those of *A. cylindrica*, and have been recorded in detail by Miller & Lang (1968). These workers describe the deposition of dense fibrillar layers on the cell surface with the concomitant accumulation of quantities of granular material identified as cyanophycin which has been shown to consist of a co-polymer of aspartate and arginine of molecular weight around 30000 (Simon, 1971). The photosynthetic thylakoids remain in the akinete, but in a contorted form (Plate 2, fig. 4). Miller & Lang (1968) paid particular attention to the germination of akinetes, which was achieved by transference to fresh medium. The new filament or germling seen in Plate 2, fig. 5 has shrunk from the spore envelope prior to envelope breakage. Carbohydrate- and lipid-containing particles are particularly numerous but the cyanophycin granules of the akinete appear to have been utilised by the developing cells.

It is not clear from the literature whether the akinete contains one, or more, resting bodies. From the electron micrographs of *Cylindrospermum* sp. it would appear that only one new filament emerges. Light-microscopy frequently reveals akinetes that have split and the release of contents from akinetes of *Anabaena cylindrica* would indicate the presence of many spore-like bodies. The quantitative determination of DNA, *in vivo*, by the fluorescent procedure shows that akinetes of *Anabaena baltica* contain more DNA than do vegetative cells and Ueda (1971) concludes that they may be polyploid.

Isolated akinetes from *A. cylindrica* possess a cell wall with a carbohydrate composition similar to that of heterocysts (Table 1), have an apparently identical carotenoid complement to that of vegetative cells and, like heterocysts, have a reduced amount of phycocyanin and chlorophyll (Fay, 1969a). A preparation of akinetes, free of vegetative cells and heterocysts, was achieved by careful extrusion of filaments in a French Pressure Cell and differential centrifugation in sucrose. Isolated akinetes

Table 1. *Composition of walls of vegetative cells, heterocysts and akinetes of* Anabaena cylindrica (Dunn & Wolk, 1970)

Cell type		Vegetative	Heterocysts	Akinete
Total analysis (%)	carbohydrate	18	62	41
	lipid	3	15	11
	amino compounds	65	4	24
Polysaccharide constituents (%)	glucose	35	73	76
	mannose	50	21	17
	galactose	5	3	3
	xylose	8	4	4
	fucose	2	0	0

were able to fix carbon dioxide in the light at a much slower rate than the vegetative cells, although respiratory release of carbon dioxide was slightly greater by akinetes. Akinetes did not fix nitrogen as measured by ^{15}N assimilation nor effect the reduction of acetylene (Fay, 1969b).

A detailed analysis of the induction of akinete formation by *Anabaena cylindrica* has been carried out by Wolk (1965a, 1966, 1970). In this organism akinetes develop adjacent to an already formed heterocyst (Plate 3, fig. 1). Wolk (1966) describes experiments which support the views of early workers that this association is not random. Cultures were agitated and the degree of akinete formation in broken filaments compared with intact filaments. Akinete formation by cells adjacent to heterocysts was reduced by agitation, whilst the ability of other vegetative cells to develop akinetes was not so affected. The adjustment of growth medium to provide low phosphate and relatively high calcium concentration enabled Wolk (1965a) to considerably increase the degree of akinete production. The inclusion of calcium glucuronate (23 mM) increased the number of akinetes from short chains of 1–4 to longer chains containing seven or more akinetes in a row.

FINE STRUCTURAL DIFFERENCES BETWEEN VEGETATIVE CELLS AND HETEROCYSTS

The most obvious feature of heterocysts is the elaboration of the outer layers of the cell into a thick, complex structure that is evident in the light microscope but requires electron microscopy for detailed analysis. Wildon & Mercer (1963) noted that the two-layered outer envelope of *Nostoc muscorum* heterocysts was absent from vegetative cells and that the vegetative cell wall became thicker in heterocysts, especially around the pore. A uniform, non-fibrous outer envelope has been described in heterocysts of *Chlorogloea fritschii* (Whitton & Peat, 1967) although the pore structures were less apparent than in other species (Lang, 1965). The outer envelope has been examined in most detail in heterocysts from *Anabaena cylindrica* where first two (Lang, 1965), then three (Lang & Fay, 1971), components have been identified. An outer fibrous layer of irregular thickness surrounds a broad homogeneous layer; a laminated layer forms the innermost section of the outer envelope and this is most evident towards the pore where it becomes associated with the outer layer of the cell wall. The two outer layers of the envelope are seen clearly in Plate 3, fig. 2, but the inner laminated region is not apparent in this glutaraldehyde–osmium treated section and is most clearly seen after glutaraldehyde–permanganate fixation (Lang & Fay, 1971). An earlier picture of Wildon & Mercer (1963), used to show the nature of the pore septum, employed glutaraldehyde–permanganate and clearly shows the inner laminated layer of the heterocyst outer envelopment described by Lang & Fay (1971) (Plate 3, fig. 3). A freeze-etched preparation of the pore region of an isolated, unfixed heterocyst, also shows the laminated ultrastructure of the inner layer where the surface of the heterocyst has been destroyed (Winkenbach, Wolk & Jost, 1972). The four-layered cell wall of heterocysts, comparable to that of vegetative blue-green algae, is seen in Plate 3, fig. 2; the inner layer being designated L_I and the outer L_{IV} according to the terminology employed for vegetative cell walls (see Drews, 1972). Around the connection of the heterocyst to the adjacent vegetative cell the L_{II} layer of the heterocyst wall thickens in both *Anabaena cylindrica* (Lang & Fay, 1971) and *Nostoc muscorum* (Wildon & Mercer, 1963).

The development of the heterocyst envelope separates the newly-forming heterocyst from the adjoining vegetative cell and produces the characteristic pore through which the only contact between the two cells is maintained. The plasmalemmata of the cells are in close alignment and are transversed by inter-cellular connections across the pore septum; these have been termed 'microplasmodesmata' by Lang & Fay (1971) to indicate the uncertainty as to whether they are truly tubular and open (Plate 3, fig. 3). This picture of the pore region of an *Anabaena* sp. (Wildon & Mercer, 1963) also shows clearly the osmiophilic plug material which has been observed in heterocysts from several species.

The cytoplasmic differences between the mature heterocyst and a vegetative cell of *Anabaena cylindrica* are illustrated in Plate 3, fig. 4 which shows a cross-section of both types of cell connected via the heterocyst pore. In the later stages of heterocyst formation from a vegetative cell there is a loss of granular inclusion bodies which are thought to be composed of polyphosphate and of glycogen. The loss of storage material has been followed cytochemically during heterocyst formation (see Kale & Talpasayi, 1969). Ribosomes are present in developing and mature heterocysts. The thylakoids of the vegetative cell undergo marked changes during heterocyst formation in all species that have been examined. They become contorted and develop into a reticulated system that, in *Anabaena cylindrica* at least, extends throughout the heterocyst cell (Plate 3, fig. 2). Concentric whorls of newly-formed membranes that are connected with the honey-comb region of distorted membranes adjacent to the pore region have been described by Lang & Fay (1971) who suggest that considerable *de novo* membrane synthesis accompanies heterocyst formation. The honey-comb region is seen clearly at the base of the osmiophilic plug in *Anabaena* sp. (Plate 3, figs. 2, 4). Lang & Fay (1971) observed that the close stacking of membranes in the region is facilitated by the lack of phycobilisomes and glycogen granules and suggest that some membrane structures in heterocysts may differ quantitatively from those of vegetative cells.

After Feulgen-staining, nucleoplasmic areas appear similar in both heterocysts and vegetative cells and autoradiographs indicate that tritiated thymidine is incorporated and located in a similar area in both cell types (Leak, 1965). In a quantitative study of DNA *in situ* by fluorescence estimation, after staining with fluorochrome coriphosphin, Ueda (1971) noted a fluorescence from heterocysts that differed qualitatively from that from vegetative cells of *Anabaena baltica*.

BIOCHEMICAL CHARACTERISTICS OF HETEROCYSTS

In any system comprising more than one cell type the separation of individual cell types is a crucial stage in the understanding of their biochemistry. The breakage of heterocysts from the containing filaments of *Anabaena cylindrica* may be readily achieved by extrusion from a pressure vessel such as a French Cell or by application of controlled ultrasonic treatment. Subsequently, heterocysts may be isolated and washed free of vegetative cell fragments by low speed centrifugation (Fay & Walsby, 1966; Wolk, 1968). The observation that filaments of blue-green algae were lysed by lysozyme but that heterocysts and akinetes were not, led to the use of this enzyme in heterocyst isolation. The recognition that isolation by physical means, although yielding apparently intact heterocyst populations, may cause the partial loss of contents has prompted many workers to adopt an isolation procedure based on lysozyme digestion of the vegetative cells. Fay & Lang (1971) have examined in detail ultrastructural damage to heterocysts of *Anabaena cylindrica* isolated by various procedures and conclude that only isolation by lysozyme treatment yielded intact heterocysts.

The respiratory activity of heterocysts mechanically isolated from *Anabaena cylindrica* was measured manometrically by Fay & Walsby (1966) and found to be some 30 % greater than that of the intact filament. Lysozyme-isolated heterocysts from the same organism had less respiratory activity than did the intact filament when measured in an oxygen electrode system, as was the NADPH oxidase activity of broken heterocysts (S. Bradley & N. G. Carr, unpublished results). The oxidation of reduced pyridine nucleotides has been linked to ATP synthesis in disintegrated heterocyst preparations (Scott & Fay, 1972). Photosynthetic assimilation of carbon dioxide into isolated heterocysts of *A. cylindrica* could not be demonstrated by Fay & Walsby (1966) and the conclusion that heterocysts did not fix carbon dioxide was supported by radioautographic evidence using intact filaments (Wolk, 1968; Stewart, Haystead & Pearson, 1969). By measuring the proportion of radioisotope from [^{14}C]sodium bicarbonate in vegetative cells and heterocysts, using radioautography, Wolk (1968) was able to demonstrate the flow of carbon material from the former to the latter. This movement of fixed carbon dioxide was not reabsorption of material excreted into the growth medium by the vegetative cells and presumably took place via the septum separating heterocyst from vegetative cell.

For some years cytochemical analysis has been applied to the development of heterocysts and to attempts to gain an understanding of their biochemistry (Kale & Talpasayi, 1969). The absence of polyphosphate granules was noted by Talpasayi (1963) who later showed that the loss of this storage material occurred early in heterocyst development (Talpasayi & Bahal, 1967). The intracellular reduction of 2,3,5-triphenyl tetrazolium chloride (TTC) by *Anabaena* sp. has been known for some time (Drawert & Tischer, 1956). Kale, Bahal & Talpasayi (1970) have noted that formazan crystals, from the reduction of TCC appeared more rapidly in mature heterocysts of *Anabaena ambigua* than in vegetative cells. Some attention has been directed toward documenting the 'reducing activity' of heterocysts and in equating this activity with the process of nitrogen fixation, perhaps even that of nitrogenase itself. Stewart *et al.* (1969) observed that photographic film was rapidly darkened on contact with heterocysts and showed that these cells from *A. cylindrica* and *A. flos-aquae* contained greater TCC reducing activity than vegetative cells. They noted that the deposition of formazan

crystals within the heterocyst inhibited acetylene reduction. Thus it is
clear that heterocysts are capable of acting in a reducing capacity and
that this may be demonstrated by the application of TCC or silver
nitrate. The exact relationship of this process to nitrogen fixation is less
apparent since the time course of TCC reduction does not always
coincide with the presence of mature heterocysts and, more unfortunately, the reduction of TCC appears to proceed with filaments that have
been held at 100° for 5 min, which would indicate the absence of an
enzyme-linked process (S. Bradley & N. G. Carr, unpublished results).
The presence of ascorbic acid in heterocysts was deduced from cytochemical reduction of silver nitrate (Talpasayi, 1967) but no concentration of this material relative to vegetative cells and non-heterocystous
blue-green algae was indicated after extraction and estimation by the
diphenolindophenol procedure of Roe (1954). Early cytochemical
evidence had suggested that the outer structures of the heterocyst, but
not the vegetative cell, contained cellulose. The scepticism regarding the
presence of this polymer in heterocysts was supported by the chemical
analysis of vegetative heterocyst and akinete walls, in which no evidence
of cellulose was obtained (Dunn & Wolk, 1970). The main difference
between vegetative and heterocyst walls of A. cylindrica lies in the much
greater proportion of carbohydrate, and consequent smaller proportion
of amino compounds in the heterocyst wall (Table 1). The high proportion of amino compounds in the akinete wall distinguishes it from that of
the heterocyst. However, the close similarity in sugar composition of the
walls of the two differentiated cell types suggest the possession of a
common polysaccharide material not present in vegetative cell walls.
From the labelling patterns obtained after a pulse of [^{14}C]bicarbonate
applied early in the heterocyst development sequence, Dunn, Simon
& Wolk (1971) concluded that the muramic acid component of the
heterocyst was derived from the vegetative cell wall whilst a
hexosamine fraction arose de novo during differentiation. Presumably
this corresponds to the outer envelope described in electron micrographs.

One of the most discussed features of heterocysts, at least of Anabaena
sp., has been their lack of the biliprotein accessory photosynthetic
pigments of the phycocyanin type. These can account for significant
(10–20 %) proportions of the cell protein; the lack of phycocyanin in
heterocysts accounts for their pale, green appearance – quite distinct
from a normal vegetative cell. Phycocyanin is located, at least in part, in
discrete bodies called phycobilisomes (Gantt & Conti, 1969) that are
distributed in an ordered fashion on the photosynthetic thylakoids, the

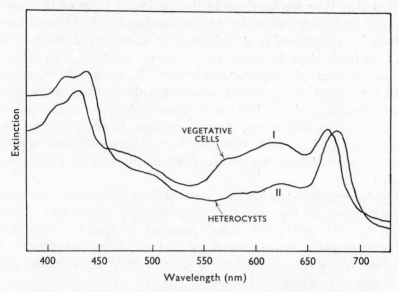

Fig. 1. Spectra of vegetative cells and heterocysts of *Anabaena cylindrica*.

absence of these structures being another aspect of the thylakoid re-arrangement in heterocyst development discussed above. The other photopigments appear to be present in heterocysts, although the chlorophyll content is rather less than in vegetative cells; the major carotenoid pigments are present although in a slightly altered proportion (Fay, 1969*a*; Winkenbach *et al.* 1972). The spectrum of a total extract from vegetative cells and heterocysts of *A. cylindrica* illustrates this point (Fig. 1).

Other lipid-soluble components show more marked differences. Walsby & Nichols (1969) showed that heterocysts of *Anabaena cylindrica* did not contain the four polyunsaturated acid-containing acyl lipids present in the vegetative cells (mono- and di-galactosyldiglyceride, phosphatidyl glycerol and a sulpholipid). Heterocysts did, however, contain a glycoside and an acyl lipid which did not contain linolenic acid, both of which were absent from vegetative cells. Subsequent work (Bryce, Welti, Walsby & Nichols, 1972) has shown that the heterocyst glycoside consists of a single hexose unit bound to a C_{26} trihydric alcohol, the specific hydroxyl group involved in the glycosidic linkage not being known. The major alcohol moiety was 1,3,25-tri-hydroxyhexacosane and this was glycosidically linked to either glucose or galactose. The function of these glycolipids in the metabolism of heterocysts is not known, although Winkenbach *et al.* (1972) have suggested that they are localised in the laminated layer of the outer envelope. Lipid species which have been associated with electron transport have been examined in *A. cylindrica* by S. Bradley, J. Pennock & N. G. Carr (unpublished results). The heterocyst contains, on a chlorophyll basis, as much plastoquinone and vitamin K_1 as does the vegetative cell but tocopherol or tocopherol quinone are absent from heterocysts. Neither vegetative cells nor heterocysts possess ubiquinone. The lack of tocopherol, or its derivatives, may be relevant to the absence from heterocysts of linolenic acid. Tocopherols have been suggested as antioxidants concerned in the protection of polyunsaturated fatty acids (see Tappel, 1962).

Poly-β-hydroxybutyrate has been isolated from filaments of *Anabaena cylindrica* grown in the presence of acetate and preliminary results indicate that this polymeric reserve material is located mainly, and possibly exclusively, in heterocysts (N. G. Carr, unpublished results). A possible role of this material in the process of nitrogen fixation and assimilation will be discussed below.

The best described metabolic feature of heterocysts, other than their being the postulated site of nitrogen fixation, is the absence of a functional photosystem II. Because phycocyanin has been shown to act as a photopigment for photosystem II, its reported absence from heterocysts (Fay, 1969*a*) led to the assumption that photosystem II was not present, although the evolution of O_2 by blue-green algae which contained much reduced phycocyanin levels had been reported (Susor, Duane & Krogmann, 1964). The usual criteria of phycocyanin absence, marked reduction of extinction in the region of 620 nm, indicates only quantitative loss and it is difficult to prove the complete absence of phycocyanin from the mature heterocyst (Fig. 1). Thomas (1970) has employed a microspectrophotometric technique which measured the pigments, *in vivo*, of individual vegetative cells and heterocysts of an *Anabaena* sp. The principal chlorophyll *a* peak in vegetative cells is at 666 nm and that of phycocyanin at 620 nm; in heterocysts the absorption due to chlorophyll is reduced and is maximal around 682 nm, whilst the phycocyanin is very reduced. The movement of the heterocyst chlorophyll peak to 682 nm indicates the absence of the short wave form of chlorophyll *a*, chlorophyll 670. The most detailed spectrophotometric evidence that heterocysts do not contain photosystem II has been provided by Donze, Haveman & Schiereck (1972) with *A. cylindrica*. Isolated heterocysts when illuminated produced an oxidation of photosystem I reaction centre P700, with bands at 705 nm and 437 nm and of a cytochrome at 420 nm and 560 nm. These workers found a high

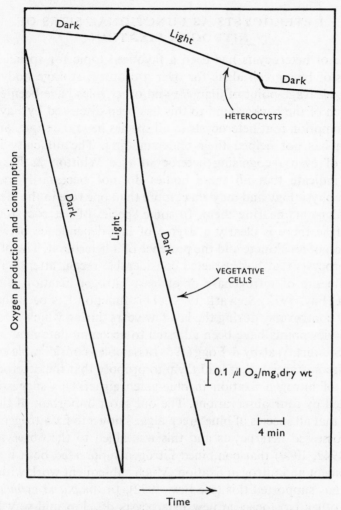

Fig. 2. Oxygen uptake by isolated heterocysts of *Anabaena cylindrica*. Note the absence of O_2 production in light indicative of absence of photosystem II.

concentration of P700 to chlorophyll molecules and concluded that photosystem II was absent. This conclusion was supported by the absence of delayed light from chlorophyll and reduced fluorescent emission from heterocysts; furthermore heterocysts did not exhibit Hill-reaction activity. The lack of O_2 evolution when heterocysts are illuminated (Fig. 2) is a direct measure of absence of photosystem II activity (Bradley & Carr, 1971).

HETEROCYSTS AS FUNCTIONAL SITES OF
NITROGEN FIXATION

The role of heterocysts has been a favoured topic for speculation by students of blue-green algae for over a century; storage body, spore-inducer, cleavage-point of filament and other roles have been explored and much of the background to this has been discussed by Fay (1972). The assumption that heterocysts in all species have the same, and only, function has not helped their understanding. The difficulty in some species of even recognising heterocysts (see Whitton & Peat, 1967) should indicate that all these bodies do not necessarily possess a uniform physiology and may enact more than one role in the life history of the filaments bearing them. In some species of heterocystous blue-green algae there is clearly a degree of interdependence between the formation of an akinete and the presence of a heterocyst. The other role for heterocysts that has attracted considerable recent attention is their being the site of nitrogenase or at least nitrogen fixation within the filament (Fay, 1972; Stewart, 1972). This function has been the centre of lively controversy during the last few years during which most of the original viewpoints have been adjusted to accommodate new evidence.

Fay, Stewart, Walsby & Fogg (1968) assembled considerable evidence, albeit circumstantial, that lead them to propose that the heterocyst was the site of nitrogen fixation in blue-green algae. This suggestion was supported by four observations. The one most important of these was the fact that all species of blue-green algae known to fix nitrogen (at that time) possessed heterocysts and this was allied to the observation by Fogg (1942, 1949) that combined nitrogen suppressed both heterocyst development and nitrogen fixation. Much subsequent work with several species has supported this (see Fay, 1972). In *Anabaena cylindrica* and certain other Nostocaceae new heterocysts develop mid-way between existing heterocysts in the filament, indicative of a gradient down the filament which Fogg (1949) suggested could be of ammonia or a related compound. In those species (*Gloeotrichia* and *Calothrix*) in which the heterocyst is basal, growth is restricted to those cells adjacent to the heterocyst. Finally, Fay *et al.* (1968) pointed to the reductive nature of nitrogen fixation and its inhibition by oxygen and that the heterocyst is cytochemically more reducing than vegetative cells. They also suggested that the reduced phycocyanin content of heterocysts may cause reduction in photosystem II activity, hence a decrease in oxygen production.

The firm identification of the heterocyst as the site of nitrogen fixation by blue-green algae was suggested by Stewart *et al.* (1969) in the light of further experimental evidence. These workers demonstrated a low, but measurable, rate of acetylene reduction by isolated heterocysts when supplied with ATP and NaS_2O_4. They further showed that under certain conditions heterocysts, but not vegetative cells, could form formazan crystals from TCC; when this was the case acetylase reduction was nearly abolished, but the filaments still fixed carbon dioxide. When intact filaments of *Anabaena cylindrica* were exposed to ^{15}N, and the proportion of fixed ^{15}N determined in both heterocysts and vegetative cells, only a small (6 %) amount was observed in the heterocyst after 30 min exposure (Stewart *et al.* 1969). It is possible that shorter exposure periods would have aided the determination of the location of N_2 fixation.

The first serious blow to this attractive theory came with the report by Wyatt & Silvey (1969) that a unicellular blue-green alga, *Gloeocapsa alpicola*, clearly unable to differentiate heterocysts, could grow in the absence of combined nitrogen and possessed the enzyme nitrogenase (measured by acetylene reduction) and that the synthesis of this was repressed by nitrate.

This was followed by the work of Smith & Evans (1970) in which, using an assay system with optimal amounts of ATP and dithionate, they measured nitrogenase and found this enzyme to be mainly (20 %) recovered in the soluble fraction. Since only vegetative cells had been broken, they concluded that the enzyme resided mainly in vegetative cells and not within the heterocyst (see also Booth, 1970). The marked oxygen sensitivity of nitrogenase was noted and no increase in specific activity of extracts found when heterocysts were broken (Smith & Evans, 1971). The latter workers had grown *Anabaena cylindrica* under conditions which minimised O_2 tension, and had starved the organism of N_2 for 48 h prior to harvesting in order to maximise nitrogenase activity.

Meanwhile Stewart & Lex (1970) had grown the non-heterocystous *Plectonema boryanum* free of combined nitrogen under conditions of low O_2 tension and measured a nitrogenase enzyme, most active under micro-aerophilic conditions, and inactive in the presence of oxygen. Maximum activity was obtained after incubation under mixtures of argon and carbon dioxide. It would appear that in this organism nitrogenase is formed and is capable of effecting nitrogen fixation in vegetative cells provided that oxygen tension is reduced. Clearly a blue-green alga carrying out photosynthesis must have some degree of aerobiosis due to the oxygen evolved during photosynthesis. It is noteworthy that Smith & Evans (1970) had used argon and carbon dioxide as a final gas phase in their cultures of *Anabaena cylindrica*, thus accounting for the synthesis of active nitrogenase in vegetative cells, this interpretation being supported by the work of Neilson, Rippka & Kunisawa (1971) with *Anabaena* sp. Wolk & Wojciuch (1971) after conducting vegetative cell breakage by sonication under hydrogen have achieved significant improvements in nitrogenase estimation in isolated heterocysts, which when incubated in light in the absence of added cofactors, reduced acetylene at rates up to 30 % those of the intact filament.

An elegant microscopic experiment has recently been described by Van Gorkom & Donze (1972) which strongly suggests that nitrogen fixation occurs in the heterocyst. When *Anabaena cylindrica* was held under hydrogen and carbon dioxide, phycocyanin was lost, not only from the heterocyst, but also from the vegetative cells (as would be expected from the results of Allen & Smith (1969) on phycocyanin as a nitrogen storage material). The resulting filaments showed none of the characteristic autofluorescence of phycocyanin when viewed microscopically. When nitrogen was readmitted to the culture as CO_2:air, gradients of phycocyanin fluorescence could be seen in the filaments; fluorescent intensity was greatest adjacent to a heterocyst and minimal at mid-point between heterocysts. When nitrogen was readmitted in CO_2:N_2 (i.e. anaerobically), the rate of phycocyanin synthesis was similar but gradients of phycocyanin concentration were rare and less pronounced.

There would appear to be considerable indirect, and some direct, evidence that heterocysts are the locations of nitrogen fixation, especially when the filament is under aerobic conditions. Certain blue-green algae can, and do, fix nitrogen without possessing heterocysts; *Plectonema boryanum* demanding semi-anaerobic growth conditions for this (Stewart & Lex, 1970), others being more tolerant of aerobic growth (Wyatt & Silvey, 1969; Rippka, Neilson, Kunisawa & Cohen-Bazire, 1971). It is not unreasonable to assume, in line with Fay *et al.* (1968) that a particular advantage of the development of the heterocyst is to provide an anaerobic environment for nitrogen fixation to occur within a photosynthesising, O_2-evolving filament.

In contrast to the considerable attention that has been devoted to measuring nitrogenase in heterocysts, very little information is available on their other biochemical activities, including those directly consequent upon nitrogen fixation. There is no direct evidence as to whether protein synthesis occurs in the mature heterocyst. It may be relevant that Van Gorkom & Donze (1972) did not observe the reappearance of phycocyanin within the heterocyst itself, but only in the adjacent vegetative cells and it is possible that a low molecular weight nitrogenous com-

PLATE 1

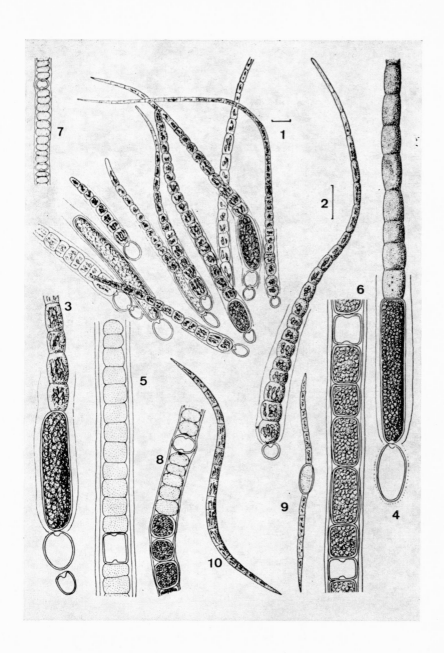

PLATE 2

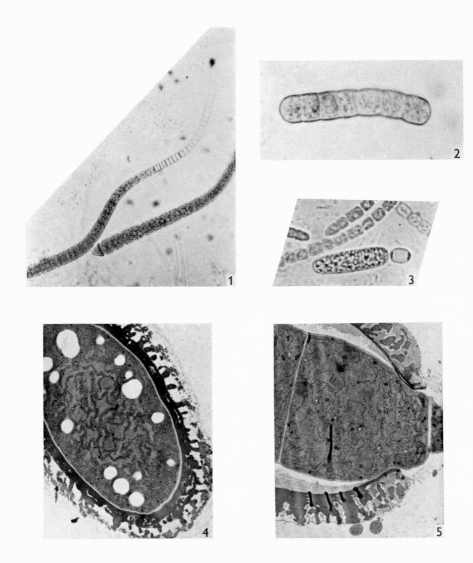

PLATE 3

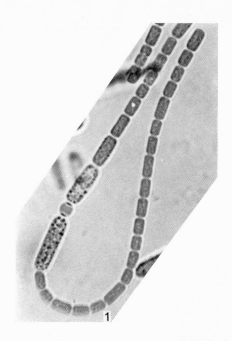

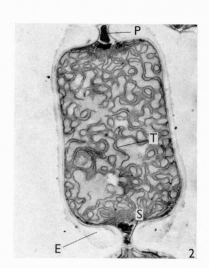

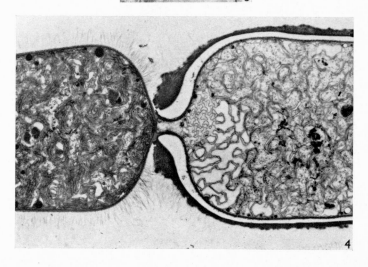

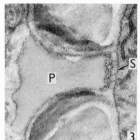

PLATE 4

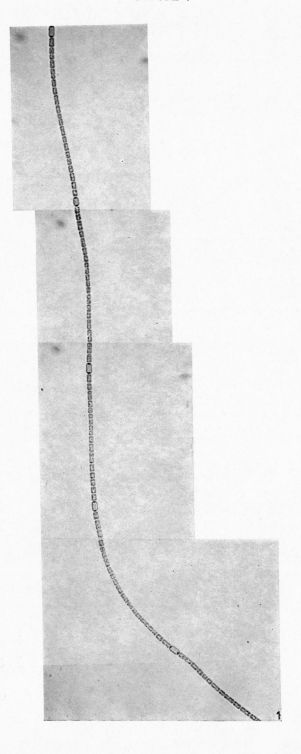

PLATE 4 (*cont.*)

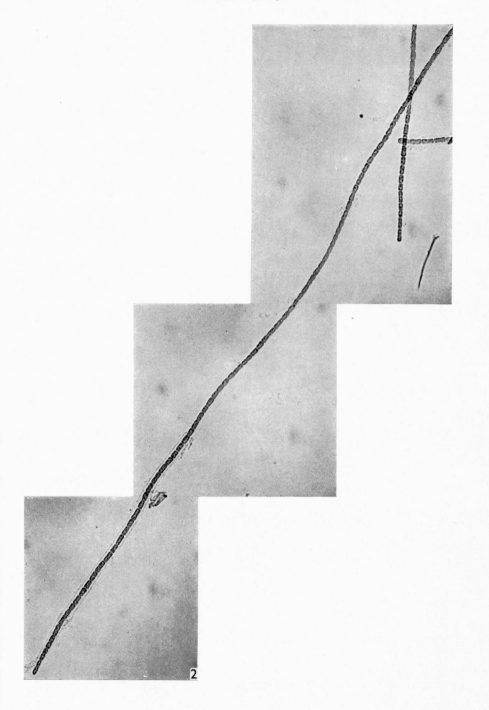

2

pound passes from heterocysts into the filament. The fact that hetero-cysts have, on occasions, been observed to germinate yielding a new filament of vegetative cells argues against their having lost permanently their protein synthesising capacity.

Recently, Scott & Fay (1972) have noted that glutamate, but not alanine, was formed in heterocyst preparations by an amination process. This implies that α-oxoglutarate is the major acceptor molecule for the ammonia produced by nitrogen fixation. Cox & Fay (1969) investigated manometrically the relationship between rates of decarboxylation and nitrogen fixation by *Anabaena cylindrica* in an attempt to identify the nature of the reductant used for nitrogen fixation, which was presum-ably passed from vegetative cell to heterocyst (Wolk, 1968). Their results, based on whole cell studies, suggested pyruvate as the source of reductant. The enzyme pyruvate:ferrodoxin oxidoreductase which cleaves pyruvate as below, has been described in detail in a non-fixing blue-green alga (Leach & Carr, 1971); it is allosterically activated by ATP.

$$CH_3COCOOH + CoA + ox.Fd \rightarrow CH_3CO.CoA + CO_2 + red.Fd.$$

This enzyme is present in *A. cylindrica* although its activity in hetero-cysts has not been unequivocally established. The activation by ATP would correlate the availability of reduced ferredoxin with the presence of ATP, which is also required for nitrogen fixation. ATP could, presumably, be formed within the heterocyst by cyclic phosphorylation. The photostimulation of nitrogen fixation has been investigated by Fay (1970) and the action spectrum of acetylene reduction indicated an involvement of photosystem I, which yields ATP.

The enzymes of the glyoxylate cycle have been detected in extracts of heterocysts of *Anabaena cylindrica* (S. Bradley, P. A. Jackson & N. G. Carr, unpublished results). This pathway would allow the assembly, from acetyl-CoA produced from pyruvate, of 4-carbon or 5-carbon acceptor molecules. Thus isocitrate could give rise to α-oxoglutarate and hence glutarate, or could be cleaved by isocitrate lyase to form succinate plus glyoxylate which would accept acetyl-CoA to replenish the 4-carbon pool. Enzymes of the glyoxylate cycle have been measured, at low activity, in other blue-green algae (Pearce & Carr, 1967) although when Hoare, Hoare & Moore (1967) followed the route of ^{14}C-labelled acetate they found no evidence of a functional glyoxylate cycle in *Anacystis nidulans*. Thus, there are indications that this pathway has a specific role in heterocysts and this is strengthened by the observation that malate dehydrogenase and malate synthetase are several-fold more

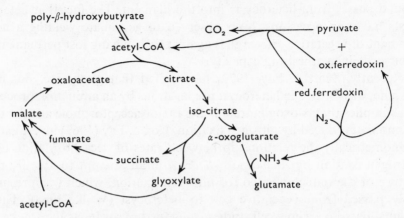

Fig. 3. Suggested intermediary metabolism in relation to nitrogen fixation
in heterocysts of *Anabaena cylindrica*.

active in heterocyst than vegetative cell-free preparations (S. Bradley, P. A. Jackson & N. G. Carr, unpublished results). The presence of poly-β-hydroxybutyrate in heterocysts, which would provide a reserve store of 2-carbon units, is consistent with the operation of a glyoxylate cycle and since poly-β-hydroxybutyrate is more reduced than acetyl-CoA this polymer would also act as a reductant store. The heterocyst could thus go some way to being able to balance the available reducing potential to the supply of amino group acceptor molecules in the form of α-oxoglutarate. The enzyme reactions thought to operate in heterocysts and to be concerned in the assimilation of the fixed nitrogen are summarised in Fig. 3.

HETEROCYST DEVELOPMENT IN *ANABAENA CYLINDRICA*

The effect of environmental variation on heterocyst production in blue-green algae has been known for some time; for example Canabaeus (1929) showed that sodium chloride apparently affected both heterocyst frequency and size. The reports by Fogg (1949, 1951) that the addition to the growth medium of substances providing an available source of combined nitrogen prevented heterocyst formation has formed the basis of much modern work. Nitrate and organic amino compounds produced transient inhibition of heterocyst formation whilst ammonium ions continued to exert their inhibition for some time. Fogg concluded that it was ammonia, or a direct metabolic product of ammonia, which prevented heterocyst formation. With the suggestion that heterocysts were the sites of nitrogen fixation (Fay *et al.* 1968; Stewart *et al.* 1969) came the idea that the product of nitrogen fixation, ammonia, diffused down the filament from the heterocyst. New heterocysts were formed only when the ammonia concentration was minimal, i.e. mid-way between existing heterocysts. Examination of heterocyst formation in several species of blue-green algae has confirmed, in principle, that formation results from the absence of ammonia (Michelson, Davis & Tischer, 1967; Ogawa & Carr, 1969; Fay, Kumar & Fogg, 1964; Talpasayi & Kale, 1967; Singh & Srivastava, 1968).

When a culture of *Anabaena cylindrica* is deprived of ammonia the regular occurrence in the filaments of a type of cell termed *proheterocyst* may be observed and in a nitrogen-fixing culture both heterocysts and proheterocysts are present. The proheterocyst is a clearly recognisable cell, greater in diameter, and often in length, than the vegetative cell, which is formed prior to the heterocyst development (Fritsch, 1951; Fogg, 1951; Lang, 1965; Kale & Talpasayi, 1969). Proheterocysts appear not to contain polyphosphate but do possess polysaccharide reserves, which are lost as they develop the characteristic cell wall of mature heterocysts (Kale & Talpasayi, 1969).

Proheterocysts occur in the filaments of *Anabaena cylindrica* which have been deprived of nitrogen sources in a sequential order, comparable to that of mature heterocysts. The standard interheterocyst cell number in that species is about ten to twenty, depending upon growth conditions (Plate 4, fig. 1). The mechanism by which this sequence is maintained has attracted some attention and it certainly presents a very simple expression of inter-cell co-operation. Wilcox (1970) observed that the mean interheterocyst distance of mature heterocysts in $-N$ medium was similar to the mean distance between presumptive heterocysts (proheterocysts) found in $+N$ medium. The transformation of proheterocysts into mature heterocysts after transfer to $-N$ medium was observed by time-lapse photography. Wilcox (1970) concluded that growth in the presence of ammonia (0.02 % NH_4Cl) allows the partial expression of heterocyst pattern and suggests that this argues against the Fogg hypothesis that ammonia (or a metabolic product of ammonia) concentration determines heterocyst sequence. However, Kulasooriya, Lang & Fay (1972) have shown thet they can obtain complete suppression of heterocyst (and proheterocyst) formation after growth of *A. cylindrica* in the presence of $(NH_4)_2HPO_4$ (10^{-3} M). Our experience also has been that concentrations of NH_4Cl (10^{-3} M) lead to a total absence of heterocysts from the filaments (Plate 4, fig. 2). Kulasooriya *et al.* (1972) have suggested the cellular C:N ratio to be the dominant influence on heterocyst formation. Thus heterocyst differentiation started when the C:N ratio became greater than 6:1, and heterocysts were entirely absent when the ratio was 4.5:1. The changes in the C:N ratio may well reflect the alteration in phycocyanin content, known to decline on nitrogen starvation (Allen & Smith, 1969) and it is a little difficult to envisage the mechanism by which the overall C:N ratio determines heterocyst frequency.

Although the correlation between lack of available nitrogen and heterocyst occurrence is impressive, the manner in which the availability of a nitrogenous source could maintain the sequence of heterocysts in an actively growing culture presents greater problems. There is no evidence that the mid-heterocystous cells of *Anabaena cylindrica* grow and divide any less rapidly than those adjacent to the heterocysts, although in some species with basal heterocysts this is not the case (Fritsch, 1945). To establish down the filament a gradient of a substance that can also serve as a precursor of amino acid and hence protein synthesis, would be to deprive the interheterocystous cells of the same concentration of that precursor available to cells adjoining heterocysts.

An interesting alternative approach to heterocyst sequencing is suggested by results of Wolk (1967), which were carried out before the nitrogen fixing role of heterocysts had been extensively discussed. In order to examine the relationship between existing heterocysts and the formation of new heterocysts within the filament, an actively growing nitrogen-fixing culture of *Anabaena cylindrica* was fragmented in a blender, and the broken filaments allowed to grow. Wolk (1967) found 1.44 times as many heterocysts in the blended culture as there were in the control experiment and concluded that the spacing of heterocysts within a filament was a result of heterocysts inhibiting vegetative cells from developing into more new heterocysts. The simultaneous differentiation of spaced heterocysts in *Anabaena azollae imbricatae* (Schwabe, 1947) and the similar phenomena observed by several workers in *Anabaena cylindrica* led Wolk (1970) to comment that the ability of one heterocyst to inhibit other heterocyst differentiation precedes the apparent ability of that heterocyst to fix nitrogen.

The experiment described in Fig. 4 was designed to examine the role of ammonia (or some other product of nitrogen fixation) on the sequential development of heterocysts. A continuous culture of *Anabaena cylindrica* was deprived of ammonia and thereby initiated into heterocyst formation. Heterocyst numbers (pro- and mature) reached a maximum at 12 h, when nitrogenase was well developed and growth had restarted, the subsequent decline in heterocyst frequency being a result of vegetative cell division. When the experiment was conducted under

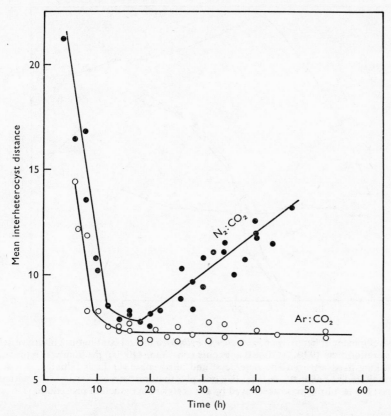

Fig. 4. An experiment, based on the continuous culture procedure (see Fig. 5), designed to examine heterocyst sequence in filaments. *Anabaena cylindrica* was grown continuously on $N_2:CO_2$ (95:5) or $Ar:CO_2$ (95:5) and the formation of total (pro- and mature) heterocysts counted.

an $Ar:CO_2$ (95:5) gas phase, the maximum number of heterocysts formed after some 14 h did not increase in the ensuing 30 h. Thus even though the ammonia concentration must have been minimal (no fixed nitrogen supplied and no nitrogen gas) there was no heterocyst development from the mid-heterocystous vegetative cells. We suggest that this experiment provides strong evidence against the 'ammonia-gradient' view of heterocyst spacing and indicates that the process of heterocyst formation itself is the pattern-controlling factor. An early product of heterocyst differentiation could be a molecule whose function is to inhibit heterocyst formation in adjacent cells. One is still left with the fact that ammonia, in growing cultures, does suppress heterocyst production. It may be that the differentiation of a vegetative cell to form a heterocyst is prevented by both ammonia (or nitrogenous product

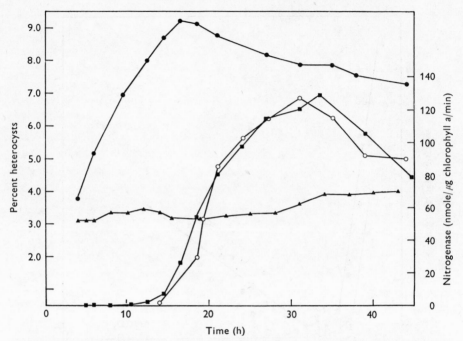

Fig. 5. Development of heterocysts in *Anabaena cylindrica* in a 5 l continuous culture vessel (mean generation time 19 h). At time 0 ammonia containing (10^{-3} M) medium was replaced by one lacking fixed nitrogen and heterocyst and nitrogenase synthesis initiated. ▲—▲, growth monitored at E680; ●—●, total (i.e. pro- plus mature) heterocysts; ○—○, mature heterocysts; ■—■, nitrogenase as assayed by the acetylenase reduction procedure.

thereof) concentration and separately by a product of adjacent, developing heterocysts.

The application of continuous culture technique to filamentous blue-green algae has been described by Bone (1971*a*, *b*) and Thomas & David (1971). The latter have examined the induction of heterocysts in *Anabaena* sp. L-31 in three steady-state conditions obtained with dilution rates of 0.21, 0.58 and 0.70 and found correlation between heterocyst formation and ammonia (formed from nitrate) release into the media at the highest dilution rate. The differentiation of heterocysts in continuous cultures of *Anabaena cylindrica* after deprivation of nitrogen source may be followed on a regular and relatively fast time base (S. Bradley & N. G. Carr, unpublished results). Within 4 h of initiation significant numbers of proheterocysts may be discerned and these reach maximum levels around 12 h, when general conversion to mature heterocysts occurs. Nitrogenase activity accompanies the formation of the mature form (Fig. 5). The establishment of a reproducible system by which a continuous culture can be switched from virtually no heterocyst to about 8 % heterocysts within a 24 h period, clearly offers a viable experimental system for the examination of the biochemistry of the differentiation process. Aliquots of such a differentiating culture were removed at time intervals following initiation, and transferred to non-heterocyst producing conditions; after 24 h the degree of heterocyst development was measured. After 5 h initiation, the culture of *A. cylindrica* will produce at least some heterocysts, even after transference to ammonia-containing media. This may be equivalent to the commit-ment point described in other differentiating systems but an examination during the first 5 h after initiation of changes in biochemical activity and of the synthesis and stability of RNA is necessary for firm comparisons to be drawn.

The germination of heterocysts, that is their dedifferentiation to yield vegetative cells, has been reported only rarely. Desikachary (1946) has described the process in two species of Rivulariaceae and Kumar (1962) has reported that heterocysts of *Camptylonema lahorense* will divide. Double heterocysts are occasionally seen in place of the usual solitary heterocyst of *Anabaena cylindrica*, but there is no evidence in the latter of division of a mature, fully-developed cell. It may be that when differentiation of the vegetative cell occurs at a particular point in the replication process, two, instead of one, heterocysts result. A detailed account of the germination of a small (3–10 %) proportion of *A. cylindrica* heterocysts has been presented by Wolk (1965*b*) who recorded photographically germination of heterocysts and found that the process was virtually dependent upon the presence of glucose and ammonium ions. A spontaneous, non-sporulating mutant clone of *Gleotrichia ghosei* has been isolated by Singh & Tiwari (1970). In this organism, ammonia does not inhibit heterocyst development and stimulates the production of hormogonia from mature heterocysts. The control of development in this mutant clone, lacking akinetes, and the relationship of this control to the parent type, shown in Plate 1, illustrates the complexity and challenge of the developmental control of three cell types in certain blue-green algae species.

The authors acknowledge with thanks permission to reproduce photographs and wish to thank Dr P. Fay and Dr C. P. Wolk for the opportunity of reading manu-scripts that were in press. Work from our own laboratory has been supported by the Science Research Council and the Medical Research Council.

REFERENCES

ALLEN, M. M. & SMITH, A. J. (1969). Nitrogen chlorosis in blue-green algae. *Archiv für Mikrobiologie*, 69, 114–20.

BONE, D. H. (1971a). Nitrogenase activity and nitrogen assimilation in *Anabaena flos-aquae* growing in continuous culture. *Archiv für Mikrobiologie*, 90, 234–41.

BONE, D. H. (1971b). Kinetics of synthesis of nitrogenase in batch and continuous culture of *Anabaena flos-aquae*. *Archiv für Mikrobiologie*, 80, 242–51.

BOTHE, H. (1970). Photosynthetische Stickstoffixierung mit einem zellfreien Extrakt aus der Blaualge *Anabaena cylindrica*. *Bericht der Deutschen Botanischen Gesellschaft*, 83, 421–32.

BOURRELLY, P. (1970). *Les Algues d'eau douce*. III. Paris: Éditions N. Boubée & Cie.

BRADLEY, S. & CARR, N. G. (1971). The absence of a functional photosystem II in heterocysts of *Anabaena cylindrica*. *Journal of General Microbiology*, 68, xiii–xiv.

BRYCE, T. A., WELTI, D., WALSBY, A. E. & NICHOLS, B. W. (1972). Monohexoside derivatives of long-chain polyhydroxy alcohols; a novel class of glycolipid specific to heterocystous algae. *Phytochemistry*, 11, 295–302.

CANABAEUS, L. (1929). Über die Heterocysten und Gasvakuolen der Blaualgen und ihre Beziehung zueinander. In *Pflanzenforschung*, vol. 13, ed. R. Kolkowitz. Jena: Fischer.

CARR, N. G. & CRAIG, I. W. (1970). The relationship between bacteria, blue-green algae and chloroplasts. In *Phytochemical Phylogeny*, ed. J. B. Harborne, pp. 119–43. London: Academic Press.

COX, R. M. & FAY, P. (1969). Special aspects of nitrogen fixation by blue-green algae. *Proceedings of the Royal Society*, B, 172, 357–66.

DESIKACHARY, T. V. (1946). Germination of the heterocysts in two members of the Rivulariaceae, *Gloeotrichia raciborskii* Wolosz, and *Rivularia mangini* Fremy. *Journal of the Indian Botanical Society*, 25, 11–17.

DONZE, M., HAVEMAN, J. & SCHIERECK, P. (1972). Absence of photosystem 2 in heterocysts of the blue-green alga *Anabaena*. *Biochimica et Biophysica Acta*, 256, 157–61.

DRAWERT, H. & TISCHER, I. (1956). Über Redox-Vorgänge bei Cyanophyceen unter besonderer Berüchsichtigung der Heterocysten. *Naturwissenschaften*, 43, 132.

DREWS, G. (1973). Fine structure and chemical composition of the cell envelope. In *The Biology of Blue-green Algae*, ed. N. G. Carr & B. A. Whitton, pp. 99–116. Oxford: Blackwell.

DUNN, J. H., SIMON, R. D. & WOLK, C. P. (1971). Incorporation of amino sugars into walls during heterocyst differentiation. *Developmental Biology*, 26, 159–64.

DUNN, J. H. & WOLK, C. P. (1970). Composition of the cellular envelopes of *Anabaena cylindrica*. *Journal of Bacteriology*, 103, 153–8.

EDELMAN, M., SWINTON, D., SCHIFF, J. A., EPSTEIN, M. T. & ZELDIN, B. (1967). Deoxyribonucleic acid of the blue-green algae (*Cyanophyta*). *Bacteriological Reviews*, 31, 315–31.

FAY, P. (1969a). Cell differentiation and pigment composition in *Anabaena cylindrica*. *Archiv für Mikrobiologie*, 67, 62–70.

FAY, P. (1969b). Metabolic activities of isolated spores of *Anabaena cylindrica*. *Journal of Experimental Botany*, 20, 100–9.

FAY, P. (1970). Photostimulation of nitrogen fixation in *Anabaena cylindrica*. *Biochimica et Biophysica Acta*, 216, 353–6.

FAY, P. (1973). The heterocyst. In *The Biology of Blue-green Algae*, ed. N. G. Carr & B. A. Whitton, pp. 238–59. Oxford: Blackwell.

Fay, P., Kumar, H. D. & Fogg, G. E. (1964). Cellular factors affecting nitrogen fixation in the blue-green alga *Chlorogloea fritschii*. *Journal of General Microbiology*, **35**, 351–60.

Fay, P. & Lang, N. J. (1971). The heterocysts of blue-green algae. 1. Ultrastructural integrity after isolation. *Proceedings of the Royal Society*, B, **178**, 185–92.

Fay, P., Stewart, W. D. P., Walsby, A. E. & Fogg, G. E. (1968). Is the heterocyst the site of nitrogen fixation in blue-green algae? *Nature, London*, **220**, 810–12.

Fay, P. & Walsby, A. E. (1966). Metabolic activities of isolated heterocysts of the blue-green alga *Anabaena cylindrica*. *Nature, London*, **209**, 94–5.

Fogg, G. E. (1942). Studies on nitrogen fixation by blue-green algae. 1. Nitrogen fixation by *Anabaena cylindrica* Lemm. *Journal of Experimental Botany*, **19**, 78–87.

Fogg, G. E. (1949). Growth and heterocyst production in *Anabaena cylindrica* Lemm. II. In relation to carbon and nitrogen metabolism. *Annals of Botany*, N.S. **13**, 241–59.

Fogg, G. E. (1951). Growth and heterocyst production in *Anabaena cylindrica* Lemm. III. The cytology of heterocysts. *Annals of Botany*, N.S. **15**, 23–5.

Fritsch, F. E. (1945). *The Structure and Reproduction of Algae*, vol. II. London: Cambridge University Press.

Fritsch, F. E. (1951). The heterocyst: a botanical enigma. *Proceedings of the Linnean Society of London*, **162**, 194–211.

Gantt, E. & Conti, S. F. (1969). Ultrastructure in blue-green algae. *Journal of Bacteriology*, **97**, 1486–93.

Geitler, L. (1932). *Cyanophyceae*. Leipzig: Akademische Verlagsgesellschaft.

Gorkom, H. J. van & Donze, M. (1972). Localization of nitrogen fixation in *Anabaena*. *Nature, London*, **234**, 231–2.

Hoare, D. S., Hoare, S. L. & Moore, R. B. (1967). The photoassimilation of organic compounds by autotrophic blue-green algae. *Journal of General Microbiology*, **49**, 351–70.

Kale, S. R., Bahal, M. & Talpasayi, E. R. S. (1970). Wall development and tetrazolium chloride reduction in heterocysts of blue-green algae, *Anabaena ambigua*. *Experientia*, **26**, 605.

Kale, S. R. & Talpasayi, E. R. S. (1969). Heterocysts: A review. *Indian Biologist*, **1**, 19–29.

Kulasooriya, S. A., Lang, N. J. & Fay, P. (1972). The heterocysts of blue-green algae, III. Differentiation and nitrogenase activity. *Proceedings of the Royal Society*, B, in press.

Kumar, H. D. (1962). Division of heterocyst in *Camptylonema lahorense* Ghose. *Revue Algologique*, N.S. **6**, 330.

Lang, N. J. (1965). Electron microscopic study of heterocyst development in *Anabaena azollae* Strasburger. *Journal of Phycology*, **1**, 127–34.

Lang, N. J. (1968). Ultrastructure of blue-green algae. In *Algae, Man and the Environment*, ed. D. F. Jackson, pp. 235–48. Syracuse: Syracuse University Press.

Lang, N. J. & Fay, P. (1971). The heterocysts of blue-green algae. II. Details of ultrastructure. *Proceedings of the Royal Society*, B, **178**, 193–203.

Lazaroff, N. (1970). Experimental control of Nostocacean development. In *Proceedings of 1st International Congress on Blue-Green Algae*, ed. T. V. Desikarchary. Madras, India.

Lazaroff, N. (1973). Photomorphogenesis and Nostocacean development. In *The Biology of Blue-green Algae*, ed. N. G. Carr & B. A. Whitton, pp. 279–316. Oxford: Blackwell.

LAZAROFF, N. & VISHNIAC, W. (1961). The effect of light on the developmental cycle of *Nostoc muscorum*, a filamentous blue-green alga. *Journal of General Microbiology*, 25, 365–74.

LAZAROFF, N. & VISHNIAC, W. (1962). The participation of filament anastomosis in the developmental cycle of *Nostoc muscorum*, a blue-green alga. *Journal of General Microbiology*, 28, 203–23.

LEACH, C. K. & CARR, N. G. (1971). Pyruvate: ferredoxin oxidoreductase and its activation by ATP in the blue-green alga, *Anabaena variabilis*. *Biochimica et Biophysica Acta*, 245, 165–74.

LEAK, L. V. (1965). Electron microscopic autoradiography incorporation of H^3-thymidine in a blue-green alga, *Anabaena* sp. *Journal of Ultrastructural Research*, 12, 135–46.

LEAK, L. V. (1967). Studies on the preservation and organisation of DNA-containing regions in a blue-green alga, a cytochemical and ultrastructural study. *Journal of Ultrastructural Research*, 20, 190–205.

MAKINO, F. & TSUZUKI, J. (1971). Absence of histone in the blue-green alga *Anabaena cylindrica*. *Nature, London*, 231, 446–7.

MICHELSON, J. C., DAVIS, E. B. & TISCHER, R. G. (1967). The effect of various nitrogen sources upon heterocyst formation in *Anabaena flos-aquae* A-37. *Journal of Experimental Botany*, 18, 397–405.

MILLER, M. M. & LANG, N. J. (1968). The fine structure of akinete formation and germination in *Cylindrospermum*. *Archiv für Mikrobiologie*, 60, 303–13.

NEILSON, A., RIPPKA, R. & KUNISAWA, R. (1971). Heterocyst formation and nitrogenase synthesis in *Anabaena* sp. A kinetic study. *Archiv für Mikrobiologie*, 76, 139–50.

OGAWA, R. E. & CARR, J. F. (1969). The influence of nitrogen on heterocyst production in blue-green algae. *Limnology and Oceanography*, 14, 342–51.

PEARCE, J. & CARR, N. G. (1967). The metabolism of acetate by the blue-green algae, *Anabaena variabilis* and *Anacystis nidulans*. *Journal of General Microbiology*, 49, 301–13.

PIERSON, B. K. & CASTENHOLZ, R. W. (1971). Bacteriochlorophylls in gliding filamentous prokaryotes from hot springs. *Nature, New Biology, London*, 233, 257.

RIPPKA, R., NEILSON, A., KUNISAWA, R. & COHEN-BAZIRE, G. (1971). Nitrogen fixation by unicellular blue-green algae. *Archiv für Mikrobiologie*, 76, 341–8.

ROBINSON, B. L. & MILLER, J. H. (1970). Photomorphogenesis in the blue-green alga, *Nostoc commune* 584. *Physiologia Plantarum*, 23, 461–72.

ROE, T. H. (1954). Chemical determination of ascorbic, dehydroascorbic and diketogluconic acids. In *Methods in Biochemical Analysis*, vol. I, ed. D. Glick, pp. 115–39. New York: Interscience Publishers.

SCHWABE, G. H. (1947). Blaualgen und Lebenstraum. II. Morphologische Reaktionen von *Anabaena azollae imbricatae*. *Acta Botanica Taiwanica*, 1, 60–82.

SCOTT, W. E. & FAY, P. (1972). Phosphorylation and amination in heterocysts of *Anabaena cylindrica*. *British Phycological Journal*, 7, 283–4.

SIMON (1971). Cyanophycin granules from blue-green alga *Anabaena cylindrica*: a reserve material consisting of copolymers of aspartic acid and arginine. *Proceedings of the National Academy of Sciences, U.S.A.* 68, 265–7.

SINGH, H. N. & SRIVASTAVA, B. S. (1968). Studies on morphogenesis in a blue-green alga. I. Effect of inorganic nitrogen sources on developmental morphology of *Anabaena doliolum*. *Canadian Journal of Microbiology*, 14, 1341–6.

SINGH, R. N. & TIWARI, D. N. (1970). Frequent heterocyst germination in the blue-green alga *Gloeotrichia glosei* Singh. *Journal of Phycology*, 6, 172–6.

SMITH, R. V. & EVANS, M. C. W. (1970). Soluble nitrogenase from vegetative cells of the blue-green alga *Anabaena cylindrica*. *Nature, London*, 225, 1253–4.

SMITH, R. V. & EVANS, M. C. W. (1971). Nitrogenase activity in cell-free extracts of the blue-green alga, *Anabaena cylindrica*. *Journal of Bacteriology*, **105**, 913–17.

STEWART, W. D. P. (1973). Nitrogen fixation. In *The Biology of Blue-green Algae*, ed. N. G. Carr & B. A. Whitton, pp. 260–78. Oxford: Blackwell.

STEWART, W. D. P. & LEX, M. (1970). Nitrogenase activity in the blue-green alga *Plectonema boryanum* Strain 594. *Archiv für Mikrobiologie*, **73**, 250–60.

STEWART, W. D. P., HAYSTEAD, A. & PEARSON, H. W. (1969). Nitrogenase activity in heterocysts of blue-green algae. *Nature, London*, **224**, 226–8.

SUSOR, W. A., DUANE, W. C. & KROGMANN, D. W. (1964). Studies on photosynthesis using cell-free preparation of blue-green algae. *Record of Chemical Progress*, **25**, 197–208.

TALPASAYI, E. R. S. (1963). Polyphosphate containing particles of blue-green algae. *Cytologia*, **28**, 76–80.

TALPASAYI, E. R. S. (1967). Localisation of ascorbic acid in heterocysts of blue-green algae. *Current Science*, **36**, 190–1.

TALPASAYI, E. R. S. & BAHAL, M. R. (1967). Cellular differentiation in *Anabaena cylindrica*. *Zeitschrift für Pflanzenphysiologie*, **56**, 100–1.

TALPASAYI, E. R. S. & KALE, K. S. (1967). Induction of heterocysts in the blue-green alga *Anabaena ambigua*. *Current Science*, **36**, 218–19.

TAPPEL, A. L. (1962). Vitamin E as the biological antioxidant. In *Vitamins and Hormones*, vol. 20, ed. R. S. Harris & I. G. Wool, pp. 493–510. New York: Academic Press.

THOMAS, J. (1970). Absence of the pigments of photosystem II of photosynthesis in heterocysts of a blue-green alga. *Nature, London*, **228**, 181–3.

THOMAS, J. & DAVID, K. A. V. (1971). Studies on the physiology of heterocyst production in the nitrogen-fixing blue-green alga *Anabaena* sp. L-31 in continuous culture. *Journal of General Microbiology*, **66**, 127–31.

UEDA, K. (1971). Die quantitative bestimmung des DNS-gehalts in den Zellen von Cyanophyceen durch fluorochromierung mit Coriphosphin. *Biochemie und Physiologie der Pflanzen*, **162**, 439–49.

WALSBY, A. E. & NICHOLS, B. W. (1969). Lipid composition of heterocysts. *Nature, London*, **221**, 673–4.

WHITTON, B. A. & PEAT, A. (1967). Heterocyst structure in *Chlorogloea fritschii*. *Archiv für Mikrobiologie*, **58**, 324–38.

WILCOX, M. (1970). One-dimensional pattern found in blue-green algae. *Nature, London*, **228**, 686–7.

WILDON, D. C. & MERCER, F. V. (1963). The ultrastructure of the heterocyst and akinete of the blue-green algae. *Archiv für Mikrobiologie*, **47**, 19–31.

WINKENBACH, F., WOLK, C. P. & JOST, M. (1972). Lipids of membranes and of the cell envelope in heterocysts of a blue-green alga. *Planta, Berlin*, in press.

WOLK, C. P. (1965a). Control of sporulation in a blue-green alga. *Developmental Biology*, **12**, 15–35.

WOLK, C. P. (1965b). Heterocyst germination under defined conditions. *Nature, London*, **205**, 201–2.

WOLK, C. P. (1966). Evidence of a role of heterocysts in the sporulation of a blue-green alga. *American Journal of Botany*, **53**, 260–2.

WOLK, C. P. (1967). Physiological basis of the pattern of vegetative growth of a blue-green alga. *Proceedings of the National Academy of Sciences, U.S.A.* **57**, 1246–51.

WOLK, C. P. (1968). Movement of carbon from vegetative cells to heterocysts in *Anabaena cylindrica*. *Journal of Bacteriology*, **96**, 2138–43.

WOLK, C. P. (1970). Aspects of the development of a blue-green alga. *Annals of the New York Academy of Sciences*, **175**, 641–7.

WOLK, C. P. & WOJCIUCH, E. (1971). Photoreduction of acetylene by heterocysts. *Planta, Berlin*, **97**, 126–34.

WYATT, J. T. & SILVEY, J. K. G. (1969). Nitrogen fixation by *Gloeocapsa*. *Science, New York*, **165**, 908–9.

EXPLANATION OF PLATES

PLATE 1

Drawing of some morphological forms found among the Nostocacae family of blue-green algae (Bourrelly, 1970).

1–3. *Gloeotrichia echinulata*. Trichomes with basal heterocysts, with (3) and without akinetes (2).

4. *Gloeotrichia pisum*.

5–6. *Nodularia implexa*, with heterocyst (5), heterocysts and akinetes (6).

7–8. *Nodularia harveyana*.

9–10. *Raphidiopsis curvata*.

PLATE 2

Fig. 1. *Rivularia biasolettiana*, × 125, field material from Slapstone Sike, Upper Teesdale. Trichome tapers from 12 to 5 μm with a basal heterocyst. (Courtesy of Drs B. A. Whitton, S. Kirkby & A. Donaldson.)

Fig. 2. Hormogonia of *Rivularia biasolettiana*, × 625, field material from Slapstone Sike, Upper Teesdale. (Courtesy of Drs B. A. Whitton, S. Kirkby & A. Donaldson.)

Fig. 3. Akinete freed from the filament but with a heterocyst still attached, *Anabaena cylindrica*, × 750.

Fig. 4. Akinete of *Cylindrospermum* sp. osmium fixation, $Ba(MnO_4)_2$ post-stain, × 4750. The dense fibrillar layer is continuous over the surface, the large clear areas are cyanophycin granules. (Miller & Lang, 1968.)

Fig. 5. Germinating akinete of *Cylindrospermum* sp. Glutaraldehyde-osmium fixation, $Ba(MnO_4)_2$ post-stain, × 3250. Division has occurred and a new cell is emerging through the outer envelope of the akinete. (Miller & Lang, 1968.)

PLATE 3

Fig. 1. A filament of *Anabaena cylindrica* (× 710) containing a heterocyst with two adjoining akinetes.

Fig. 2. Section of heterocyst of *Anabaena cylindrica* still attached to the vegetative cells after fixation with glutaraldehyde-permanganate (× 7700). The thylakoids (T) become contorted at the pole which leads to a pore (P) through the outer envelope (E); the heterocyst being separated from the adjacent vegetative cell by a septum (S). (Fay & Lang, 1971.)

Fig. 3. Pore region of a heterocyst of *Anabaena* sp. permanganate fixed × 32000. Note plasmadesmata which traverses the septum (S) and the osmiophilic plus region (P). (Wildon & Mercer, 1963.)

Fig. 4. Section of a vegetative cell (left) and heterocyst (right) of *Anabaena cylindrica*. Note the thickened outer envelope of the heterocyst and rearrangement of the thylakoids. (Lang, 1968.)

PLATE 4

Fig. 1. Sequence of heterocysts in a filament of *Anabaena cylindrica* (× 200).

Fig. 2. An ammonia-grown filament of *Anabaena cylindrica* showing complete absence of proheterocysts and heterocysts (× 200).

THE CELL CYCLE OF A EUKARYOTE

J. M. MITCHISON

Department of Zoology, University of Edinburgh,
West Mains Road, Edinburgh EH9 3JT

INTRODUCTION

A biologist reading a symposium on differentiation in micro-organisms might be excused a feeling of surprise on meeting an article about the cell cycle. The cell cycle is the basic time unit in growing cells, and there is an old-established proposition that growth and differentiation are antithetic properties of cellular systems. There is certainly truth in this proposition when comparing the steady exponential growth of a logarithmic phase culture with the specialised changes which take place in spore formation or in the later stages of slime-mould development. Does it, however, apply when we look in finer detail at the cycles of individual cells? At the structural level, most growing cells do not show any dramatic changes through the earlier stages of the cycle. But morphogenesis goes on throughout these stages as more intra-cellular organelles are made, and there are major structural changes at the end of the cycle as a cell goes through mitosis and cleavage. At the chemical level, bulk properties such as volume, dry mass, total protein and total RNA increase steadily through the cycle, following, in most but not all cells, an exponential or linear pattern. But the position is different when we examine individual species of macromolecule rather than large groups such as total protein. Nearly twenty years have passed since DNA was first found to be synthesised periodically rather than continuously, and this now appears to be a general rule for the cycle of all eukaryotic cells. More recently, this pattern of periodic synthesis at particular points in the cycle has been found with histones, immunoglobulin and many enzymes. The conclusion is that the cell cycle shows in miniature two of the most important characteristics of differentiating systems: morphogenesis, and the periodic syntheses which are the manifestations of ordered gene expression. Moreover, the cell cycle is one of the few fields of developmental biology in which there are testable hypotheses about the mechanisms which control the temporal expression of the genome (pp. 196–7).

These general arguments do not need to be carried further since they have been developed in more detail elsewhere (Mitchison, 1971, 1973). Instead, we can turn and see how far it has been possible to analyse the temporal complexity of the cycle in the fission yeast that my colleagues and I have worked on for the last sixteen years; and, in particular, examine how the patterns of enzyme synthesis illuminate the events of the cycle.

THE CELL CYCLE OF *SCHIZOSACCHAROMYCES POMBE*

The fission yeast *Schizosaccharomyces pombe* is an easy and undemanding micro-organism for physiological and biochemical work (Mitchison, 1970). It also has a good genetical background (Leupold, 1970) though chromosome mapping is not as well developed as it is in *Saccharomyces cerevisiae*. There are 6–7 chromosomes and the normal vegetative phase is haploid. The usual eukaryotic organelles are present – a single nucleus with nucleolus and nuclear membrane, mitochondria and Golgi apparatus. Nuclear division takes place within an intact nuclear membrane and with an ordered array of microtubules that looks somewhat like a mitotic spindle but may not behave like one (McCully & Robinow, 1971). The main obvious differences from a budding yeast are the shape, the mode of growth and division, and the absence of a vacuole in growing cells. The shape is like a scaled-up version of a bacterial rod, a cylinder with rounded ends about 3.5 μm in diameter and 6 μm in length at the start of the cycle. During the cycle, it grows only in length by tip growth which is predominantly at one end (Johnson, 1965). Nuclear division happens at 0.75 of the way through the cycle, and at 0.85 a septum or cell plate develops across the middle of the cell and eventually splits at cleavage to give two daughter cells. The mode of growth has two advantages for experimental work. One is that the cycle stage of an individual cell can be determined with some precision from the cell length. The second is that it is quick and easy to count in a microscope the proportion of cells containing cell plates. This cell plate index is equivalent to the mitotic index of higher cells.

Our early work on the cycle was done with single cell techniques. Growing cells were measured in an interference microscope or grain counts were made of individual fixed cells in autoradiographs after pulses of precursors. In recent years, we have largely used synchronous cultures made by a selection method (Mitchison & Vincent, 1965). This involves concentrating living cells from an exponential phase culture and then centrifuging them in a sucrose gradient. Small cells (at the start of the cycle) move slowly through the gradient and can be separated off and grown up as a synchronous culture. This technique does not give a very high degree of synchrony but it can be, and has been, used on a variety of cell types (bacteria, budding yeast and mammalian cells).

Before discussing the patterns of enzyme synthesis, it may be helpful to give a brief summary of the synthetic patterns of other macro-molecules through the cycle. Volume follows a curve of increase which is approximately exponential through the first three-quarters of the cycle but then remains constant during the last quarter of the cycle when nuclear division and cell plate formation take place. Total dry mass increases linearly in complex medium (Mitchison, 1957) and expo-nentially in minimal medium (Stebbing, 1971). Total protein and total RNA increase exponentially throughout the cycle (Mitchison & Wilbur, 1962; Mitchison & Lark, 1962; Mitchison, Cummins, Gross & Creanor, 1969; Stebbing, 1971). In contrast, DNA is synthesised during a very restricted period of the cycle (10–15 min out of a total cycle time of 150 min). There may be a difference between strains in the exact timing of this S period. In some strains at any rate, it is at the time of cell division, which gives values of G1/S/G2 as 0.2/0.1/0.7 of the cycle when G1 is taken as starting at nuclear division (Mitchison & Creanor, 1971a; see Fig. 5). These patterns of increase of the bulk properties of *Schizo-saccharomyces pombe* are broadly similar to those of the 'average' eukaryotic cell. The two main differences are the early and unusually short S period, and the continuing synthesis of RNA during nuclear division. The reason for the latter fact may be that there does not seem to be a major contraction of yeast chromosomes before mitosis (Williamson, 1966).

ENZYME SYNTHESIS DURING THE CELL CYCLE

There has been a great deal of interest in recent years about the patterns of enzyme synthesis through the cycle (reviews by Donachie & Masters, 1969; Halvorson, Carter & Tauro, 1971; Mitchison, 1971). The tech-nique is relatively simple and usually consists of a series of enzyme assays made on successive samples from a synchronous culture. The increase in activity is assumed to be a measure of enzyme synthesis and in only a few cases have measurements been made of total enzyme protein. The patterns of synthesis can be approximately classified into two main groups depending on whether synthesis is periodic or continuous, and these two groups can be further sub-divided (Fig. 1). A stable enzyme which doubles during one portion of the cycle gives a 'step' pattern which is like that of DNA synthesis in eukaryotes and has its own G1, S and G2 periods. The true length of the period of synthesis is often

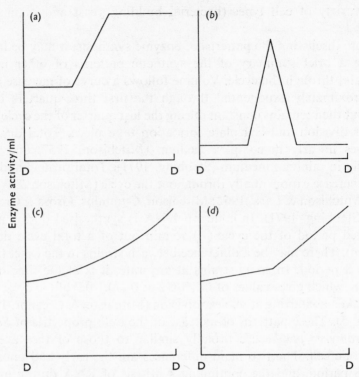

Fig. 1. Patterns of enzyme synthesis in synchronous cultures during one cell cycle. (a) Periodic, step; (b) periodic, peak; (c) continuous, exponential; (d) continuous, linear; D, cell division. (Mitchison, 1969a.)

difficult to establish since no synchronous culture is ever perfectly synchronised; thus what may be a very rapid doubling in an individual cell cycle will be spread over a period of time determined by the degree of asynchrony of the culture. An unstable enzyme gives a 'peak' pattern in which activity rises to a peak during the period of synthesis but then declines as the enzyme is broken down or inactivated. Enzymes which are synthesised continuously may follow a variety of patterns and only two of the simplest are shown in Fig. 1, an exponential curve with continuously increasing rate, and a linear pattern in which the rate doubles sharply at one point in the cycle.

Some 130 enzymes have now been assayed through the cycle in a wide range of cells from *Escherichia coli* to mammalian cells. In the majority of cases, synthesis is periodic, though there are a good number of continuous enzymes especially in mammalian cells. There are also several cases in which the classification breaks down and it is difficult to say whether synthesis is or is not periodic.

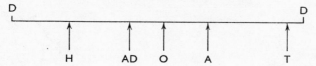

Fig. 2. Cell cycle map of the timing of five step enzymes in *Schizosaccharomyces pombe*. Each arrow gives the mid-point of the period of synthesis of an enzyme. A, aspartate transcarbamylase; AD, alcohol dehydrogenase; H, homoserine dehydrogenase; O, ornithine transcarbamylase; T, tryptophan synthetase; D, cell division. (Bostock *et al*. 1966; A. A Robinson, personal communication.)

This mixture of periodic and continuous synthesis also applies to the enzymes we have examined in *Schizosaccharomyces pombe*. Five enzymes are synthesised periodically in steps. These are aspartate transcarbamylase, ornithine transcarbamylase, tryptophan synthetase, homoserine dehydrogenase and alcohol dehydrogenase (Bostock, Donachie, Masters & Mitchison, 1966; A. A. Robinson, personal communication). Fig. 2 shows that the periods of synthesis of these enzymes are spread through the cycle and are not concentrated at any one point. Four other enzymes; sucrase, maltase, and acid and alkaline phosphatase, are synthesised continuously (Bostock *et al*. 1966; Mitchison & Creanor, 1969). The maltase pattern cannot be defined accurately since it is affected by the method of synchronisation, but the other three enzymes all follow the linear pattern with a point of rate doubling once per cycle (Fig. 1, and experimental results in Fig. 3). Although there is a good deal of scatter between different experiments, the rate doubling points of all three enzymes centre round a point about a quarter of the way through the cycle.

Before trying to interpret the control mechanisms that lie behind the step or the linear patterns, it is better to consider another way of using enzyme assays on synchronous cultures. This is to remove samples at intervals through the cell cycle and induce or derepress an enzyme. The rate at which the enzyme activity rises in the sample is the inducibility or 'potential' (Kuempel, Masters & Pardee, 1965) of that enzyme at that stage of the cycle. Most of the work on enzyme potential has been done in prokaryotes, and there the pattern of change in potential is usually similar to the pattern of synthesis of a step enzyme. The potential remains constant through most of a cycle and then doubles fairly sharply at a characteristic point. There is good reason in prokaryotes for believing that this point is at the time when the structural gene for that enzyme doubles. The clearest evidence is that the sucrase potential in *Bacillus subtilis* doubles at the same time as the sucrase-transforming capacity of the DNA (Masters & Pardee, 1965). There is other supporting evidence including the identity in spacing and order between the

7

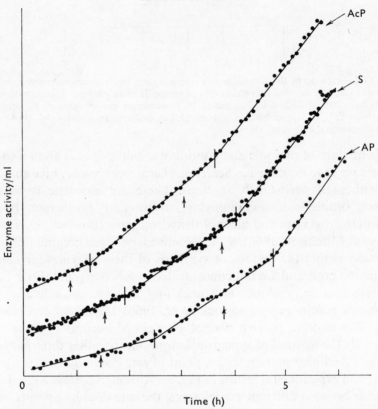

Fig. 3. Enzyme synthesis ('continuous linear') in three different synchronous cultures of *Schizosaccharomyces pombe*. AcP, acid phosphatase; S, sucrase; AP, alkaline phosphatase. Arrows mark peaks of cell plate index. Vertical lines mark points of rate change calculated by a statistical method. (Mitchison & Creanor, 1969.)

potential doubling steps in the cycle and the order of genes on the chromosome in *Escherichia coli* (Helmstetter, 1968; Pato & Glaser, 1968; Donachie & Masters, 1969).

We have applied the same technique to *Schizosaccharomyces pombe* and have measured the changes in the potential of sucrase and maltase when derepressed by lowering the glucose content of the medium (Mitchison & Creanor, 1969, 1971*a*). There are steps in potential similar to those found in prokaryotes (Fig. 4). If these steps are put onto cell cycle maps, it can be seen that the mean positions are about 0.34 and 0.46 (Fig. 5). These values have to be corrected for what we have called 'precursor delay' since there is a lag of about 15 min between the time at which protein synthesis has been inhibited by cycloheximide and the time when enzyme activity ceases to rise. It may be that this lag represents the time needed to assemble the sub-units of the enzyme or to

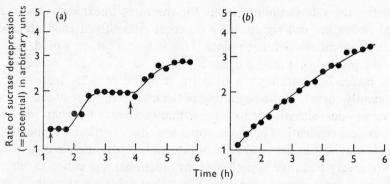

Fig. 4. (*a*) Change of sucrase potential in a synchronous culture of *Schizosaccharomyces pombe*. Each point is the slope from a sample derepressed for sucrase and then assayed at three successive times after sampling. Arrows mark peaks of cell plate index. (*b*) Change in sucrase potential in an asynchronous control culture. (Mitchison & Creanor, 1969.)

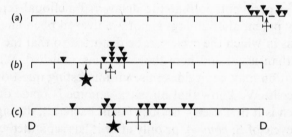

Fig. 5. Cell cycle maps of the timing in synchronous cultures of *Schizosaccharomyces pombe* of DNA synthesis and enzyme potential. Arrows with cross bars are mean values with standard errors. (*a*) DNA synthesis. Each triangle is the time of the mid-point of the rise in DNA (mid-S period) in the first cycle of a synchronous culture. (*b*) Sucrase potential. Each triangle is the time of the mid-point of rise of sucrase potential. Arrow with star is the mean value corrected for precursor delay. (*c*) Maltase potential, as above. (Mitchison & Creanor, 1971*a*.)

position it in the cell. The effect of this lag is that the real point of potential rise comes 15 min before the apparent measured rise. This is shown in Fig. 5, with the corrected mean values at 0.24 and 0.36. Because of the scatter, there is no significant difference between these mean values, but it is clear that there *is* a significant difference between them and the DNA synthesis time (S period) which is also shown in Fig. 5.

Returning now to the patterns of uninduced synthesis, we can find a connection between linear patterns and potential changes. Linear patterns have been found in bacteria as well as in *Schizosaccharomyces pombe* and they have been interpreted in a way similar to that for potential – that the rate doubling point is also at the time of doubling of the structural gene. The evidence, however, is thinner (Mitchison, 1971). In

S. pombe, the rate doubling points for the three linear enzymes mentioned above are not significantly different either from each other or from the potential doubling points. This is just what we would expect from the prokaryotic model *if* the S period was short and was at 0.25 of the cycle. But, although it is short, it is not at 0.25. Instead it is significantly earlier, at the beginning of the cycle. One way of looking at this paradoxical situation is to suggest that there may be a gap between the chemical replication of the genome and its 'functional replication' which is manifest in the doublings of rate and of potential. We have put forward a very tentative hypothesis that functional replication is related to chromosomal events since there is other evidence that there are chromosomal changes at this time (Mitchison & Creanor, 1969). But there may, of course, be other explanations, including for instance translational control and delays. One other point that is emerging from preliminary experiments is that the delay in functional replication as measured by potential doublings is not a universal phenomenon. There are situations in which the cycle can be distorted so that the S period is much later than usual and here the delay has vanished. Whatever the final explanation may be, it does raise an interesting question for other eukaryotic cells. We know that an extra genome is made during the S period. When is it first used – at once (as in bacteria), during G2 (as in the normal cycle of *S. pombe*), or only after it has separated off as a new set of chromosomes after mitosis?

CONTROL OF PERIODIC ENZYME SYNTHESIS

The evidence above suggests that the control of enzyme potential and of rate changes in linear enzymes is a gene dosage effect which is closely linked to DNA replication. The reverse is true of step enzymes (for convenience, I shall restrict the discussion to these enzymes, but the argument applies equally well to unstable peak enzymes). Not only are the steps spread throughout the cycle rather than being associated with the S period (Fig. 2, also Tauro, Halvorson & Epstein, 1968), but they will also continue in the absence of DNA synthesis, as we shall see later. The two current theories, therefore, of the control of periodic enzyme synthesis do not link it directly to DNA replication. I shall describe them both briefly but a fuller discussion is given in recent reviews about enzymes in the cell cycle (Donachie & Masters, 1969; Halvorson *et al.* 1971; Mitchison, 1971).

The first theory can be called 'oscillatory repression' and has been put forward by Donachie, Goodwin, Masters and Pardee. It is primarily concerned with enzymes subject to end-product repression. A system such as this is subject to negative feed-back and it will oscillate under the right conditions producing periodic enzyme synthesis. There is no inherent reason why the period of oscillation should have the same frequency as the cell cycle, so it has to be assumed that there is an entrainment mechanism which locks in the oscillations with some other event in the cycle.

The second theory, produced by Halvorson and his colleagues, can be called 'linear reading' or 'sequential transcription'. In essence, it suggests that genes are transcribed in the same order as their linear sequence on a chromosome. One or more RNA polymerases would move along the genome during the cycle, so that a gene would only be transcribed or induced for a short period once in the cycle. The most persuasive evidence in support of this is a series of enzymes that show steps in synthesis which mirror their genetic order both in the normal cell cycle of *Bacillus subtilis* (Masters & Pardee, 1965) and after spore germination (Kennett & Sueoka, 1971).

In arguments about these theories, the main contribution from our work on *Schizosaccharomyces pombe* so far is to show that some enzymes at any rate are not subject to the linear reading mechanism. These are the linear enzymes shown in Fig. 3. Not only are they (and maltase) synthesised continuously but also two of them can be derepressed throughout the cycle. It has been suggested that continuous enzymes are under the control of several genes which are transcribed at different points in the cycle. If so, it might be difficult in practice to distinguish between a continuous curve and one made up of a number of small steps. But this does not explain why there is both a point of rate doubling between two linear segments and also a point of potential doubling.

We cannot yet decide between these theories, and it may be that both of them are partially correct. They are stimulating and testable, and one of them, linear reading, is particularly interesting as providing a rationale for the gene order in a chromosome; though, of course, we do not know the reason, if any, why enzyme steps are needed during a cycle. It is also worth remembering that they are the only theories of temporal control in a differentiating system.

THE 'GROWTH CYCLE' AND THE 'DD CYCLE'

I want now to turn to another hypothesis that has been important in some of our recent work on *Schizosaccharomyces pombe*. It is not strictly concerned with the reasons for periodic synthesis, but it does

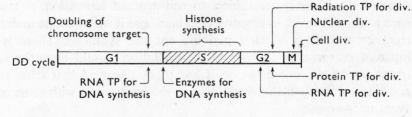

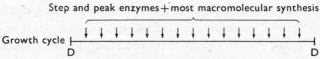

Fig. 6. The 'DNA-division cycle' and the 'growth cycle'. TP, transition point; D, division. (Mitchison, 1971.)

illustrate the relations between enzyme steps and other events of the cycle.

The passage of a cell through the cycle can be analysed in terms of movement past a series of cycle 'markers' (Mitchison, 1969b). Markers can be morphological, like the stages of mitosis; or biochemical, like periodic synthesis. They can also be physiological, like 'transition points' which mark a time before which an inhibitor (e.g. of cell division) is effective and after which it is ineffective. No doubt many of these transition points are dependent on underlying biochemical events such as the completion of the synthesis of a protein required for division.

The importance of markers is that it enables us to identify some of the causal relations in the cycle. If by distortion or blocking of the cycle, the cell passes through one set of markers but does not pass through another set, then the causal connection within the sets is closer than it is between them. The hypothesis that I want to put forward, and for which the evidence will be produced later, is that many of the existing cycle markers can be assigned to one or other of two sequences, or cycles since they span the complete cell cycle. These can be called the 'DNA-division cycle' or 'DD cycle', and the 'growth cycle' (Fig. 6). The main markers in the DD cycle are the start and finish of the S period, the stages of mitosis, nuclear division and cell division. Other markers may be the protein, RNA and radiation transition points for division, the RNA transition point for DNA synthesis, and the point in late G1 where the chromosome targets for radiation double. In addition, the DD cycle should include the synthesis of nuclear histones, the synthesis of the enzymes associated with DNA replication, and the synthesis of the proteins, structural and enzymic, which are associated with division, e.g. the proteins of the mitotic apparatus or the gene products whose temperature-sensitive controls have been shown in the elegant experiments of Hartwell (1971). The reasons for these assignations have been discussed more fully elsewhere (Mitchison, 1971). The growth cycle, on the other hand, contains most of the macromolecular synthesis that is involved in doubling the mass of the cell during the cycle. Until recently, there have been few markers available for this cycle, but we can now assign to it as markers most of the periodic enzymes, step and peak. The existence of the growth cycle is more doubtful than that of the DD cycle and it may well turn out to be composed of a number of independent sequences rather than a single sequence; but we can assume it to be one sequence for the purposes of the discussion here.

BLOCKING THE DD CYCLE AFTER SELECTION SYNCHRONY

If there are two cycles, a DD cycle and a growth cycle, which are partially independent, it should be possible to take a synchronous culture and dissociate the two cycles. We have done this by blocking the DD cycle in two ways and showing that the enzyme steps which mark the growth cycle continue unaffected (A. A. Robinson, personal communication).

One way is to block DNA synthesis by use of an inhibitor. A synchronous culture, made with our normal method of selection synchrony by gradient separation, was split into two, one part being a control, and the other part being treated with hydroxyurea at one hour after setting up (Fig. 7). The effect of the hydroxyurea was to allow the forthcoming division to take place at its normal time, two hours after the start of the experiment, but to inhibit the S period which normally takes place at the time of division. The cells subsequently recovered the ability to synthesise DNA, even in the presence of the hydroxyurea, but the second division was delayed by two and a half hours compared to the control. Yet assays of ornithine transcarbamylase showed that there was practically no difference in the step pattern of synthesis between the treated culture and the control. Substantially similar results were obtained with aspartate transcarbamylase and with alcohol dehydrogenase. This is not the only case where the patterns of enzyme synthesis have continued unaffected by a DNA block. It has also been found with *Bacillus subtilis* (Masters & Donachie, 1966) and with mammalian cells (Churchill & Studzinski, 1970; Gelbard, Kim & Perez, 1969; Klevecz, 1969).

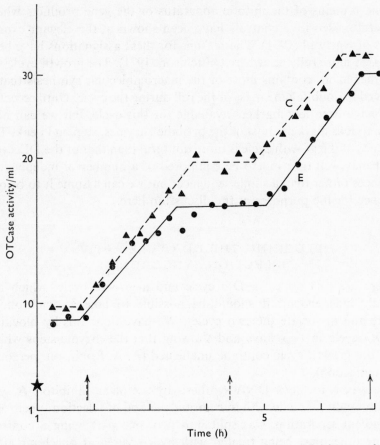

Fig. 7. Effect of hydroxyurea on ornithine transcarbamylase activity in a synchronous culture of *Schizosaccharomyces pombe*. E, Experimental culture to which hydroxyurea (to give 8 mM) was added one hour after starting the culture (star). C, Control culture without hydroxyurea. Plain arrows mark cell plate index peaks in E, dashed arrows mark cell plate index peaks in C. (A. A. Robinson, personal communication.)

The second way is to use another inhibitor, mitomycin C. Separate experiments showed that mitomycin C is not in itself an inhibitor of DNA synthesis in *Schizosaccharomyces pombe* but it does block nuclear division, possibly by causing cross-links in DNA which can in time be repaired. If it is applied to a synchronous culture half an hour after it has been set up, unlike hydroxyurea it prevents the subsequent division (Fig. 8). The succeeding S period is also blocked – a fact which underlines the normal interdependence of the main events of the DD cycle. But, as with hydroxyurea, the synthetic pattern of ornithine transcarbamylase is scarcely affected. The rise and the short subsequent plateau in activity in the control is mirrored in the mitomycin C-treated

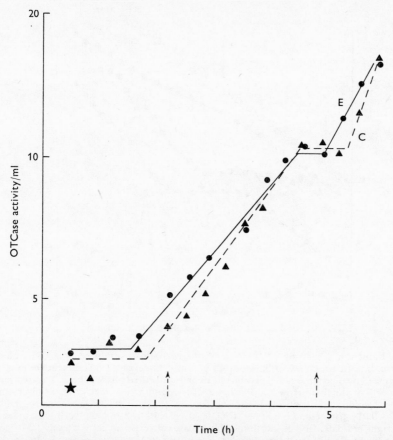

Fig. 8. Effect of mitomycin C on ornithine transcarbamylase activity in a synchronous culture of *Schizosaccharomyces pombe*. E, Experimental culture to which mitomycin C (to give 100 μg/ml) was added half an hour after starting the culture (star). C, Control culture without mitomycin C. Dashed arrows mark cell plate index peaks in C. (A. A. Robinson, personal communication.)

culture. In the latter culture, the cells which have been division-inhibited grow to unusual lengths. As with hydroxyurea, similar results were obtained with aspartate transcarbamylase and alcohol dehydrogenase.

INDUCTION SYNCHRONY BY INHIBITION OF
DNA SYNTHESIS

Another illustration of the relations between the DD cycle and the growth cycle comes from our recent work on induction synchrony (Mitchison & Creanor, 1971*b*). The method of *selection* synchrony described earlier does not appear to distort the synthetic patterns of the

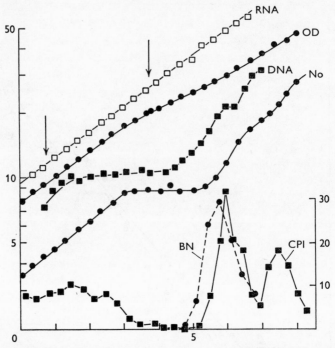

Fig. 9. Induction synchrony of *Schizosaccharomyces pombe* produced by a DNA block. Abscissa, time in hours; left ordinate, arbitrary units. Effect of a 3 h treatment with deoxy-adenosine (2 mM) on various cell parameters of an exponential phase culture. The treatment commences at the first arrow and finishes at the second arrow. RNA, one arbitrary unit (AU) = 1 μg/ml. OD, optical density at 595 nm, 1 AU = 0·0255 absorbance. DNA, 1 AU = 0.01 μg/ml. No, cell number, 1 AU = 0.715 × 10⁶ cells/ml. CPI, cell plate index as % on right ordinate. BN, proportion of binucleate cells without cell plates as 1.5 × % on right ordinate. Results from three similar experiments. (Mitchison & Creanor, 1971*b*.)

cycle but it does give a low yield since most of the cells in the original culture are thrown away. We have also, however, developed a method of *induction* synchrony in which all the cells of an asynchronous culture are induced to divide in synchrony. The yield is much higher but there are definite distortions of the growth patterns. The technique is to inhibit DNA synthesis for about a generation time and then reverse the inhibition. This method has been used extensively with higher cells (references in Mitchison, 1971) but only occasionally with lower eukaryotes (Villadsen & Zeuthen, 1970).

If a normal asynchronous culture of *Schizosaccharomyces pombe* is treated for three hours with deoxyadenosine, DNA synthesis is inhibited after a lag of about one hour and does not recommence until half an hour after the end of the inhibitor pulse (Fig. 9). The DNA doubles to reach a short plateau at the time of the first synchronous division and then rises again. The cell plate index drops during the pulse from about 10 % (the normal value in an asynchronous culture) to zero. It remains low for the first one and a half hours after the pulse and then rises to a peak value of about 30 % just before the first synchronous division. This value is about the same as that found with selection synchrony. The second cell plate peak, however, occurs after only one and a half hours, 60 % of the normal cycle time either in asynchronous cultures or after selection synchrony. One result of this shortened cycle is that there is practically no plateau in the cell number curve between the two synchronous divisions. The curve for binucleate cells without cell plates shows that nuclear division takes place at the normal time, a little before the appearance of the cell plate.

All these events are associated with the DD cycle, and they are consistent with the hypothesis that cells will continue to move round the DD cycle until they come to the S period when they are blocked from any further progress. At the time when the DNA block becomes fully effective (one hour after the start of the pulse), most of the cells are in G2 since this phase occupies 70 % of the cycle. These cells will continue round the DD cycle and divide until they reach the next S period, and it is their divisions that presumably account for the increasing cell number and continuing cell plate index until nearly the end of the pulse. When the block is removed, all the cells move synchronously into the S period and then into the following division.

The success of this technique of synchronisation depends on the fact that the DD cycle has been blocked for a generation and then released. But the results with hydroxyurea show that events of the growth cycle will continue even though the DD cycle has been blocked. The prediction would therefore be that this method of synchronising division will *not* synchronise the growth cycles in the culture. We have two sets of results that show that this prediction appears to be fulfilled.

Growth in RNA is totally unaffected by a deoxyadenosine pulse, and optical density is slightly affected only after the end of the pulse (Fig. 9). The result is that the cells are larger than normal at the first synchronous division. This is shown in the histograms of the length of dividing cells (with cell plates) in Fig. 10. Compared to the normal situation in an asynchronous culture, the cells are not only 70–80 % larger but they are also twice as variable. This is just what would be expected if the whole spread of cells at all stages of the cycle in an asynchronous culture continued to grow *without* dividing until this time, and there was no synchronisation of their growth cycles. Fig. 10 also shows two other situations. One of them is at the first synchronous division in a culture

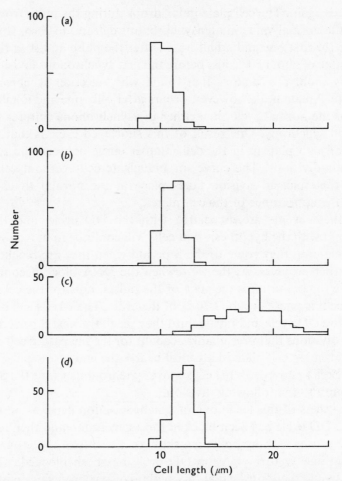

Fig. 10. Histograms of length of cells with cell plates (dividing cells) in cultures of *Schizo-saccharomyces pombe*. (a) From an asynchronous culture; (b) from the first synchronous division after selection synchrony by gradient centrifugation; (c) from the first synchronous division after induction synchrony with deoxyadenosine; (d) from the second synchronous division after induction synchrony. (Mitchison & Creanor, 1971b.)

synchronised by selection, where there is little difference from the asynchronous culture. The other situation is at the second synchronous division after induction synchrony. It shows that the cells are shorter and less variable than they were at the first division, and have largely returned to normal.

A more critical test of whether or not the growth cycle remains un-synchronised is to see what happens to enzymes that show periodic synthesis after selection synchrony. We have followed the activity of aspartate and ornithine transcarbamylase and the results are given in

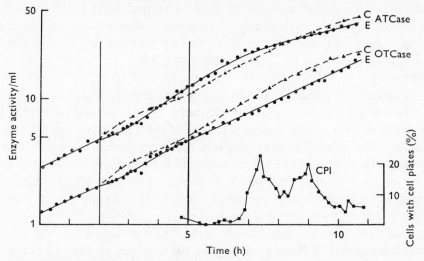

Fig. 11. Effect of induction synchrony on the activity of two step enzymes in *Schizosaccharomyces pombe*. ATCase, activity of aspartate transcarbamylase; OTCase, activity of ornithine transcarbamylase. Activity was measured for 2 h in an asynchronous exponential phase culture. The culture was then split. One half was kept as an asynchronous control (C). The other half was synchronised with a 3 h pulse of 2 mM deoxyadenosine (E). Activities were measured in both halves. CPI, cell plate index in E. (C. H. Sissons, J. M. Mitchison & J. Creanor, personal communication.)

Fig. 11. These are both step enzymes after selection synchrony but they do not show this behaviour after induction synchronisation. There are some fluctuations compared to the control, especially during the pulse, but there are no significant steps during the synchronous divisions.

Although most step enzymes would not be expected to show their steps after induction synchrony by a DNA block, the situation with enzyme potential would be different. If, as has been suggested earlier, there is a fairly close link between potential increase and DNA replication, then potential should stay constant during the DNA block and increase fairly sharply at or after the time when replication starts again. We are testing this now, and preliminary results indicate that sucrose potential does indeed show this stepwise behaviour.

CONCLUSIONS

At first sight, these results might suggest that this technique of induction synchrony should be dropped since it distorts the synthetic patterns of the cycle. But, in fact, it remains a powerful tool because it can be used in combination with selection synchrony to assign events to either one of the two cycles. If an event is only synchronised by selection, it should be part of the growth cycle, while if it is synchronised by both methods, it should be part of the DD cycle. As examples, the former applies to the step enzymes above, and the latter to enzyme potential. In any case, induction synchrony will remain useful as a method of studying events like mitosis or DNA synthesis where the large yield may be important and there is no particular concern about the growth cycle.

The concept of the two separable cycles is no more than a working hypothesis whose main novelty is the use of the step enzymes as markers in a process of growth which until recently was regarded as a continuous process. As has been said, the growth cycle is almost certainly more than one causal sequence, and, although there are theories of its temporal control, they are probably too simple. We also need to know much more about the relations between the two cycles. Although it seems to be possible for growth to continue while the DD cycle is blocked, how long will it continue? Will the growth cycle, for example, go round a second time during a DNA block? Then there is the question of the mechanism which must function in normal growth to lock the two cycles together. It is striking how rapidly the abnormal situation at the first synchronous division after induction synchrony is rectified by unusually short succeeding cycles both in *Schizosaccharomyces pombe* and in heat-synchronised *Tetrahymena* (Fig. 11; Zeuthen, 1964). Perhaps an excess of 'division proteins' are synthesised during the induction period, but that is another story. The thing that is clear is that we are only at the beginning of the process of unravelling a temporal complexity in the cell cycle which matches the spatial complexity of the cell.

REFERENCES

BOSTOCK, C. J., DONACHIE, W. D., MASTERS, M. & MITCHISON, J. M. (1966). Synthesis of enzymes and DNA in synchronous cultures of *Schizosaccharomyces pombe*. *Nature, London*, **210**, 808-10.

CHURCHILL, J. R. & STUDZINSKI, G. P. (1970). Thymidine as synchronising agent. III. Persistence of cell cycle patterns of phosphatase activities and elevation of nuclease activity during inhibition of DNA synthesis. *Journal of Cell Physiology*, **75**, 297-304.

DONACHIE, W. D. & MASTERS, M. (1969). Temporal control of gene expression in bacteria. In *The Cell Cycle. Gene–Enzyme Interactions*, ed. G. M. Padilla, G. L. Whitson & I. L. Cameron, pp. 37-76. New York and London: Academic Press.

GELBARD, A. S., KIM, J. H. & PEREZ, A. G. (1969). Fluctuation in deoxycytidine monophosphate deaminase activity during the cell cycle in synchronous populations of HeLa cells. *Biochimica et Biophysica Acta*, **182**, 564-6.

HALVORSON, H. O., CARTER, B. L. A. & TAURO, P. (1971). Synthesis of enzymes during the cell cycle. *Advances in Microbial Physiology*, **6**, 47-106.

HARTWELL, L. H. (1971). Genetic control of the cell division cycle in yeast. IV. Genes controlling bud emergence and cytokinesis. *Experimental Cell Research*, **69**, 265–76.

HELMSTETTER, C. E. (1968). Origin and sequence of chromosome replication in *Escherichia coli* B/r. *Journal of Bacteriology*, **95**, 1634–41.

JOHNSON, B. F. (1965). Autoradiographic analysis of regional wall growth of yeast, *Schizosaccharomyces pombe*. *Experimental Cell Research*, **39**, 613–24.

KENNETT, R. H. & SUEOKA, N. (1971). Gene expression during outgrowth of *Bacillus subtilis* spores. The relationship between gene order on the chromosome and temporal sequence of enzyme synthesis. *Journal of Molecular Biology*, **60**, 31–44.

KLEVECZ, R. R. (1969). Temporal order in mammalian cells. I. The periodic synthesis of lactate dehydrogenase in the cell cycle. *Journal of Cell Biology*, **43**, 207–19.

KUEMPEL, P. L., MASTERS, M. & PARDEE, A. B. (1965). Bursts of enzyme synthesis in the bacterial duplication cycle. *Biochemical and Biophysical Research Communications*, **18**, 858–67.

LEUPOLD, U. (1970). Genetical methods for *Schizosaccharomyces pombe*. In *Methods in Cell Physiology*, vol. 4, ed. D. M. Prescott, pp. 169–77. New York and London: Academic Press.

MCCULLY, E. K. & ROBINOW, C. F. (1971). Mitosis in the fission yeast *Schizosaccharomyces pombe*: a comparative study with light and electron microscopy. *Journal of Cell Science*, **9**, 475–507.

MASTERS, M. & DONACHIE, W. D. (1966). Repression and control of cyclic enzyme synthesis in *Bacillus subtilis*. *Nature, London*, **209**, 476–9.

MASTERS, M. & PARDEE, A. B. (1965). Sequence of enzyme synthesis and gene replication during the cell cycle of *Bacillus subtilis*. *Proceedings of the National Academy of Sciences, U.S.A.* **54**, 64–70.

MITCHISON, J. M. (1957). The growth of single cells. I. *Schizosaccharomyces pombe*. *Experimental Cell Research*, **13**, 244–62.

MITCHISON, J. M. (1969a). Enzyme synthesis in synchronous cultures. *Science, New York*, **165**, 657–63.

MITCHISON, J. M. (1969b). Markers in the cell cycle. In *The Cell Cycle. Gene–Enzyme Interactions*, ed. G. M. Padilla, G. L. Whitson & I. L. Cameron, pp. 361–72. New York and London: Academic Press.

MITCHISON, J. M. (1970). Physiological and cytological methods for *Schizosaccharomyces pombe*. In *Methods in Cell Physiology*, vol. 4, ed. D. M. Prescott, pp. 131–65. New York and London: Academic Press.

MITCHISON, J. M. (1971). *The Biology of the Cell Cycle*. London: Cambridge University Press.

MITCHISON, J. M. (1973). Differentiation in the cell cycle. In *The Cell Cycle in Development and Differentiation. First Symposium of the British Society for Developmental Biology* (in press). London: Cambridge University Press.

MITCHISON, J. M. & CREANOR, J. (1969). Linear synthesis of sucrase and phosphatases during the cell cycle of *Schizosaccharomyces pombe*. *Journal of Cell Science*, **5**, 373–91.

MITCHISON, J. M. & CREANOR, J. (1971a). Further measurements of DNA synthesis and enzyme potential during cell cycle of fission yeast *Schizosaccharomyces pombe*. *Experimental Cell Research*, **69**, 244–7.

MITCHISON, J. M. & CREANOR, J. (1971b). Induction synchrony in the fission yeast *Schizosaccharomyces pombe*. *Experimental Cell Research*, **67**, 368–74.

MITCHISON, J. M., CUMMINS, J. E., GROSS, P. R. & CREANOR, J. (1969). The uptake of bases and their incorporation into RNA during the cell cycle of *Schizo-*

saccharomyces pombe in normal growth and after a step-down. *Experimental Cell Research*, **57**, 411–22.

MITCHISON, J. M. & LARK, K. G. (1962). Incorporation of ³H-adenine into RNA during the cell cycle of *Schizosaccharomyces pombe*. *Experimental Cell Research*, **28**, 452–5.

MITCHISON, J. M. & VINCENT, W. S. (1965). Preparation of synchronous cell cultures by sedimentation. *Nature, London*, **205**, 987–9.

MITCHISON, J. M. & WILBUR, K. M. (1962). The incorporation of protein and carbohydrate precursors during the cell cycle of fission yeast. *Experimental Cell Research*, **26**, 144–57.

PATO, M. & GLASER, D. A. (1968). The origin and direction of replication of the chromosome of *Escherichia coli* B/r. *Proceedings of the National Academy of Sciences, U.S.A.* **60**, 1268–74.

STEBBING, N. (1971). Growth and changes in pool and macromolecular components of *Schizosaccharomyces pombe* during the cell cycle. *Journal of Cell Science*, **9**, 701–17.

TAURO, P., HALVORSON, H. L. & EPSTEIN, R. L. (1968). Time of gene expression in relation to centromere distance during the cell cycle of *Saccharomyces cerevisiae*. *Proceedings of the National Academy of Sciences, U.S.A.* **59**, 277–84.

VILLADSEN, I. S. & ZEUTHEN, E. (1970). Synchronisation of DNA in *Tetrahymena* populations by temporary limitation of access to thymine compounds. *Experimental Cell Research*, **61**, 302–10.

WILLIAMSON, D. H. (1966). Nuclear events in synchronously dividing yeast cultures. In *Cell Synchrony*, ed. I. L. Cameron & G. M. Padilla, pp. 81–101. New York and London: Academic Press.

ZEUTHEN, E. (1964). The temperature-induced division synchrony in *Tetrahymena*. In *Synchrony in Division and Growth*, ed. E. Zeuthen, pp. 99–158. New York: Interscience.

ASCOSPORE FORMATION IN YEAST

M. TINGLE, A. J. SINGH KLAR, S. A. HENRY AND H. O. HALVORSON

Rosentiel Basic Medical Sciences Research Center,
Brandeis University, Waltham, Massachusetts 02154

INTRODUCTION

The life cycle of heterothallic yeasts makes them particularly suited for a study of differentiation at the single cell level. In *Saccharomyces cerevisiae*, haploid cells grow vegetatively by mitotic division. Zygote formation occurs when cells of the opposite mating type fuse followed by nuclear fusion. Zygotes reproduce mitotically by budding. Meiosis and sporulation can be induced by changing the medium to one which is nutritionally deprived, usually a nitrogen-free medium. Following two meiotic divisions and sporogenesis in the diploid cell, four haploid ascospores are produced within an ascus. Strains differ in their ability to sporulate. In sporulating strains, ascospore formation is completed in about twenty-four hours. Upon transfer to a growth medium, germination ensues, leading again to mitotic division.

The meiotic cycle and sporogenesis prove an attractive system for studying the differentiation process. Dissection of differentiation requires simultaneous investigations of both the biochemical and the cytological changes, as well as genetic analysis of the process. Yeasts are becoming increasingly popular objects for experimentation, since they can be manipulated as prokaryotic organisms but are biochemically very closely related to higher eukaryotic cells. Although the cytological studies are as yet incomplete (i.e. chromosomes being difficult to observe) yeasts have many of the structures common to higher cells (McCully & Robinow, 1971; Moens & Rapport, 1971a; Guth, Hashimoto & Conti, 1972). These include a nucleus surrounded by a typical nuclear membrane, a nucleolar region, a mitotic spindle, mitochondria, plasma membrane, centriole, ribosomes, vacuoles and a cell wall. Genetic studies including tetrad analysis have shown that yeasts undergo a meiotic cycle similar to that in higher organisms, which includes chromosome replication, pairing of homologous chromosomes, crossing over at the four strand stage and separation of the chromosomes to the four daughter nuclei.

What factors control the production of mature ascospores which are morphologically, genetically, and biochemically distinct from the diploid cells? Does this process involve many genes or are few events involved which lead to a co-ordinated change in biosynthetic activity? Based on the experience gained over the last fifteen years on sporulation in *Bacillus*, the view that a relatively simple initiating event can cause the transition has slowly emerged (discussed elsewhere in this Symposium). Although many mutations affecting bacterial sporulation have been isolated, only a few of these are known to involve spore specific components. It is not unlikely that a similar phenomenon may be involved in the formation of yeast ascospores. Recently, mutations in loci controlling meiosis and sporulation have been isolated in yeast. As discussed below, these results suggest that a large number of different genetic loci may control this developmental sequence. The question still remains how many unique structural elements are involved in sporulation.

Sporulation in yeast, like that in bacteria, usually occurs in cells facing starvation. The capacity of a yeast cell to sporulate is related to the cell age and its previous growth conditions (Haber & Halvorson, 1972). Thus, sporulation ability is in part governed by the shifting metabolism of a starving cell. This phenomenon is reminiscent of the development of components leading to a natural survival in the establishment of the cryptobiotic state. In the dormant state, the selective conservation of endogenous materials plays a very important role. One might therefore expect that the induction of enzymes which mobilize endogenous reserves selectively, or the synthesis of metabolites which protect the loss of key enzyme activity are essential to the process of spore formation.

NUCLEAR CYTOLOGY DURING SPORULATION

Beware that you do not lose the substance by grasping at the shadow. (AESOP.)

Sporulation in *Saccharomyces cerevisiae* has been the object of numerous electron microscopic studies (Hashimoto, Conti & Naylor, 1958; Hashimoto, Gerhardt, Conti & Naylor, 1960; Mundkur, 1961; Osumi, Sando & Miyake, 1966; Marquardt, 1963; Lynn & Magee, 1970; Engels & Croes, 1968; Moens, 1971; Moens & Rapport, 1971a, b; Black & Gorman, 1971; Peterson, Gray & Ris, 1972). Cellular structures such as microtubules, spindle plaques, synaptinemal complexes and the nuclear membranes have been difficult to preserve or resolve. Recently, improvements in fixation, serial sectioning, and freeze-etching, have led to a detailed picture of cytological events of meiosis and sporulation for *S. cerevisiae* and, to a somewhat lesser degree, for *Hansenula wingei*.

To understand the control of meiosis and sporulation in yeast, correlations must be made between the temporal sequence of well defined cytological stages and biochemical and physiological events. There are difficulties in attempting to draw such parallels. For example, the time required for completion of sporulation varies somewhat from strain to strain, different conditions have been employed and in a single sporulation culture, individual cells may enter a given stage of development several hours apart.

The work of Moens (1971) and Moens & Rapport (1971a, b) offers perhaps the best available reconstruction of the cytological events during meiosis and sporulation in *Saccharomyces cerevisiae*. These authors accomplished their reconstruction of these events by serial sectioning of OsO₄-fixed cells collected at intervals during sporulation. From the events reported by Moens and others, the cytological stages are defined in Fig. 1 and outlined in Table 1. The timing of stages listed in Table 1 must be considered approximate since it is compiled from data obtained in different laboratories and represents the time at which approximately 50 % of the sporulating cells have reached this level of development.

Based on the observations by Moens & Rapport, the stages can be summarized as follows: Stage I is the period after transfer to sporulation medium when a single indistinct spindle plaque with few microtubules is observed (Fig. 1a). Stage II is the period from the duplication of the spindle plaques (Fig. 1b) to the time at which the spindle plaques have moved to face each other for the first meiotic division (Fig. 1c). Stage III consists of the period of nuclear elongation during the first meiotic division (Fig. 1d). Stage IV includes replication of the plaques for the second division (Fig. 1e), the second division and concurrent early development of the ascospores (Fig. 1g, h). Stage V is the maturation period of the ascospores (Fig. 1i). A scanning electron microscopic photograph of a mature ascus and ascospores is seen in Plate 1. Some of the differences between mature spores and vegetative cells are summarized in Table 2.

Moens and Rapport have described these events in detail: During the first 4–5 h after transfer to sporulation medium, the single spindle plaque is indistinct and has few microtubules associated with it (stage I). After 8 h, many cells have two plaques, with microtubules projecting into the nucleus (stage II). Some cells have two plaques connected by a bridge. The mode of replication of the plaques is uncertain. The plaques in later stages face each other and are found with varying distances between them. As the plaques separate (stage III), the nucleus elongates but does not divide. The nuclear envelope apparently remains intact

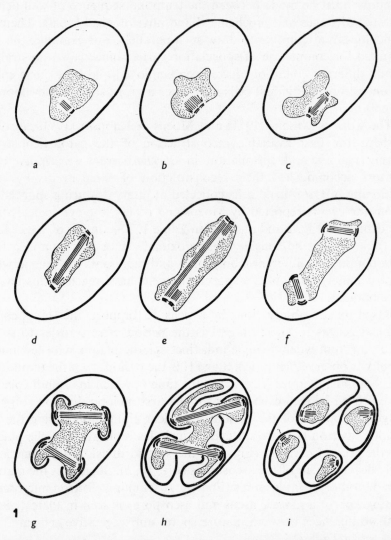

Fig. 1. Schematic representation of the nuclear events during meiosis and spore formation of *Saccharomyces cerevisiae*. (*a*) Stage I. Spindle plaque has not duplicated and few micro-tubules are associated with it. (*b*) Stage II. Spindle plaque has duplicated; more micro-tubules are observed. (*c*) End of stage II. Spindle plaques have migrated to face each other but have not yet begun to separate. (*d*) Stage III. Spindle plaques are separating during the first meiotic division. Nucleus is elongating with its membrane intact. (*e*) Beginning of stage IV. Spindle plaques have duplicated at the end of the first meiotic division. (*f*) Stage IV. Spindle plaques have migrated to face each other for the second meiotic division. (*g*) Stage IV. Second meiotic division with 'prospore wall' shown. Spore wall formation beginning. (*h*) Stage IV. Second division and spore wall formation are almost completed. Nuclear membrane is still intact. (*i*) Stage V. Four spores are visible in the ascus. Spore maturation is in progress. (Moens & Rapport, 1971*a*.)

Table 1

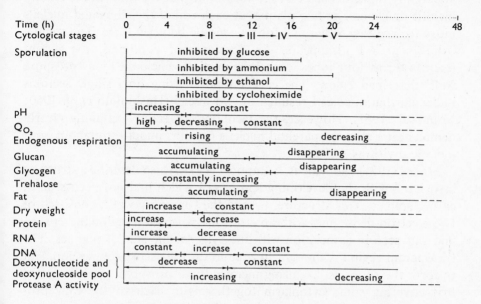

Time (h)	0	4	8	12	16	20	24	48
Cytological stages	I		II	III	IV	V		
Sporulation	inhibited by glucose							
	inhibited by ammonium							
	inhibited by ethanol							
	inhibited by cycloheximide							
pH	increasing		constant					
Q_{O_2}	high	decreasing		constant				
Endogenous respiration	rising				decreasing			
Glucan	accumulating			disappearing				
Glycogen	accumulating			disappearing				
Trehalose	constantly increasing							
Fat	accumulating				disappearing			
Dry weight	increase		constant					
Protein	increase	decrease						
RNA	increase	decrease						
DNA	constant	increase	constant					
Deoxynucleotide and deoxynucleoside pool	decrease		constant					
Protease A activity	increasing			decreasing				

Table 2. *Morphological and biochemical properties of ascospores of Saccharomyces cerevisiae**

STRUCTURE

Ascospores are generally spherical with a smooth surface and usually without colour. Two to four spores are present in each ascus. The ascus wall does not lyse rapidly upon maturity.

CHEMICAL COMPOSITION

Ascospores have a higher carbohydrate content and contain more glucan, mannan and trehalose than vegetative cells. The lipid content of ascospores is higher than vegetative cells. Ascospores contain less RNA and ribonucleotides and although higher in proline contain reduced amounts of protein and amino acids than vegetative cells. Spore wall contains an antigen which is immunologically distinct from those present in vegetative cell wall.

SURFACE PROPERTIES

Ascospore walls are thicker and stronger than those of vegetative cells. Spores contain more lipids which may account for the fact that spores are acid fast, hydrophobic and preferentially migrate into non-aqueous solvents.

RESISTANCE

Ascospores are slightly more resistant to heat, to the toxic effects of alcohol and to the action of cell wall lytic enzymes such as glusulase.

* For general review see Phaff, Miller & Mrak, 1966 or Fowell, 1969.

throughout the process. At maximum separation, the spindle plaques apparently replicate (stage IV) to form four plaques prior to the second meiotic division. The two plaques at each end of the elongated nucleus move to face each other and form a spindle between them. The spindle from meiosis I disappears. As development continues, the parent nucleus forms four lobes, each lobe partially enclosed by a 'prospore wall' described below. The earlier reports using only single sections indicated that meiosis I resulted in two nuclei (Hashimoto *et al.* 1960; Mundkur, 1961). However, the technique of serial sectioning clearly demonstrates that the parental nucleus remains intact throughout both meiotic divisions.

Moens (1971) and Moens & Rapport (1971*a*) also describe a structure associated with the developing spore wall which had not been seen in earlier studies. This structure, termed the 'prospore wall', consisted of two electron-dense lines with an electron-transparent region in between and was seen in association with the cytoplasmic side of plaques. The transparent region is thought to expand during ascospore development to form the spore wall. During meiosis II, a 'prospore wall' is seen at first covering some cytoplasm together with each of the four protuberances from the parent nucleus. Material from the parent nucleus apparently moves into the spore nucleus and cytoplasm flows under the developing spore wall. The process continues until the spore is completely formed. During the maturation stage (stage V) the nucleus and plaques become separated from the spore wall and the plaque gradually takes the less distinct appearance typical of vegetative cells. Moens (1971) feels that the use of $KMnO_4$ as a fixative in earlier studies prevented the observation of the 'prospore wall' and thus resulted in confusion over the identity of the membrane associated with spore wall formation.

Recently, even finer resolution of the structure of the meiotic spindle plaques has been reported employing the technique of spheroplasting prior to fixation with formaldehyde–glutaraldehyde and OsO_4 (Peterson *et al.* 1972). In these preparations the nuclear membranes are especially well preserved. The meiosis I spindle plaque was located in a region of the nuclear membrane which had the appearance of a large nuclear pore. The appearance of the meiosis II plaques is shown in Plate 2. In this plate the structure termed the 'prospore wall' (2*c*, arrow) by Moens & Rapport (1971*a*) is clearly visible.

Synaptinemal complexes have also been observed in *Saccharomyces cerevisiae* (Engel & Croes, 1968; Moens & Rapport, 1971*b*). The occurrence of the complexes was found to be maximal some 9–10 h after introduction into sporulation medium (stage II), several hours preceding the first division (Engels & Croes, 1968).

Moens & Rapport (1971b) described synaptinemal complexes consisting of an electron transparent region with an opaque central zone, bounded by two dense regions. Several such complexes were often observed in a single structure termed the 'polycomplex' located at the edge of the nucleus. A structure not found in mitotically dividing cells, referred to as the 'polycomplex body', was often found to contain the polycomplex. Moens and his co-worker suggested that the polycomplex body may play a role in polycomplex formation. By the time of spindle elongation during the first meiotic division, polycomplexes are no longer visible.

Meiosis and sporulation in the yeast *Saccharomyces cerevisiae* has also recently been investigated using the freeze-etch technique as well as thin sectioning (Guth *et al.* 1972). These workers observed nuclear pores with spindle fibres sometimes extending through the pores. They confirmed the conclusion that the nucleus elongates without separation and dissolution of nuclear wall during the first division. With the second division, they observed four bilamellar structures, which they term the 'forespore membranes', located near the four buds from the parent nucleus. The 'forespore membrane' corresponds to the 'prospore wall' described by Moens (1971). Although Guth *et al.* (1972) were unable to substantiate the division of the spindle plaques described by Moens (1971) and Moens & Rapport (1971a), the basic outlines of meiosis by the two studies are in agreement.

Cytological studies are more useful to cell physiologists when structural changes can be correlated with biochemical activities in the cell. For example, R. F. Illingworth, A. H. Rose & A. Beckett (personal communication) employed electron microscopy in conjunction with a study of lipid synthesis during sporulation. More frequently light microscopy is employed in conjunction with physiological studies (Pontefract & Miller, 1962; Sando & Miyake, 1971; Croes, 1967a).

Stage II involving spindle plaque duplication cannot be resolved by light microscopy but stage III, the first meiotic division, stage IV, the second meiotic division and early ascospore development and stage V, spore maturation can be readily defined. Using Giemsa stain to follow the nuclear divisions, Pontefract & Miller (1962) reported the appearance of 'chromosome-like bodies' after 10 h in sporulation media. They also observed both first and second meiotic divisions. Most investigators have failed to visualize chromosomes in *Saccharomyces cerevisiae* but Tamaki (1965) reported seeing 18 bivalents prior to the first meiotic division. However, Tamaki's observations have not been confirmed by other investigators. Employing Giemsa staining Croes (1967a) correlated nuclear events during meiosis with a number of biochemical properties.

The cytology of sporulation in other yeasts, including *Hansenula* which has 'hat-shaped' spores, has also been studied. Its large chromosomes can easily be visualized by light microscopy during the early stages of meiosis (Stock & Black, 1970). As in *Saccharomyces cerevisiae*, the nuclear membrane remains intact throughout meiosis as the nucleus invaginates first in one plane and then in a second perpendicular plane producing four lobes which become the spore nuclei (Black & Gorman, 1971). These workers concluded that the membranes enclosing the meiotic products and responsible for laying down spore wall and forming the cytoplasmic membrane are derived from endoplasmic reticulum. Some workers studying *S. cerevisiae* reached a similar conclusion (Lynn & Magee, 1970), but this has most recently been disputed (Moens & Rapport, 1971a) as they reported that the spore wall develops from the 'prospore wall'.

MEIOTIC DNA REPLICATION

When a diploid yeast cell undergoes meiosis and sporulation DNA replication occurs and a diploid '2n' nucleus is converted to the '4n' state. Eventually, a haploid set of chromosomes is incorporated into each of the four ascospores.

Several laboratories have reported that the DNA concentration per cell as measured colorimetrically with the diphenylamine reagent doubles during sporulation (Croes, 1966; Esposito, Esposito, Arnaud & Halvorson, 1969; Roth & Lusnak, 1970; Sando & Miyake, 1971). DNA synthesis occurs discontinuously and is usually initiated 4–7 h after introducing the cells into sporulation medium and is completed by T_{10}. The percentage increase in DNA concentration is related to the number of cells in the culture which undergo sporulation.

Sando & Miyake (1971) followed changes in the DNA precursor pools during sporulation. Oligodeoxynucleotide content decreased during stage I with the lowest value at T_6. Following this the pools increased until T_{12} and thereafter decreased continuously. On the other hand, deoxynucleotides and deoxynucleosides decreased continuously from T_0 to T_{12}.

Darland (1969) using a large yeast cell fraction isolated from stationary phase cultures by centrifugation through renografin gradients, reported sporulation in the absence of net DNA synthesis. In these same cells a

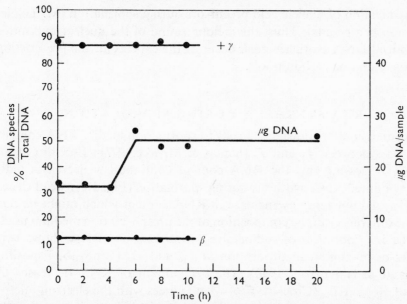

Fig. 2. DNA duplication during sporulation. Whole cell DNA was isolated from the diploid *Saccharomyces cerevisiae* S41 at intervals during sporulation as described by Sena (1972). Samples were ultracentrifuged to equilibrium in CsCl to separate the DNA into nuclear (α, γ) and mitochondrial (β) bands.

The area percentage of each DNA band was calculated from densitometer tracings of the ultraviolet light photographs. These percentages were reproducible to within less than 5 % of one another. (Sena, Ph.D. Thesis, University of Wisconsin, 1972.)

limited incorporation of [³H]adenine into alkali-stable hot TCA-soluble material was observed from 3–6 h into sporulation which could be attributed to either repair synthesis or replication of mitochondrial DNA. Since meiotic DNA synthesis is required for sporulation, this increase must have occurred in the resting culture prior to introduction into sporulation medium.

Yeasts contain both nuclear and mitochondrial DNA (mDNA). In vegetatively growing cells of *Saccharomyces lactis* and *S. cerevisiae* both nuclear and mDNA have been reported to be individually synchronized but replicate at different times during the cell cycle (Smith, Tauro, Schweizer & Halvorson, 1968; Cottrelle & Avers, 1970). However, Williamson & Moustacchi (1971) employing different cultural conditions found that in *S. cerevisiae*, mDNA replication was continuous over most, if not all, of the cell cycle. This difference has yet to be resolved. On the other hand, in *S. cerevisiae* mDNA (β) and nuclear DNA (α and γ) replication is linked during sporulation (Fig. 2).

Experiments designed to study the mode and timing of DNA replication during meiosis and sporulation have been hampered by the lack of permeation of nucleic acid precursors during sporulation (M. Tingle, unpublished results). Thus, the meiotic round of the nuclear and mitochondrial DNA synthesis must occur at the expense of the pre-existing nitrogen and/or carbon pools.

RNA SYNTHESIS DURING SPORULATION

Immediately after transferring cells to sporulation media, RNA content increases between T_4 and T_6 (Sando & Miyake, 1971; Esposito et al. 1969). Following this the RNA content continuously decreases indicating limited RNA synthesis during sporulation (Fig. 3). Instead Croes (1967a) did not find any increase in RNA content which might be due to the differences in the composition of the presporulation medium used.

The incorporation of radioactive precursors into RNA shows two peaks of maximum incorporation at T_{10} and T_{25} (Esposito, Esposito, Arnaud & Halvorson, 1970). The net amount of RNA decreases to about 50 % at T_{24} (Croes, 1967b) which agrees with the electron micrographs by Mundkur (1961) showing a dramatic decrease in cytoplasmic ribosomes during sporulation. The cytoplasm becomes less basophilic during this period.

Because the appropriate inhibitors are not available it has not been possible to test directly if RNA synthesis is required for yeast sporulation. Radioactive isotope labelling experiments have demonstrated that ribosomal, transfer and heterogeneous RNA species are synthesized during sporulation but no difference in RNA species have been detected between sporulating and non-sporulating cells (Darland, 1969) except 20S RNA. Accumulation of a 20S RNA species has been demonstrated (Kadowaki & Halvorson, 1971a, b) which has recently been identified as an 18S rRNA precursor (Sogin & Haber, 1972). From the kinetics of rRNA synthesis, it is evident that rRNA accumulation in the sporulating cell is slow compared to that in the vegetative cell.

PROTEIN SYNTHESIS DURING SPORULATION

Protein synthesis is required for meiosis and spore formation. Cycloheximide inhibits the sporulation process up to the stage where mature ascospores are observed (Esposito et al. 1969; Darland, 1969). The net amount of protein increases during stage I (2–6 h into sporulation) and declines thereafter (Esposito et al. 1969; Croes, 1967a; Sando & Miyake, 1971). The maximum increase in protein content varies from 10 to 35 % depending on the composition of the presporulation medium.

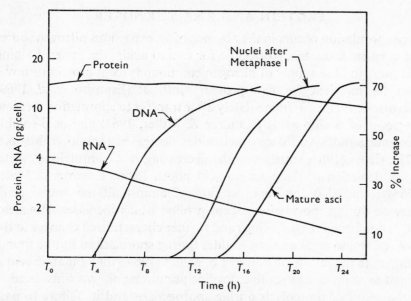

Fig. 3. Percentage increase in DNA, mature asci and nuclei after metaphase I and average protein and RNA content of the yeast cell during meiosis and ascus formation. (Croes, 1967a.)

Incorporation of [^{14}C]amino acids into hot TCA-precipitable material has shown that two periods of maximum protein synthesis occur during sporulation (Esposito *et al.* 1969). The first period occurs between T_4 and T_7 (stage I) and the second period from T_{23} to T_{27} (stage V – ascospore formation). Since proteins synthesized during sporulation have not been characterized, there is no evidence at present for sporulation-specific proteins. On the other hand, ascospores contain an immunologically distinct antigen absent in vegetative cells which has been tentatively identified as a protein structure component (Snider & Miller, 1966).

There is some evidence that mitochondrial protein synthesis is also required during sporulation. Puglisi & Zennaro (1971) observed that erythromycin, an inhibitor of mitochondrial protein synthesis, prevents sporulation in erythromycin-sensitive strains but not in erythromycin-resistant strains. Erthryomycin did not affect respiration in sporulating cultures of either sensitive or resistant strains at low concentrations (10 mg/ml). Similarly, P. Tauro (personal communication) observed that sporulation is sensitive to chloramphenicol, an inhibitor of mitochondrial but not cytoplasmic protein synthesis in yeast.

PROTEIN AND RNA TURNOVER

Since sporulation occurs in the absence of an exogenous nitrogen source and even in mutants auxotrophic for amino acids, the internal amino acid pool and the supply of nitrogen compounds from protein turnover must be sufficient for new protein synthesis (Esposito *et al.* 1969; Halvorson, 1958*a*, *b*). Immediately after transfer to sporulation medium the pools of amino acids (Ramirez & Miller, 1964) and acid-soluble deoxynucleosides and deoxynucleotides decreased (Sando & Miyake, 1971). Croes (1967*a*) attributed the decreasing rate of protein synthesis to the depletion of these amino acid pools. Later, Ramirez & Miller (1964) found that the concentration of almost all the amino acids decrease during sporulation except proline which increases. Rousseau (1972) confirmed these findings and further characterized changes in the pools of amino acids and nucleotides during sporulation. But he found, contrary to the results of Ramirez & Miller (1964), that glutamic acid is almost as abundant as proline. Proline accumulation was considered to be the result of proteolysis during sporogenesis and its failure to pass through the cell wall into the medium (Ramirez & Miller, 1964). The failure of a protease negative strain to complete sporulation (Chen & Miller, 1968) can be interpreted as further support for the endogenous origin of precursors for macromolecule synthesis.

The nucleotide pool is greatly reduced during sporulation (Abdel-Wabab, Miller, Gabriel & Hoffmann-Ostenhof, 1961; Miller & Hoffmann-Ostenhof, 1964). The concentration of nucleotide di- and triphosphates, and uridine diphosphate glycosides decreased during sporulation and this was suggested to be due to their use as precursors for spore wall formation (Miller & Hoffmann-Ostenhof, 1964). Also, Rousseau (1972) observed that almost all of the nucleotides and nucleosides decrease during sporulation except CMP, AMP and UDP-peptides. Significantly, the level of GTP, UDP and NAD is very low in spores (Fig. 4).

It is well established that protein and RNA turnover dramatically increase in resting cells. The rate of protein turnover is 23 times greater in resting cells than in logarithmically growing yeast cells (Halvorson, 1958*a*). Similar increases have been observed in bacteria induced by starvation or adaptation to utilize new substrates (Mandelstam, 1960). To estimate protein turnover during yeast sporulation, vegetatively growing cells were labelled with radioactive amino acid, then transferred to unlabelled sporulation medium and the decrease in radioactivity in the protein fraction measured.

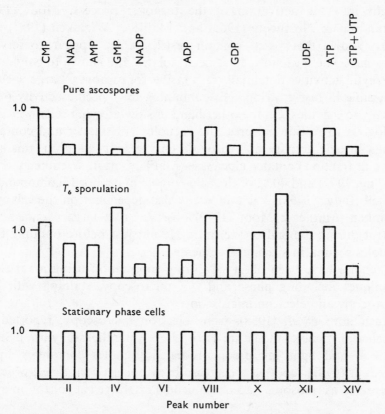

Fig. 4. Nucleotide pool analysis of some stages of sporulation of *Saccharomyces cerevisiae*. Stationary growth in YEP medium, T_6 sporulation, 6 h in sporulation medium and mature ascospores. (Rousseau, Ph.D. Thesis, University of Wisconsin, 1972.)

During sporulation the radioactivity in proteins labelled during vegetative growth represents a minimum estimate of protein turnover since recycling of the amino acid pool for new protein synthesis and the amino acid pools during sporulation have not been included in the estimate of turnover (Esposito *et al.* 1969). Similarly, in bacterial sporulation Monro (1961) found that about 80 % of the prelabelled vegetative proteins are turned over during sporulation of *Bacillus thuringensis*.

There is very little information available on the mechanism of protein and RNA turnover in sporulating yeast although the proteases present in vegetative cells are well characterized (Hayashi, Oka, Doi & Hata, 1968*a*, *b*; Doi, Hayashi & Hata, 1969). Chen & Miller (1968) reported that protease A increases during sporulation until stage V after which it declines rapidly. Activation of turnover requires ATP (Halvorson, 1962). Cytological studies with vegetatively growing cells has provided a possible insight into the activation of the turnover process. Moor (1967), Sentandreu & Northcote (1969) and Matile & Wiemken (1967) have clearly shown that yeasts contain vesicles which appear to serve as lysosomes. These vacuoles have been isolated and shown to contain the hydrolytic activities potentially responsible for turnover. Large vacuoles are visible in fast-growing cells containing low specific activity of the hydrolytic enzymes. On the other hand in slow-growing cells, which are analogous to resting cultures, the vacuoles are smaller and contain a higher specific activity of hydrolytic enzymes. In addition, this same vacuolar fraction contains glucanase (Matile, Cortat, Wiemken & Frey-Wysling, 1971) and 40 % of the acid-soluble phosphorus compounds of the cell (Indge, 1968). Vacuole size is also dependent on the cell cycle. Wiemken, Matile & Moor (1970) observed that large vacuoles were present during the end of the cell cycle but just before budding these vacuoles shrunk in size and fragmented.

Vacuoles have been observed during sporulation using a variety of techniques including phase and UV microscopy, staining with light microscopy and electron microscopy.

Hashimoto *et al.* (1958), using electron microscopy, reported the presence of vacuoles in mature ascospores. A parallel study using Sudan Black B and light microscopy revealed vacuoles containing lipids. Mundkur (1961) reported a decrease in cytoplasmic particles, which he interpreted as ribosomes, accompanied by extensive vacuolization of the cell.

Soon after transfer to acetate sporulation medium, Pontefract & Miller (1962), using phase microscopy, reported that cells increased in size and large vacuoles were observed in unstained cells. Small granules began to appear which increased in size and were easily seen just before spore formation. Croes (1967a) reported that the large central vacuoles seen in vegetative cells appear to fragment into numerous smaller vacuoles and that the process of vacuolization may be well underway in stage I.

Fowell (1969) commented that considerable confusion is caused by the presence of both vacuoles and smaller granules during sporulation; the granules containing glycogen and fat. According to the criteria used by Fowell, in the various reports it is unclear which are 'vacuoles' and which are 'granules'. Even if these two groups of cytoplasmic organelles could be adequately distinguished, it is uncertain if each group is homogeneous. Svihla, Dainko & Schlenk (1964) studied the fate of the vacuoles during sporulation using ultraviolet. By microscopy and following incorporation of UV-absorbing S-adenosyl-methionine into the vacuoles, he found that these are abolished and their contents released into the intersporal space. The structures described by others as vacuoles later in the sporulation process may well be the 'granules' of Fowell's terminology.

It is of interest that during sporulation one of the prominent cytological features is the decrease in size and fragmentation of the vacuoles. Presumably, the proteases and nucleases are released into the cytoplasm and contribute to the active turnover during spore formation. The mechanism for stabilization of vacuoles may well be a critical factor in determining the potential for spore formation.

RESPIRATORY ACTIVITY DURING SPORULATION

Yeast sporulation is dependent upon oxidative metabolism. The oxidation rate of an exogenous carbon source such as acetic acid is high during stage I, falls rapidly during stages II and III and then remains constant during stages IV and V (Croes, 1967a). On the other hand, the endogenous oxidation rate rises continuously during stages I to IV and then decreases during stage V (Vezinhet, Arnaud & Galzy, 1971). Esposito et al. (1969) found that during sporulation in acetate medium about 62 % of the assimilated acetate is respired, 22 % remains in the soluble pool, and 16 % is incorporated into lipids, protein, nucleic acids and all other components.

Miyake, Sando & Sato (1971) observed that the activity of tricarboxylic acid cycle enzymes increase during stage I. They found that isocitrate lyase activity increased markedly in a sporogenic strain but not in an asporogenic strain. The rate of production of $^{14}CO_2$ from [1-^{14}C]acetate is more than from [2-^{14}C]acetate implying that the glyoxylate cycle predominates during sporulation. This result is in contradiction to Croes (1967b) who suggested that sporulation was triggered by an insufficiency of the glyoxylate cycle. Further support for the involvement of the glyoxylate pathway comes from the effect of CO_2 on sporulation. It has long been recognized that low concentrations of CO_2 are essential for sporulation whereas high concentrations are inhibitory (Adams & Miller, 1954; Bright, Dixon & Whymper, 1949; Miller, 1971). Of particular interest was the finding by Bettelheim & Gay (1963) that when glyoxylate was included in the acetate sporulation medium, the system was no longer dependent upon CO_2.

Undoubtedly most of the energy and/or carbon needs for sporulation are derived from exogenous sources. Possible endogenous energy reserves will be discussed later. Vezinhet et al. (1971) and Pontefract &

Miller (1962) observed that the endogenous respiration of cells transferred to acetate sporulation medium increased during stage IV. It is interesting to note that glycogen and fat increased in amount during the same stage. Once endogenous respiration reaches a maximum, glycogen and fat start disappearing from the sporulating cells. Also, lipids (Eaton, 1960) do not serve as substrates for endogenous respiration in mitotically growing yeast.

METABOLISM OF STORAGE CARBOHYDRATES

In yeast and many other micro-organisms carbohydrates are the primary energy reserves. As pointed out by Wilkinson (1959), these reserves, which generally provide for improved survival or other biological advantage, accumulate when the energy supply exceeds that required for growth and are utilized when exogenous energy becomes limiting for growth and maintenance. For example, during sporulation in *Clostridium* (Strasdine, 1972), intracellular glucan accumulates at the time of exhaustion of nutrients from the medium. Glucan breakdown and the subsequent endogenous fermentation precedes sporulation.

The volume and dry weight of sporulating cells increases during stages I and II (Croes, 1967a). Roth (1970) estimated that 67 % of the dry weight increase is due to the increase in carbohydrate content. The insoluble carbohydrates increase 3–4-fold to the middle of stage II whereas trehalose increases continuously throughout sporulation leading to an 18-fold increase in the ascospores. Pontefract & Miller (1962), from cytological studies, could not detect glycogen in vegetative cells but reported evidence for its appearance during stage I. The glycogen increased until early stage V and then declined. This was confirmed by chemical analysis by Galzy & Bizeau (1966) who also noted a parallel between glycogen breakdown and a rise in glycogen phosphorylase activity. Pazonyi & Markus (1955) found in addition to trehalose, considerable glucan and mannan accumulation during sporogenesis.

Recently, Kane, Roth & Erwin (1972) have assayed the glycogen–glucan and mannan content of sporulating and non-sporulating strains of *Saccharomyces cerevisiae*. In sporulating cells the glycogen–glucan fraction increased 8-fold during stages I and II and remained constant until stage IV, at which point it decreased. A similar increase in glycogen–glucan content was observed in non-sporulating cells in acetate medium but no decrease during the period of spore formation was seen. Thus, the glycogen–glucan fraction may serve as a carbon or energy source for spore formation. Mannan synthesis increased (prior to ascospore

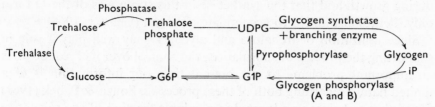

Fig. 5. The pathway for the metabolism of
trehalose and of glycogen in yeast

formation) in both sporulating and non-sporulating cells. The authors suggested that the mannan present in ascospore cell walls may be obtained from pre-existing mannans in cell walls.

In vegetative cells glycogen and trehalose are the primary carbohydrate reserves. The storage of these reserve carbohydrates is dependent on the metabolic state of the cells. Glycogen reserve may occupy up to 20 % of the dry weight of the vegetative cell and trehalose some 6–10 %. The glycogen content increases when glucose is fermented in nitrogen-low medium or on transfer from anaerobic to aerobic conditions while trehalose content increases on transfer from anaerobic to aerobic conditions (Suomalainen & Oura, 1971). In nitrogen-limited as well as glucose-limited media, the accumulation of glycogen and trehalose is strongly dependent upon the growth rate. The accumulation increases with decreasing growth rate (Kuenzi & Fiechter, 1972). M. T. Kuenzi (personal communication) found in chemostat experiments a correlation between the glycogen and trehalose content and the ability to sporulate. The highest content of glycogen and trehalose were observed in slow-growing chemostat cultures. In these cultures, 5 % of the cells contained spores. In cells growing at slightly higher growth rates, none of the cells sporulated, although the carbohydrate contents were high, however, on removal of the exogenous carbohydrate 5 % of these cells sporulated.

The pathway for trehalose and glucan metabolism (Fig. 5) in yeast has been recently reviewed by Manners (1971). Two forms of glycogen synthetase, a glucose 6-phosphate (G6P) independent (I) and a G6P dependent synthetase (D), have been reported (Rothman & Cabib, 1970) and these are interconvertible depending upon the concentration of G6P. During vegetative growth, the increase in glycogen before the onset of the stationary phase of growth was correlated with the rapid increase in the I form of the synthetase. Mutants unable to synthesize trehalose and glycogen have been reported in *Saccharomyces cerevisiae* (Chester & Byrne, 1968). Rothman & Cabib (1970) isolated a similar mutant and observed that this mutant failed to accumulate glycogen

8

during growth and that the synthetase in this strain was of the D form only. Petites also failed to accumulate glycogen (Chester, 1968).

Since utilization of trehalose and glycogen may well play a role in supporting the process of sporulation, the control over its breakdown is an important problem. At the present time our information is of a limited nature regarding both of these processes. Souza & Panek (1968) found that trehalase was located in a soluble fraction of the vegetative cells whereas trehalose was in the particulate fraction. The enzyme was present at all times during the cell cycle suggesting that compartmentalization occurs.

In yeast, glycogen is degraded by glycogen phosphorylase (for review see Preiss, 1969). The binding of glucose 1-phosphate (G1P) to phosphorylase is competitively inhibited by UDPG and G6P. These findings suggest that recycling of G1P to UDPG and to glycogen may be important in the overall regulation of glycogen degradation in yeast (Sagardia, Gotay & Rodriguez, 1971). Fosset, Muir, Nielson & Fischer (1971) have further characterized yeast phosphorylase. These workers separated two forms of the enzyme, a glycogen phosphorylase A, which was present as a phosphorylated dimer of the enzyme, molecular weight of 250 000, and a dephosphorylated B tetramer form of the enzyme with a molecular weight of 390 000. They further reported an ATP-dependent protein kinase which phosphorylates phosphorylase B and converted it to the A form. G6P was an effective inhibitor of both forms of the enzyme; the inhibition was non-competitive with both G1P and glycogen. In growing yeast cultures the high G6P levels observed during the periods of maximal glycogen synthesis (Rothman & Cabib, 1969) would inhibit phosphorylase B activity completely and lead to activation of glycogen synthetase. The inhibition by G6P could be overcome by conversion of phosphorylase B into the A form. Thus, G6P may play a dual regulatory role. The exact relations between the effectors, such as G6P and the metabolism of carbohydrate reserves are as yet not understood.

It seems reasonably clear that carbohydrate reserves must be mobilized to support carbon requirements for spore formation. An unravelling of the changing regulatory controls operative during both starvation and commitment to spore formation will be difficult and undoubtedly occupy the interest of researchers in this field for many years. The recent report of the effect of cyclic AMP on sporulation (N. Yanagishima, personal communication) is extremely interesting in this regard. They have shown that cyclic AMP reverses inhibition of sporulation by glucose. At the moment, only limited insights into regulation of carbohydrate utilization are available. Of most value to an understanding for spore formation is, (1) the metabolism of resting cells and (2) a comparison of the behaviour of sporogenic and asporogenic strains placed under sporulation conditions. These controls provide some basis for identifying the gross sequential events involved in cell metabolism which are associated with the process of spore formation.

LIPID SYNTHESIS

Cytologically fat globules are visible in vegetative cells by staining with Sudan Black B. Pontefract & Miller (1962) found that after stage II, the fat content started to increase and this increase continued to stage V. In early stage V, masses of fat deposits can be seen beside the maturing spore wall. Late in stage V, little fat can be seen in any of the epiplasm outside the spores but inside the spores there are small globules. Esposito *et al.* (1969) also observed the lipid synthesis during sporulation and found that there are two periods of lipid synthesis extending from stages I to III and then late in stage V.

R. F. Illingworth, A. H. Rose & A. Beckett (personal communication) studied the change in lipid composition in sporulating cells. They found that total lipids increased to four times the amount found in vegetative stationary cells. The increase could be accounted for primarily by synthesis of the triglycerides and sterol esters and to a smaller extent by phospholipids. The phospholipid and sterol compositions were relatively constant throughout sporulation, but there was a significant increase in the relative proportion of unsaturated fatty acids. These workers confirmed the finding of Esposito *et al.* (1969) that there are two major periods during which [^{14}C]acetate is incorporated into total lipids. There is, however, evidence that the first period of lipid synthesis after exposure to acetate sporulation medium is not specific to sporulation. S. Henry & H. O. Halvorson (unpublished results) found that the rate of incorporation of [^{14}C]acetate and ^{32}P into lipids during stages I through IV was virtually identical in both sporulating diploid cells and non-sporulating haploids. The relative proportion of the various lipid classes synthesized was comparable in both haploids and diploids. Haploids and diploids both underwent extensive vacuolization under these conditions. The haploids, however, did not initiate the second period of synthesis which begins at about the time the first ascospores are being formed (stage IV). The lipids synthesized in diploids at this time are primarily neutral lipids.

CAPACITY OF CELLS TO SPORULATE

Upon introduction into sporulation medium yeast cells have two possibilities: either to grow mitotically or to undergo sporulation. Not all cells in a population sporulate: cells which sporulate discontinue vegetative growth, whereas cells which fail to sporulate continue vegetative growth. Based on bud formation and enlargement, cells which sporulated ceased vegetative growth from the time they were introduced into sporulation medium whereas cells which did not produce ascospores exhibit limited vegetative growth (Haber & Halvorson, 1972). Mother and daughter cells differ in their sporogenic capacity. Although they appear to be nearly identical, mother and daughter cells can be differentiated by fluorescence staining of a chitin 'bud scar' found only on the mother cell (Yanagita, Yagisawa, Oishi, Sando & Suto, 1970a). Daughter cells (cells without a bud scar) sporulate poorly when compared with the sporulation of mother cells (Yanagita et al. 1970a). Sando & Miyake (1971) using cells from late stationary phase, found at the end of sporulation essentially only two types of cells: asci containing spores and unsporulated cells in which the nucleus was still in a premeiotic state. Recent studies (Haber & Halvorson, 1972) have shown that the capacity of a cell to sporulate was dependent on the stage of the vegetative cell cycle. Single cells recently formed from cell scission sporulated poorly, cells in the middle of the vegetative cell cycle had increased sporulation capacity, and cells in the last third of the cell cycle sporulated abundantly.

Haber & Halvorson (1972) postulated that sporulation capacity may be limited by the availability of one or more essential compounds which fluctuate periodically through a cell cycle. Yanagita, Oishi & Ishiko (1970b) found that respiratory activity was 1.3 times higher and polyphosphate content was almost three times greater in the adult cells than in the infant cells. Another plausible explanation for the cell cycle dependence of sporulation is the existence of specific functions which could be induced only during a limited period of the vegetative cell cycle. Another factor affecting the ability of cells to sporulate is their content of RNA and protein at the time they are introduced into sporulation medium (Croes, 1967b).

One prerequisite for maximum sporulation is the preadaptation of the cells to respiratory growth. Cells harvested during log-phase fermentative growth are least able to sporulate while those cells which have grown to stationary phase and have changed from fermentative to oxidative metabolism are fully capable of sporulation (Croes, 1967a, b;

Esposito *et al.* 1969). Cells growing in acetate medium are capable of sporulation when harvested in log phase (Roth & Halvorson, 1969).

To what extent is the process controlled by the expression of functions specific for the new specialized state? Most cellular modifications are initiated by changes in the levels of nutrients, biosynthetic intermediates and end products with pre-existing and competing enzyme pathways. Sporulation has been viewed as the product of a modified cell cycle under unbalanced growth in bacteria by Hitchins & Slepecky (1969) and in yeast (J. E. Haber & H. O. Halvorson, 1972, unpublished results).

INDUCTION OF SPORULATION

Now, as we have made sure that, in twice twenty-four hours time, animalcules appear, which have reached their full growth; and...we imagine that they have come forth from little round particles. (A. van Leeuwenhoek, letter 96.)

One of the least understood processes in sporulation is the mechanism whereby yeast cells which are potentially capable of sporulation are diverted from mitotic to meiotic growth. This induction step is influenced by a wide spectrum of environmental factors (for review see Fowell, 1969), the most prominent of which are the absence of nitrogen and the presence of a non-fermentable carbon source in the sporulation medium (Miller, 1963b).

From the studies of Miller (1963b) it is clear that the frequency of sporulation is low when both nitrogen and carbon sources are readily available for metabolism. Sporulation can be suppressed by the addition of rapidly metabolizable sugars such as glucose, fructose, mannose, sucrose and maltose but not by compounds which are incapable of supporting growth.

The addition of ammonium ions to the sporulating medium also inhibits sporulation although it does not prevent a large number of nuclei from entering the first stage of meiosis (Miller, 1963a; Miller & Hoffmann-Ostenhof, 1964). Gosling & Duggan (1971) demonstrated that ammonium ions inhibit both the increase in glyoxylate cycle enzymes and the adaptation of yeast to acetate utilization. Miller (1957) found that dihydroxyacetone as well as acetate was an effective carbon source for sporulation. Moreover, sporulation in dihydroxyacetone, unlike that in acetate, is not inhibited by ammonium sulphate. Croes (1967b) has proposed that the enhanced rate of protein synthesis following exposure to sporulation medium occurs in the absence of equivalent production of most amino acids. The glyoxylate cycle is required for the production of amino acids during acetate metabolism (Kornberg,

1965); the decreased amount of amino acids points to the insufficiency of this cycle during sporulation. The finding that the inhibition of vegetative growth on transfer to sporulation medium can be overcome by addition of a small amount of glyoxylate (Bettelheim & Gay, 1963) supports this hypothesis.

Exogenous acetate is not required throughout sporulation. Darland (1969) showed that cells could be transferred from acetate sporulation medium to distilled water at the end of stage I and still complete maximal sporulation. Exposure to acetate for as little as 5–10 min resulted in an eight-fold increase in sporulation above control exposed only to distilled water.

COMMITMENT

In the beginning...the earth was void and empty... (GENESIS i. 1)

Commitment in yeast is analogous to the more extensively studied problem in bacterial sporulation. One of the first reports was that of Saito (1916) in *Saccharomyces carlsbergensis* who found that when cells were exposed to sporulation conditions for a critical time they could no longer be converted back to mitotic growth by the addition of nutrients. In *Bacillus* it is clear that there is no single point of commitment for the whole sporulation process. As pointed out by Mandelstam (1971) 'the cells after initiation in a deficient medium become successively committed to one sporulation event after another'. It is not as yet clear whether the same phenomenon occurs in yeast sporulation.

In yeast sporulation inhibitory compounds differ in the stage in which commitment is expressed. For example, Darland (1969) found that about 7 h before the appearance of mature asci, cells can no longer be prevented from sporulation by the addition of glucose. Similarly, about 3–4 and 7–8 h before the appearance of mature asci, cells become insensitive to inhibition of sporulation by ammonium chloride and ethanol (A. J. S. Klar, personal communication; Miller, 1971) respectively. The primary limitation in these experiments is that they determine only the period in which sporulation sensitivity is lost. Kirsop (1954) and Ganesan, Holter & Roberts (1958) concluded that commitment occurred at about T_{10} (stage II). Transferring the cells to growth medium after this time did not stop sporulation. Environmental conditions such as pH affect the time of commitment, e.g. as shown by Fowell (1967) that at high pH value, cells are committed at T_5 (stage I).

GENETIC CONTROL OF SPORULATION

In heterothallic yeasts, the two alleles of the mating type locus, a and α, control mating as well as meiosis (Roman, Phillips & Sands, 1955). In order for a diploid to sporulate it must be heterozygous for the mating type alleles a and α. Diploids of the genotype α/α or a/a do not sporulate and are blocked in stage I (Roth & Halvorson, 1969; Roth & Lusnak, 1970; Roth, 1972). While a typical early increase in mass was observed in these cells, no indication of DNA replication, recombination or nuclear division was observed. Apparently, the earliest detectable event which is blocked in a/a or α/α strains is pre-meiotic DNA replication.

Further genetic analysis of sporulation in yeast has been hampered by some severe methodological limitations. First, the mutants had to be induced in a haploid strain and then converted into a homoallelic diploid to allow expression of the mutation. Second, since genetic analysis in yeast is accomplished by studying the meiotic products, mutants blocked in meiosis and sporulation could not be characterized genetically unless they were in some way conditional. Thirdly, the isolation and selection of sporulation deficient mutants blocked in a specific event in meiosis or ascospore formation was not possible since few sporulation specific components have been identified. Recently, however, several selection systems have been devised to overcome some of these difficulties and sporulation mutants blocked in the whole developmental sequence have been isolated in several yeasts. In *Schizosaccharomyces pombe* sporulation deficient mutants appear spontaneously at a frequency of 0.3 % (see Fincham & Day, p. 309). Esposito, Frink, Bernstein & Esposito (1972) calculated that 48 ± 27 loci are specifically involved in meiotic and sporulation function in *Saccharomyces cerevisiae*. Therefore, it is apparent that there are a number of unique genes governing sporulation which do not affect growth.

Bresch, Miller & Egel (1968) were the first to isolate mutants in sporulation and meiosis in *Schizosaccharomyces pombe*. In this yeast sporulation specific mutants can be obtained in homothallic strains permitting the expression of recessive alleles. Furthermore, asporogenic colonies can be distinguished from sporogenic clones by treatment with iodine vapor (Bresch *et al.* 1968). Over 300 mutants have been classed in 24 complementation groups, three of which are linked to each other and are adjacent to the mating locus. Thus far dominant mutations have not been isolated. Cytological studies, to distinguish the point of the defect in meiosis or sporulation, indicate that 4 genes are concerned with meiotic divisions while 18 are involved in later stages of ascospore development.

Esposito & Esposito (1969) isolated temperature-sensitive mutants of *Saccharomyces cerevisiae* which sporulated normally at the permissive temperature (25°) but were blocked in the formation of asci at the restrictive temperature (34°). Genetic analysis of sporulation specific mutants was carried out at the permissive temperature.

The mutations were originally produced in a homothallic strain of yeast carrying the diploidizing genes HO_α and HM_1. These genes cause the conversion of either the a or α locus to the opposite mating type. Haploid ascospores then fuse with their sister cells after a few mitotic divisions. The diploids formed are homoallelic at all loci except the mating type. Mutations were produced in the haploid ascospore and the diploids subsequently formed were tested for sporulation. Therefore, all mutations were homoallelic in the diploid state and were expressed whether dominant or recessive. From complementation tests with recessive mutants 11 separate cistrons have been identified (Esposito *et al.* 1972). All the recessive mutants complete either the first or second nuclear division as measured by the Giemsa stain. Three dominant sporulation mutants have been found but complementation tests were not reported. Two of the mutants do not complete nuclear division while the third remains in the mononucleate state. One mutant, Spo-98, produces aneuploid ascospores which fail to germinate at the permissive temperature (Esposito *et al.* 1972).

Relatively little biochemical information is available for these meiotic mutants. Of the mutants isolated by Esposito & Esposito (1969), only Spo-1, Spo-2 and Spo-3 have been studied. Spo-1 and Spo-2 do not increase their DNA to the parental level during sporulation. None of the mutants are defective in protein or RNA synthesis at 34 °C and all mutants retain normal activities for protein turnover (Esposito *et al.* 1970). Esposito *et al.* (1970) investigated the physiological effects of three temperature-sensitive mutations. From temperature shift experiments the temporal period of expression of the mutated function can be identified. The three mutants displayed different temperature-sensitive periods indicating that the specific mutational lesions affected different gene functions in sporulation.

Roth & Fogel (1971) devised a selection technique for isolating mutants deficient in meiotic recombination. Mutations of this kind could represent alterations in the synthesis of deoxynucleotide precursors, DNA polymerases or ligases. This method requires a parental strain disomic for chromosome III and heteroallelic for a locus on this

Table 3. *Classes of mutations possible during stages of meiosis and sporulation*

Stage	Types of mutations possible	Existing mutants
Vegetative growth	Altered vegetative function; sporulation aberrant or at a reduced frequency	*S. cerevisiae* petite mutations (Ephrussi, 1953) *S. cerevisiae* UV-sensitive mutants (Resnick, 1969) *S. cerevisiae* his 1 mutants (Littlewood & Fink, 1972)
I	Mutation affecting meiotic DNA replication	
II	Mutations preventing spindle plaque duplication or migration, microtubule formation pairing of homologous chromosomes, formation of synaptinemal complexes, meiotic recombination	*S. cerevisiae* rec mutants (Roth & Fogel, 1971) *S. cerevisiae* Spo 1–Spo 11 (Esposito *et al.* 1972) *S. pombe* me I–1–me I–4 (Bresch *et al.* 1968)
III	Mutations affecting nuclear or spindle elongation	*S. pombe* me II–1 (Bresch *et al.* 1968)
IV	Mutations affecting spindle plaque duplication, disappearance of meiosis I spindle, migration of spindle plaques, formation of meiosis II spindle, centromere separation, nuclear elongation 'prospore wall' and early spore wall formation	*S. pombe* Spo 1–Spo 10 (Bresch *et al.* 1968)
V	Mutations affecting maturation of ascospores resulting in spores with abnormal walls, metabolic deficiencies or poor germination properties	

chromosome (e.g. leu_{2-1}/leu_{2-2}). The disomic strain is mutagenized with ethyl methane sulfonate and the clones are tested for reduced interallelic recombination as measured by a reduction in leu^+ colonies. Roth & Fogel (1971) isolated over 91 mutants apparently blocked in early meiotic events.

A summary of the meiotic and sporulation pathway and the stage at which various yeast mutants are blocked is shown in Table 3. The majority of sporulation-deficient mutants in *Schizosaccharomyces pombe* are blocked in ascospore development (Bresch *et al.* 1968), while the most frequent class in *Saccharomyces cerevisiae* correspond to mutants which are unable to complete meiosis. This may be due to a bias in isolation of mutant colonies. In *S. cerevisiae* mutants are usually classified as non-sporulating strains by the microscopic determination of the presence or absence of asci. This procedure may leave undetected a group of mutants in the later stages of ascospore maturation (stage V). In *S. pombe* iodine vapour can be used to distinguish between wild type and mutant colonies. This technique may allow the detection of stage V mutants.

In addition to the biochemical information ultrastructural studies may also be employed to characterize the altered functions in the mutant strains. Although individual chromosomes cannot be seen in yeast, recent technological advances in staining and sectioning of sporulating yeast cells have led to the definition of meiotic spindles, spindle plaques, microtubules, nucleolus and other structures associated with ascospore formation (Moens, 1971; Peterson *et al.* 1972). Impairments in any of the processes necessary for chromosome separation should result in sporulation deficiencies. For example Spo-3, which remains primarily mononucleate and expresses its temperature-sensitive function halfway through the meiotic cycle (Esposito *et al.* 1970), exhibits abnormal meiosis (Plate 3). R. Gray (personal communication) compared thin sections of wild type and Spo-3 cells at various stages of sporulation at the non-permissive temperature. Meiosis in Spo-3 appears normal for at least 8 h and then remains arrested in stage II. In some cases the spindle plaques remain adjacent rather than migrating to face each other (Fig. 1). Also evident in the mutant was the unusual number of parallel tubule structures ranging from 60 Å to 80 Å in diameter and in parallel bundles up to 2000 Å in width.

ALTERNATIVE GENETIC APPROACHES

It is difficult to determine the deficiency of a particular mutant unless one selects for an alteration in a known function. Since the mutants which have been isolated thus far are pleiotrophic and affect the entire complex developmental sequence, assignment of a definite function to each mutant gene may prove difficult. Even when a missing function can be identified there is no guarantee that the structural gene for that function is mutated.

In this section we suggest some alternative genetic approaches for sporulation studies.

Dominant mutations

Of all the sporulation specific mutants obtained only three have been shown to be dominant (Esposito *et al.* 1972). Yet these mutations are more likely to affect regulatory or structural components. Dominant mutants may also display altered transcriptional and translational controls or they may result in subunit alterations in the structures of membranes, spindles and microtubules. Furthermore, such mutations could be isolated easily since they can be induced in diploids and selected directly by their ability to sporulate at a restrictive temperature.

Conditional mutants which are resistant to inhibitors or which escape suppression of sporulation

Erythromycin specifically inhibits mitochondrial protein synthesis in vegetatively growing yeast (Lamb, Clark-Walker & Linnane, 1968). Recently Puglisi & Zennaro (1971) have shown that sporulation is also inhibited by erythromycin. The isolation of erythromycin-resistant conditional mutants specifically affecting sporulation would be of interest in determining the mitochondrial contribution to sporulation.

Sporulating cultures of yeast are sensitive to exogenously added glucose, or ammonium ion or cyclic AMP (Miller, 1957; Miller, 1963b; N. Yanagishima, personal communication). Mutants which escape such inhibition may also increase our knowledge of the metabolic controls governing sporulation. N. S. Grewal & J. Miller (personal communication) found some strains which produce diploid spores preferentially and also escape inhibition by glucose at concentrations up to 1 %.

In animal cells, colchicine inhibits cell division directly by binding to subunits of the mitotic spindle (Borisy & Taylor, 1967). In the fission yeast *Schizosaccharomyces pombe* colchemid has been shown to arrest vegetative cell growth, but only at very high concentrations compared to mammalian cells (5×10^{-3} M) (Lederberg & Stetten, 1970). Mutants resistant to colchemid were capable of ascospore formation in colchemid. Under restricted vegetative growth conditions *Saccharomyces cerevisiae* has also been shown to be inhibited by colchemid but again only at high concentrations (1–10 mM) (J. E. Haber, J. Peloquin, H. O. Halvorson & G. S. Borisy, unpublished results). Binding studies with labelled colchemid suggest that the drug's targets were microtubule subunits. Isolation of colchicine and colchemid-sensitive yeast strains would be of great value in searching for any possible sporulation-specific mutants.

Strains producing aberrant asci

Although appearance of some two-spored asci in sporulating cultures is affected by the age of the cells in the division cycle (Haber & Halvorson, 1972), the mechanism of this regulation is not understood. N. S. Grewal & J. Miller (personal communication) have recently described a strain of *Saccharomyces cerevisiae* which typically forms two-spored asci. Genetic and biochemical analysis of these asci may serve to increase our knowledge on the meiotic process itself, particularly in regard to nuclear spindle orientation.

Asci containing more than four spores are occasionally observed in sporulating cultures. Whether they arise from an altered meiotic cycle or post-meiotic vegetative replication is not as yet clear. A particularly interesting example was observed by B. Littlewood & C. R. Fink (personal communication) who discovered that some *Saccharomyces cerevisiae* strains which contain mutations near or at the feed-back site in hi_1 gene also produce multiple-spored (8–16) asci. From genetic and cytological studies they have shown that this results from a post-meiotic division. This mutation thus appears to affect both meiotic division and the number of nuclei per diploid cell.

Altered ascospores

A different approach will be required to isolate stage V mutants. Conditional mutants could be isolated in which the ascospores formed at a restrictive temperature were more sensitive or resistant to osmotic pressure, heat or to the action of alcohol. Such mutants should include those with altered spore wall components. Similarly, mutants which fail to germinate could also reflect modifications in the spore wall.

Mutants defective in a known vegetative cell function which also display reduced sporulation capacity

In yeast, mutants in specific *vegetative* cell functions or growth often display reduced sporulation when crossed with a wild-type strain. In most cases this reduction is the result of a secondary mutation(s) which can be eliminated by repeated outcrossing with wild-type strains. However, some mutants still display reduced sporulation even after extensive outcrossing. Diploids heterozygous for the mutant phenotype show usually reduced sporulation but the effect is usually more marked in homoallelic diploids.

A frequent example of this type of mutation is the petite mutation in yeast. Diploids must be respiratory-sufficient to go through meiosis. Thus, diploids homozygous for a petite character do not form asci while diploids heterozygous for the wild-type and petite characters are able to sporulate (Ephrussi, 1953).

Radiation-sensitive mutants provide a second example. A number of mutants sensitive to ultraviolet light and ionizing radiation during vegetative growth have been isolated (Cox & Parry, 1968; Resnick, 1969; Puglisi, 1967). In diploids homozygous for these mutations, sporulation ability was significantly reduced. These mutants may be blocked in recombination and thus unable to proceed through meiosis. Comparable types of mutants have been found in *Neurospora crassa*, where the lysine-5 mutation prevents the normal ripening of ascospores carrying it (Stadler, 1956).

Analysis of mutants with known biochemical defects may be more informative than analysis of sporulation-specific mutants in which it is difficult to determine the primary effects of the respective mutations. For example, an enzyme may have a different function or stability during vegetative growth than during sporulation. Also, sporulation may be more sensitive to a particular modification in type or amount of a given enzyme. Mutations in enzymes which are involved in catabolism and/or required for the metabolism of endogenous substrates could be more detrimental for endotrophic sporulation than for vegetative growth. A mutation which reduces enzyme activity may also alter the rate of turnover of an enzyme. Even though sporulation is characterized by increased proteolytic activity relatively little is known about selective digestion of proteins during mitosis or meiosis. Chen & Miller (1968) have described a strain with low protease activity which fails to sporulate; however, a genetic analysis correlating the two effects was not undertaken.

THIS AND THAT

When thou hast done; Thou has not done. For I have more.
(JOHN DONNE, 'A Hymn to God the Father'.)

As is evident in this review, sporulation in yeast is complex and involves the overlapping events of starvation, meiosis and ascospore formation. These parallel processes make it difficult to design genetic and biological experiments on individual stages. This difficulty can be minimized if the overall developmental sequence can be biologically dissected. That is, if the processes of starvation, meiosis and spore formation can each be studied independently. There are numerous examples in the literature which suggest that this separation is possible in yeast. First, ascospores are sometimes observed in cultures growing in complete medium and in chemostat cultures. Although this has not received much attention in yeast, a similar phenomenon has been well studied in bacteria (Dawes & Mandelstam, 1970). Second, spore formation, in the absence of meiosis, would generate asci containing diploid spores. The recent report from N. S. Grewal & J. Miller (unpublished) of three strains producing asci containing exclusively two spores with a diploid DNA content suggests this possibility. Third, the complete separation of meiosis and spore formation has not yet been accomplished. The behaviour of some of the temperature-sensitive mutants reported in this review, as well as the characterization of strains in which meiosis is followed by a number of mitotic divisions (B. Littlewood & C. R. Fink, personal communication) hold promise that this separation may be biologically possible.

We would like to thank Drs R. Gray, P. Rousseau and E. Sena for permission to quote unpublished data.

This work was supported by a Public Health Service grant Al-1459, a National Science Foundation grant B-1750 and by a National Institute of Health Postdoctoral Fellowship held by one of us (S.A.H.).

REFERENCES

ABDEL-WABAB, M. F., MILLER, J. J., GABRIEL, O. & HOFFMANN-OSTENHOF, O. (1961). Zur Kenntnis des Phosphatstoffwechsels der Hefe, V. Mitt: Weitare Untersuchungen über freie Nucleotide in Saureeutraden aus verschiedenen Stoffwechselbedingungen unterworfener Hefe. *Monatshefte für Chemie und verwandte Teile anderer Wissenschaften*, **92**, 22–30.

ADAMS, A. M. & MILLER, J. J. (1954). Effect of gaseous environment and temperature on ascospore formation in *Saccharomyces cerevisiae* Hansen. *Canadian Journal of Botany*, **32**, 320–34.

BETTELHEIM, K. A. & GAY, J. L. (1963). Acetate-glyoxylate medium for sporulation of *Saccharomyces cerevisiae*. *Journal of Applied Bacteriology*, **26**, 224–31.

BLACK, S. H. & GORMAN, C. (1971). The cytology of *Hansenula*. III. Nuclear segregation and envelopment during ascosporogenesis in *Hansenula wingei*. *Archiv für Mikrobologie*, **79**, 231–48.

BORISY, G. G. & TAYLOR, E. W. (1967). The mechanism of action of colchicine. Colchicine binding to sea urchin eggs and the mitotic apparatus. *Journal of Cell Biology*, **34**, 535–48.

BRESCH, C., MILLER, G. & EGEL, R. (1968). Genes involved in meiosis and sporulation of a yeast. *Molecular and General Genetics*, **102**, 301–6.

BRIGHT, T. B., DIXON, P. A. & WHYMPER, J. W. T. (1949). Effect of ethyl alcohol and carbon dioxide on sporulation of baker's yeast. *Nature, London*, **164**, 544.

CHEN, A. W. & MILLER, J. J. (1968). Proteolytic activity of intact yeast cells during sporulation. *Canadian Journal of Microbiology*, **14**, 957–63.

CHESTER, V. E. (1968). Heritable glycogen storage deficiency in yeast and its induction by ultraviolet light. *Journal of General Microbiology*, **51**, 49–56.

CHESTER, V. E. & BYRNE, M. J. (1968). Carbohydrate composition and UDP-glucose concentration in a normal yeast and a mutant deficient in glycogen. *Archives of Biochemistry and Biophysics*, **127**, 556–62.

COTTRELLE, S. F. & AVERS, C. J. (1970). Evidence for mitochondrial synchrony in synchronous cell cultures of yeast. *Biochemical and Biophysical Research Communications*, **38**, 973–80.

COX, B. S. & PARRY, J. M. (1968). The isolation, genetics and survival characteristics of ultraviolet light-sensitive mutants in yeast. *Mutation Research*, **6**, 37–55.

CROES, A. F. (1966). Duplication of DNA during meiosis in baker's yeast. *Experimental Cell Research*, **41**, 452–4.

CROES, A. F. (1967a). Induction of meiosis in yeast. I. Timing of cytological and biochemical events. *Planta, Berlin*, **76**, 209–26.

CROES, A. F. (1967b). Induction of meiosis in yeast. II. Metabolic factors leading to meiosis. *Planta, Berlin*, **76**, 227–37.

DARLAND, G. K. (1969). The physiology of sporulation in *Saccharomyces cerevisiae*. Ph.D. Thesis, University of Washington.

DAWES, I. W. & MANDELSTAM, J. (1970). Sporulation of *Bacillus subtilis* in continuous culture. *Journal of Bacteriology*, **103**, 529–35.

DOI, E., HAYASHI, R. & HATA, T. (1967). Purification of yeast proteinases. II. Purification and some properties of yeast proteinase C. *Agricultural and Biological Chemistry*, **31**, 160–9.

EATON, R. E. (1960). Endogenous respiration of yeast. I. The endogenous substrate. *Archives of Biochemistry and Biophysics*, **88**, 17–25.

ENGELS, F. & CROES, A. (1968). The synaptinemal complex in yeast. *Chromosoma, Berlin*, **25**, 104–6.

EPHRUSSI, B. (1953). *Nucleo-cytoplasmic Relations in Microorganisms*. Oxford: Clarendon Press.

ESPOSITO, M. S. & ESPOSITO, R. E. (1969). The genetic control of sporulation in *Saccharomyces*. I. The isolation of temperature-sensitive sporulation-deficient mutants. *Genetics*, **61**, 79–89.

ESPOSITO, M. S., ESPOSITO, R. E., ARNAUD, M. & HALVORSON, H. O. (1969). Acetate utilization and macromolecular synthesis during sporulation of yeast. *Journal of Bacteriology*, **100**, 180–6.

ESPOSITO, M. S., ESPOSITO, R. E., ARNAUD, M. & HALVORSON, H. O. (1970). Conditional mutants in meiosis in yeast. *Journal of Bacteriology*, **104**, 202–10.

ESPOSITO, R. E., FRINK, N., BERNSTEIN, P. & ESPOSITO, M. S. (1972). The genetic control of sporulation in *Saccharomyces*. II. Dominance and complementation of mutants of meiosis and spore formation. *Molecular and General Genetics*, **114**, 241–8.

FINCHAM, J. R. S. & DAY, P. R. (1971). In *Fungal Genetics*, 3rd ed. *Botanical Monographs*, vol. 4, p. 39. Oxford and Edinburgh: Blackwell.

FOSSET, M., MUIR, L. W., NIELSEN, L. D. & FISCHER, E. H. (1971). Purification and properties of yeast glycogen phosphorylase a and b. *Biochemistry*, **10**, 4105–13.

FOWELL, R. R. (1967). Factors controlling the sporulation of yeasts. II. The sporulation phase. *Journal of Applied Bacteriology*, **30**, 450–74.

FOWELL, R. R. (1969). Sporulation and hybridization of yeasts. In *The Yeasts*, vol. 1, *Biology of Yeasts*, ed. A. H. Rose & J. S. Harrison, pp. 303–83. London and New York: Academic Press.

GALZY, P. & BIZEAU, C. (1966). Action de l'acide acétique sur la sporulation de *Saccharomyces cerevisiae* Hansen. *Comptes Rendus de la Société de Biologie*, **160**, 176–8.

GANESAN, A. T., HOLTER, H. & ROBERTS, C. (1958). Some observations on sporulation in *Saccharomyces*. *Compte rendu des travaux du Laboratoire de Carlsberg*, **13**, 1–6.

GOSLING, J. P. & DUGGAN, P. F. (1971). Activities of tricarboxylic acid cycle enzymes, glyoxylate cycle enzymes and fructose disphosphate in baker's yeast during adaptation to acetate oxidation. *Journal of Bacteriology*, **106**, 908–14.

GUTH, E., HASHIMOTO, T. & CONTI, S. (1972). Morphogenesis of ascospores in *Saccharomyces cerevisiae*. *Journal of Bacteriology*, **109**, 869–80.

HABER, J. E. & HALVORSON, H. O. (1972). Cell cycle dependency of sporulation in *Saccharomyces cerevisiae*. *Journal of Bacteriology*, **109**, 1027–33.

HALVORSON, H. O. (1958a). Intracellular protein and nucleic acid turnover in resting yeast cells. *Biochimica et Biophysica Acta*, **27**, 255–66.

HALVORSON, H. O. (1958b). Studies on protein and nucleic acid turnover in growing cultures of yeast. *Biochimica et Biophysica Acta*, **27**, 267–76.

HALVORSON, H. O. (1962). In *The Function and Control of Intracellular Protein Turnover in Microorganisms in Amino Acid Pools*, ed. J. T. Holden, pp. 633–45. Amsterdam, London and New York: Elsevier.

HASHIMOTO, T., CONTI, S. & NAYLOR, H. (1958). Fine structure of microorganisms. III. Electron microscopy of resting and germinating ascospores of *Saccharomyces cerevisiae*. *Journal of Bacteriology*, **76**, 406–16.

HASHIMOTO, T., GERHARDT, P., CONTI, S. & NAYLOR, H. (1960). Studies on the fine structure of microorganisms. V. Morphogenesis of nuclear and membrane structures during ascospore formation in yeast. *Journal of Biophysical and Biochemical Cytology*, **7**, 305–17.

HAYASHI, R., OKA, Y., DOI, E. & HATA, T. (1968a). Activation of intracellular proteinases of yeast. I. Occurrences of inactive precursors of proteinases B and C and their activation. *Agricultural and Biological Chemistry*, **32**, 359–66.

HAYASHI, R., OKA, Y., DOI, E. & HATA, T. (1968b). Activation of intracellular proteinases of yeast. II. Activation and some properties of pro-proteinase C. *Agricultural and Biological Chemistry*, **32**, 367–73.

HITCHINS, A. D. & SLEPECKY, R. A. (1969). Bacterial sporulation as a modified procaryotic cell division. *Nature, London*, **223**, 804–7.

INDGE, K. J. (1968). Polyphosphates of the yeast cell vacuole. *Journal of General Microbiology*, **51**, 447–55.

KADOWAKI, K. & HALVORSON, H. O. (1971a). Appearance of a new species of ribonucleic acid during sporulation in *Saccharomyces cerevisiae*. *Journal of Bacteriology*, **105**, 826–30.

KADOWAKI, K. & HALVORSON, H. O. (1971b). Isolation and properties of a new species of ribonucleic acid synthesized in sporulating cells of *Saccharomyces cerevisiae*. *Journal of Bacteriology*, **105**, 831–6.

KANE, S. M., ROTH, R. M. & ERWIN, J. A. (1972). Polysaccharide metabolism during yeast sporulation. *Abstracts of the Annual Meeting of the American Society for Microbiology*, p. 184.

KIRSOP, B. H. (1954). Studies in yeast sporulation. I. Some factors influencing sporulation. *Journal of the Institute of Brewery*, **60**, 393–6.

KORNBERG, H. L. (1965). The co-ordination of metabolic routes. In *Function and Structure in Microorganisms. Symposium of the Society of General Microbiology*, **15**, 343–68.

KUENZI, M. T. & FIECHTER, A. (1972). Regulations of carbohydrate composition of *Saccharomyces cerevisiae* under growth limitation. *Archiv für Mikrobiologie*, **84**, 254–65.

LAMB, A. J., CLARK-WALKER, G. D. & LINNANE, A. W. (1968). The biogenesis of mitochondria. 4. The differentiation of mitochondrial and cytoplasmic protein synthesizing systems *in vitro* by antibiotics. *Biochimica et Biophysica Acta*, **161**, 415–27.

LEDERBERG, S. & STETTEN, G. (1970). Colchemid sensitivity of fission yeast and the isolation of colchemid-resistant mutants. *Science, New York*, **168**, 485–7.

LYNN, R. R. & MAGEE, P. T. (1970). Development of the spore wall during ascospore formation in *Saccharomyces cerevisiae*. *Journal of Cell Biology*, **44**, 688–92.

MCCULLY, E. & ROBINOW, C. (1971). Mitosis in the fission yeast *Schizosaccharomyces pombe*: A comparative study with light and electron microscopy. *Journal of Cell Science*, **9**, 475–507.

MANDELSTAM, J. (1960). The intracellular turnover of protein and nucleic acids and its role in biochemical differentiation. *Bacteriological Reviews*, **24**, 289–308.

MANDELSTAM, J. (1971). Recurring patterns during development in primitive organisms. In *Control Mechanisms of Growth and Differentiation. Symposia of the Society for Experimental Biology*, **25**, 1–26.

PLATE 1

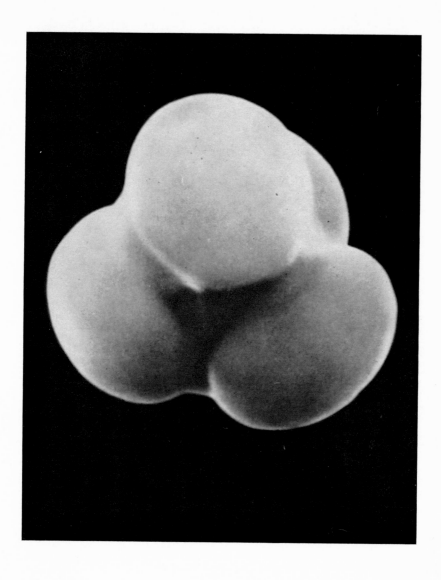

(*Facing p. 240*)

PLATE 2

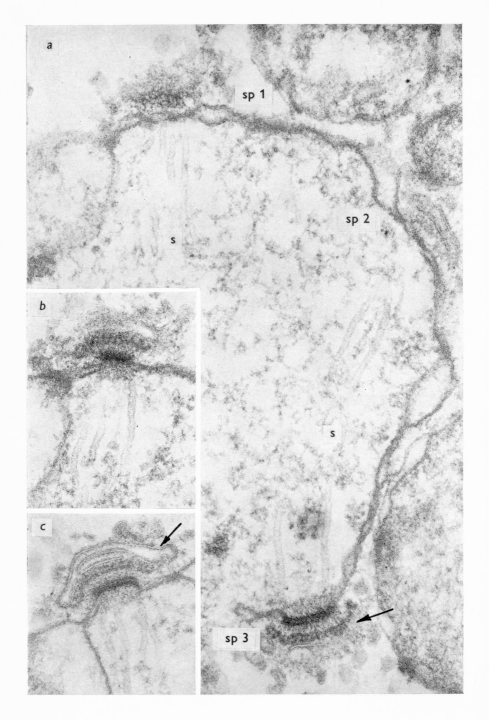

PLATE 3

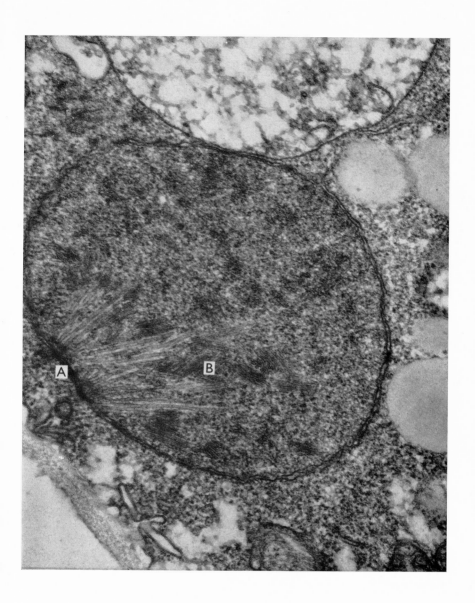

MANNERS, D. J. (1971). The structure and biosynthesis of storage carbohydrates in yeasts. In *The Yeasts*, vol. 2, ed. A. H. Rose & J. S. Harrison, pp. 419–40. London and New York: Academic Press.

MARQUARDT, H. (1963). Elektronenoptische Untersuchungen über die Ascosporenbildung bei *Saccharomyces cerevisiae* unter cytologischem und cytogenetischen Aspelet. *Archiv für Mikrobiologie*, 46, 308–20.

MATILE, P., CORTAT, M., WIEMKEN, A. & FREY-WYSLING, A. (1971). Isolation of glucanase-containing particles from budding *Saccharomyces cerevisiae*. *Proceedings of National Academy of Sciences, U.S.A.* 68, 636–40.

MATILE, P. & WIEMKEN, A. (1967). The vacuole as the lysosome of the yeast cell. *Archiv für Mikrobiologie*, 56, 148–55.

MILLER, J. J. (1957). The metabolism of yeast sporulation. II. Stimulation and inhibition by monosaccharides. *Canadian Journal of Microbiology*, 3, 81–90.

MILLER, J. J. (1963a). Determination by ammonium of the manner of yeast nuclear division. *Nature, London*, 196, 214–15.

MILLER, J. J. (1963b). The metabolism of yeast sporulation. V. Stimulation and inhibition of sporulation and growth by nitrogen compounds. *Canadian Journal of Microbiology*, 9, 259–77.

MILLER, J. J. (1971). Some recent observations on spore formation and germination in *Saccharomyces*. *Spectrum*, 1, 73–9.

MILLER, J. J. & HOFFMANN-OSTENHOF, O. (1964). Spore formation and germination in *Saccharomyces*. *Allegemeine Mikrobiologie, Morphologie, Physiologie und Oekologie de Mikroorganismen*, 4, 273–95.

MIYAKE, S., SANDO, N. & SATO, S. (1971). Biochemical changes in yeast during sporulation. II. Acetate metabolism. *Development, Growth and Differentiation*, 12, 285–95.

MOENS, P. (1971). Fine structure of ascospore development in the yeast *Saccharomyces cerevisiae*. *Canadian Journal of Microbiology*, 17, 507–10.

MOENS, P. & RAPPORT, E. (1971a). Spindles, spindle plaques and meiosis in the yeast *Saccharomyces cerevisiae* (Hansen). *Journal of Cell Biology*, 50, 344–61.

MOENS, P. B. & RAPPORT, E. (1971b). Synaptic structures in the nuclei of sporulating yeast, *Saccharomyces cerevisiae* (Hansen). *Journal of Cell Science*, 9, 665–77.

MONRO, R. E. (1961). Protein turnover and the formation of protein inclusions during sporulation of *Bacillus thuringensis*. *Biochemical Journal*, 81, 225.

MOOR, H. (1967). Endoplasmic reticulum as the initiator of bud formation in yeast. *Archiv für Mikrobiologie*, 57, 135–46.

MUNDKUR, B. (1961). Electron microscopical studies of frozen-dried yeast. III. Formation of the tetrad in *Saccharomyces*. *Experimental Cell Research*, 25, 24–40.

OSUMI, M., SANDO, N. & MIYAKE, S. (1966). Morphological changes in yeast cell during sporogenesis. *Japan Women's University Journal*, 13, 70–9.

PAZONYI, B. & MARKUS, L. (1955). The fractionation and measurement of carbohydrates in relation to the life cycle of yeast. *Agrokemia es Talajtan, Budapest*, 4, 225–35.

PETERSON, J., GRAY, R. & RIS, H. (1972). Meiotic spindle plaques in *Saccharomyces cerevisiae*. *Journal of Cell Biology*, 53, 837–41.

PHAFF, H. J., MILLER, M. W. & MRAK, E. M. (1966). *The Life of Yeasts*. Cambridge, Mass.: Harvard University Press.

PONTEFRACT, R. D. & MILLER, J. J. (1962). The metabolism of yeast sporulation. IV. Cytological and physiological changes in sporulating cells. *Canadian Journal of Microbiology*, 8, 573–84.

PREISS, J. (1969). The regulation of the biosynthesis of α-1,4-glucans in bacteria and plants. In *Current Topics in Cellular Regulation*, vol. 1, ed. B. L. Horecker & E. R. Stadtman, pp. 125–60. New York: Academic Press.

PUGLISI, P. P. (1967). Genetic control of radiation sensitivity in yeast. *Radiation Research*, **31**, 856–66.

PUGLISI, P. P. & ZENNARO, E. (1971). Erythromycin inhibition of sporulation in *Saccharomyces cerevisiae*. *Experientia*, **27**, 963–4.

RAMIREZ, C. & MILLER, J. J. (1964). The metabolism of yeast sporulation. VI. Changes in amino acid content during sporogenesis. *Canadian Journal of Microbiology*, **10**, 623–31.

RESNICK, M. A. (1969). Genetic control of radiation sensitivity in *Saccharomyces cerevisiae*. *Genetics*, **62**, 519–31.

ROMAN, H., PHILLIPS, M. M. & SANDS, S. M. (1955). Studies on polyploid *Saccharomyces cerevisiae*. I. Tetraploid segregation. *Genetics*, **40**, 546–61.

ROTH, R. (1970). Carbohydrate accumulation during the sporulation of yeast. *Journal of Bacteriology*, **101**, 53–7.

ROTH, R. (1972). Coordinate genetic control of replication and recombination during meiosis in yeast. In *Abstracts of the Annual Meeting of the American Society for Microbiology*, p. 73.

ROTH, R. & FOGEL, S. (1971). A system selective for yeast mutants deficient in meiotic recombination. *Molecular and General Genetics*, **112**, 295–305.

ROTH, R. & HALVORSON, H. O. (1969). Sporulation of yeast harvested during logarithmic growth. *Journal of Bacteriology*, **98**, 831–2.

ROTH, R. & LUSNAK, K. (1970). DNA synthesis during yeast sporulation: genetic control of an early development event. *Science, London*, **168**, 493–4.

ROTHMAN, L. B. & CABIB, E. (1969). Regulation of glycogen synthesis in the intact yeast cell. *Biochemistry*, **8**, 3332–41.

ROTHMAN, L. B. & CABIB, E. (1970). Two forms of yeast glycogen synthetase and their role in glycogen accumulation. *Proceedings of National Academy of Sciences, U.S.A.* **66**, 967–74.

ROUSSEAU, P. (1972). Germination of yeast spores in *Saccharomyces cerevisiae*. Ph.D. Thesis, University of Wisconsin.

ROUSSEAU, P., HALVORSON, H. O., BULLA, L. A. & JULIAN, G. S. (1972). Germination and outgrowth of single spores of *Saccharomyces cerevisiae* viewed by scanning electron and phase microscopy. *Journal of Bacteriology*, **109**, 1232–8.

SAGARDIA, F., GOTAY, I. & RODRIGUEZ, M. (1971). Control properties of yeast glycogen phosphorylase. *Biochemical and Biophysical Research Communications*, **42**, 829–35.

SAITO, K. (1916). Chemical conditions for the development of the reproductive organs of some yeasts. *Journal of the College of Science, Imperial University of Tokyo*, **39**, 1–70.

SANDO, N. & MIYAKE, S. (1971). Biochemical changes in yeast during sporulation. I. Fate of nucleic acids and related compounds. *Development, Growth and Differentiation*, **12**, 273–83.

SENA, E. P. (1972). DNA replication in yeast. Ph.D. thesis, University of Wisconsin.

SENTANDREU, R. & NORTHCOTE, D. H. (1969). The formation of buds in yeast. *Journal of General Microbiology*, **55**, 393–8.

SMITH, D., TAURO, P., SCHWEIZER, E. & HALVORSON, H. O. (1968). The replication of mitochondrial DNA during the cell cycle in *Saccharomyces lactis*. *Proceedings of the National Academy of Sciences, U.S.A.* **60**, 936–42.

SNIDER, I. J. & MILLER, J. J. (1966). A serological comparison of vegetative cell and ascus walls and the spore coat of *Saccharomyces cerevisiae*. *Canadian Journal of Microbiology*, **12**, 485–8.

SOGIN, S. J. & HABER, J. E. (1972). Identification of the nature of RNA in *Saccharomyces cerevisiae*. *Federation of American Societies for Experimental Biology*, **31**, 1356.

SOUZA, N. O. & PANEK, A. D. (1968). Location of trehalase and trehalose in yeast cells. *Archives of Biochemistry*, **125**, 22–8.

STADLER, D. T. (1956). A map of linkage group XI of *Neurospora*. *Genetics*, **41**, 528–43.

STOCK, D. & BLACK, S. (1970). Meiosis in *Hansenula holstii* and *Hansenula wingei*. *Journal of General Microbiology*, **64**, 365–72.

STRASDINE, G. A. (1972). The role of intracellular glucan in endogenous fermentation and spore maturation in *Clostridium botulinum*. *Canadian Journal of Microbiology*, **18**, 211–17.

SUOMALAINEN, H. & OURA, E. (1971). Yeast nutrition and solute uptake. In *The Yeasts*, vol. 2, ed. A. H. Rose & J. S. Harrison, pp. 3–74. New York and London: Academic Press.

SVIHLA, G., DAINKO, J. L. & SCHLENK, F. (1964). Ultraviolet microscopy of the vacuole of *Saccharomyces cerevisiae* during sporulation. *Journal of Bacteriology*, **88**, 449–56.

TAMAKI, H. (1965). Chromosome behavior at meiosis in *Saccharomyces cerevisiae*. *Journal of General Microbiology*, **41**, 93–8.

VEZINHET, F., ARNAUD, A. & GALZY, P. (1971). Étude de la respiration de la levure au course de la sporulation. *Canadian Journal of Microbiology*, **17**, 1179–84.

WIEMKEN, A., MATILE, PH. & MOOR, H. (1970). Vacuolar dynamics in synchronously budding yeast. *Archiv für Mikrobiologie*, **70**, 89–103.

WILKINSON, J. F. (1959). The problem of energy storage compounds in bacteria. *Experimental Cell Research*, supplement 7, 111–30.

WILLIAMSON, D. H. & MOUSTACCHI, E. (1971). The synthesis of mitochondrial DNA during the cell cycle in the yeast *Saccharomyces cerevisiae*. *Biochemical and Biophysical Research Communications*, **42**, 195–201.

YANAGITA, T., YAGISAWA, M., OISHI, S., SANDO, N. & SUTO, T. (1970a). Sporogenic activities of mother and daughter cells in *Saccharomyces cerevisiae*. *Journal of General Applied Microbiology*, **16**, 347–50.

YANAGITA, T., OISHI, S. & ISHIKO, S. (1970b). Sporulation in relation to yeast cells in relation to their 'division age'. *Meeting on Genetic and Biochemical Regulation of Dormancy, Kyoto, Japan*.

EXPLANATION OF PLATES

PLATE 1

Scanning electron micrographs (× 5500) of an ascus of *Saccharomyces cerevisiae* containing four ascospores. (Rousseau *et al*. 1972.)

PLATE 2

(*a–c*) Meiosis II plaques in profile. (*a*) Three of the four plaques present in a meiosis II nucleus are visible in this section. The components of spindle plaque 3 (sp 3) are most clearly seen. The outer zone of the plaque (arrow) is larger than in meiosis I and it is noticeably denser than the inner zone. The material comprising the region of low electron opacity between the outer and central zones of this plaque, as well as the outer zone itself, appears to be oriented in parallel arrays perpendicular to the surface of the nucleus. Only edges of spindle plaques 1 (sp 1) and 2 (sp 2) are viewed in this section. Microtubules comprising both spindles (s) are present (× 73000). (*b*, *c*) Spindle plaque 1 as observed in the next two sections (× 73000). Flattened, membranous vesicles (arrow) pressed to the outer zone of the plaque can be observed in (*c*). (Peterson *et al*. 1972.)

PLATE 3

The nucleus of a cell of *Saccharomyces cerevisiae* mutant Spo-3 which has been exposed to sporulation medium for 23 h at the restrictive temperature. The meiosis spindle plaque (*A*) has duplicated but has failed to migrate. Bundles of microfibrils (*B*) can be seen in the nucleoplasm. (× 51000). (Gray, unpublished data.)

FUNDAMENTAL ASPECTS OF
HYPHAL MORPHOGENESIS

S. BARTNICKI-GARCIA

Department of Plant Pathology, University of California,
Riverside, California 92502

The key to the hypha lies in the apex. (N. F. ROBERTSON.)

INTRODUCTION

This article is, first, a synopsis of certain aspects of the structure and function of fungal hyphae which may have a bearing on hyphal morphogenesis and, second, a speculative excursion into the mechanism of apical growth. This is an attempt to understand how a hyphal tube is produced and the emphasis is therefore on apical growth – the process by which a fungal hypha elongates. For reviews on various aspects of hyphal growth relevant to this article see Robertson (1961, 1965a, b, 1968), Park & Robinson (1966a), Burnett (1968), Green (1969), Larpent (1970) and Brody (1972).

For the most part, hyphal morphology is dictated by the shape of the hyphal wall, consequently we will be primarily concerned with understanding the apical genesis of a cylindrical wall. For simplicity, we shall assume that the apical dome wall is hemispherical and that the lateral walls are cylindrical (Fig. 1). Almost always hyphae depart somewhat from this shape and have a more pointed dome, slightly conical lateral walls, and a more or less meandering longitudinal axis.

The fact that fungal hyphae and sporangiophores grow apically has long been established by various techniques including application of external markers (Burgeff, 1915; Castle, 1958; Grove, Bracker & Morré, 1970), measurement of distances between tips, septa and branches (Smith, 1923; Butler, 1961) and the elegant method of staining with fluorescent antibodies (Marchant & Smith, 1968). None of these studies, however, could establish whether the structural elements of the wall were actually inserted at the apex. In fact, indirect evidence, based on the response of hyphal tips to experimental manipulation, led to the suggestion that maximum wall synthesis occurred, not at the very tip, but in an annular band located immediately behind a non-extensible apical cap (Burnett, 1968). More recently, however, it has been conclusively established, by autoradiographic labelling of hyphae of different fungi, that microfibrillar constituents are preferentially deposited in the hyphal

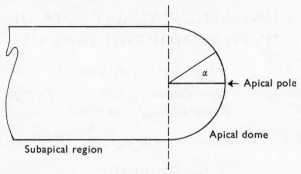

Fig. 1. Hyphal tip parameters.

apex and that maximum incorporation occurs at the apical pole (Bartnicki-Garcia & Lippman, 1969; Katz & Rosenberger, 1970, 1971; Gooday, 1971).

CYTOLOGY OF THE HYPHAL TIPS

Spitzenkörper and apical vesicles

Cytologists have long been searching hyphal tips of fungi for internal structures that might offer clues to the mechanism of apical growth. As far back as 1924, Brunswik had already found a plausible candidate for the cytological basis of apical growth – a densely staining, apical body or 'Spitzenkörper', seen in the tips of fixed hyphae of *Coprinus* spp. Because of its position, he suggested that the Spitzenkörper was connected with apical growth. Three decades later, Girbardt (1955), using phase contrast microscopy, confirmed the existence of the Spitzen-körper in the growing tips of a number of Basidiomycetes and Asco-mycetes. In these fungi, the Spitzenkörper appears as a refractive, roughly spherical body with a diffuse outline; it is located in close proximity to the apical pole. Significantly, the Spitzenkörper is seen in the tips of actively growing hyphae only – it vanishes when growth is experimentally arrested and re-forms just before growth is resumed (Girbardt, 1955, 1957; McClure, Park & Robinson, 1968). Moreover, changes in the direction of hyphal growth could be correlated with *prior* changes in the position of the Spitzenkörper (Girbardt, 1957). Thus, a strong case can be made for a prominent role of this organelle in apical growth.

By means of electron microscopy, McClure *et al.* (1968) concluded that the Spitzenkörper in *Aspergillus niger* was a conglomeration of small cytoplasmic vesicles. The peculiar abundance of vesicles in the hyphal tips of fungi has been verified in other laboratories throughout the world (Brenner & Carroll, 1968; Hemmes & Hohl, 1969; Girbardt, 1969; Grove et al. 1970; Grove & Bracker, 1970). By far the most precise studies on apical organization have been those from the laboratories of Girbardt in Jena and Bracker at Purdue University. Their investigations have demonstrated the seemingly universal occurrence of a large number of vesicles in the hyphal apex of growing fungi and, at the same time, they have shown some distinct differences in the organization of the apical vesicles in fungi belonging to different taxonomic classes (Plate 1). According to Girbardt (1969), there are two types of Spitzenkörper: a spheroidal one characteristic of the higher fungi and a 'cap-like' (crescentic) Spitzenkörper found in the Zygomycetes. His studies on the Basidiomycete, *Polystictus versicolor*, led to a three-dimensional reconstruction of the internal structure of a hyphal tip (Plate 2, fig. 1). In this model, the spheroidal Spitzenkörper is seen as a rather compact cluster of vesicles of two radically different sizes.

Grove & Bracker (1970) recognized three types of apical organization. In addition to the two types mentioned above for higher fungi and Zygomycetes, a third type was described for Oömycetous fungi (see Plate 1 and Fig. 2). They concluded that the Spitzenkörper seen with the optical microscope does not correspond to the agglomeration of large cytoplasmic vesicles revealed by electron microscopy, but represents a specialized central region within the vesicle cluster. This region, which only occurs in the hyphal tips of higher fungi, could be either vesicle-free and with an electron-dense core, or it may contain an aggregate of smaller vesicles (Plate 1). Grove & Bracker proposed that the term, Spitzenkörper, applies to this central region only (Fig. 2).

One important contribution emanating from these cytological studies is the realization that the cytoplasm of the apical dome consists mainly of apical vesicles to the near or total exclusion of other organelles such as nuclei, mitochondria and ribosomes (Fig. 3) (Zalokar, 1959; McClure et al. 1968; Girbardt, 1969; Grove et al. 1970; Grove & Bracker, 1970). Thus, the array of well-known functions associated with the latter organelles becomes of secondary importance to the main problem at hand: to establish a pattern of wall construction that will continuously transform the hemispherical surface of the apical dome wall into a cylindrical surface. There is compelling evidence that the apical vesicles play an important role as part of a secretory system (exocytosis) carrying materials and enzymes for the growth of the plasmalemma and the cell wall. The studies of McClure et al. (1968), Brenner & Carroll (1968), Girbardt (1969) and, particularly, Grove et al. (1970), point to

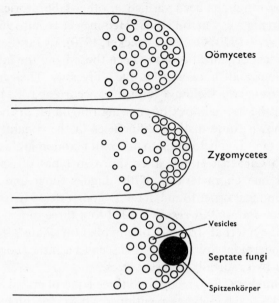

Fig. 2. The three types of organization of apical vesicles
in fungi according to Grove & Bracker (1970).

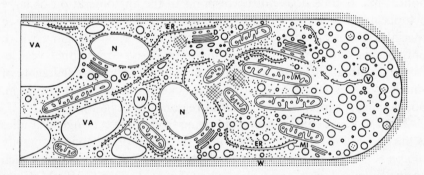

Fig. 3. Distribution of organelles in a hypha of *Pythium ultimum*. D = dictyosome;
ER = endoplasmic reticulum; M = mitochondrion; MI = microbody; N = nucleus;
R = ribosome; V = cytoplasmic vesicle; VA = vacuole. (Grove, Bracker & Morré, 1970.)

the dynamic nature of the apical vesicles which are continuously being
produced by the Golgi apparatus (or an equivalent generator of vesicles),
migrate to the apex and eventually coalesce with the apical plasmalemma.
And, in so doing, they simultaneously accomplish two essential functions:
they extend the plasmalemma surface and discharge enzymes (Girbardt,
1969) and/or material needed for cell wall formation (Grove *et al.* 1970).
The exact function of the vesicles must await biochemical characteriza-

tion. However, in another apically growing system, the pollen tube, there is evidence that apical vesicles contain polysaccharides for the amorphous layer of the pollen tube wall as well as the enzymatic potential for cellulose synthesis (van der Woude, Morré & Bracker, 1971).

Apical corpuscle

In germinating sporangiospores of *Mucor rouxii*, a single, electron-dense organelle ($\sim 0.2 \mu$m), the apical corpuscle (Plate 2, fig. 2) was seen intimately associated with the apical wall of the incipient germ tubes (Bartnicki-Garcia, Nelson & Cota-Robles, 1968 *a*). It was suggested that the organelle had a role in germ tube emission and its continued apical growth. Subsequent studies have revealed structures similar to the apical corpuscle in germinating spores of *Botrytis fabae* (Richmond & Pring, 1971), *B. cinerea* (Gull & Trinci, 1971), and *Phytophthora parasitica* (Bartnicki-Garcia, 1969). The spores of *Aspergillus nidulans* (Weisberg & Turian, 1971) and *Microsporum gypseum* (Page & Stock, 1971) show peripheral structures resembling the apical corpuscle which these authors categorized as lomasomes and lysosomes respectively. These findings tend to support the presumed role of the 'apical corpuscle' in germ tube emergence; however, its occurrence in apices of mature hyphae has not been substantiated (Girbardt, 1969; Grove & Bracker, 1970) and therefore its function in apical growth of hyphae remains questionable.

HYPHAL WALL ARCHITECTURE

General

Hyphal walls of fungi have a chemically complex nature (Aronson, 1965). They usually contain two or more polysaccharides as main constituents plus some protein and lipid. The chemical composition of the wall can vary drastically from one taxonomic class to the next (Bartnicki-Garcia, 1968). Consequently, the biochemical machinery needed for apical growth must vary accordingly. Yet some considerations of the architecture of young hyphal walls suggest certain points of similarity. Thus, a dense network of microfibrils constitutes the skeletal support of all hyphal walls examined. These microfibrils appear to be embedded in an amorphous matrix (Aronson, 1965). Significantly, studies on the structure of young hyphal walls of fungi (via shadow cast specimens of isolated walls) reveal a seemingly universal dual texture: a distinctly microfibrillar texture on the inner face and a non- or only vaguely fibrillar (granular, amorphous) appearance on the outer surface (Tamaki, 1959; Carbonell & Rodriguez, 1968; Hunsley & Burnett,

1970; Tokunaga & Bartnicki-Garcia, 1971*b*; Fultz & Woolf, 1972). In the walls of some Oömycetes, microfibrils reach the outer surface where they can be seen together with amorphous material (Pao & Aronson, 1970). It is debatable whether the dual texture of the hyphal wall originates from two truly distinct layers or whether it represents a graded zonation of a single wall which is predominantly or exclusively microfibrillar on the innermost face and mainly or entirely amorphous on its outermost surface. There is evidence for a multiplicity of layers in some hyphal walls, particularly in older portions (e.g. Hunsley & Burnett, 1970).

An important, but unresolved point for understanding hyphal wall morphogenesis, is the relative role of the two structural elements of the wall: namely, microfibrils and amorphous matrix. In this connection, the germinating cyst of *Phytophthora palmivora* provides an interesting insight into the architectural design of the hyphal wall. The hyphal (germ tube) wall arises directly from the cyst wall. The cyst wall is only a microfibrillar envelope around a spherical cell and has no permanent or visible coat of amorphous matrix (Tokunaga & Bartnicki-Garcia, 1971*b*). The germ tube, on the other hand, possesses from the base to the tip (Plate 3) the typical dual texture of a hyphal wall: amorphous outside, microfibrillar inside. Evidently, a microfibrillar skeleton is essential for both the cyst and the germ tube walls. During germination the fungus not only extends the microfibrillar fabric but also adds an extra coating of amorphous β-1,3-glucan to the developing germ tube wall. This occurs even though the cyst is germinating in the absence of nutrients and is dependent entirely on its own reserves. Then one would not expect the formation of superfluous structures and one must strongly suspect that the amorphous outer coat is an indispensable component of the germ tube wall. Whether it simply confers needed rigidity to the cylindrical wall or whether it participates more actively in the determination of hyphal morphogenesis remains unknown. Similar changes in wall structure occur in the germinating cyst of *Allomyces macrogynus* (Fultz & Woolf, 1972), a fungus with a different wall chemistry.

Apical dome wall

The evidence presented by Hunsley & Burnett (1970) and that of our own work (see Plate 3) indicate no major differences in the architecture of the dome wall compared to the lateral walls immediately behind. The inner microfibrillar network extends from the very tip to the oldest parts of the hyphal wall (contrary to Marchant's (1966) claim) and, in most fungi, it is completely covered by a continuous outer coat of amorphous material. In cross section, the dome wall appears similar to the subapical walls though there may be transient changes in thickness of the dome wall (bumps), probably caused by vesicle discharges (Grove & Bracker, 1970).

Microfibril orientation and lamination have been of considerable importance in studies of the growth and architecture of plant cell walls (e.g. Roelofsen, 1959). However, except for the sporangiophore of *Phycomyces*, they have not commanded similar attention in the fungi. Hunsley & Burnett (1970) concluded that apical microfibrils were not oriented in any predominant direction. On the other hand, Aronson & Preston (1960) reported that the microfibrils in the rhizoid tips of *Allomyces macrogynus* tend to lie in a preferred longitudinal orientation. This is also the case in the germ tube apex of *Phytophthora palmivora* (see Plate 3). Aronson & Preston (1960) presented evidence for lamellae with differently oriented microfibrils and expressed the possibility of changes in microfibril orientation during growth. But much more work is needed before the significance of microfibrillar orientation in hyphal morphogenesis can be established. Likewise, there is evidence showing other changes in wall architecture as the hypha ages, including differential thickening of the wall layers, acquisition of new layers (Hunsley & Burnett, 1970; Manocha & Colvin, 1968) and increases in the diameter of the microfibrils (Hunsley & Burnett, 1968). Whether these are merely manifestations of secondary wall growth of little or no importance in hyphal morphogenesis or whether they represent the end result of fundamental changes initiated at the growing tip, are questions that cannot be presently answered.

For the time being, it would be reasonable to conclude that the hyphal wall growth mechanism must be similar, in principle, throughout the fungi and, as Hunsley & Burnett (1970) indicated, it must form microfibrils at the apex together with amorphous materials in such a way that a more or less coaxial sequence of polymers can be derived.

Apical pore

Strunk (1963) found an apical pore at the tip of hyphal walls of *Polystictus versicolor*. The rim of this pore consists of circularly oriented microfibrils. Her claim was challenged by Scurfield (1967) and later by Hunsley & Burnett (1970) who found no evidence for a pore in the apical walls of several fungi. These three authors suggested that the apical pores were in reality septal orifices from hyphal wall fragments broken in such a way that they had an exposed cross-wall at one end. However, she was able to refute this criticism by employing undisrupted hyphae (Strunk, 1968). There is, nevertheless, the possibility that it may have been artificially formed during handling, particularly since hyphal tips have the tendency to burst (through an opening) invisible by light microscopy when subjected to environmental changes (see p. 256). Because of its location and structural differentiation, a genuine apical pore would most certainly have a profound implication in apical growth; consequently, it seems essential to ascertain its true nature. Even if it turns out to be an artifact of autolysis, its formation could mark a particularly weak area in the apical wall of potentially key importance in apical growth (see p. 258).

BIOPHYSICS OF APICAL GROWTH

Gradient of wall expansion

An essential attribute of apical growth is the restriction of surface expansion to the terminal region of the cell, yet restriction *per se* does not necessarily give rise to a tubular form; various other shapes are possible (Green, 1969). Hence, Green has pleaded that a morphogenetic response be dissected into its two components of surface behaviour: rate and directionality of expansion. In tip growth, a dome-shaped zone generates a cylindrical surface at its base. Seemingly, surface expansion in this dome region is isotropic and the essential basis for apical growth is the maintenance of a steep descending gradient in the rate of wall expansion within the apical dome (Green, 1969).

The surface behaviour of tip-growing cells has been considered by Reinhardt (1892), de Wolff & Houwink (1954), Castle (1958), and Green & King (1966). Assuming that the dome is hemispherical, that growth is radially symmetrical, and that surface expansion is equal in all directions, the gradient in local relative rate of expansion area must decrease as a cosine function of α (see Fig. 1) (Green, 1969); i.e. maximum wall expansion occurs at the apical pole ($\alpha = 0°$) decreasing to zero at the base of the dome ($\alpha = 90°$). This, however, is an idealized approximation. Hyphal tips are usually tapered and therefore the rate of wall expansion probably decreases more gradually over a longer distance.

Turgor pressure

Turgor appears to be an indispensable component of hyphal elongation (Robertson, 1968). The osmotic pressure in hyphal tips is considerably higher than that of the surrounding medium (Adebayo, Harris & Gardner, 1971) and normal extension growth depends on this differential (Park & Robinson, 1966b). Hyphal tip elongation ceases by application of hypertonic solutions (Robertson, 1959; Park & Robinson, 1966*b*) and resumes when the turgor is restored.

The exact role of turgor pressure in hyphal growth has not been clarified. It may be needed only to maintain the plasmalemma tightly appressed against the apical wall and thus permit the flow of materials to the apical wall. However, even though unequivocal proof is lacking, it seems reasonable to believe that turgor pressure provides the driving force that extends the apical wall at the same time that new wall mass is added. It is, however, unlikely that turgor alone could expand the wall and the need for concurrent active lysis, or enzymatic loosening of the wall fabric, has been frequently invoked (Green, 1969; Park & Robinson, 1966*a*).

BIOCHEMISTRY OF APICAL GROWTH

Wall biosynthesis

Site of wall deposition

Autoradiographic studies (in which growing hyphae are given a brief pulse of a tritiated wall precursor) have demonstrated that the structural polymers of hyphal walls (chitin, chitosan, β-glucans and galactose-containing polymers) are preferentially *deposited* in the apex (Plate 4, fig. 1) (Bartnicki-Garcia & Lippman, 1969; Katz & Rosenberger, 1970, 1971; Gooday, 1971). The maximum rate of incorporation of radioactive precursors per unit area occurs at, or within, 1 μm of the apical pole, and decreases sharply over a short distance corresponding approximately to the length of the apical dome (Fig. 4). (A residual, slowly declining gradient of wall deposition persists in the tubular portion of the hypha and probably accounts for increases in girth, thickness or microfibril diameter (see p. 251).) These findings indicate the existence in the apical dome of a sharp, descending, presumably radial, gradient of wall synthesis centered at or very near the apical pole. Inasmuch as wall synthesis probably parallels wall expansion (provided wall thickness and density are conserved during growth), these findings support the prediction of a steep gradient of wall expansion in the apical dome as the basis for apical growth (see p. 252). The values obtained for the relative rates of wall synthesis per unit area approximate the theoretical values for the relative rate of surface expansion calculated as a cosine function of α (Fig. 4). This agreement, however, must be considered only tentative since the small area of the apical dome and the limitations of the autoradiographic procedures employed allowed only three measurements of wall incorporation for the entire apical dome.

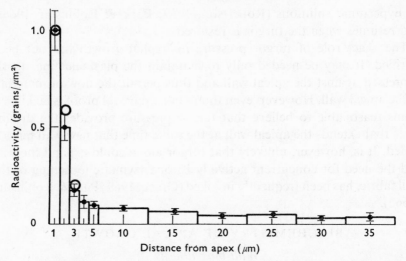

Fig. 4. Rates of wall synthesis along young hyphae. Rates of synthesis (●) were measured autoradiographically in hyphal segments 1 μm long at the top, and 10 μm long elsewhere. In these hyphae, the apical dome is quasihemispherical and has a radius close to 3 μm. Points are mean values with standard error markers for $n = 25$. (Bartnicki-Garcia & Lippman, 1969.) For comparison, the theoretical values for the relative rate of area expansion in the dome region are also shown (○). These are the relative values of cosine α (Fig. 1) measured at the midpoint of each of three equal segments of a hemispherical dome.

Sites of synthesis of cell wall polymers

To understand how the biochemical machinery of apical growth operates, it is necessary to know where cell wall polymers are made, particularly the subcellular sites where polymerization and/or assembly are accomplished. Although the experiments described in the preceding section convincingly demonstrate the sites of final *deposition* of certain sugar monomers (N-acetylglucosamine, glucose and galactose), they do not necessarily represent the sites where these sugars were polymerized. Thus, these wall polysaccharides could have been synthesized somewhere in the endomembrane system and then transported to their final destination in the wall by the apical vesicles (or other mechanisms). Alternatively, wall polymers may be largely or totally assembled *in situ*.

Various pieces of evidence support the concept that the microfibrillar skeleton of the wall is synthesized *in situ*, either on the outer surface of the plasmalemma or in the wall fabric itself:

(i) Cell fractionation studies of hyphae of *Phytophthora cinnamomi* (Wang & Bartnicki-Garcia, 1966), *Mucor rouxii* (McMurrough, Flores-Carreón & Bartnicki-Garcia, 1971) and *Neurospora crassa* (Mishra & Tatum, 1972) showed that the sugar nucleotide transferases for the synthesis of β-1,3-β-1,6-glucan, chitin and β-1,3-glucan, respectively, were present mainly or exclusively in the hyphal wall fraction. In *M. rouxii*, about 85 % of the recoverable activity of chitin synthetase was associated with the cell walls and the remainder was mostly bound to particles sedimenting between 10^4 and $10^5 \times g$. Probably the former represents enzyme at the site of operation while the latter is nascent enzyme (McMurrough *et al.* 1971) or enzyme *en route* to its destination.

(ii) Autoradiographic studies showed that chitin synthetase was localized preferentially in the apical region of hyphal walls of *Mucor rouxii*. Young germlings were freed of cytoplasm by freeze-thawing (McMurrough *et al.* 1971) or disrupted by glass beads (unpublished results) and incubated with uridine diphosphate–*N*-acetyl-[^{3}H]-D-glucosamine. The resulting apical pattern of labelling disclosed the site of operation of chitin synthetase and was similar to the apical pattern of wall deposition obtained by exposing the living hyphae to the free sugar, *N*-acetyl-[^{3}H]-D-glucosamine (Plate 4, fig. 2). In the disupted cells, only about one half of the hyphae showed chitin synthetase activity at the tips, and the reason for this is not entirely clear. It could be an indication of intermittent wall synthesis, but it could also be an artifact, particularly in view of the finding that the pattern of apical deposition of chitin in living hyphae of *Aspergillus nidulans* is readily lost by osmotic shock or metabolic inhibitors (Katz & Rosenberger, 1971).

(iii) In cells devoid of a wall, e.g. zoospores of *Phytophthora palmivora*, there is no chemical evidence of preformed microfibrils in the cytoplasm even though the cell is capable of rapidly laying down a microfibrillar wall (Tokunaga & Bartnicki-Garcia, 1971*a*, *b*; M. C. Wang & S. Bartnicki-Garcia, unpublished results).

(iv) From electron microscopic images, Bracker (1971) concluded that the wall fibrils were not carried in the apical vesicles of *Gilbertella persicaria*. These vesicles contained amorphous material of an appearance similar to the amorphous matrix of the cell wall. This situation is analogous to that in pollen tubes (van der Woude *et al.* 1971).

Pending biochemical characterization of apical vesicles and determination of the degree of prefabrication of various wall components in the cytoplasm, it may be tentatively concluded that cell wall polymer formation takes place both in the cytoplasm and in the wall itself. The microfibrillar network is assembled, if not entirely polymerized, *in situ*, whereas the matrix material is probably preformed internally and need only be anchored to the wall.

Cell wall lysis

The role of lytic (plasticizing, loosening or softening) enzymes in cell wall growth of bacteria (Higgins & Shockman, 1971) and plants (Morré & Eisinger, 1968; Ray, 1969) has been extensively discussed. Direct proof for the participation of lytic enzymes in apical growth of fungal hyphae is not available, but there is a body of circumstantial evidence which suggests or supports their participation, for instance:

(i) Enzymes capable of degrading cell wall polymers (proteases and β-1,3-glucanases) have been isolated from the hyphal walls of *Neurospora crassa* (Mahadevan & Mahadkar, 1970); a wall-bound β-1,3-endoglucanase was implicated in cell wall extension in *Schizosaccharomyces pombe* (Barras, 1972).

(ii) There is a positive correlation between the level of cell wall lytic enzymes and apical growth phenomena such as the frequency of branching in the wild type *v.* a colonial mutant of *Neurospora crassa* (Mahadevan & Mahadkar, 1970), the onset of germination in conidia in *Neurospora crassa* (Mahadevan & Rao, 1970) and *Microsporum gypseum* (Page & Stock, 1971), and the hormone-induced branching in *Achlya* (Thomas & Mullins, 1967). It is not certain, however, whether these activities are pertinent only to the initiation of apical growth or to its sustenance.

(iii) Fungal hyphae have the intrinsic capacity for executing a complete but carefully controlled dissolution of their hyphal tips as is unmistakably clear during anastomosis and clamp connection.

(iv) The hyphal tips of many fungi are highly susceptible to bursting when subjected to environmental changes and they even undergo occasional spontaneous bursting. In a study of the conditions resulting in apical bursting (Bartnicki-Garcia & Lippman, 1972b) we confirmed and extended earlier observations made by Robertson (1958, 1959) and Park & Robinson (1966b) on the ability of fungi to burst when flooded with distilled water or dilute acid solutions. The hyphal tips of colonies of *Mucor, Rhizopus, Neurospora, Aspergillus*, etc., but not *Schizophyllum* or *Phytophthora*, burst readily, in a matter of seconds, when their colonies are flooded with distilled water (Plate 4, fig. 3) or when the temperature of incubation is raised rapidly. The bursting of hyphal tips of *Mucor rouxii* upon addition of distilled water is not merely an osmotic phenomenon. By adjusting cultivation conditions it is possible to grow hyphae whose tips do not burst when flooded with distilled water, yet burst readily when flooded with a variety of dilute solutions of acids (any acid solution less than 0.1 M, pH below 4–5), certain critical

PLATE 1

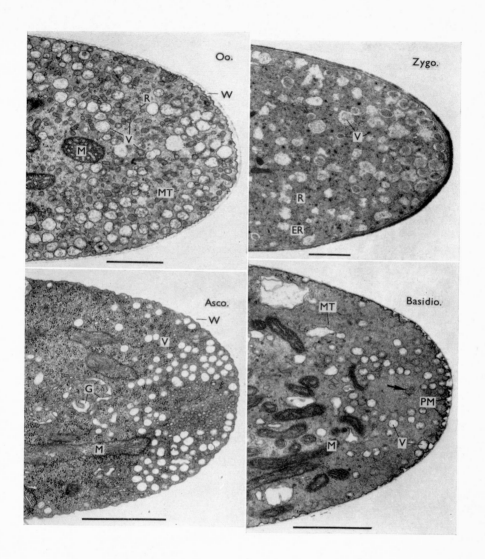

(Facing p. 256)

PLATE 2

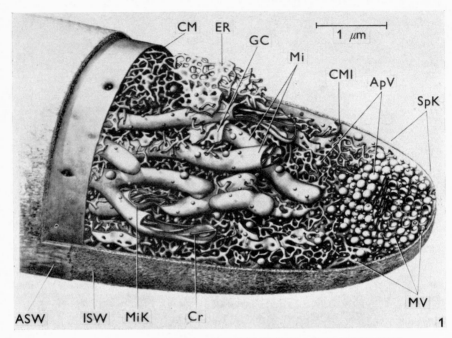

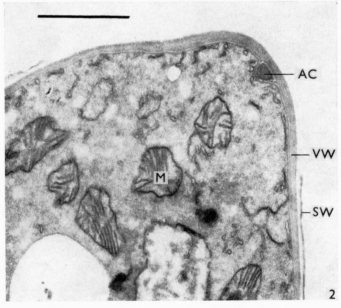

PLATE 3

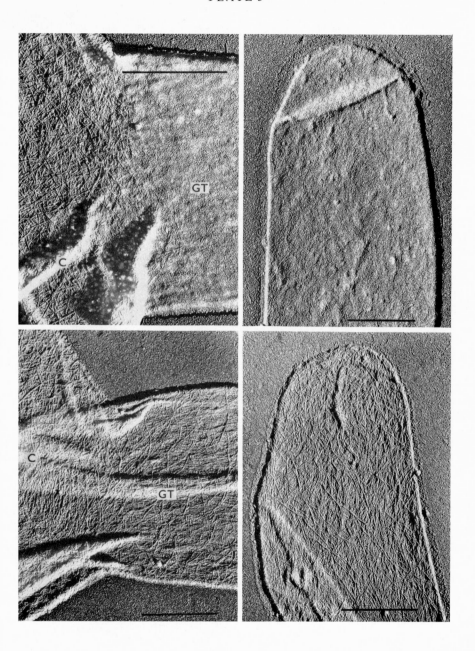

PLATE 4

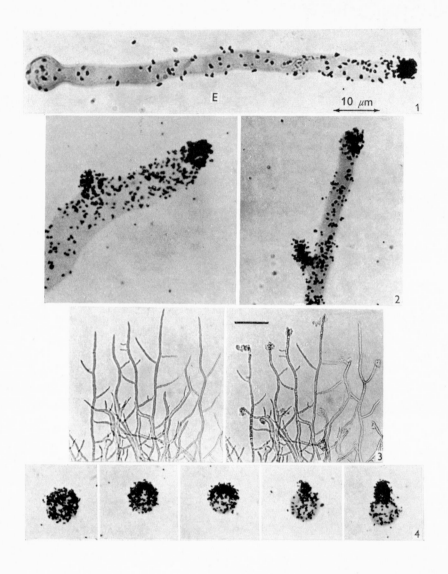

E

10 μm

1

2

3

4

concentrations of neutral salts (e.g. 0.05 M-NaCl but not 0.01 M-NaCl) as well as some detergents (0.1 mM-sodium dodecyl sulphate), chelating agents (0.001 M-EDTA), alcohols (10 % ethanol), etc.

It is unlikely that these solutions caused bursting due to the sudden drop in external osmotic pressure since the hyphal walls of the fungus are able to withstand the flooding with distilled water without bursting. Also, the temperature coefficient of apical bursting is too high for a physical explanation of the phenomenon. Consequently, it becomes necessary to postulate that the solutes in the water must have elicited some active chemical weakening of the wall. The variety of substances inducing bursting precludes an explanation based on a common chemical action on the cell wall. Therefore, we must conclude that these substances somehow increased the activity of enzymes involved in weakening the cell wall. This then becomes circumstantial proof for the existence of a high potential of lytic activity in the apexes of ordinary hyphae examined when they were actively engaged in elongating themselves (Bartnicki-Garcia & Lippman, 1972b). The lytic enzymes are probably carried in vesicles, as shown for the β-1,3-glucanase activity of *Saccharomyces cerevisiae* (Matile, Cortat, Wiemken & Frey-Wyssling, 1971). The autolytic response may be triggered by an excessive discharge of these vesicles or, indirectly, by inhibition of the counteracting process of wall synthesis (see below).

Control of apical wall growth
Delicate balance between wall synthesis and wall lysis

If, as is commonly believed, lytic enzymes are essential for wall growth, it follows that a harmonious balance must exist between the processes of wall synthesis and wall lysis to assure the orderly growth of the cell wall. Too much synthesis would cause thickening of the apical wall and arrestment of elongation; too much lysis would lead to bursting (or perhaps swelling) of the apex. The observation that hyphal tips tend to burst rather readily (see p. 256) when disturbed by a number of treatments (some of them rather harmless) may be taken as an indication that the postulated balance between wall lysis and wall synthesis does exist and that it is more or less precariously set, particularly in rapidly growing hyphae, and can be easily shifted in the direction of lysis. This imbalance can result from either an increase in lytic activity or a decrease in wall synthesis. This conclusion is supported by the findings that application of an inhibitor of chitin synthetase (polyoxin D) or of an enzyme with chitinase activity (lysozyme) both cause the same effect on *Mucor rouxii*: bursting of the tips (Bartnicki-Garcia &

Lippman, 1972*a*, *b*; unpubl. mat.). Likewise, sorbose, an inhibitor of wall-glucan synthetase (Mishra & Tatum, 1972), causes apical disintegration in hyphae of *Neurospora crassa* (Rizvi & Robertson, 1965).

To maintain harmony in apical wall metabolism and consequently uniformity of shape, it was postulated (Bartnicki-Garcia & Lippman, 1969) that the gradient of wall synthesis in the hyphal apex is accompanied by a parallel gradient of wall lysis. Such parallel distribution may not be difficult to account for, if both types of enzymes are delivered to the tip by vesicles. Presumably, they may arrive together in the same vesicle or they may come in different vesicles having similar migration patterns.

Control of apical wall metabolism

Undoubtedly, one of the most fundamental aspects of apical growth is the spatial regulation of wall metabolism (synthesis + lysis). What determines a maximum rate of wall metabolism (and hence growth) at the apical pole and its sharp drop to near cessation at the base of the apical dome?

In Robertson's (1965*b*) scheme, apical growth is the result of two separate processes: formation of plastic wall at the apex and the rigidification (setting) of the wall beyond the margin of the apical dome to prevent further extension. Park & Robinson (1966*a*) subscribed to this view but in addition they postulated that the plasticity of the apex was due to 'a balance between the lysis and forging of bonds between intersecting microfibrils'. They also proposed that rigidification was accomplished by a second wall-building mechanism. Although proof is lacking, I prefer not to invoke a rigidification mechanism mainly because the gradients of wall synthesis and wall lysis advocated herein (see below) would automatically account for the near cessation of wall growth at the base of the apical dome and there would be no need to introduce such an additional process.

These descending gradients in wall synthesis and wall lysis are probably created by two conditions: (i) the continuous discharge of packets of enzymes and certain wall precursors in an apical pattern (maximum at the pole, minimum at the base of the apical dome), and (ii) a finite function or life for these enzymes so that they cease to operate by the time the point in the dome surface at which they were initially secreted has become part of the lateral wall. The first condition is probably met by the pattern of migration and fusion of the apical vesicles carrying the necessary enzymes and wall precursors. The second condition may be satisfied, particularly for the lytic enzymes, by assuming that these enzymes are unstable or are actively destroyed by proteases known to be associated with the wall. In this connection, it is worth noting that proteins (enzymes) involved in softening cell walls of higher plants are unstable (Cleland, 1970; Morré & Eisinger, 1968). In the case of wall synthetases, their activity may possibly terminate not through inactivation but by depletion of some key substrate.

Another possibility for regulation of apical growth was pointed out by J. L. Reissig (personal communication) who had earlier reported that a galactosamine-rich polysaccharide isolated from *Neurospora crassa* inhibited its own growth (Reissig & Glasgow, 1971). This material is probably part of the glycoprotein reticulum of *Neurospora crassa* hyphal walls described by Hunsley & Burnett (1970). Significantly, this reticulum seems to be absent from the apex and makes its appearance subapically. A possible function for this growth inhibitory polymer might be to terminate the wall growth activity initiated in the apical dome.

AN INTEGRATED MODEL FOR THE MECHANISM OF APICAL GROWTH IN FUNGI

The preceding information can serve as a basis for formulating a model of the mechanism of apical growth in fungi. Given our present state of knowledge, such a model, while highly hypothetical, is at the same time essential to embark on the long process of elucidating the genesis of a fungal hypha. Instead of postulating a model of wall growth designed primarily for apical growth, it seems more meaningful to formulate a general, unitary model of cell wall growth* based on the assumption that wall growth results from the cumulative action of minute hypothetical units of wall growth. And, then, treat apical growth (as well as other growth types) as patterns of distribution of these units of wall growth. The model proposed may be regarded as an updated version of the scheme of mosaic growth of the plant cell wall (Frey-Wyssling & Stecher, 1951) described by Green & Chapman (1955) in terms of a mosaic of minute 'growth events'.

The conceptualization of a unit of cell wall growth may not be just a mathematical abstraction (cf. Castle, 1958) but may have physical reality; a unit may turn out to be the amount of growth obtained from the discharge of a single vesicle (or a minimum combination of different vesicles) carrying components essential for wall growth.

* For the present purpose, cell wall growth refers to an increase in surface area of the wall without any appreciable change in the thickness or density of the wall.

Unitary model of cell wall growth

The model is an attempt to integrate all the essential components needed for cell wall growth into a single, hypothetical, minimal unit (Fig. 5). For simplicity, the model considers a wall with only two components: an amorphous substance and a microfibrillar skeleton. The production of a unit of wall growth is depicted in Fig. 5 as follows. (A) Lytic enzymes from a cytoplasmic vesicle are secreted into the wall. (B) These enzymes attack the microfibrillar skeleton by splitting either inter- or intramolecular bonds. (C) The dissociated (broken or thinned-out) microfibrils can no longer withstand the high turgor pressure and become stretched out or separated from one another with the consequent increase in surface area of the wall. (No attack on the outer amorphous components is shown, but this might prove to be necessary.) (D) Microfibril-synthesizing enzymes operating in the wall itself or on the outer surface of *new* plasmalemma (formed by exocytosis) rebuild the microfibrils by producing new chains or by extending old ones (or broken ones). In this scheme, the synthetases are secreted via vesicles, whereas the soluble precursors for microfibril synthesis are assumed to be transported across the plasmalemma (via lipid intermediates?). Vesicles containing amorphous wall material, in a largely or entirely preformed state, deposit their contents against the wall. Given the high turgor pressure of the cell, the vesicular contents would be forced through the microfibrillar fabric, and most of it would reach the outer surface of the wall where it would somehow be firmly anchored. (E) In this manner the cell wall has expanded one unit area without losing its overall properties including the coaxial arrangement (differential layering) of wall polymers (see p. 251). In Fig. 5, growth is depicted in steps for the sake of clarity; most likely all these processes take place simultaneously. Also, although the two enzymes and the matrix substance are all shown arriving in separate vesicles, it is possible that these components may be carried in one or two types of vesicle only.

Apical growth: polar distribution of wall growth units

Briefly, apical growth may be viewed as the result of a tendency of the postulated units of cell wall growth to concentrate around a point on the cell surface which is or becomes the apical pole of a hyphal tube; so long as this polarity is preserved, the growing cell will extend in a tubular fashion. If the density of these wall growth units around the apical pole were to decrease, radially, as a cosine function (see p. 252), the idealized shape of a hypha shown in Fig. 1 would be obtained.

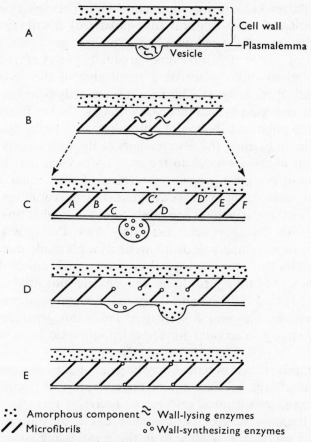

:·:· Amorphous component ≈ Wall-lysing enzymes
// Microfibrils   °₀° Wall-synthesizing enzymes

Fig. 5. Hypothetical representation of the events in a
unit of cell wall growth. (See text for explanation.)

The actual shape and diameter of the hypha would be determined by
the spatial distribution of the wall growth units and by the relative
ratios of biosynthetic and lytic activities in these units. Such a complex
interaction would be readily affected by a wide assortment of environmental factors and by mutations in various loci, all of which can cause
drastic changes in hyphal morphology. (For examples of environmental
and mutational effects see Bartnicki-Garcia & Nickerson (1962) and
Brody (1972) respectively.)

The germination of asexual spores of *Mucor rouxii* may be used as an
example to illustrate the presumed origin of hyphal growth. Initially, a
spore germinates by growing into a large spherical cell (Bartnicki-
Garcia, Nelson & Cota-Robles, 1968b). A vegetative wall is synthesized
de novo, under the spore wall, by a uniformly disperse pattern of wall

synthesis (Bartnicki-Garcia & Lippman, 1969). This spherical cell wall probably arises from a uniform (non-polarized) distribution of wall growth units around the cell periphery. Before a germ tube protrudes, the pattern of wall synthesis becomes gradually polarized until most of the wall synthesis takes place on a small area of the surface of the spherical cell (Plate 4, fig. 4). This area subsequently becomes the apical dome of the emerging hyphal tube. Accordingly, the key factor in apical growth is the polar distribution of cell wall growth units. Since vesicles may contain, in essence, the determinants of the wall growth units, the displacement of these vesicles to the apex poses what may be the ultimate question in apical growth – the cause of their polar migration. There is no evidence that vesicles are guided by microtubules or microfilaments, for these structures are rarely found in hyphal tips (Grove & Bracker, 1970). Since vesicles ostensibly lack the means for self-propulsion, there remains electrophoresis as a plausible mechanism to convey vesicles from their subapical origins to their apical destination (Jaffe, 1968). According to Jaffe, the electric potential that exists between the apex and the subapical regions of *Neurospora crassa* hyphae (as measured by Slayman & Slayman, 1962) can generate a current sufficiently strong to account for the electrophoretic movement of the vesicles.

Possibly, this electric potential may be generated by reactions ensuing from the discharge of the vesicles themselves, thus forming a self-perpetuating electrochemical gradient. Conceivably this gradient may be started by the same vesicles that maintain it. Yet, the obligate period of unpolarized spherical growth seen in the germinating spore of *Mucor rouxii*, prior to germ tube emission, suggests that a special entity, a differentiated vesicle or another structural element, may be needed to initiate polarized wall growth. One such candidate may be the apical corpuscle or another analogous structure found in germinating spores of fungi (see p. 249).

REFERENCES

ADEBAYO, A. A., HARRIS, R. F. & GARDNER, W. R. (1971). Turgor pressure of fungal mycelia. *Transactions of the British Mycological Society*, 57, 145–51.

ARONSON, J. M. (1965). The cell wall. In *The Fungi*, vol. 1, ed. G. C. Ainsworth & A. S. Sussman, pp. 49–76. New York and London: Academic Press.

ARONSON, J. M. & PRESTON, R. D. (1960). The microfibrillar structure of the cell walls of the fungus *Allomyces. Journal of Biophysical and Biochemical Cytology*, 8, 247–56.

BARRAS, D. R. (1972). A β-glucan endohydrolase from *Schizosaccharomyces pombe* and its role in cell wall growth. *Antonie van Leeuwenhoek*, 38, 65–80.

BARTNICKI-GARCIA, S. (1968). Cell wall chemistry, morphogenesis, and taxonomy of fungi. *Annual Review of Microbiology*, 22, 87–108.

BARTNICKI-GARCIA, S. (1969). Cell wall differentiation in the Phycomycetes. *Phytopathology*, **59**, 1065–71.

BARTNICKI-GARCIA, S. & LIPPMAN, E. (1969). Fungal morphogenesis: cell wall construction in *Mucor rouxii*. *Science, New York*, **165**, 302–4.

BARTNICKI-GARCIA, S. & LIPPMAN, E. (1972a). Inhibition of *Mucor rouxii* by Polyoxin D: Effects on chitin synthetase and morphological development. *Journal of General Microbiology*, **71**, 301–9.

BARTNICKI-GARCIA, S. & LIPPMAN, E. (1972b). The bursting tendency of hyphal tips of fungi: presumptive evidence for a delicate balance between wall synthesis and wall lysis in apical growth. *Journal of General Microbiology*, **73**, in press.

BARTNICKI-GARCIA, S., NELSON, N. & COTA-ROBLES, E. (1968a). A novel apical corpuscle in hyphae of *Mucor rouxii*. *Journal of Bacteriology*, **95**, 2399–402.

BARTNICKI-GARCIA, S., NELSON, N. & COTA-ROBLES, E. (1968b). Electron microscopy of spore germination and cell wall formation in *Mucor rouxii*. *Archiv für Mikrobiologie*, **63**, 242–55.

BARTNICKI-GARCIA, S. & NICKERSON, W. J. (1962). Nutrition, growth and morphogenesis of *Mucor rouxii*. *Journal of Bacteriology*, **84**, 841–58.

BRACKER, C. E. (1971). Cytoplasmic vesicles in germinating spores of *Gilbertella persicaria*. *Protoplasma*, **72**, 381–97.

BRENNER, D. M. & CARROLL, G. C. (1968). Fine-structural correlates of growth in hyphae of *Ascodesmis sphaerospora*. *Journal of Bacteriology*, **95**, 658–71.

BRODY, S. (1972). Metabolism, cell walls and morphogenesis. In *Cell Differentiation*, ed. S. Coward. New York and London: Academic Press. (In press.)

BRUNSWIK, H. (1924). Untersuchungen über Geschlechts und Kernverhaltnisse bei der Hymenomyzetengattung *Coprinus*. In *Botanische Abhandlungen*, vol. 5, ed. K. Goebel, pp. 1–152. Jena: Gustav Fisher.

BURGEFF, H. (1915). Untersuchungen über Variabilität, Sexualität und Erblichkeit bei *Phycomyces nitens* Kuntze. *Flora*, N.F. **108**, 353–488.

BURNETT, J. H. (1968). *Fundamentals of Mycology*. New York: San Martin's Press.

BUTLER, G. M. (1961). Growth of hyphal branching systems in *Coprinus disseminatus*. *Annals of Botany*, N.S. **25**, 341–52.

CARBONELL, L. M. & RODRIGUEZ, J. (1968). Mycelial phase of *Paracoccidioides brasiliensii* and *Blastomyces dermatitidis*: an electron microscope study. *Journal of Bacteriology*, **96**, 533–43.

CASTLE, E. S. (1958). The topography of tip growth in a plant cell. *Journal of General Physiology*, **41**, 913–26.

CLELAND, R. (1970). Protein synthesis and wall extensibility in the *Avena* coleoptile. *Planta, Berlin*, **95**, 218–26.

DE WOLFF, P. M. & HOUWINK, A. L. (1954). Some considerations on cellulose fibril orientation in growing cell walls. *Acta Botanica Neerlandica*, **3**, 396–7.

FREY-WYSSLING, A. & STECHER, H. (1951). Das Flächenwachstum der pflanzlichen Zellwände. *Experientia*, **7**, 420–1.

FULTZ, S. A. & WOOLF, R. A. (1972). Surface structure in *Allomyces* during germination and growth. *Mycologia*, **64**, 212–18.

GIRBARDT, M. (1955). Lebendbeobachtungen an *Polystictus versicolor* (L.). *Flora*, **142**, 540–63.

GIRBARDT, M. (1957). Der Spitzenkörper von *Polystictus versicolor* (L.). *Planta, Berlin*, **50**, 47–59.

GIRBARDT, M. (1969). Die Ultrastruktur der Apikalregion von Pilzhyphen. *Protoplasma*, **67**, 413–41.

GOODAY, G. W. (1971). An autoradiographic study of hyphal growth of some fungi. *Journal of General Microbiology*, **67**, 125–33.

GREEN, P. B. (1969). Cell morphogenesis. *Annual Review of Plant Physiology*, **20**, 365–94.

GREEN, P. B. & CHAPMAN, G. B. (1955). On the development and structure of the cell wall in *Nitella*. *American Journal of Botany*, **42**, 685–93.

GREEN, P. B. & KING, A. (1966). A mechanism for the origin of specifically oriented textures in development with special reference to *Nitella* wall texture. *Australian Journal of Biological Sciences*, **19**, 421–37.

GROVE, S. N. & BRACKER, C. E. (1970). Protoplasmic organization of hyphal tips among fungi: vesicles and Spitzenkörper. *Journal of Bacteriology*, **104**, 989–1009.

GROVE, S. N., BRACKER, C. E. & MORRÉ, D. J. (1970). An ultrastructural basis for hyphal tip growth in *Pythium ultimum*. *American Journal of Botany*, **59**, 245–66.

GULL, K. & TRINCI, A. P. J. (1971). Fine structure of spore germination in *Botrytis cinerea*. *Journal of General Microbiology*, **68**, 207–20.

HEMMES, D. E. & HOHL, H. R. (1969). Ultrastructural changes in directly germinating sporangia of *Phytophthora parasitica*. *American Journal of Botany*, **56**, 300–13.

HIGGINS, M. L. & SHOCKMAN, G. D. (1971). Procaryotic cell division with respect to wall and membranes. *Critical Reviews in Microbiology*, **1**, 29–72.

HUNSLEY, D. & BURNETT, J. H. (1968). Dimensions of microfibrillar elements in fungal walls. *Nature, London*, **218**, 462–3.

HUNSLEY, D. & BURNETT, J. H. (1970). The ultrastructural architecture of the walls of some hyphal fungi. *Journal of General Microbiology*, **62**, 203–18.

JAFFE, L. F. (1968). Localization in the developing *Fucus* egg and the general role of localizing currents. *Advances in Morphogenesis*, **7**, 295–328.

KATZ, D. & ROSENBERGER, R. F. (1970). The utilization of galactose by an *Aspergillus nidulans* mutant lacking galactose phosphate–UDP glucose transferase and its relation to cell wall synthesis. *Archiv für Mikrobiologie*, **74**, 41–51.

KATZ, D. & ROSENBERGER, R. F. (1971). Hyphal wall synthesis in *Aspergillus nidulans*: effect of protein synthesis inhibition and osmotic shock on chitin insertion and morphogenesis. *Journal of Bacteriology*, **108**, 184–90.

LARPENT, J. P. (1970). Problèmes posés par le développement des systèmes filamenteux. *Physiologie Végétale*, **8**, 335–47.

McCLURE, W. K., PARK, D. & ROBINSON, P. M. (1968). Apical organization in the somatic hyphae of fungi. *Journal of General Microbiology*, **50**, 177–82.

McMURROUGH, I., FLORES-CARREÓN, A. & BARTNICKI-GARCIA, S. (1971). Pathway of chitin synthesis and cellular localization of chitin synthetase in *Mucor rouxii*. *Journal of Biological Chemistry*, **246**, 3999–4007.

MAHADEVAN, P. R. & MAHADKAR, U. R. (1970). Role of enzymes in growth and morphology of *Neurospora crassa*: cell-wall-bound enzymes and their possible role in branching. *Journal of Bacteriology*, **101**, 941–7.

MAHADEVAN, P. R. & RAO, S. R. (1970). Enzyme degradation of conidial wall during germination of *Neurospora crassa*. *Indian Journal of Experimental Biology*, **8**, 293–7.

MANOCHA, M. S. & COLVIN, J. R. (1968). Structure of the cell wall of *Pythium debaryanum*. *Journal of Bacteriology*, **95**, 1140–52.

MARCHANT, R. (1966). Wall structure and spore germination in *Fusarium culmorum*. *Annals of Botany*, **30**, 821–30.

MARCHANT, R. & SMITH, D. G. (1968). A serological investigation of hyphal growth in *Fusarium culmorum*. *Archiv für Mikrobiologie*, **63**, 85–94.

MATILE, P., CORTAT, M., WIEMKEN, A. & FREY-WYSSLING, A. (1971). Isolation of glucanase-containing particles from budding *Saccharomyces cerevisiae*. *Proceedings of the National Academy of Sciences, U.S.A.* **68**, 636–40.

MISHRA, N. C. & TATUM, E. L. (1972). Effect of L-sorbose on polysaccharide synthetases of *Neurospora crassa*. *Proceedings of the National Academy of Sciences, U.S.A.* **69**, 313–17.

MORRÉ, D. J. & EISINGER, W. R. (1968). Cell wall extensibility: its control by auxin and relationship to cell elongation. In *Biochemistry and Physiology of Plant Growth Substances*, ed. F. Wightman & G. Setterfield. Ottawa: Runge Press.

PAGE, W. J. & STOCK, J. J. (1971). Regulation and self-inhibition of *Microsporum gypseum* macroconidia germination. *Journal of Bacteriology*, **108**, 276–81.

PAO, V. M. & ARONSON, J. M. (1970). Cell wall structure of *Sapromyces elongatus*. *Mycologia*, **62**, 531–41.

PARK, D. & ROBINSON, P. M. (1966a). Aspects of hyphal morphogenesis in fungi. In *Trends in Plant Morphogenesis*, ed. E. G. Cutter, pp. 27–44. London: Longmans, Green.

PARK, D. & ROBINSON, P. M. (1966b). Internal pressure of hyphal tips of fungi, and its significance in morphogenesis. *Annals of Botany*, N.S. **30**, 425–39.

RAY, P. M. (1969). The action of auxin on cell enlargement in plants. *Developmental Biology Supplement*, **3**, 172–205.

REINHARDT, M. O. (1892). Das Wachsthum der Pilzhyphen. *Jahrbucher für Wissenschaftliche Botanik*, **23**, 479–566.

REISSIG, J. L. & GLASGOW, J. E. (1971). Mucopolysaccharide which regulates growth in *Neurospora*. *Journal of Bacteriology*, **106**, 882–9.

RICHMOND, D. V. & PRING, R. J. (1971). Fine structure of germinating *Botrytis fabae* Sardiña conidia. *Annals of Botany*, **35**, 493–500.

RIZVI, S. R. H. & ROBERTSON, N. F. (1965). Apical disintegration of hyphae of *Neurospora crassa* as a response to L-sorbose. *Transactions of the British Mycological Society*, **48**, 469–77.

ROBERTSON, N. F. (1958). Observations on the effect of water on the hyphal apices of *Fusarium oxysporum*. *Annals of Botany*, N.S. **22**, 159–73.

ROBERTSON, N. F. (1959). Experimental control of hyphal branching and branch form in hyphomycetous fungi. *Journal of the Linnean Society of London Botany*, **56**, 207–11.

ROBERTSON, N. F. (1961). Mycology. In *Contemporary Botanical Thought*, ed. A. M. MacLeod & L. S. Cobley. Edinburgh: Oliver and Boyd.

ROBERTSON, N. F. (1965a). The fungal hypha. *Transactions of the British Mycological Society*, **48**, 1–8.

ROBERTSON, N. F. (1965b). The mechanism of cellular extension and branching. In *The Fungi*, vol. 1, ed. G. C. Ainsworth & A. S. Sussman, pp. 613–23. New York and London: Academic Press.

ROBERTSON, N. F. (1968). The growth process in fungi. *Annual Review of Phytopathology*, **6**, 115–36.

ROELOFSEN, P. A. (1959). *The Plant Cell Wall*. Berlin-Nikolassee: Gebrüder Borntraeger.

SCURFIELD, G. (1967). Apical pore in fungal hyphae. *Nature, London*, **214**, 740–1.

SLAYMAN, C. L. & SLAYMAN, C. W. (1962). Measurement of membrane potentials in *Neurospora*. *Science, New York*, **136**, 876–7.

SMITH, J. H. (1923). On the apical growth of fungal hyphae. *Annals of Botany*, **37**, 341–3.

STRUNK, C. (1963). Über die Substruktur der Hyphenspitzen von *Polystictus versicolor*. *Zeitschrift für allgemeine Mikrobiologie*, **3**, 265–74.

STRUNK, C. (1968). Zur Darstellung des Apicalporus bei *Polystictus versicolor*. *Archiv für Mikrobiologie*, **60**, 255–61.

TAMAKI, T. (1959). Electron microscopical studies of *Microsporum japonicum* and *Trichophyton interdigitale*. *Shikoku Acta Medica, Japan*, **15**, 252–7.

THOMAS, D., DES, J. & MULLINS, J. T. (1967). Role of enzymatic wall-softening in plant morphogenesis: hormonal induction in *Achlya*. *Science, New York*, **156**, 84–5.

TOKUNAGA, J. & BARTNICKI-GARCIA, S. (1971*a*). Cyst wall formation and endogenous carbohydrate utilization during synchronous encystment of *Phytophthora palmivora* zoospores. *Archiv für Mikrobiologie*, **79**, 283–92.

TOKUNAGA, J. & BARTNICKI-GARCIA, S. (1971*b*). Structure and differentiation of the cell wall of *Phytophthora palmivora*: cysts, hyphae and sporangia. *Archiv für Mikrobiologie*, **79**, 293–310.

VAN DER WOUDE, W. J., MORRÉ, D. J. & BRACKER, C. E. (1971). Isolation and characterization of secretory vesicles in germinated pollen of *Lilium longiflorum*. *Journal of Cell Science*, **8**, 331–51.

WANG, M. C. & BARTNICKI-GARCIA, S. (1966). Biosynthesis of β-1,3- and β-1,6-linked glucan by *Phytophthora cinnamoni* hyphal walls. *Biochemical and Biophysical Research Communications*, **24**, 832–7.

WEISBERG, S. H. & TURIAN, G. (1971). Ultrastructure of *Aspergillus nidulans* conidia and conidial lomasomes. *Protoplasma*, **72**, 55–67.

ZALOKAR, M. (1959). Growth and differentiation of *Neurospora* hyphae. *American Journal of Botany*, **46**, 602–10.

EXPLANATION OF PLATES

PLATE 1

Comparative cytology of hyphal apices of an Oömycete (*Pythium aphanidermatum*), a Zygomycete (*Gilbertella persicaria*), an Ascomycete (*Aspergillus niger*) and a Basidiomycete (*Armillaria melleae*). W = wall, V = vesicle, PM = plasmalemma, M = mitochondria, ER = endoplasmic reticulum, G = Golgi cisternae, R = ribosomes, MT = microtubule. Arrow in the Basidiomycete apex points to a small densely staining core, Marker = 1 μm. (Grove & Bracker, 1970.)

PLATE 2

Fig. 1. Model of the apical region of *Polystictus versicolor*. ApV = apical vesicle; ASW = outer layer of cell wall; CM = plasmalemma; CMI = plasmalemma invagination; Cr = mitochondrial crista; ER = endoplasmic reticulum; GC = Golgi cisterna; ISW = inner fibrillar layer of cell wall; Mi = mitochondria; MiK = mitochondrial membrane; MV = microvesicle; SpK = Spitzenkörper. (Girbardt, 1969.)

Fig. 2. Apical corpuscle (AC) in the incipient germ tube of a germinating spore of *Mucor rouxii*. VW = vegetative wall, SW = spore wall. Marker = 1 μm. (Bartnicki-Garcia, Nelson & Cota-Robles, 1968*a*.)

PLATE 3

Architecture of the hyphal wall: germ tube of *Phytophthora palmivora*. On the left are views of the base of the germ tube wall (GT) emerging from the cyst (C). On the right are the corresponding views of their apexes. Upper pictures show the texture of the outer surface of the wall. Lower pictures show the inner microfibrillar skeleton exposed by removal of the outer layer with an exo-β-1,3-glucanase. (Tokunaga & Bartnicki-Garcia, 1971*b*.)

PLATE 4

Fig. 1. Apical pattern of cell wall construction in an anaerobically grown hypha of *Mucor rouxii* exposed for 5 min to *N*-acetyl-[³H]-D-glucosamine and then chemically extracted to remove cytoplasm. (Bartnicki-Garcia & Lippman, 1969.)

Fig. 2. Autoradiographic localization of sites of wall deposition (*in vivo*) and sites of operation of chitin synthetase (*in vitro*) in *Mucor rouxii*. Left: sites of wall deposition (chitin + chitosan) in living hyphae of *M. rouxii* incubated with a brief pulse of [³H]GlcNAc. Right: chitin synthetase sites in hyphal walls of *M. rouxii* incubated with UDP-[³H]-GlcNAc. (McMurrough *et al.* 1971.)

Fig. 3. Bursting of hyphal tips of *Mucor rouxii* by flooding a colony with distilled water. Left: immediately after addition of water. Right: one minute later. Marker = 100 μm. (Barnicki-Garcia & Lippman, 1972*b*.)

Fig. 4. Progressive stages in the polarization of cell wall synthesis during spore germination of *Mucor rouxii*. From an anaerobic culture (N$_2$) exposed to [^{3}H]-*N*-acetyl-D-glucosamine for 5 min and processed for autoradiography as described by Bartnicki-Garcia & Lippman (1969). Grains represent chitin and chitosan polymers. (Previously unpublished.)

DIFFERENTIATION IN THE MUCORALES

G. W. GOODAY

Department of Biochemistry, University of Aberdeen, Scotland

INTRODUCTION

The choice between the development of sexual or asexual reproductive cells in the Mucorales is determined by the presence or absence of sexual interaction with a compatible partner:

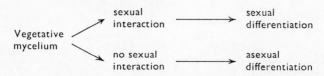

This article is concerned with the control of these developmental sequences. Although the primary mechanisms of sex determination still remain obscure, recent advances in our understanding of the hormonal control of the intercellular reactions are described. The major role of the cell wall in both developmental sequences is discussed. Most of the examples are taken from members of the Mucoraceae, considered a primitive family of the Mucorales by Benjamin (1959).

A complete life cycle of a member of the Mucorales, illustrated by *Mucor mucedo* (Fig. 1), is rarely encountered, since in nature compatible mating types rarely meet, and in the laboratory the sexually formed zygospores are very difficult to germinate. The mycelium of a member of the Mucorales is typically composed of unbranched coenocytic hyphae growing on a simple nutrient-rich substrate. A short distance behind the growing hyphal tips the vegetative mycelium differentiates to give sporangiophores. The resultant sporangia contain the asexual sporangiospores which when released readily germinate to give more vegetative mycelium. However, if the hyphae encounter others of opposite mating type, sporangiophores are not produced, and sexual differentiation predominates with the production of characteristic sexual hyphae, the zygophores, which are formed nearer to the hyphal tips. Compatible zygophores are mutually attractive, growing towards each other (zygotropism), fusing in mated pairs and rapidly swelling at their tips to give the progametangia. These are delimited by cross walls to give gametangia supported by suspensor cells, and the central fusion

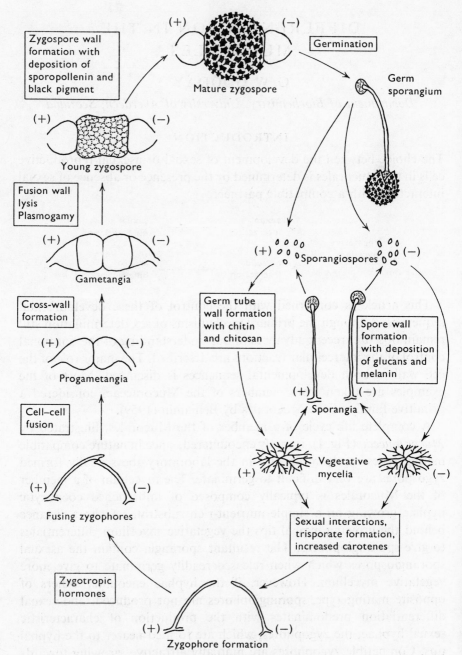

Fig. 1. Annotated life cycle of *Mucor mucedo*.
For further details, see text.

wall breaks down to allow plasmogamy, giving a large cell with nuclei from each mating type. A very thick black wall develops around the maturing zygospore. After a dormant period the zygospore may germinate to give a germ sporangium and so complete the life cycle. The nuclear events within the zygospore will not concern us here, save to say that they involve karyogamy and meiosis.

SEXUAL REPRODUCTION

Sexuality in the Mucorales

In the last century, zygospore formation in the Mucorales was known to be sporadic and unpredictable. Some rare isolates readily formed zygospores. Others quickly lost this ability, but most isolates never formed zygospores. This confused situation was clarified in the classic paper by Blakeslee (1904) in which he described the two forms of mating systems in these fungi. A minority of species were homothallic, mating between adjacent hyphae of the same mycelium, while the majority of species were heterothallic, with zygospore formation occurring only between individuals of two mating types, designated $(+)$ and $(-)$, of the same species. He also described the phenomenon of interspecific sexual reactions in which opposite mating types of different species will sometimes interact to allow some recognisable preliminary sexual differentiation to occur where the two cultures meet. These interactions allow the mating type, $(+)$, $(-)$ or neutral (i.e. showing no observable sexual reaction with the test strains used), of all isolates of Mucorales to be determined by mating them with strains related back to those of *Rhizopus nigricans* used in Blakeslee's original work.

In an extensive survey using many natural isolates of 36 species, Blakeslee, Cartledge, Welch & Bergner (1927) showed that both mating types were widely distributed (393 $(+)$ and 470 $(-)$) but that 235 isolates were neutral. These same authors and Blakeslee & Cartledge (1927) found no case of a heterothallic strain being a sexual intergrade, i.e. all remained true to type, and were never sometimes $(+)$ and sometimes $(-)$ depending on the mating partner. However, interspecific reactions were also seen between two homothallic species, or between homothallic and heterothallic species. As discussed later, some homothallic species seem to have a more $(+)$ or more $(-)$ behaviour.

The intensities of the interspecific reactions vary, but most commonly they are confined to zygophore formation with or without zygotropism and progametangium formation. Satina & Blakeslee (1930) comment that septum formation to delimit a gametangium was rarely observed, and the formation of an azygospore (a parthenogenetic zygospore from only one gametangium) was a very rare occurrence.

There are a few records of the interspecific reaction progressing as far as plasmogamy, with resultant zygospore production. Saito & Naganishi (1915) record hybrids between *Mucor* species, but Blakeslee (1920) suggests they might have been using morphologically different strains, and not different species. Callen (1940) records hybrids between the homothallic *Rhizopus sexualis* and *Rhizopus*, *Mucor* and *Absidia* species, but this has not been confirmed, and it must be very difficult to determine the exact origin of all hyphae in such matings. Burgeff (1925) succeeded in germinating a hybrid zygospore of *Phycomyces blakesleeanus* and *P. nitens*, but here again there must be some doubt as to the exact designation of the strains used. The occurrence of hybrid zygospores must remain not proven until they can be obtained from genetically defined strains and the products of their germination can be analysed. Certainly, Blakeslee & Cartledge (1927) and Satina & Blakeslee (1929, 1930) found no hybrid spores in a very large number of interspecific matings.

The study of interspecific reactions is important as it has defined the levels of control of the different stages of sexual reproduction (see Table 2). Thus Burnett (1953) has analysed some of the data of Blakeslee & Cartledge (1927) on interspecific reactions between *Mucor* and *Absidia* species, and has found no significant difference in intensity, as measured by visual inspection, between matings within the genera or between the genera. Clearly, all stages up to fusion, wall lysis and plasmogamy are common to all the species from the many genera participating in interspecific reactions.

Control by hormones

The presence or absence of the single chemical, trisporic acid, controls the formation of zygophores. If trisporic acid is added to unmated cultures of either (+) or (−) *Mucor mucedo*, sporangiophore production is suppressed, and zygophores are formed instead.

Trisporic acid, being a diffusible chemical controlling sexual differentiation, is thus clearly a sex hormone as the term is defined by Raper (1960). The history of the identification of the sex hormone activity of trisporic acid has been reviewed recently (Van den Ende & Stegwee, 1971; Gooday, 1972) and will not be repeated here in detail.

Trisporic acid can be bio-assayed by pipetting a sample into a well just in front of the advancing hyphal tips of (+) or (−) cultures of *Mucor mucedo* growing over nutrient agar plates. Some hours later zygophores will be seen and these can readily be counted. The zygophores have a characteristic pointed appearance, and grow in irregular spirals. They are formed nearer to the hyphal tips than the sporangiophores would have been. The zygophores and their vegetative mycelium are a much brighter orange than the mycelium in untreated cultures, due to an increase in carotene content (Gooday, 1968a). Thomas & Goodwin (1967) have shown an 80-fold increase of β-carotene in the $(-)$ strain of *Blakeslea trispora* when trisporic acid is added. These authors also report an increase in sterol content (chiefly ergosterol) in mated cultures of *B. trispora*.

However, apart from the formation of the zygophores themselves, the effects on isoprenoids, and a lack of a detectable effect on respiration (in *Mucor mucedo*, G. W. Gooday, unpublished material; in *Blakeslea trispora*, Feofilova & Taptykova, 1971), virtually nothing is known of the biochemical action of trisporic acid. A recent observation indicating another metabolic difference between mated and unmated cultures is that gallic acid accumulates in unmated but not in mated cultures of *B. trispora* (J. D. Bu'Lock, personal communication). This deserves further attention, particularly as sporangiophores of *Phycomyces blakesleeanus* have a high gallic acid content (Dennison, 1959).

The structure of trisporic acid C, the major component of the mixture of trisporic acids obtained from mated cultures of the Mucorales, is given in Table 1. In the bio-assay, free *cis*- and *trans*-trisporic acids C and B (the related ketone) are about equally active on both $(+)$ and $(-)$ mycelium, and trisporic acid A (with no oxygen in the side chain) is less active. Chemical methylation of trisporic acids B and C greatly attenuates their activity on $(+)$ *Mucor mucedo*, but has little effect on their activity on the $(-)$ strain. Trisporol C (the related neutral alcohol, Austin, Bu'Lock & Drake, 1970) and a related C_{18} ketone have no detectable activity on the $(-)$ strain, but have some activity on the $(+)$ strain (Bu'Lock, Drake & Winstanley, 1972).

The trisporic acids are biosynthesised as metabolites of β-carotene, via cleavage to give retinal, subsequent loss of a C_2 fragment, and a series of oxidations (Austin, Bu'Lock & Gooday, 1969; Austin *et al.* 1970; Bu'Lock, Austin, Snatze & Hruban, 1970). Bu'Lock & Winstanley (1971b) have shown that barbiturate will stimulate trisporic acid synthesis in mated fermentations of *Blakeslea trispora*, possibly by stimulating mixed-function oxygenase systems. Unmated fermentations of both $(+)$ and $(-)$ *B. trispora* will convert trisporol C to trisporic acids, but the $(+)$ is much more efficient. Conversion of added $[^{14}C]$-C_{18} ketone to trisporol C is very rapid in mated cultures, and can be detected in (+) but not in (−) fermentations of *B. trispora* (D. Quarrie, D. Taylor & J. D. Bu'Lock, personal communication). Trisporol C and the C_{18} ketone are probably intermediates in the biosynthetic sequence from retinal to trisporic acids, and so at least some of the intermediate enzymes of this pathway are constitutively present in detectable amounts in unmated cultures of *B. trispora*.

Burgeff (1924) first showed that physical contact is not necessary for hormone production, since the two mating types of *Mucor* will interact through a permeable membrane separating them to give rise to zygophore production. Plempel (1960, 1963) obtained the first reliable cell-free hormone preparations from culture filtrates of mated *M. mucedo*. He suggested a scheme in which two hormones ('gamones'), one produced by each mating type, were active on the opposite mating types, and were produced in response to 'progamones' from the opposite mating types. Recent work has shown that the one hormone, trisporic acid, is produced by both mating types and is active on both mating types. However, no detectable trisporic acid is produced in unmated cultures, and the nature of the metabolic co-operation whereby trisporate synthesis is initiated in mated cultures is still obscure. Diverse schemes have been considered and rejected in turn (Van den Ende, 1968; Gooday, 1968a; Sutter & Rafelson, 1968; Sutter, 1970; Van den Ende, Wiechmann, Reyngoud & Hendriks, 1970; Van den Ende & Stegwee, 1971). Most of these schemes involved cross-feeding of substrate levels of intermediates or of the appropriate enzymes, and this has not been detected in culture filtrates of *Blakeslea trispora*. Instead, experiments on hormone production by *B. trispora*, following the mixing together of mycelium of each mating type that had been pre-grown separately, have shown that both mating types contribute about equally to the final yield of trisporic acid (G. Marsman, quoted by Van den Ende & Stegwee, 1971). Protein synthesis is required in both mating types for this *de novo* synthesis (Bu'Lock & Winstanley, 1971a). There is now growing evidence that trisporic acid synthesis (albeit very much lower than in mated cultures and apparently very much dependent on culture conditions) can be detected in unmated cultures of both mating types of *B. trispora* in contact with culture filtrates of opposite mating type (R. P. Sutter; H. Van den Ende, personal communications). Thus a possible model involves a continual mutual interchange of very small amounts of uncharacterised agents (corresponding to Plempel's 'progamones') to induce the trisporate pathway in both mating types (Bu'Lock & Winstanley, 1971a). If this scheme is correct, these postulated factors should give rise to zygophore production in unmated strains via the formation of trisporic acid, but this has not yet been detected. As discussed later, a spontaneous zygophore-producing strain of (+) *M. mucedo* has been described which could have been constitutively producing trisporic acid. A renewed search for such strains could provide valuable support for this model.

The earlier suggestions of volatile substances being involved in the induction of zygophores (Blakeslee, 1904; Burgeff, 1924) have not been substantiated, and Köhler (1935) has shown that Burgeff's results can be explained by heterokaryosis of the strains (see later). However, evidence for the involvement of volatile metabolites in zygospore formation in the homothallic *Rhizopus sexualis* has been obtained by Hepden & Hawker (1961) who found that inhibition of zygospore production by low temperatures can be counteracted by a stream of air from cultures grown at higher temperatures. In view of such unexplained observations and of the rapidly growing list of physiologically active volatile fungal metabolites (Hutchinson, 1971), it is still possible that the factors involved in the induction of trisporate synthesis could be active either in a vapour phase or a liquid phase in different circumstances. As we have to propose the presence of two uncharacterised volatile substances to account for zygotropism (see later), the same substances could conceivably act at different concentrations as trisporate inducers and zygotropic attractants. We have an example of such a system in the different activities of antheridiol in *Achlya*, giving antheridial branching at low concentrations and chemotropism at higher concentrations (Barksdale, 1969).

Trisporic acid has been chemically identified from mated cultures of *Mucor mucedo*, *Phycomyces blakesleeanus*, *Blakeslea trispora*, (+) cultures of *B. trispora* and the homothallic *Zygorhynchus moelleri* (Van den Ende, 1968; Gooday, 1968*b*; Austin *et al.* 1969; Sutter, 1970; Sutter, Dehaven & Whitaker, 1972). The mutual induction of zygophores that occurs in the many interspecific sexual reactions suggests that trisporic acid is the sex hormone throughout the Mucorales. It will doubtless be identified from other species, but as it is active at very low concentrations (10^{-8} M in the bio-assay and certainly much lower *in vivo*), it may not always accumulate in detectable amounts. *M. mucedo*, the species in which Blakeslee originally discovered heterothallism, gives a yield of only 5×10^{-7} M, but as its zygophores are very distinctive it has always been the species of choice for the hormone bio-assays. *B. trispora*, the species from which trisporic acids were first characterised (Caglioti *et al.* 1964) gives very high yields (5×10^{-4} M, Bu'Lock & Winstanley, 1971*b*), and so has been used for the recent work to elucidate the biochemical mechanisms of the sexual collaboration leading to trisporic acid production. However, it must be remembered that this species may not necessarily exhibit a typical hormone situation, and any scheme proposed for the sexual interactions of *B. trispora* must also be tested with *M. mucedo* and other species. As Burnett (1965a) has discussed, the mating system in the Mucorales does not appear to operate very efficiently in natural environments. Zygospore formation is rare, zygospores are loath to germinate, and neutral strains are widespread. Thus there is possibly little selective pressure towards maintaining a tightly-controlled sexual interaction once compatible mating types have met. *B. trispora* could represent a situation in which hormone production is to some extent out of control. Nevertheless, the original initiation of trisporic acid synthesis by interacting compatible mycelia of heterothallic species seems to be very efficiently controlled, as we have not yet been able to mimic this experimentally in unmated cultures. Thus the sexual interactions ensure that zygophore induction and the concomitant suppression of sporangiophores only occur in close proximity to a mating partner, where the chances of mating are high.

The role of hormones in sexual reproduction in homothallic species is unclear. Trisporic acid has not been detected as a metabolite of homothallic species growing alone, but, as mentioned above, it is active in such low concentrations that it may not accumulate in detectable amounts. We can assume that it is involved because interspecific sexual reactions between two homothallic species and between homothallic and heterothallic species are well documented (Satina & Blakeslee, 1930), and *Blakeslea trispora* (+) and the homothallic *Zygorhynchus moelleri* interact to give detectable trisporate synthesis (Van den Ende, 1968). Zygospore formation can sometimes be induced in apparently sterile homothallic strains by interspecific reactions (e.g. Schipper, 1971). Some homothallic strains react with (+), others with (−), and others with both (+) and (−) heterothallic strains (Satina & Blakeslee, 1930), so although they are presumably constitutive for the necessary hormone production, some species have a genetic constitution to complement trisporate synthesis in (−) strains, and others in (+) strains.

Trisporic acid is produced as a typical secondary metabolite, accumulating late in mated fermentations of *Blakeslea trispora* (Bu'Lock, Drake & Winstanley, 1969; Bu'Lock & Winstanley, 1971b). In mated *Mucor mucedo* grown from mixed spore suspensions on agar plates, it is not detected until at least a day after inoculation. Its synthesis by *M. mucedo* on agar plates can be prevented by adverse conditions, such as low oxygen tension, but it can be detected about a day after return to atmospheric oxygen tension (Gooday, 1968a). In common with many other fungal sporulation systems some minimum vegetative growth has to occur (the metabolic 'trophophase' of Bu'Lock, 1967) before differentiation can commence in the metabolic 'idiophase'. This picture of the hormone being a secondary metabolite, being formed a day after the start of vegetative growth, may seem at variance with the observation that the young areas just behind the hyphal tips appear to be the site of production and response to the hormones when two compatible *Mucor* colonies meet (as illustrated in Plate 1). However, these young areas are connected with all of the coenocytic mycelium behind, and as Trinci (1971) has so clearly shown, a non-septate hypha allows a large peripheral growth zone (e.g. 8.66 mm in *Rhizopus stolonifera*) to contribute to vegetative apical growth, and so when the two colonies meet, two large peripheral growth zones containing older mycelium could participate in the sexual interactions.

Comparison with other fungal hormone systems

Despite many examples of hormonal control of interactions between two cells during sexual reproduction in fungi (reviewed by Raper, 1960; Machlis, 1966; Barksdale, 1969), only two other hormones have been characterised in detail. Table 1 compares trisporic acid with antheridiol and sirenin, sex hormones of the water moulds, *Achlya* spp. and *Allomyces* spp. Both antheridiol and sirenin are produced by the female cells without the need for stimulation by the male. Hormone B, giving rise to oögonial formation in the females, is only produced by the male *Achlya* plant in response to antheridiol, and so the control of its formation by the presence or absence of compatible mycelium may be compared with the control of trisporic acid production in the Mucorales. Trisporic acid, antheridiol and sirenin are all produced in very small quantities (apart from the *Blakeslea trispora* fermentation), are all active at very low concentrations, and are all specific to the species producing them or to closely related species. They are all concerned with intercellular communication between two cells, antheridiol and sirenin involving attraction of the male by the female, and trisporic acid giving rise to zygophores which are then mutually attractive.

Trisporic acid, antheridiol and hormone B all help to control stages in the orderly sequences of events culminating in sexual fusion in *Mucor* and in *Achlya*. The work by Barksdale (1969) is particularly instructive as she has shown that antheridiol can evoke a sequence of different responses from recipient male mycelium of *Achlya*. Antheridiol induces branching, delimitation of antheridia with accompanying meiosis, chemotropism, and release of hormone B. The response to antheridiol varies according to the concentration of antheridiol, the nutritional status of the mycelium, and the strain of fungus used.

Table 1. *Fungal sex hormones*

Hormone	Mucor Trisporic acid C	Achlya Antheridiol	Allomyces Sirenin
Molecular formula	$C_{18}H_{26}O_4$	$C_{29}H_{42}O_5$	$C_{15}H_{24}O_2$
Molecular weight	306	470	236
Structure			
Optimal yield of hormone (molar concentrations)	5×10^{-4} (*B. trispora*) 5×10^{-7} (*M. mucedo*)	6×10^{-9} (*A. bisexualis*)	1×10^{-6} (*A. arbuscula*/ *A. javanicus* hybrid)
Sensitivity of bioassay (molar conc.)	1×10^{-8}	1×10^{-11}	1×10^{-10}
Specificity of production and of activity	Many families in the Mucorales	*Achlya* spp.	*Allomyces* spp.
Control of synthesis	Interaction between $(+)$ and $(-)$ mating types gives synthesis by $(+)$ and $(-)$ mycelia	Synthesised by female mycelium	Synthesised by female motile gametes
Morphogenetic action of hormone	$(+)$ and $(-)$ mycelia produce zygophores; sporangiophore production prevented	Male mycelium branches; antheridia delimited with meiosis; chemotropism of antheridia	Chemotaxis of male gametes
Biochemical action of hormone	Increase in carotenoids and sterols	Increase in cellulase; production of hormone B	
References	Van den Ende, 1968; Austin *et al.* 1969; Bu'Lock & Winstanley, 1971*b*; Gooday, 1972	Thomas & Mullins, 1967; Barksdale, 1969	Machlis, Nutting, Williams & Rapoport, 1966; Nutting, Rapoport & Machlis, 1968

Zygotropism

The meeting of compatible zygophores is controlled by zygotropism. There appears to be no biological parallel to this mutual attraction between two cells. Two compatible zygophores grow steadily towards one another from distances of up to 2 mm apart, eventually to meet and fuse (Plates 1 and 2). The original description by Blakeslee (1904) is exactly applicable to these Plates: ' . . . a mutual attraction which may be termed zygotactic, is exercised between the zygophoric hyphae belonging to opposite mycelia and they may be seen gradually to approach each other. . . Two minutes before contact occurred, and while the hyphae were separated by a distance equal to about a third of their width, very slight protrusions were observed on the sides mutually facing, seemingly as if the forces which were drawing the filaments laterally had effected a bulging of the delicate walls at their growing points.' In contradiction to the account then (and sometimes still) in textbooks, Blakeslee emphasised that: 'The progametes in all cases are, from the very first, mutually adherent, and by their enlargement push apart the zygophoric hyphae from which they have originated.'

We know virtually nothing of the mechanism of zygotropism despite such elegant work as that by Banbury (1955) who has published photographs clearly showing attraction between two populations of zygophores of opposite mating type on adjacent agar blocks. Volatile hormones must surely be responsible for zygotropism, but the apparent necessity of having two complementary substances, both specific to transmitter and receiver, suggests a very unusual situation.

Zygophores of *Mucor mucedo* are not phototropic, but their asexual equivalents, the sporangiophores, are strongly phototropic. It is possible that zygotropism has the same cellular mechanism as phototropism, but is controlled by a chemo-receptor instead of a photo-receptor.

Plempel (1960, 1962) has provided preliminary evidence for the involvement of readily oxidized volatile substances in zygotropism but despite many attempts this has not yet been confirmed by such techniques as gas–liquid chromatography, and this intriguing phenomenon remains to be elucidated.

Carotenogenesis

Most observers of sex in the Mucorales have commented on an increase in carotene content of the reacting mycelium (reviewed by Hesseltine, 1961; Burnett, 1965b). Trisporic acid was characterised as a carotenogenic agent (Caglioti *et al.* 1964) before its hormone activity was suspected. As has been previously discussed, it is now clear that

β-carotene is a precursor of trisporic acid biosynthesis and that tri-sporic acid itself then acts to stimulate β-carotene production. The net result is the formation of zygophores with a high carotene content. We have recently been able to suggest a function for this increased caroteno-genesis in *Mucor mucedo*. Radio-tracer experiments have shown that during sexual reproduction carotene can be oxidatively polymerised to give sporopollenin (G. V. Gooday, P. Fawcett, D. Green & G. Shaw, unpublished material) in a similar manner to the polymerisation of carotenoids in anthers of higher plants to give sporopollenin in pollen grains (Shaw, 1970). In *M. mucedo*, the sporopollenin is found as a component (about 1 %) of the zygospore wall, and so the polymerisa-tion, carotene to sporopollenin, is a further example of the process discussed by Bu'Lock (1961) of a microbial secondary metabolite being incorporated as a structural component during sporogenesis. The *Mucor* sporopollenin is chemically similar to higher plant sporopollenins, and is virtually identical to an artificial polymer produced by catalysed oxida-tion of β-carotene (Shaw, 1970). Like other sporopollenins, that from *Mucor* is extremely resistant to biological and chemical attack and so must give added protection to the zygospores.

Zygospore germinations and the possibility of heterokaryosis

The zygospores are truly resting spores, are very difficult to germinate, and so are good examples of the 'memnospores' of Gregory (1966). They show a constitutive dormancy (as defined by Sussman & Halvor-son, 1966), often for about three months, and even then seem little affected by a wide range of external conditions (Blakeslee, 1906; Hocking, 1967). Their percentage germination is often low (e.g. less than 1 % for *Mucor hiemalis*, Gauger, 1965). Germinations have not yet been recorded for many species (e.g. *Rhizopus sexualis*).

The most common pattern of germination is heterothallic, with all the germ-spores from one zygospore being of one mating type (Blakeslee, 1906; Gauger, 1965). However, mycelia from these germ-spores are often sexually inactive for some time after germination, later becoming either $(+)$ or $(-)$, and germ-spores sometimes give sexual hetero-karyons (Blakeslee, 1906; Cerdá-Olmedo, quoted in Bergman *et al.* 1969) like those prepared artificially by Burgeff (1914). These sexual heterokaryons show a wide range of behaviour, usually not producing zygospores, but sometimes showing characteristics of sexually reacting mycelium, such as increased carotenogenesis, and production of zygophore-like hyphae, the 'pseudophores'. They appear to be less stable than heterokaryons isogenic for mating type, as they break down to give the two mating types, sometimes as sectors in the same mycelium. Gauger (1966) investigated cultures from germ-spores of *Mucor hiemalis* that produced azygospores and found that they were unstable, breaking down to normal (+) and (−) strains on repeated single spore isolations. Heterokaryosis, although apparently not arising in the Mucorales through hyphal fusions, may be more widespread than is suspected. The sporangiospores and germ-spores have variable numbers of nuclei, and there has been little determined effort to obtain genetically homogeneous stock cultures for much of the work described here.

Köhler (1935), in a brilliant genetical analysis, showed that the strains used by Burgeff (1924) and later by Verkaik (1930) in their studies on sexuality in *Mucor mucedo* were heterokaryons, with a mixture of nuclei carrying different morphological characters. Burgeff and Verkaik had experienced difficulties in reproducing experiments, particularly those involving the induction of zygophores. Köhler obtained several different apparently homozygous strains showing different morphological characters by selection from the 'wild types'. Certain culture conditions (e.g. alkaline medium in the dark) favoured occasional spontaneous production of zygophores, particularly in one of his (+) strains. He repeatedly subcultured to select for zygophore production by plucking zygophores and allowing them to regenerate mycelium, and obtained a variety 'Laniger' which spontaneously produced zygophores. This was not due to sexual heterokaryosis (as in 'pseudophore' production by sexual heterokaryons), with (+) and (−) nuclei in the same mycelium, as then this strain should be sexually attenuated. However, 'Laniger' retained its characteristics on repeated sub-culturing, and when crossed with an apparently homozygous (−) culture, it produced zygospores as abundantly as its 'wild type' heterokaryotic parent mycelium. The zygophores of 'Laniger' had all the characteristics of ordinary zygophores, including a high carotene content. However, when Köhler illuminated these cultures, these zygophores quickly changed their appearance to give dwarf sporangia. As noted by Burgeff (1924), zygophores can also turn into sporangiophores at the edges of sexually interacting areas of 'wild-type' (+) and (−) mycelium, and if trisporic acid is put in the centre of an unmated culture the zygophores formed furthest away will later become sporangiophores (G. W. Gooday, unpublished material). Thus a sufficient concentration of trisporic acid can initiate zygophore production and maintain the zygophores, but at lower concentrations the zygophores formed can later revert to sporangia, presumably as the trisporate concentration falls below a critical level.

Köhler suggested that 'Laniger' was a homozygous (+) strain in which the production of zygophores was a primary character which could be overcome by the effect of light. He crossed 'Laniger' with an apparently homozygous (−) strain and analysed the progeny. All the 'Laniger' characteristics segregated linked with the (+) mating type. As mentioned earlier germ-spores usually give rise to mycelium that is sexually immature for several generations, and Köhler observed that 'Laniger' characters were not expressed until the mycelium was sexually mature.

It seems clear that the spontaneous production of zygophores and carotene accumulation of 'Laniger' could have been due to constitutive trisporic acid biosynthesis to give physiologically active levels of the hormone. Köhler's work shows us the contribution that genetical analysis could make to an understanding of the breeding system of the Mucors. Considerable progress has been made recently with the genetics of carotenoid and phototropic mutants of *Phycomyces blakesleeanus* (Heisenberg & Cerdá-Olmedo, 1968; Bergman *et al.* 1969), and it is hoped that these techniques will yield information on the control of sexuality in the Mucors.

THE ROLE OF THE CELL WALL IN DIFFERENTIATION
Cell wall composition

The value of cell wall composition as a character in studying phylogenetic and ontogenetic relationships in fungi is now firmly established (Bartnicki-Garcia, 1968, 1970). The vegetative cell walls of *Mucor* have chitin (poly *N*-acetyl-D-glucosamine), chitosan (poly D-glucosamine) and polyphosphates (presumably associated with the chitosan) as characteristic components (e.g. *M. rouxii*, Bartnicki-Garcia, 1968; *M. ramannianus*, Jones, Bacon, Farmer & Webley, 1968). Poly D-glucuronides have also been identified as major components (Bartnicki-Garcia & Reyes, 1968). This composition contrasts strongly with that of other major groups of fungi such as Ascomycetes and Basidiomycetes which have glucans and chitin as major components.

The walls of the sporangiophores and yeast cells of *M. rouxii* (Bartnicki-Garcia, 1968) and of the arthrospores (formed by fragmentation of vegetative hyphae) of *M. ramannianus* (Jones *et al.* 1968) have the same components as the vegetative mycelium but in differing proportions. Although not analysed in detail, chitin and chitosan are components of the zygospore wall of *M. mucedo* (G. W. Gooday, unpublished) and the warty outer layer is impregnated with a black pigment and sporopollenin (see above).

However, the sporangiospore walls are totally different in composition, being chiefly of glucan, with a greatly reduced amino-sugar content, and with a deposit of melanin (Bartnicki-Garcia, 1968; Jones *et al.* 1968). Other fungi, such as *Aspergillus phoenicis* (Bloomfield & Alexander, 1967), show quantitative differences between spore and hyphal walls, although these have similar qualitative compositions. Bartnicki-Garcia (1970) has put forward the intriguing suggestion that the *Mucor* sporangiospore wall composition is an ontogenetic recapitulation of cell wall phylogeny, for as well as the chemical discontinuity there is also the structural discontinuity at spore formation between the spore wall and the sporangial wall, and at spore germination between the spore wall and the germ-tube wall (see below). The sporangiospore wall would represent an ancient trait, suggesting a relationship with the chytridomycete water moulds, and the cell wall produced on germination would be the more advanced type characteristic of the Zygomycetes.

Protective functions can be ascribed to the melanin and sporopollenin in the spore walls as melanin can inhibit lysis by polysaccharases (Bull, 1970) and sporopollenin is extremely resistant to chemical and biological degradation (Shaw, 1970).

Cell wall fusions

Vegetative intra-specific cell–cell fusions, which are so common in Ascomycetes and Basidiomycetes, are very rare in the Mucorales (Buller, 1933). That it is the cell walls which prevent fusion of the vegetative hyphae is shown by the experiments creating viable artificial heterokaryons. If two drops of protoplasm from two cut hyphae of different mycelia of the same fungus are mixed they can regenerate a wall and then grow normally (Weide, 1939; Heisenberg & Cerdá-Olmedo, 1968). Heterokaryons can also be formed inside a hypha by injecting cytoplasm from another hypha with a syringe, or by inserting half a sporangiophore inside another one (Burgeff, 1914; Bergman *et al.* 1969). The only cell–cell fusion observed in *Mucor mucedo* is that of the compatible zygophores during sexual reproduction. This fusion is a rapid process, for as soon as they meet (Plates 1 and 2) zygophores have to be torn apart to be separated. Hawker & Gooday (1969) have examined early stages of fusion with the electron microscope, and find that the two fusion walls immediately lose their separate identities and appear as one wall.

The specificity of the sexual fusion must lie in components of the cell walls of the zygophores which are not present in the walls of the intermingling vegetative hyphae below. The description of specific cross-reacting glycoproteins on the surfaces of sexually agglutinating yeast cells of *Hansenula wingei* (Crandall & Brock, 1968) suggests that a search for sex-specific agglutinins in *Mucor* could be rewarding.

However, the surface recognition between zygophores is sex-specific and not species-specific, for imperfect sexual reactions between different species of Mucorales involving fusion between (+) and (−) strains are well documented (see above). The suggestion by Burgeff (1924) that the fusion of some mycoparasitic members of the Mucorales and their Mucoraceous hosts could represent an imperfect sexual reaction (i.e. that the parasitism is sex-specific) was not borne out by extensive investigations by Satina & Blakeslee (1926), who did however find in some cases there was a tendency for an increase in the extent of parasitism when host and parasite were of different mating types.

Cell wall formation

The hyphae of the Mucorales are typically coenocytic, growing apically and not forming cross walls. There are exceptions, and Benjamin (1959) regards septa in vegetative hyphae as a characteristic of the more advanced families. However, cross walls are formed in vegetative mycelium in response to injury, or to cut off older portions of hyphae (Buller, 1933), during the formation of some forms of differentiated cells such as chlamydospores (intercalary resting cells), arthrospores, or some sporangiophores (e.g. *Pilobolus* spp.). Although formed centripetally, these septa usually do not have central pores (except in some of the more advanced families; Benjamin, 1959), such as are found in Ascomycetes and Basidiomycetes. Thus the septa effectively seal off areas of the hyphae, preventing protoplasmic streaming and passage of nuclei and other organelles. However, this need not mean a total severing of connection, for plasmodesmata are present in the septa delimiting the gametangia (Hawker, Gooday & Bracker, 1966; Hawker & Gooday, 1967; Hawker & Beckett, 1971), and could well be present in other septa.

The formation of the columellar wall and sporangiospore walls during maturation of the sporangiophore of *Gilbertella persicaria* has been beautifully illustrated by Bracker (1966, 1968). Columellar cleavage to delimit the sporangium is nearly complete when the sporogenous region is in mid-cleavage, and is completed before late cleavage. The cleavage vesicles are identical in appearance to those that delimit the spores, but the wall is quite distinct from the spore walls, as it is fibrillar in section and extends down inside the sporangiophore wall, to taper to an end. The *de novo* formation of the walls of the sporangiospores involves the encapsulation of small groups of nuclei by the proliferating cleavage vesicles in the sporangium. The cleavage vesicles have small electron dense granules over their inner surfaces, which become the outer surfaces of the developing spores. These granules coalesce into rods to form a network which envelops the spore protoplasts and becomes thickened with a two-layered appearance. A much thicker secondary spore wall is then laid down between this and the spore plasmalemma. With this elaborate development it is not surprising that the spore walls are totally different in chemical composition to the sporangiophore walls (see above). The spores of *Gilbertella* are ornamented with ribbon-like appendages at both ends, and these are formed before the spores are delimited, and eventually become attached to the outermost layer of the spore coat. At present we have no knowledge of the chemical nature of the different parts of the spore wall.

On germination, sporangiospores actively swell to become large spherical cells (e.g. from $4.5 \times 5.5 \,\mu$m to about $14 \,\mu$m diameter in *Rhizopus arrhizus*; Ekundayo & Carlile, 1964). During this swelling a completely new wall of randomly arranged fibrils in an amorphous matrix is laid down inside the original spore wall (e.g. *Rhizopus* spp., Hawker & Abbott, 1963; Ekundayo, 1966; *Mucor rouxii*, Bartnicki-Garcia, Nelson & Cota-Robles, 1968; *Gilbertella persicaria*, Bracker & Halderson, 1971). This new wall grows out to become the germ tube, and is in structural continuity with the vegetative mycelium. Thus during germination we can visualise the transition from the glucan-rich spore wall to the chitin–chitosan vegetative wall.

The delimitation of the gametangia from the suspensor cells during sexual reproduction has been described for *Rhizopus sexualis* (Hawker & Gooday, 1967; Hawker & Beckett, 1971). The two septa form by the centripetal coalescence of vesicles and eventually seal off the gametangia, apart from the plasmodesmata referred to above. Extensive endoplasmic reticulum and vesicles are associated with this process and with the subsequent thickening to give what will eventually be the end walls of the mature zygospore.

The thickened zygospore wall is formed by the layering down of material inside the original gametangial walls, which remain as torn fragments on the outside of the mature spore (Hawker & Gooday, 1968; Hawker & Beckett, 1971). The two major layers are the outer very thick warty coating and the inner translucent elastic layer which is laid down inside the outer layer with a thin 'smoothing layer' between them, acting as a template. The thick outer layer is heavily pigmented, is probably the site of sporopollenin in *Mucor mucedo*, and gives histo-

chemical reactions for chitin and chitosan. The inner layer gives histochemical reactions for chitin and is strongly birefringent in polarised
light (G. W. Gooday, unpublished material). The inner layer is im-
permeable to chemical fixatives and so probably helps to explain the
failure to induce germination experimentally.

Cell wall lysis

It seems obvious that controlled enzymic lysis of cell walls must occur
during morphogenesis of fungal hyphae, although it is very difficult to
obtain direct evidence for it. Some recent examples include: the involvement of lysosomes in the extreme case of 'deliquescence' of the gills of
the ink-caps, *Coprinus* spp. (Iten & Matile, 1970), and in the budding of
yeast cells (Matile, Cortat, Wiemken & Frey-Wyssling, 1971); the big
increase in cellulase content in male hyphae of *Achlya* responding to
antheridiol by branching (Thomas & Mullins, 1967); the controlled
enzymic degradation of complex septa to allow migration of nuclei
during heterokaryotisation in Basidiomycetes that has been investigated
in *Schizophyllum commune* (Niederpruem & Wessels, 1969; Janszen &
Wessels, 1970). In the Mucorales, enzymic wall lysis is possibly involved
in such processes as spore germination and sporangiospore release, and
it is almost certainly involved during sexual reproduction in the lysis of
the fusion wall that separates the gametangia.

Fine-structural studies of conjugating *Rhizopus sexualis* (Hawker &
Gooday, 1969; Hawker & Beckett, 1971) and *Phycomyces blakesleeanus*
(Sassen, 1965) show that as the gametangial septa are being completed
the fusion wall is being totally broken down, from the centre outwards.
During this dissolution the plasmalemma next to the wall appears
wrinkled, and the vesicles and lomasomes become associated with it.
Sassen (1965) claimed the involvement of chitinase in this dissolution,
but this awaits confirmation as his assay was for N-acetyl-D-glucos-
aminidase and not for chitinase, and he also found this enzyme in the
vegetative mycelium.

The process of fusion wall dissolution, and hence plasmogamy,
unlike that of zygophore fusion, apparently does not occur in interspecific sexual reactions (see earlier), and so is species specific. These
two processes must be controlled by different mechanisms (see
Table 2).

Table 2. *Control of sexual reproduction in Mucorales*

Stage	Specific to Mucorales	Specific to mating type	Specific to species	Possible site of specificity	Possible no. of genes involved
(1) Production of trisporate 'inducers'	.	+	.	Enzymes of secondary metabolism	Few
(2) Response to 'inducers' by production of trisporate	.	+	.	Synthesis of new enzymes	Few
(3) Response to trisporate by zygophore production	+	.	.	Differentiation of vegetative mycelium	Many
(4) Zygotropism	.	+	.	Growth of zygophore tip	Few
(5) Zygophore fusion	.	+	.	Mutual recognition at cell-wall surface	Few
(6) Gametangial formation	+	.	.	Control of septum formation and cell-wall differentiation	Many
(7) Lysis of fusion wall	.	.	+	Mutual release of enzymes at cell surface	Few
(8) Plasmogamy	.	.	+	Mutual recognition at plasmalemma surface	Few
(9) Zygospore development	+	.	.	Differentiation to give new cell wall	Many
(10) Karyogamy	.	.	+	Recognition between nuclei	Few

Compiled from: Blakeslee, 1904; Burgeff, 1924; Satina & Blakeslee, 1929, 1930; Austin *et al.* 1969; Bu'Lock & Winstanley, 1971a.

CONCLUSION

The chief concern of this article has been a consideration of the remarkable intercellular co-ordination of differentiation shown when sexual reproduction supplants asexual reproduction in the Mucorales. Sexual reproduction involves a sequence of interactions between two cultures. Some of these interactions can take place in interspecific matings, but the complete sequence giving rise to the zygospore only occurs in intraspecific matings. A summary of the specificity and levels of control of these interactions is given in Table 2. At all stages the two mating types behave in a complementary fashion, and we are still continuing the search started by A. F. Blakeslee in 1904, and carried on by him and his colleagues for many years, to find the true difference between them. The steps in the sequence that are mating-type specific and those that are species specific plausibly involve few genes, whereas those which are common to all interacting members of the Mucorales involve complete morphogenetic developments, and are almost certainly controlled by large numbers of genes. The limited amount of genetical analysis that has been done indicates that the mating type is controlled by a pair of allelomorphs at one locus, but there is no evidence that a group of genes is not involved, either as a chromosome segment or controlled by a 'super gene' (Bergman *et al.* 1969). Table 2 requires four different activities of the mating type determinant and although it is possible that the one gene controls all these it seems more likely that several genes are involved.

The hormonal control of sexual reproduction in the Mucorales enables the formation of the sexual cells to be regulated so that they are only formed in close proximity to a compatible mating partner. This is clearly important, as the zygophores are formed at the expense of the sporangiophores. Thus the formation and action of trisporic acid precludes sporangiospore production and so prevents the normal means of dispersal of these fungi. It might be said that as it is produced by both mating types and is active on both mating types, trisporic acid cannot be thought of as a hormone involved in the intercellular co-ordination of sexual reproduction. However, there is some evidence that trisporic acid is not translocated along hyphae to any great extent from where it is synthesised (G. W. Gooday, unpublished material) but instead it is released to diffuse through the growth medium. Therefore, it can co-ordinate the activities of hyphae some distance from those actually involved in the primary sexual interactions. The end result is a wide zone of zygophores, on either side of the line of meeting of the

PLATE 1

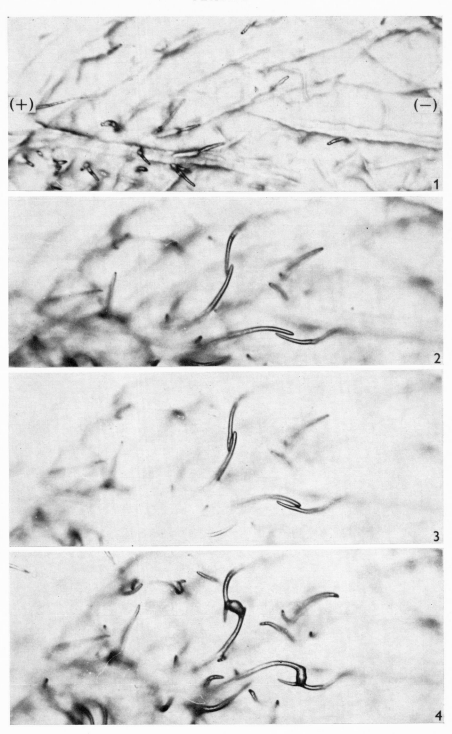

(*Facing p. 288*)

PLATE 2

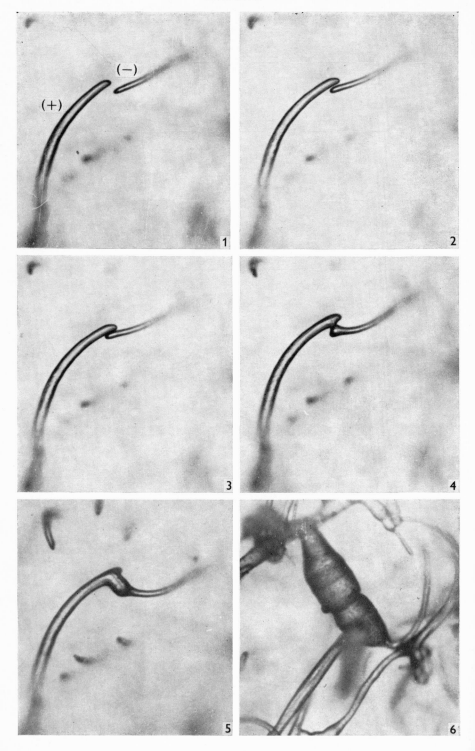

two cultures, that are able to participate in zygospore production by arching over from both sides and meeting in the middle. The formation and action of trisporic acid can then be represented as follows:

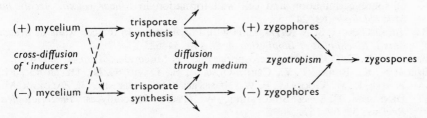

Thus the zygophores, with the distinctive characters of their high carotene content, their zygotropism, and their ability to fuse in compatible pairs, are only formed when there is a good chance of completion of the sexual process to give zygospores. The four events: (1) initiation of trisporate synthesis, (2) resultant zygophore formation, (3) zygotropism, and (4) zygophore fusion, form an orderly sequence in which the two interacting fungi are co-ordinated by intercellular chemical communication: (1) by the interchange of the trisporate 'inducers', (2) by the diffusion of trisporic acid itself, (3) by the interchange of zygotropic hormones, and (4) by the mutual cell–cell recognition at the two cell surfaces.

If we look at other fungal hormone systems as well, we see that both *Mucor* and *Achlya* use sex hormones to co-ordinate the differentiation of the cells of both mating partners before they are in physical contact, and *Mucor*, *Achlya* and *Allomyces* all use sex hormones to bring cells into physical contact. The central role of diffusible chemicals in the control of these sequences suggests that these may be examples of phenomena of cell–cell interactions of much wider biological significance.

REFERENCES

AUSTIN, D. J., BU'LOCK, J. D. & DRAKE, D. (1970). The biosynthesis of trisporic acids from β-carotene via retinal and trisporol. *Experientia*, **26**, 348–49.

AUSTIN, D. J., BU'LOCK, J. D. & GOODAY, G. W. (1969). Trisporic acids: Sexual hormones from *Mucor mucedo* and *Blakeslea trispora*. *Nature, London*, **223**, 1178–9.

BANBURY, G. H. (1955). Physiological studies in the Mucorales. III. The zygotropism of zygophores of *Mucor mucedo* Brefeld. *Journal of Experimental Botany*, **6**, 235–44.

BARKSDALE, A. W. (1969). Sexual hormones of *Achlya* and other fungi. *Science, New York*, **166**, 831–7.

BARTNICKI-GARCIA, S. (1968). Cell wall chemistry, morphogenesis, and taxonomy of fungi. *Annual Review of Microbiology*, **22**, 87–108.

BARTNICKI-GARCIA, S. (1970). Cell wall composition and other biochemical markers in fungal phylogeny. In *Phytochemical Phylogeny*, ed. J. B. Harborne, pp. 81–103. New York and London: Academic Press.

BARTNICKI-GARCIA, S., NELSON, N. & COTA-ROBLES, E. (1968). Electron microscopy of spore germination and cell wall formation in *Mucor rouxii*. *Archiv für Mikrobiologie*, **63**, 242–55.

BARTNICKI-GARCIA, S. & REYES, E. (1968). Polyuronides in the cell walls of *Mucor rouxii*. *Biochimica et Biophysica Acta*, **170**, 54–62.

BENJAMIN, R. K. (1959). The merosporangiferous Mucorales. *Aliso*, **4**, 321–433.

BERGMAN, K., BURKE, P. V., CERDÁ-OLMEDO, E., DAVID, C. N., DELBRÜCK, M., FOSTER, K. W., GOODELL, E. W., HEISENBERG, M., MEISSNER, G., ZALOKAR, M., DENNISON, D. S. & SHROPSHIRE, W. (1969). Phycomyces. *Bacteriological Reviews*, **33**, 99–157.

BLAKESLEE, A. F. (1904). Sexual reproduction in the Mucorineae. *Proceedings of the American Academy of Arts and Sciences*, **40**, 205–319.

BLAKESLEE, A. F. (1906). Zygospore germinations in the Mucorineae. *Annales Mycologici*, **4**, 1–28.

BLAKESLEE, A. F. (1920). Sexuality in Mucors. *Science, New York*, **51**, 375–82 and 403–9.

BLAKESLEE, A. F. & CARTLEDGE, J. L. (1927). Sexual dimorphism in Mucorales. II. Interspecific reactions. *Botanical Gazette*, **84**, 51–7.

BLAKESLEE, A. F., CARTLEDGE, J. L., WELCH, D. S. & BERGNER, A. D. (1927). Sexual dimorphism in Mucorales. I. Interspecific reactions. *Botanical Gazette*, **84**, 27–50.

BLOOMFIELD, B. J. & ALEXANDER, M. (1967). Melanins and resistance of fungi to lysis. *Journal of Bacteriology*, **93**, 1276–80.

BRACKER, C. E. (1966). Ultrastructural aspects of sporangiospore formation in *Gilbertella persicaria*. In *The Fungus Spore*, ed. M. F. Madelin, pp. 39–60. London: Butterworths.

BRACKER, C. E. (1968). The ultrastructure and development of sporangia in *Gilbertella persicaria*. *Mycologia*, **60**, 1016–67.

BRACKER, C. E. & HALDERSON, N. K. (1971). Wall fibrils in germinating sporangiospores of *Gilbertella persicaria* (Mucorales). *Archiv für Mikrobiologie*, **77**, 366–76.

BULL, A. T. (1970). Inhibition of polysaccharases by melanin: Enzyme inhibition in relation to mycolysis. *Archives of Biochemistry and Biophysics*, **137**, 345–56.

BULLER, A. H. R. (1933). *Researches on Fungi*, vol. 5. London: Longmans, Green.

BU'LOCK, J. D. (1961). Intermediary metabolism and antibiotic synthesis. *Advances in Applied Microbiology*, **3**, 293–342.

BU'LOCK, J. D. (1967). *Essays in Biosynthesis and Microbial Development*. New York and London: John Wiley and Sons.

BU'LOCK, J. D., AUSTIN, D. J., SNATZKE, G. & HRUBAN, L. (1970). Absolute configuration of trisporic acids and the stereochemistry of cyclization in β-carotene biosynthesis. *Journal of the Chemical Society*, D, 255–6.

BU'LOCK, J. D., DRAKE, D. & WINSTANLEY, D. J. (1969). Growth of *Blakeslea trispora* relative to carotene, sterol, and cofactor production. *Journal of General Microbiology*, **58**, xi–xii.

BU'LOCK, J. D., DRAKE, D. & WINSTANLEY, D. J. (1972). The trisporic acid series of fungal sex hormones: structures, specificity and transformations. *Phytochemistry*, **11**, 2011–18.

BU'LOCK, J. D. & WINSTANLEY, D. J. (1971*a*). Carotenoid metabolism and sexuality in Mucorales. *Journal of General Microbiology*, **68**, xvi–xvii.

BU'LOCK, J. D. & WINSTANLEY, D. J. (1971b). Trisporic acid production by *Blakeslea trispora* and its promotion by barbiturate. *Journal of General Microbiology*, **69**, 391–4.

BURGEFF, H. (1914). Untersuchungen über Variabilität, Sexualität und Erblichkeit bei *Phycomyces nitens* Kunze I. *Flora*, **107**, 259–316.

BURGEFF, H. (1924). Untersuchungen über Sexualität und Parasitisimus bei Mucorineen. I. *Botanische Abhandlungen*, **4**, 1–135.

BURGEFF, H. (1925). Über Arten und Artkreuzung in der Gattung Phycomyces Kunze. *Flora*, **118/119**, 40–6.

BURNETT, J. H. (1953). A study of the breeding systems in lower plants. D.Phil. Thesis, University of Oxford.

BURNETT, J. H. (1965a). The natural history of recombination systems. In *Incompatibility in Fungi*, ed. K. Esser & J. R. Raper, pp. 98–113. Berlin: Springer-Verlag.

BURNETT, J. H. (1965b). Functions of carotenoids other than in photosynthesis. In *Chemistry and Biochemistry of Plant Pigments*, ed. T. W. Goodwin, pp. 381–403. New York and London: Academic Press.

CAGLIOTI, L., CAINELLI, G., CAMERINO, B., MANDELLI, R., PRIETO, A., QUILICO, A., SALVATORI, T. & SELVA, A. (1964). Sulla constituzione degli acidi trisporici. *La Chimica e l'Industria, Milano*, **46**, 961–6.

CALLEN, E. O. (1940). Morphology, cytology and sexuality of the homothallic *Rhizopus sexualis* (Smith) Callen. *Annals of Botany*, **4**, 791–818.

CRANDALL, M. A. & BROCK, T. D. (1968). Molecular basis of mating in the yeast *Hansenula wingei*. *Bacteriological Reviews*, **32**, 139–63.

DENNISON, D. S. (1959). Gallic acid in Phycomyces sporangiophores. *Nature, London*, **184**, 2036.

EKUNDAYO, J. A. (1966). Further studies on germination of sporangiospores of *Rhizopus arrhizus*. *Journal of General Microbiology*, **42**, 283–91.

EKUNDAYO, J. A. & CARLILE, M. J. (1964). The germination of sporangiospores of *Rhizopus arrhizus*; spore swelling and germ-tube emergence. *Journal of General Microbiology*, **35**, 261–9.

FEOFILOVA, E. P. & TAPTYKOVA, S. D. (1971). Effect of trisporic acids on respiration of the (−) strain of *Blakeslea trispora*. *Mikrobiologiya*, **40**, 495–500.

GAUGER, W. L. (1965). The germination of zygospores of *Mucor hiemalis*. *Mycologia*, **57**, 634–41.

GAUGER, W. L. (1966). Sexuality in an azygosporic strain of *Mucor hiemalis*. I. Breakdown of the azygosporic component. *American Journal of Botany*, **53**, 751–5.

GOODAY, G. W. (1968a). Hormonal control of sexual reproduction in *Mucor mucedo*. *New Phytologist*, **67**, 815–21.

GOODAY, G. W. (1968b). The extraction of a sexual hormone from the mycelium of *Mucor mucedo*. *Phytochemistry*, **7**, 2103–5.

GOODAY, G. W. (1972). Fungal sex hormones. *Biochemical Journal*, **127**, 2–3P.

GREGORY, P. H. (1966). The fungus spore: what it is and what it does. In *The Fungus Spore*, ed. M. F. Madelin, pp. 1–14. London: Butterworths.

HAWKER, L. E. & ABBOTT, P. McV. (1963). An electron microscope study of maturation and germination of sporangiospores of two species of *Rhizopus*. *Journal of General Microbiology*, **32**, 295–8.

HAWKER, L. E. & BECKETT, A. (1971). Fine structure and development of the zygospore of *Rhizopus sexualis* (Smith) Callen. *Philosophical Transactions of the Royal Society of London*, B, **263**, 71–100.

HAWKER, L. E. & GOODAY, M. A. (1967). Delimitation of the gametangia of *Rhizopus sexualis* (Smith) Callen: an electron microscope study of septum formation. *Journal of General Microbiology*, **49**, 371–6.

HAWKER, L. E. & GOODAY, M. A. (1968). Development of the zygospore wall in *Rhizopus sexualis* (Smith) Callen. *Journal of General Microbiology*, **54**, 13–20.

HAWKER, L. E. & GOODAY, M. A. (1969). Fusion, subsequent swelling and final dissolution of the apical walls of the progametangia of *Rhizopus sexualis* (Smith) Callen: an electron microscope study. *New Phytologist*, **68**, 133–40.

HAWKER, L. E., GOODAY, M. A. & BRACKER, C. E. (1966). Plasmodesmata in fungal cell walls. *Nature, London*, **212**, 635.

HEISENBERG, M. & CERDÁ-OLMEDO, E. (1968). Segregation of heterokaryons in the asexual cycle of *Phycomyces*. *Molecular and General Genetics*, **102**, 187–95.

HEPDEN, P. M. & HAWKER, L. E. (1961). A volatile substance controlling early stages of zygospore formation in *Rhizopus sexualis*. *Journal of General Microbiology*, **24**, 155–64.

HESSELTINE, C. W. (1961). Carotenoids in the fungi Mucorales. *United States Department of Agriculture, Technical Bulletin*, no. 1245, 33 pp.

HOCKING, D. (1967). Zygospore initiation, development and germination in *Phycomyces blakesleeanus*. *Transactions of the British Mycological Society*, **50**, 207–20.

HUTCHINSON, S. A. (1971). Biological activity of volatile fungal metabolites. *Transactions of the British Mycological Society*, **57**, 185–200.

ITEN, W. & MATILE, P. (1970). Role of chitinase and other lysosomal enzymes of *Coprinus lagopus* in the autolysis of fruiting bodies. *Journal of General Microbiology*, **61**, 301–9.

JANSZEN, F. H. A. & WESSELS, J. G. H. (1970). Enzymic dissolution of hyphal septa in a Basidiomycete. *Antonie van Leeuwenhoek*, **36**, 255–7.

JONES, D., BACON, J. S. D., FARMER, V. C. & WEBLEY, D. M. (1968). Lysis of cell walls of *Mucor ramannianus* Möller by a *Streptomyces* sp. *Antonie van Leeuwenhoek*, **34**, 173–82.

KÖHLER, F. (1935). Genetische Studien an *Mucor mucedo* Brefeld. *Zeitschrift für induktive Abstammungs- u. Vererbungslehre*, **70**, 1–54.

MACHLIS, L. (1966). Fungal sex hormones. In *The Fungi*, vol. 2, ed. G. C. Ainsworth & A. S. Sussman. New York and London: Academic Press.

MACHLIS, L., NUTTING, W. H., WILLIAMS, M. W. & RAPOPORT, H. (1966). Production, isolation and characterisation of sirenin. *Biochemistry, Easton*, **5**, 2147–52.

MATILE, P., CORTAT, M., WIEMKEN, A. & FREY-WYSSLING, A. (1971). Isolation of glucanase-containing particles from budding *Saccharomyces cerevisiae*. *Proceedings of the National Academy of Sciences, U.S.A.* **68**, 636–40.

NIEDERPRUEM, D. J. & WESSELS, J. G. H. (1969). Cytodifferentiation and morphogenesis in *Schizophyllum commune*. *Bacteriological Reviews*, **33**, 505–35.

NUTTING, W. H., RAPOPORT, H. & MACHLIS, L. (1968). The structure of sirenin. *Journal of the American Chemical Society*, **90**, 6434–8.

PLEMPEL, M. (1960). Die zygotropische Reaktion bei Mucorineen. I. *Planta, Berlin*, **55**, 254–8.

PLEMPEL, M. (1962). Die zygotropische Reaktion der Mucorineen. III. *Planta, Berlin*, **58**, 509–20.

PLEMPEL, M. (1963). Die chemischen Grundlagen der Sexualreaktion bei Zygomyceten. *Planta, Berlin*, **59**, 492–508.

RAPER, J. R. (1960). The control of sex in fungi. *American Journal of Botany*, **47**, 794–808.

SAITO, K. & NAGANISHI, H. (1915). Bemerkungen zur Kreuzung zwischen verschiedenen *Mucor*-Arten. *Botanical Magazine, Tokyo*, **29**, 149–54.

SASSEN, M. M. A. (1965). Breakdown of the plant cell wall during the cell-fusion process. *Acta Botanica Neerlandica*, **14**, 165–96.

SATINA, S. & BLAKESLEE, A. F. (1926). The Mucor parasite *Parasitella* in relation to sex. *Proceedings of the National Academy of Sciences, U.S.A.* **12**, 202–7.

SATINA, S. & BLAKESLEE, A. F. (1929). Criteria of male and female in bread moulds (Mucors). *Proceedings of the National Academy of Sciences, U.S.A.* **15**, 735–40.

SATINA, S. & BLAKESLEE, A. F. (1930). Imperfect sexual reactions in homothallic and heterothallic Mucors. *Botanical Gazette*, **90**, 299–311.

SCHIPPER, M. A. A. (1971). Induction of zygospore production in *Mucor saximontensis*, an agamic strain of *Zygorhynchus moelleri. Transactions of the British Mycological Society*, **56**, 157–8.

SHAW, G. (1970). Sporopollenin. In *Phytochemical Phylogeny*, ed. J. B. Harborne, pp. 31–58. New York and London: Academic Press.

SUSSMAN, A. S. & HALVORSON, H. O. (1966). *Spores, Their Dormancy and Germination*. New York and London: Harper and Row.

SUTTER, R. P. (1970). Trisporic acid synthesis in *Blakeslea trispora. Science, New York*, **168**, 1590–2.

SUTTER, R. P., DEHAVEN, R. N. & WHITAKER, J. P. (1972). Trisporic acids in *Phycomyces blakesleeanus. Bacteriological Abstracts* (in press).

SUTTER, R. P. & RAFELSON, M. E. (1968). Separation of β-factor synthesis from stimulated β-carotene synthesis in mated cultures of *Blakeslea trispora. Journal of Bacteriology*, **95**, 426–32.

THOMAS, D. M. & GOODWIN, T. W. (1967). Studies on carotenogenesis in *Blakeslea trispora*. I. General observations on synthesis in mated and unmated strains. *Phytochemistry*, **6**, 355–60.

THOMAS, D. & MULLINS, J. T. (1967). Role of enzymatic wall softening in plant morphogenesis: hormonal induction in *Achlya. Science, New York*, **156**, 84–5.

TRINCI, A. P. J. (1971). Influence of the width of the peripheral growth zone on the radial growth rate of fungal colonies on solid media. *Journal of General Microbiology*, **67**, 325–44.

VAN DEN ENDE, H. (1968). Relationship between sexuality and carotene synthesis in *Blakeslea trispora. Journal of Bacteriology*, **96**, 1298–303.

VAN DEN ENDE, H. & STEGWEE, D. (1971). Physiology of sex in Mucorales. *Botanical Review*, **37**, 22–36.

VAN DEN ENDE, H., WIECHMANN, A. H. C. A., REYNGOUD, D. J. & HENDRIKS, T. (1970). Hormonal interactions in *Mucor mucedo* and *Blakeslea trispora. Journal of Bacteriology*, **101**, 423–8.

VERKAIK, C. (1930). Über das Entstehen von Zygophoren von *Mucor mucedo* (+) unter Beeinflüssung eines von *Mucor mucedo* (−) abgeschiedenen Stoffes. *Proceedings. Koniklijke Nederlandse akademie van Wetenschappen*, C33, 656–8.

WEIDE, A. (1939). Beobachtungen an Plasmaexplantaten von Phycomyces. *Archiv für experimentelle Zellforschung*, **23**, 299–337.

EXPLANATION OF PLATES

PLATE 1

Zygospore formation, zygotropism, and fusion in *Mucor mucedo*. The (+) culture is growing from the left and the (−) from the right, meeting in the centre. All × 80.

Fig. 1. Zygophore formation. 3 h after original meeting of the two cultures.

Fig. 2. Zygotropism of two pairs of zygophores. 90 min later.

Fig. 3. Fusion of the two pairs of zygophores. 5 min later.

Fig. 4. Formation of progametangia. 30 min later.

PLATE 2

Zygotropism and fusion of a pair of zygophores of *Mucor mucedo*. The (+) zygophore is growing from the left and the (−) from the right. All × 150.

Fig. 1. The two zygophores just before meeting, having attracted each other by zygotropism.

Fig. 2. 4 min later. Note asymmetry of tips just before contact.

Fig. 3. 1 min later. Contact of zygophores.

Fig. 4. 10 min later. Swelling at points of contact.

Fig. 5. 18 min later. Progametangium formation.

Fig. 6. 12 h later. Progametangia and suspensors delimited and much enlarged. Lysis of fusion wall probably in progress.

DIFFERENTIATION IN THE ASPERGILLI

J. E. SMITH AND J. G. ANDERSON

Department of Applied Microbiology, Strathclyde University,
Glasgow, Scotland

INTRODUCTION

The genus *Aspergillus* was first defined by Micheli (1729) to include microscopic fungi possessing erect stalks and spore heads. The tendency of the spore chains or columns to radiate from the central stalk created a pattern not unlike the *aspergillium* used by priests and this led Micheli, himself a priest, to give the generic name to this group of fungi.

The aspergilli have a worldwide distribution and are undoubtedly one of the most ubiquitous of all groups of fungi. The ability to utilize an enormous variety of organic materials for growth has made the aspergilli possibly the greatest contaminants of man-made and natural organic products, particularly in warm, humid atmospheres (Smith, 1969). Certain aspergilli are also responsible for primary and secondary infections in man and animals while the conidia of some species may be allergenic when inhaled or otherwise contacted (Austwick, 1965, 1966). Many aspergilli elaborate mycotoxins during growth as contaminants of cereal and other food products and such toxins may be excreted into the substrate on which they are growing or be retained within the mould cells and released only with the rupturing of the mycelium during processing or consumption (Wogan & Pong, 1970; Semeniuk, Harshfield, Carlson, Hesseltine & Kwolek, 1971).

The aspergilli are not entirely destructive in nature and several species have had their metabolic powers harnessed for commercial exploitation. *Aspergillus niger* has been used for the commercial production of citric, gluconic and gallic acids while strains of *A. oryzae* have long been used for saccharification of rice starch in the production of saké and similar potable liquors. Strains of *A. oryzae* are used in the manufacture of soy sauce and diastase enzyme powders.

Thus the aspergilli are a group of fungi that exert a profound influence on mankind. Furthermore, as filamentous eukaryotic organisms, they represent an important link between the truly unicellular state and the multicellular condition of higher organisms. It must be anticipated that a fuller understanding of the factors which regulate growth and development in the aspergilli could well be of major importance and relevance not only to fundamental studies on differentiation in eukaryotes but also to specific industrial processes utilizing these organisms.

Thom & Raper (1945) recognised 86 species of *Aspergillus* in their monograph of the genus. The later monograph by Raper & Fennell (1965) further expanded this list to well over 100, and additional species are regularly being found. Sexual reproduction has been described for a few groups of the aspergilli in particular in some species within the *Aspergillus glaucus*, *A. fisheri*, *A. fumigatus* and *A. nidulans* series. For detailed studies on the morphology of the ascosporic aspergilli the comprehensive treatise by Raper & Fennell (1965) should be consulted. Sexually-reproducing aspergilli have been classified in the Ascomycetes in the Order Eurotiales while the non-ascosporic aspergilli are classified in the Deuteromycetes or Fungi Imperfecti in the form order Moniliaceae (Alexopoulos, 1962). It was considered by Thom & Raper (1945) that the generic name *Aspergillus* should be applied to all of these fungi, whether or not an ascosporic stage was produced, as they considered that the finding and describing of the sexual stage merely completed the characterization of a fungus already well-known.

The fruit body in this group is called a *cleistothecium* and the enclosed ascospores are released by rupturing of the cleistothecial wall. Mature cleistothecia may contain up to 100000 asci (Fig. 1). Little information is available on the physiology and biochemistry of this phase of the life cycle. Species which have been studied genetically include *Aspergillus niger* (Pontecorvo, Roper & Forbes, 1953), *A. nidulans* (Pontecorvo, Roper, Hemmons, MacDonald & Bufton, 1953), *A. oryzae*, *A. sojae* (Ishitani, Ikeda & Sakaguchi, 1956), *A. fumigatus* (Stromnaes & Garber, 1963), and *A. amstelodami* (Lewis, 1969), although only *A. nidulans* and *A. amstelodami* have been subjected to more orthodox genetical analysis utilizing sexual reproduction (Fincham & Day, 1971).

The life cycle is initiated by the germination of the conidium. The germ tube becomes a hypha which elongates by deposition of new protoplasm at the hyphal tip, and branches develop below the growing apices. An extensive mycelium develops which constitutes the thallus or vegetative phase of the life cycle. Although the mycelium is septate a coenocytic condition prevails and cytoplasmic circulation occurs through small central pores in the septae.

Asexual reproduction of the aspergilli in surface cultures is characterized by the formation of complex thick-walled conidiophores or stalks which grow erect and perpendicular from the colonizing mycelium (Fig. 1). In submerged culture conidiophores are reduced in length but are otherwise normal and grow out from mycelial pellets or from units

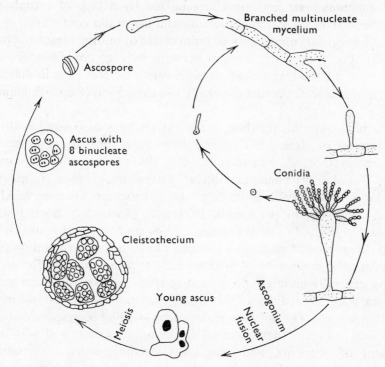

Fig. 1. Life cycle of *Aspergillus nidulans* (after Fincham & Day, 1971).

of filamentous mycelium (Vezina, Singh & Sehgal, 1965; Galbraith & Smith, 1969b; Anderson & Smith, 1971a). A novel microcycle conidiation has recently been obtained with several *Aspergillus* species in which the conidiophore develops directly from an enlarged conidium without an intervening mycelial stage (Anderson & Smith, 1971b, 1972).

THE CONIDIOSPORE

Fungal spore germination represents a simple, easily studied example of eukaryotic differentiation. While germination has been defined by Manners (1966) as the formation of a germ tube from the dormant spore Sussman (1966) has considered that any measurable irreversible change can be the accepted criterion of germination. For the purpose of this review the term germination is used in its most widely accepted sense as referring to the processes and changes occurring during the resumption of development of a resting structure and its transformation to a morphologically different structure (Allen, 1965).

Conidia may be considered as both the beginning and the end of the developmental cycle and are characterized by a lack of cytoplasmic movement, reduced metabolic activity and low water content (Gregory, 1966). *Aspergillus* conidia may be uninucleate or multinucleate and there is invariably a close relationship between average conidium size and number of nuclei. Uninucleate conidia rarely exceed 4 μm in diameter while multinucleate conidia are rarely less than 5 μm (Raper & Fennell, 1965).

The physiological and fine structural changes associated with the germination of *Aspergillus* conidia have been the subject of several investigations: *A. niger* (Yanagita, 1957, 1964; Sussman & Halvorson, 1966; Tanaka & Yanagita, 1963*a*; Tsukahara, 1968); *A. nidulans* (Border & Trinci, 1970; Bainbridge, 1971; Florance, Denison & Allen, 1972); *A. fumigatus* (Campbell, 1971); *A. giganteus* (Trinci, Peat & Banbury, 1968); *A. oryzae* (Tanaka, 1966) and *A. parasiticus* (Grove, 1972). Most conidia undergo a swelling phase during germination prior to germ tube emergence. Conidia which require nutrients for germination generally undergo more swelling than conidia which can germinate in water. Yanagita (1957) distinguished two phases within the swelling process. The initial or endogenous swelling corresponded to the small size increase that occurred in deionised water, and was independent of nutrients, and exogenous swelling which was nutrient dependent. Since the latter phase is clearly a period of active metabolism and growth and not a passive increase in size by imbibition the term spherical growth may be more applicable to this part of germination (Bartnicki-Garcia, Nelson & Cota-Robles, 1968). An increase in incubation temperature above that normally used in conidial germination studies (25–30°) was observed to have marked effects on the morphology of germination of *A. niger* conidia (Anderson & Smith, 1972). The increase in conidium diameter of × 3 which occurred during normal germination of *A. niger* conidia at 30° was within the range described by Fletcher (1969) for a variety of fungal spores while the increase of × 5.6 at 44° was unusual (Table 1). During this process an increase in spore surface area of × 30 occurred, yet the wall remained thick, clearly showing that wall extension was accompanied by wall synthesis. The full implication of these giant conidia will be discussed later.

In *Aspergillus nidulans* the cell wall of dry, dormant conidia is three-layered. After hydration, two new layers appear in the conidial wall and at germination the germ tube wall is continuous with the newly-formed innermost layer (Florance *et al.* 1972). There is a localization of organelles on the side of the conidium from which the germ tube arises and as

Table 1. *Comparison of spore diameter, spore surface, and spore volume of unswollen spores and large SG spores of Aspergillus niger produced after 48 h cultivation at 44°*

	Spore diameter (μm)	Spore surface area (μm^2)	Spore volume (μm^3)
(A) Unswollen spores	3.5	38.5	22.4
(B) Large SG spores	19.6	1207	3933
Ratio B/A	5.6	31.4	175.6

the germ tube emerges the apex has a multivesicular appearance. Apical vesicles have also been demonstrated in the germ tubes of *A. parasiticus* (Grove, 1972). In this report the vesicles could only be detected when extreme care was taken to avoid prefixation stresses or injury to the sensitive apical cytoplasm, and it was suggested that the failure to detect these vesicles in numerous earlier studies could be attributed to this factor. Although one or occasionally two germ tubes grow out at the time of germination, multiple germ tube formation can occur under certain conditions (J. G. Anderson & J. E. Smith, unpublished results). Nuclear studies have been carried out on conidia of several species of *Aspergillus* during germination (Baker, 1945; Sakaguchi & Ishitani, 1952; Yanagita, 1964; Campbell, 1971; Weijer & Weisberg, 1966; Rosenberger & Kessel, 1967, 1968; Robinow & Caten, 1969). Mitochondria are normally present in dormant conidia and during germination an increase in numbers and change in size and shape have been reported (Hawker, 1966; Sussman, 1966). A general feature of most dormant conidia is a sparse endoplasmic reticulum which rapidly increases during germination. Partial synchrony of conidial germination as judged by germ tube emergence has been achieved in *A. nidulans* (Bainbridge, 1971).

Conidia of *Aspergillus niger* have been shown to require glucose, phosphate, L-alanine or L-proline and carbon dioxide for complete germination (Rippel & Bortels, 1927; Vakil, Rao & Bhattacharyya, 1961; Yanagita, 1957). The fixation of carbon dioxide is common among germinating fungal spores and in *A. niger* the fixed carbon rapidly enters the general biosynthetic activities of the conidium and becomes widely distributed (Yanagita, 1963). Mandels (1963) has suggested that the apparent need of many fungal spores for such substances may not be nutritional but that they function as activators breaking dormancy. *A. fumigatus* can germinate in the absence of external nutrients and this could imply that dormancy in this spore can be broken by a non-nutrient factor and that the process of germination can be supported by metabolism of internal food reserves.

In *Aspergillus fumigatus* lipids have been implicated as storage products utilized during germination (Campbell, 1971) while in *A. giganteus* glycogen has been identified (Trinci *et al.* 1968). Polyols and trehalose are of widespread occurrence in dormant spores and are important reserve materials although control devices governing their metabolism remain obscure (Aitken & Niederpruem, 1970). Dormant conidia of *A. oryzae* (Horikoshi, Iida & Ikeda, 1965) and *A. niger* (Sumi, 1928; Takebe, 1960) contain large amounts of mannitol relative to trehalose. Results with *A. oryzae* (Horikoshi & Ikeda, 1966) suggest that since trehalose is inhibited by mannitol this enzyme may only function after mannitol depletion.

An abrupt increase in the rate of oxygen uptake at germ tube formation was reported for *Aspergillus niger* (Yanagita, 1957) and *A. oryzae* conidia (Terui & Mochizuki, 1955). Sussman (1966) considered that an initial glycolytic metabolism followed by a more aerobic type may be widespread in fungal spore germination. With *A. niger* conidia Bhatnagar & Krishnan (1960*a, b*) found that phosphatases and enzymes of the Embden–Meyerhof–Parnas (EMP) pathway were absent in dormant conidia but appeared during germination. Since ATP is required for pre-germination synthetic reactions it has been suggested that ATP could be synthesized from the energy released by the simple hydrolysis of poly-metaphosphates and phospholipids present in most spores (Gottlieb, 1966). Chemical analysis of *A. nidulans* (Shepherd, 1957) and *A. niger* (Nishi, 1961) conidia during germination would appear to confirm this hypothesis.

Macromolecular synthesis occurs early in spore germination (Gottlieb, 1966). In *Aspergillus niger* (Yanagita, 1963; Nishi, 1961) RNA is synthesized early in the germination cycle well in advance of DNA. Staples, Syamananda, Kao & Block (1962) have demonstrated early incorporation of ^{14}C into protein during germination of *A. niger* conidia. In *A. nidulans* chemical analysis of synchronously germinating conidia indicated RNA, protein and DNA synthesis at 30, 150 and 180 min, respectively, after initiation of germination (Bainbridge, 1971). The fact that RNA synthesis occurred before protein synthesis in *A. niger* (Yanagita, 1963) has conflicted with other suggestions that germination in fungal spores may be initiated at the translational rather than at the transcriptional level (Hollomon, 1969; Van Etten, 1969). Horikoshi *et al.* (1965) have shown that the nucleotide composition of rRNA and the sedimentation coefficients of the ribosomes from dormant and germinated conidia of *A. oryzae* are similar. Horikoshi & Ikeda (1968, 1969) have shown that the protein synthesis activity and thermal stability of conidial ribosomes are less than those from vegetative mycelium.

If the apparatus required for protein synthesis was present and potentially active in ungerminated conidia but the DNA repressed, the transcription of mRNA would be prevented. Tanaka, Kogane & Yanagita (1965) showed that DNA remained stable during spore germination of *Aspergillus oryzae* and Kogane & Yanagita (1964) showed the DNA of dormant and germinated conidia to have similar physical and chemical properties. This work does not show directly that the DNA in conidia is repressed, only that it seems to remain stable during germination. An ability to conserve mRNA in an unused but readily available form would supersede a requirement for transcription of mRNA on germination and permit protein synthesis in ungerminated spores. The occurrence of long-lived mRNA in eukaryotic cells has been discussed by Mandelstam (1969).

In a medium that supports rapid growth the replication of nuclei in each germ tube of *Aspergillus nidulans* can be highly synchronized (Kessel & Rosenberger, 1968). During hyphal tip elongation in *A. nidulans* there is a wave of nuclear divisions followed at about 30 min intervals by the formation of a series of incomplete septa (Clutterbuck, 1970). The germ tubes of *A. nidulans* have been shown to grow exponentially in length from their inception (Trinci, 1969) while the germ tube specific growth rate in *A. wentii* and *A. niger* has been shown to be greater than the specific growth rate of vegetative hyphae in submerged culture (Trinci, 1971*a*).

THE VEGETATIVE GROWTH PHASE

The somatic growth phase of the aspergilli is normally well-developed, profusely branched, septate, and hyaline, and the cells are usually multinucleate. The partitioning septa have single central pores which are initially open but may become plugged during ageing. In hyphal regions where the septal pores remain open, migration of all inclusions and organelles including nuclei is possible and in these regions the mycelium is functionally coenocytic. Studies have been made of the fine structure of vegetative hyphae of *Aspergillus niger* (Tanaka & Yanagita, 1963*a*; Tsukahara & Yamada, 1965) and *A. fumigatus* (Campbell, 1970). In a recent study on *A. nidulans* Bainbridge, Bull, Pirt, Rowley & Trinci (1971) observed that the hyphal wall was composed of up to four concentric layers.

Bartnicki-Garcia (1968) has stressed the important role of the cell wall in the structure and behaviour of fungi and rightly considers that all morphological developments in the fungi can be reduced to a question of cell wall morphogenesis. Chemically, fungal cell walls are 80–90 % polysaccharide with the remainder largely composed of protein and lipid. The aspergilli belong to the chitan–glucan category in the fungal cell wall subdivision proposed by Bartnicki-Garcia (1968). The walls of *Aspergillus niger* contain *N*-acetylglucosamine, glucose, mannose and galactose (Bloomfield & Alexander, 1967; Ruiz-Herrera, 1967; Johnson, 1965). The main components in the cell wall of *A. nidulans* are glucose and *N*-acetylglucosamine with minor quantities of mannose, galactosamine, galactose, glucuronic acid, protein and lipid (Bull, 1970; Zonneveld, 1971). The high glucan plus chitin content and the presence of galactose and galactosamine in the cell walls of *A. nidulans* is typical of Ascomycete fungi (Bull, 1970). Loss of mechanical strength of the hyphal walls of *A. nidulans* was reported by Katz & Rosenberger (1970) by a mutation which lead to the formation of hyphal walls lacking chitin. Protoplasts can now be readily obtained from many fungi including *Aspergillus nidulans* (Peberdy & Gibson, 1971; Peberdy, 1972) and under suitable conditions protoplasts of *A. nidulans* can revert to the filamentous morphological form from which they were produced. The process is complex but undoubtedly offers exciting prospects for studies on wall synthesis and on the factors controlling the differentiation of cellular form.

Fungal hyphae grow by deposition of material at the tip although the details of this process are not yet clearly understood. The zone at the hyphal apex is rich in cytoplasmic vesicles and nearly devoid of other cell components. These vesicles obviously contain wall building materials and are considered to be incorporated at the expanding surface. Using *Aspergillus niger* McClure, Park & Robinson (1968) originally suggested that the cluster of cytoplasmic vesicles seen by electron microscopy in the hyphal apex corresponded to the densely staining or refractive spheroid body called the Spitzenkörper which can be observed by light microscopy in living and fixed hyphal tips. However, Grove & Bracker (1970) now consider that the Spitzenkörper correlates with a smaller specialized zone that occurs within the cluster of apical vesicles and which sometimes contains microvesicles, tubules and a few ribosomes.

Hyphal extension by the apical growth process results in the characteristic cylindrical shape of the hyphae and represents the normal mode of development in mycelial fungi. In common with most other filamentous fungi growing hyphae of *Aspergillus nidulans* will actively incorporate *N*-acetylglucosamine almost exclusively at the tip and thus inferring that wall synthesis is limited to the apex of growing hyphae (Katz & Rosenberger, 1971). However, in the presence of cycloheximide there was an increase in the amount of labelled precursor in subapical regions of the hyphae and a reduction at the tip. The effect could be reversed by removing the inhibitor and also did not appear to be due to chitin turnover. Similar effects were also caused by osmotic shock. Both treatments also produced hyphae with abnormally large numbers of branches and septa.

Abnormalities of hyphal morphogenesis, particularly spherical cell formation, have frequently been observed in the aspergilli, apparently occurring as a response to one of several conditions unfavourable to growth. In *Aspergillus nidulans* large isodiametric cells were observed in hyphae growing under conditions of low oxygen tension (Carter & Bull, 1971). Cohen, Katz & Rosenberger (1969) have described a mutant of *A. nidulans* which produces walls with reduced chitin content when grown at high temperatures. Under these conditions the conidia germinate into irregular-shaped hyphae with swollen sections. Manganese deficiency has been reported to result in the production of large yeast-like cells from the mycelium of *A. parasiticus* by a process of arthrosporal morphogenesis (Detroy & Ciegler, 1971). Under conditions of low pH Bent & Morton (1963) observed large swollen cells in a number of fungi including *A. niger* and *A. flavus*. Other observations of the formation of globose cells in these species have been made (Testi-Camposano, 1959; Leopold & Seichertova, 1959; Wildman, 1966; Childs, Ayres & Koehler, 1971). Elevated temperatures have been shown to produce marked morphological changes in the conidia and newly-formed hyphae of *A. niger* (Anderson & Smith, 1972). At 30° hyphae were thin and relatively unbranched; at 41° hyphal swelling and excessive branch formation occurred; at 43° a thick multiseptate form was produced by a process resembling arthrospore formation; at 44° there was complete inhibition of apical growth although spherical growth of the conidia could occur. These morphological changes which are shown in Plate 1, figs. 1–4 were interpreted as an increasingly severe inhibition of the apical growth process. Austwick (1965) has described a series of changes in the hyphal morphology of *A. fumigatus* which occur during the course of an infection and which are possibly characteristic for all pathogenic aspergilli. Wide globose- or oval-celled hyphae are found at the centre of acute lesions and these represent the germinated conidium and its primary hyphae. Narrow, unbranched hyphae which may be straight or spiral and septate or non-septate are then produced. The next hyphal form is densely branched and septate and this type grows around the periphery of chronic lesions. The final form is the normal vegetative hyphae which rapidly invade dead or weakened tissue. While these changes probably reflect the degree of resistance of the host to infection nothing is known of the factors regulating the morphological form of the hyphae in the tissues (Austwick, 1965).

Aspergillus fungi have been grown on solid or liquid surface culture and in submerged liquid culture. On static culture a heterogeneous growth is produced which includes aerial, surface and submerged mycelia each with its particular physiological condition (Smith & Galbraith, 1971).

Various studies (Yanagita & Kogane, 1962, 1963a; Nagasaki, 1968a, b) on surface colonies of Aspergillus niger have demonstrated that cytological characteristics and physiological activities of the hyphae vary from the periphery to the inside portions. From differential localization of enzymes, wall thickness and intracellular structure of the hyphae it appears that hyphal cells show a characteristic differentiation depending on their location in the colony. Autoradiographic studies of nucleic acid and protein metabolism in individual hyphae of A. nidulans (Nishi, Yanagita & Maruyama, 1968) and A. niger (Fencl, 1970) clearly demonstrate the heterogeneous nature of the hyphal filament. This differentiation of cellular activities along the hyphal filament is probably an inevitable consequence of the apical growth process.

Trinci & Banbury (1967) have shown that individual hyphae of Aspergillus nidulans grow at a constant rate, i.e. their length increases linearly with time. In a subsequent study on the kinetics of branch formation Trinci (1970b) reported that lateral branch formation does not significantly affect the growth rate of the parent hyphae whereas apical branching is preceded by a slight deceleration (18 %) in growth rate. Apical and lateral branches did not immediately grow at the same rate as their parent hyphae. A short adaptation period of 14–44 min elapsed before the branches attained a constant linear growth rate. Recently, Trinci (1971b) has presented a model for the colonial growth of A. nidulans in which the radial growth rate of the colony is dependent on the length of the leading hyphae spanning the peripheral growth zone of the colony and the specific growth rate of these hyphae. For A. nidulans the radial growth rate could be divided into a short exponential phase, a phase of declining growth rate and a phase of constant growth rate (linear growth). The inoculum size did not influence the colony radial growth rate of A. nidulans (Trinci & Gull, 1970) whereas the concentration of glucose in the medium had a marked effect (Trinci,

1969). The effect of decreasing substrate water potential on the turgor pressure and colony growth rate of *A. wentii* has been examined by Adebayo, Harris & Gardner (1971). Mycelial turgor pressure ranged from 12–18 bar but showed no consistent relationship to growth rate or to substrate water potential. Jinks (1957, 1959*a*, 1966) has clearly demonstrated with *A. nidulans* and *A. glaucus* that cytoplasmic as well as nuclear determinants influence fungal growth rate. Differences in the rates of growth between sub-cultures of these fungi maintained by conidia or hyphal tip transfers were attributed to an unequal distribution of cytoplasmic elements which arose during spore formation or which occurred among different hyphal tips isolated from the same colony. In submerged culture the growth kinetics of *A. nidulans* are dependent on the growth form. Trinci (1969) and Carter & Bull (1969) have shown that diffuse filamentous mycelium grows exponentially. Individual unbranched hyphae grow at a constant rate and the ability of mycelial hyphae to show exponential growth kinetics in submerged culture has been attributed to branch formation (Trinci & Banbury, 1967). Growth of *A. nidulans* as large pellets results in cube root growth kinetics (Trinci, 1970*a*) indicating that parts of the pellets are either not growing or are growing under sub-optimum conditions. Indeed, cytological observations of the internal structure of *A. niger* pellets (Yanagita & Kogane, 1963*b*; Clark, 1962) demonstrate the heterogeneous nature of large pellets. A detailed description and model of the internal structure of *A. nidulans* pellets has been presented by Trinci (1970*a*).

The mechanisms involved in pellet formation in *Aspergillus* spp. have been discussed by Galbraith & Smith (1969*a*) and Trinci (1970*a*). While pellet formation appears to be initiated by the aggregation of germinating conidia or small mycelial clumps the mechanism of the aggregation process is not known. Of the various factors known to affect the aggregation process the pH value of the medium with *A. niger* (Galbraith & Smith, 1969*a*) and with *A. nidulans* the use of a large inoculum (Trinci, 1970*a*) and addition of desoxycholate to the medium (Dorn & Rivera, 1966) have been successfully used to induce growth forms suitable for biochemical and kinetic studies.

Because of the heterogeneous nature of the mycelial growth in surface culture most physiological studies have been made on submerged batch culture either in flasks or under the more exacting conditions of fermenter culture. Organisms growing in batch culture are exposed to an ever-changing environment and for this reason it is difficult to obtain exact information on the effect of the environment on the organism. In practice, most batch studies have been concerned with the influence of the environments on secondary metabolite formation or on the induction of new growth forms.

The work of Pirt and his associates using continuous-flow fermenters has opened up a novel way of analysing fungal development, in particular the vegetative phase. In continuous-flow culture, growth can occur in an unchanging environment and it is now possible to achieve a steady state with filamentous fungi. The physiology of *Aspergillus nidulans* has been examined under conditions of batch and glucose-limited chemostat culture and the effect of different steady-state growth rates and dissolved oxygen tensions (DOT) examined (Carter & Bull, 1969). During exponential growth in batch culture and at near maximum specific growth rates and also at low DOT levels in chemostat culture there was a greatly increased activity of the hexosemonophosphate (PP) pathway of carbon catabolism. When the glucose feed was reduced to $1.5 \times$ the maintenance ration or stopped completely there was a rapid drop in the respiratory activity of the mycelium and virtually all the glucose was catabolized by the EMP pathway (Bainbridge *et al.* 1971). In these starved cultures the autolysis of some hyphae provided the substrate for the maintenance or growth of the remaining hyphae. There was no evidence that autophagy provided substrates for endogenous metabolism. The mean length of hyphal segments and the degree of branching were independent of the dissolved oxygen tension (Carter & Bull, 1971).

Continuous culture may also represent an ideal system for studying the biogenesis of mitochondria in eukaryotic organisms. Highly active and intact mitochondria have previously been extracted from *Aspergillus niger* (Watson & Smith, 1967) and *A. oryzae* (Watson, Paton & Smith, 1969).

ASEXUAL DIFFERENTIATION
General morphology

Although the change from the vegetative condition to the asexual reproductive stage occurs biochemically at an early stage during vegetative growth (Smith & Valenzuela-Perez, 1971) the first morphological indication that a major phase of differentiation is beginning is the enlargement of certain cells in the mycelium. Such cells, termed *foot-cells* produce a single erect thick-walled stalk cell, the conidiophore, as a branch perpendicular to the long axis of the cell. Foot-cells may be submerged in the substratum or arise from aerial hyphae and their formation has been studied in *Aspergillus clavatus* (Klein, 1944). Conidiophores are generally unbranched and non-septate though some branched conidiophores have been shown in *A. glaucus* and *A. wentii*

(Raper & Fennell, 1965). The structure and composition of the characteristically thickened conidiophore walls has been studied by Castle (1945) and Farr (1954). Tanaka & Yanagita (1963b) have studied the fine structure of conidiophore development in *A. niger*. The wall of the foot-cell is considerably thicker than the wall of the distal end of the conidiophore and is composed of not less than eight layers of different densities. In some species, including *A. nidulans*, conidiophores may show superficial, wart-like concretions of dried pigment on their outer surfaces.

During elongation and growth the conidiophore tip is pointed but then enlarges into a bulbous head, the vesicle. This swelling is accompanied by extensive secondary thickening of the vesicle wall which may obtain a thickness of 1–2 μm in *Aspergillus giganteus* (Trinci *et al.* 1968). As the multinucleate vesicle develops a large number of fertile cells or sterigmata are produced over its surface either parallel and clustered in terminal groups or radially from the entire surface. The position of the sterigmata cause the characteristic shape of the sporing head when viewed in low power microscopy.

The sterigmata may occur as one series or as a double series with each primary sterigma bearing a cluster or several secondary sterigmata or phialides at the apex. From the phialides the conidia are developed and in still air conditions long chains of conidia will develop. Because of the abundance of conidia production in all aspergilli their colour (black, brown, yellow, or green) gives the characteristic colour to a mature culture. Hughes (1953) has defined the phialide as a unicellular structure which is normally terminal on a single or branched conidiophore. Phialides are typically bottle-shaped and the conidia develop as endogenous phialospores, formed in basipetal succession inside the phialide by transverse septation, and are extruded at the tip (Tubaki, 1958). Cytoplasmic continuity occurs for some time in a chain of conidia until the terminal conidia assume the general size and shape characteristic of the species. During this maturation the conidium lays down a secondary wall within the primary wall, which separates it completely from the parent cell. It is generally considered that the vesicle supplies the initial complement of one or more nuclei to the phialide (Yuill, 1950). A single nucleus is present in the phialide of *Aspergillus niger* (Tanaka & Yanagita, 1963b), *A. giganteus* (Trinci *et al.* 1968) and *A. nidulans* (Clutterbuck, 1969a). Diploid strains of *A. nidulans* have larger phialides than haploid strains; however the dimensions of the conidiophore and vesicle are unaffected by ploidy (Clutterbuck, 1969a).

Culture systems for conidiation studies

In the studies on conidiation in the aspergilli, as in studies on other forms of microbial differentiation, particular emphasis has been placed on the development and refinement of culture techniques. In many instances improvements in understanding of differentiation processes have rapidly followed the introduction of culture techniques which have increased the measure of control over morphogenesis. The use of defined media for growth, sophisticated culture vessels and the application of the continuous flow culture technique have allowed a precise analysis of the environmental factors which affect morphogenesis. While such studies represent an essential aspect of morphogenetic studies the elucidation of the biochemical mechanisms of the environmental effects is of greater fundamental importance. Studies on the biochemistry of differentiation in micro-organisms represent, in fact, a study of the summary of the differentiation of cell populations and consequently the use of culture methods which provide homogeneous conditions and the development of synchronous culture techniques are important aspects of these studies.

Growth and differentiation under static subaerial conditions represents the 'natural' mode of development of the aspergilli. Heterogeneity which is an inherent characteristic of the filamentous form is further magnified by overcrowding and compaction of hyphae within the macroscopic colonies which are formed under these conditions. While surface colonial growth is convenient for most genetical studies and for studies on the morphology of development it is unsuitable for certain biochemical studies. The particular problems associated with biochemical studies on differentiation in filamentous organisms are more fully discussed later. Because of these difficulties efforts have been made to improve on the conventional surface method and now several techniques are available for the induction of conidiation under defined homogeneous conditions. A description of the various culture methods ranging from the conventional surface method to the more recent innovations will be given together with a description of the ways in which these methods have been applied for studies of conidiation in the aspergilli. Reference will be made to the biochemical studies that have been carried out using each culture method although the results of these studies are dealt with separately (see Biochemistry of conidiation).

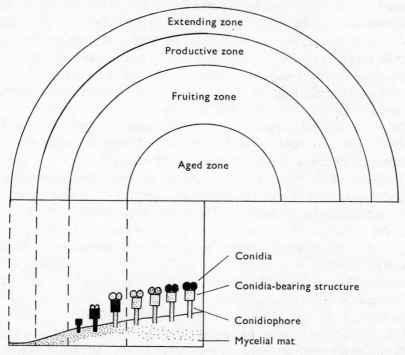

Fig. 2. Schematic diagram showing the four concentric differentiation zones in the surface colony of *Aspergillus niger*. The extent of conidiophore maturation in each of the zones is shown by the localization of basophilic substances (represented as dots of varying density). These substances are highly concentrated in the mycelium of the extending zone and in the early stages of conidiophore development. As the conidiophores mature the basophilic substances progressively move from the conidiophores into the conidia (Yanagita & Kogane, 1962).

Conidiation in surface culture

Growth on the surface of a suitable medium is the traditional culture method for filamentous fungi. Few of the aspergilli require vitamins or other growth factors so that simple synthetic culture media can normally be used. However, the nature and composition of the medium produces marked variations in growth character and in the colouration and morphology of the conidial heads.

Inoculation of the fungus at a point on a solid nutrient medium results in the formation of a circular colony which is characterized by zones of differing morphology and metabolic activity. Four concentric differentiation zones have been described for conidiating colonies of *Aspergillus niger* (Fig. 2) (Yanagita & Kogane, 1962, 1963*a*). The onset of conidiation in these colonies begins some distance behind the extending margin and there occurs a progressive change from immature conidiophores in the outer regions of the colony to aged conidiophores in the centre (Fig. 2). A similar description of the surface colonies of *A. awamori* has been given by Megee, Kinoshita, Fredrickson & Tsuchiya (1970). On the basis of the finite number of differentiation states or zones within the colony these authors have developed a useful mathematical model of biochemical differentiation and product formation for moulds cultivated in surface culture and which is also valid for certain submerged cultivations. In some aspergilli (see Raper & Fennell, 1965) zonation takes the form of fairly regular concentric rings of alternately crowded and sparse conidial structures.

Zonation effects which are characteristic of colony development from a point inoculum can be eliminated by spraying the inoculum evenly over the surface of the medium. The hyphal mats grown in this way for varying periods of time have been found to correspond in their physiological and biochemical features to the different zones in the circular colony (Yanagita & Kogane, 1963a). This method of inoculation has been effectively used with *Aspergillus niger* to produce large quantities of mycelium and conidiophores of more uniform age for studies on biochemical changes during differentiation (Yanagita & Kogane, 1963a, b; Nagasaki, 1968a, b). In these studies the fungus was grown on a cellophane membrane placed on the surface of the solidified medium to simplify the harvesting of intact hyphal mats. A technique involving lyophilization of the mat followed by brushing was also used to selectively harvest only the aerial mycelium and conidiophores for biochemical studies. Clutterbuck (1972) in a recent study of a laccase enzyme associated with conidial pigmentation in *Aspergillus nidulans* has employed a method in which the conidial inoculum is mixed with semi-solid medium and is then deposited as a shallow layer on a base of solidified medium. The mycelial mat so formed could be readily peeled away from the underlying agar. Conidiation in these cultures was controlled by a layer of cellophane placed over the surface of the medium (under which growth remained vegetative) and which when removed allowed a rapid and uniform development of conidiophores over the surface of the mycelial mat.

In vertical section the *Aspergillus* colony consists of a lower dense mycelial layer which penetrates to a greater or lesser extent into the substratum and an upper layer composed of loosely interwoven aerial hyphae. The foot cells may occur in either or both of these layers. The conidiophore grows upwards away from the medium and final differentiation of the conidial heads takes place in an aerial environment. Usually, conditions are such that complete maturation occurs although certain adverse conditions can drastically affect gross morphology and the fine structure of morphogenesis. In *A. janus* (Raper & Thom, 1944) two different types of conidial head are produced and the ratio of these types is strongly influenced by temperature. Abnormalities such as proliferations of the sterigmata and enlarged multinucleate conidia have been observed in *A. repens* and *A. echinulatus* grown at temperatures above optimum or in high humidity (Thielke, 1958; Thielke & Paravicini, 1962). Anaerobic conditions lead to conidiophore proliferations in *A. niger* (Miller & Anderson, 1961) and sterigmata proliferations considered as dedifferentiations induced by respiratory deficiency were observed in *A. amstelodami* (Bleul, 1962). A relative humidity of 100 % allows conidiophore development but inhibits vesicle formation in *A. giganteus* (Trinci & Banbury, 1967). Reduction of the relative humidity to 95 % allows vesicle formation and this technique has been used to induce the synchronous development of the conidial heads for studies on the fine structure of conidia formation (Trinci *et al.* 1968).

The growth and differentiation of some species of *Aspergillus* are influenced by illumination. In *A. janus* (Tylutki, 1952) the type of conidial head produced and in *A. giganteus* (Raper & Fennell, 1965) the size of the heads and the length of the conidiophore are affected by an interaction between temperature and light. Light has a differential effect on the formation of conidiophores and sclerotia in *A. paradoxus* (Raper & Fennell, 1965) and *A. japonicum* (Heath & Eggins, 1965). A differential effect of light on the production of cleistothecia and conidia occurs with *A. ornatus* (Raper, Fennell & Tresner, 1953). Curran (1971) has made a detailed study of sporulation in some members of the *A. glaucus* group in response to illumination and osmotic pressure. Light enhanced conidial production in all of the fungi studied and in several darkness enhanced cleistothecial formation. Conidial head formation was less inhibited at high osmotic pressures than was cleistothecial production.

The most detailed studies of morphogenetic effects of light have recently been made on *Aspergillus giganteus*. In this species light induces carotenogenesis, conidiophore extension and conidiophore orientation (Trinci & Banbury, 1968). The conidiophores were positively phototrophic to unilateral stimulation with light of wavelength below 520 nm, the phototrophic response being restricted to the apical 240 μm of the conidiophore. Since photoinduction of carotenogenesis and conidiophore extension occurred when cultures were illuminated under anaerobic conditions it was suggested that the presence of oxygen is not essential during the photochemical reaction. Also for photoinduction to occur the light must fall directly upon the mycelium, there being little or no transmission of the inductive stimulus from illuminated to adjacent non-illuminated parts of the same mycelium (Trinci & Banbury, 1969). From experiments in which the intensity of light and the frequency of illumination were varied it was suggested that 'low energy' photo-chemical reactions were involved in which light only served as a trigger to a chain reaction which maintains growth and carotenogenesis.

While environmental factors can modify the morphology of the conidial apparatus the correct genetic potential is prerequisite for normal conidiation. The aspergilli are characterized by a tendency for spontaneous mutation and variability and one of the most common expressions of this is the development, in the surface colonies, of sectors which have growth or reproductive characters different from the parent colony. At least some of this inherent variation can be attributed to the phenomenon of heterokaryosis. A detailed description of various alterations in conidiophore morphology and conidial pigmentation as a result of mutation or heterokaryosis has been provided by Raper & Fennell (1965).

For many aspergilli there is a sharp zone of demarcation between two adjacent colonies, growth apparently ceasing at this junction. Since conidiation is usually hastened at the margins of these zones of inter-colony inhibition the use of a two or three colony plate has been recommended as a culture technique for the examination and identification of *Aspergillus* spp. (Raper & Fennell, 1965). Recently, morphological (fluffy) variants of *A. nidulans* have been described which are capable of infiltrating and progressively covering the entire surface of a non-fluffy *A. nidulans* colony (Dorn, Martin & Purnell, 1967). Morphologically the fluffy strains fail to differentiate normally. Conidiation is delayed or is completely absent and in most cases the strains are self sterile while a few of the more severe forms are cross-sterile as well. Subsequently Dorn (1970) has isolated and characterized nine fluffy mutants which displayed varying degrees of delayed differentiation. With these mutants there appeared to be an inverse correlation between the extent of invasion and the level of differentiation, i.e. less differentiated mutants invade more rapidly. From a study of their genetical properties four *flu* loci and two extracistronic suppressor loci were tentatively identified. The results of Dorn *et al.* (1967) also suggest that the involvement of extrachromosomal factors must be considered for at least some of the fluffy strains.

For certain genetical studies it is advantageous to handle large numbers of discrete colonies on individual culture plates. MacKintosh & Pritchard (1963) working with *Aspergillus nidulans* have examined various plating conditions which minimize the spreading nature of surface colonial growth. It was found that sodium desoxycholate induced micro-colonies of *A. nidulans* about 3 mm in diameter and consequently that much higher plating densities could be used than was previously possible. For replica plating micro-colonies they suggested the use of damp instead of dry velveteen since the technique in its original form was unsuitable for fungi with dry spores. These techniques have since been widely used in genetical studies.

The micro-colonial growth method was employed by Clutterbuck (1969*b*) in a study involving a mutational analysis of conidial development in *Aspergillus nidulans*. In this study, which was confined to the processes following conidiophore initiation, various mutations which led to alterations in the conidial apparatus were described and classified. Only two mutant types were found which were totally lacking in conidia. Other mutants described were those with modified conidia, mutants with disturbed conidiophore or sterigma morphology and mutants affected in their conidiophore pigmentation. In a subsequent quantitative survey of conidiation mutants in *A. nidulans* Martinelli & Clutterbuck (1971) estimated that about 1000 genes occur in which mutation leads to a large or small effect on conidiation. Most of the mutants were blocked before conidiophore formation while many others failed at ill-defined stages of conidiation and it was suggested that these failures were due to the gradual build-up of metabolic deficiencies. Growth tests on the conidiation mutants showed that the mutant defect in 85 % of the conidiation mutants was not confined to conidiation and these were regarded as generally defective rather than specifically defective in conidiation. Consequently it was estimated that of the approximate 1000 genes which affect conidiation only 45 to 150 represent loci which are specifically involved in conidiation.

Hereditable extrachromosomal factors are also known to participate in the development and differentiation of fungi. A considerable contribution to present understanding of this phenomenon has come from studies on the growth, conidiation and pigmentation of surface colonies of extrachromosomal variants of *Aspergillus nidulans* and *A. glaucus* (see Jinks, 1966). Variations from wild type characteristics can be shown to be under the control of an extrachromosomal system by the failure to behave in a Mendelian manner in matings or in the failure to segregate with known genetic differences in the resolution of a heterokaryon through asexual transfers (Jinks, 1959*b*). Several studies have shown the involvement of the extrachromosomal system in conidiation. In *A. nidulans* the red mutant (Jinks, 1958; Arlett, Grindle & Jinks, 1962) shows reductions in conidiophore and conidia production while mycelial mutants (Roper, 1958) are blocked in the development of both sexual and asexual spores. In *A. glaucus* a progressive loss of developmental capacity accompanies ageing and this involves only changes in the extrachromosomal system (Mather & Jinks, 1958; Jinks, 1966). Continuous propagation by vegetative cells or asexual spores (but not sexual spores) results first in a loss of the sexual stage, then loss of the asexual stage and finally cessation of growth.

Conidiation in submerged culture

The surface culture method which has proved adequate for studies on the environmental and genetic factors affecting conidiophore development has severe limitations for certain nutritional and biochemical studies on conidiation. The physical nature of the mycelial mat creates an unavoidable degree of physiological variation within the culture. Submerged agitated culture gives more homogeneous conditions particularly if growth is in the filamentous rather than the pellet form. However, in submerged conditions most filamentous fungi, including the aspergilli, remain entirely vegetative and this is consistent with the finding that most differentiated structures are characteristic of aerial mycelium in the surface colonies. The nature of the inhibition in submerged conditions may be complex and probably involves one or a combination of several factors such as reduction in oxygen tension, changes in the physical nature of the hyphal walls associated with submergence and direct contact of the potential sporophore cells with inhibitory factors in the medium.

Conidiation of *Aspergillus* species can be induced in submerged shake culture by manipulation of the culture medium (Vezina *et al.* 1965; Galbraith & Smith, 1969*b*). For *A. ochraceus* Vezina *et al.* (1965) found that a high salt concentration together with a correct nutrient balance was required for maximum conidiation. From the exacting conditions required for this development it was considered that a precise balance was necessary between the medium and the physico-chemical nature of the environment. In submerged agitated culture it is possible to obtain precise information on the relationship between the nutritional status of the medium and the onset of conidiation. Analysis of medium changes during growth of *A. nidulans* (Carter & Bull, 1969) and *A. niger* (Galbraith & Smith, 1969*b*) in submerged culture has shown that conidiation occurs at the end of the rapid phase of growth after the exhaustion of the limiting nutrient. The studies on *A. niger* by Galbraith & Smith (1969*b*) have provided the most detailed information on the induction of submerged conidiation in *Aspergillus* species. When glucose was the limiting nutrient conidia induction was affected by the type of nitrogen source. Conidiation did not occur in the presence of excess ammonium ions in spite of glucose exhaustion although nitrate had no such inhibitory effect. Acetate, pyruvate, tricarboxylic acid cycle intermediates, glyoxylate and most amino acids were able to overcome ammonium ion inhibition of conidiation. On the basis of these findings it was possible to compare the biochemical events in differentiating and non-differentiating mycelia by utilizing culture media which differed initially in only one factor. For most studies either α-oxoglutarate (Galbraith & Smith, 1969*b*) or glutamate (Valenzuela-Perez & Smith, 1971; Smith & Valenzuela-Perez, 1971; Smith, Valenzuela-Perez & Ng, 1971) have been used as the 'inducing' compounds.

This technique provides control over the induction of conidiation at the end of the growth phase in submerged shake culture but does not influence further maturation of the conidiophores. More precise control over both the induction of conidiophores and their subsequent development has now been obtained with *Aspergillus niger* by using a fermenter culture system involving sequential medium replacement (Anderson & Smith, 1971*a*). This technique is dependent on the finding that cultural conditions can be introduced which selectively favour each stage of conidiophore development. On this basis four major developmental stages were characterized for conidiophore maturation in *A. niger*. The conidiophore morphology corresponding to each of these stages during the sequential replacement of media is depicted in Fig. 3. The first morphological stage of conidiation, foot cell formation (Plate 2, fig. 1), was induced by growth in a medium with nitrogen as the limiting nutrient (LN medium). Foot cell induction did not occur in a high nitrogen (HN) medium in which both the nitrogen and carbon sources were exhausted simultaneously nor did it occur in LN medium when the oxygen supply to the culture was reduced to a level which just maintained vegetative growth. This latter condition is designated low nitrogen, low oxygen (LNO) medium.

The second stage of development, conidiophore elongation, occurred in LN medium in the absence of exogenous nitrogen although continued metabolism of the carbon source (glucose) was required. Under these nutritional conditions no further differentiation of the conidiophore takes place. Replacement of the culture to a new medium containing a nitrogen source and a tricarboxylic acid cycle intermediate as a carbon source effectively induces the third stage, vesicle and phialide formation. Although the conidial apparatus appears to be fully

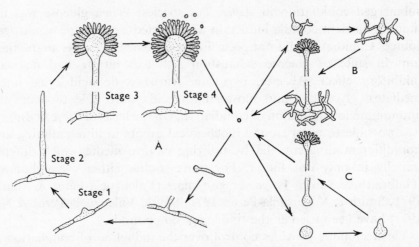

Fig. 3. Summary of induced morphogenetic sequences leading to conidiophore development in *Aspergillus niger* under submerged agitated conditions.

(A) Sequence of morphological changes in replacement fermenter culture (Anderson & Smith, 1971*a*). Stage 1, conidiophore initiation; stage 2, conidiophore elongation; stage 3, vesicle and phialide formation; stage 4, conidiospore production.

(B, C) Forms of microcycle conidiation (Anderson & Smith, 1971*b*). In B a branched mycelial system and a mature conidiophore are produced from an enlarged conidium. Treatment consists of incubation at 41° for 15 h followed by 30° for 12 h.

In (C) a mature conidiophore is produced from an enlarged conidium in the complete absence of vegetative development. Treatment consists of incubation at 44° for 48 h followed by 30° for 15 h.

developed at this stage no conidiospore production occurs from the phialides (Plate 2, fig. 2). This final fourth stage of development was most effectively induced by replacement to a medium in which glucose was the carbon source and nitrate the nitrogen source. If the culture is not replaced after stage three extensive proliferations of the phialides occur similar to those previously described under adverse subaerial conditions.

This control of the successive structural changes by media replacement results in the synchronous maturation of the conidiophores. Studies have been carried out on esterase activity (Lloyd, Anderson, Smith & Morris, 1972) and carbon catabolism (Ng, Smith & Anderson, 1972; Smith & Ng, 1972) using this system of synchronized conidiophore development. The latter studies also involved a comparison of glucose catabolism during foot cell formation in LN medium and during the maintained vegetative growth phase in HN and LNO media.

Growth in submerged agitated batch culture clearly allows a much more precise examination of the effects of medium composition on differentiation than the static surface growth method. Nevertheless, the interpretation of differentiation responses to particular culture conditions is complicated in the batch system by the transient nature of the environmental conditions during growth. In the batch culture system it is not possible, for example, to determine whether conidiation results from nutrient limitation or from the limitation of growth rate imposed by this condition. Chemostat culture on the other hand permits a study of microbial populations at various growth rates and under various metabolic steady states.

While a substantial number of studies have been made on organic acid and enzyme production by *Aspergillus* species under continuous flow conditions there have been few studies on morphological differentiation using this technique. Carter & Bull (1971) have examined the morphology of *A. nidulans* during culture in glucose-limited chemostat culture at various dissolved oxygen tensions. Although differentiated conidiophores were not produced, free conidia appeared in the medium at dissolved oxygen tensions below 3.5 mm of mercury. Also under maintenance or starvation conditions no parts of the conidial apparatus of *A. nidulans* were identified although structures thought to be conidia were observed (Bainbridge *et al.* 1971). Recently studies have also been made of the morphology of *A. niger* in carbon limited and nitrogen limited chemostat culture (Ng, Smith & McIntosh, 1973). Although conidiation was difficult to achieve under glucose limitation the results were sufficient to show close agreement with the results of Righelato, Trinci, Pirt & Peat (1968) with *Penicillium chrysogenum*. With this fungus maximum conidiation occurred at growth rates between zero and the critical growth rate above which steady state vegetative growth prevailed. When citrate was supplied as the carbon source limitation of this nutrient gave a much higher degree of conidiation than was obtained under glucose limitation. Under ammonium limitation with citrate as the carbon source there was no conidiation. When nitrate was used as a limiting nitrogen source conidiophore initiation but not maturation occurred. These experiments demonstrate that conidiation in *A. niger* can be controlled by growth rate only in particular growth media. Conidiation in this fungus appears to be determined by an interaction between growth rate and the nature of the carbon and nitrogen sources in the culture medium.

The studies on conidiation in filamentous fungi using chemostat culture are in close agreement with the experiments with *Bacillus subtilis* in which sporulation could be induced on departure from logarithmic growth in batch culture or at sufficiently high dilution rates in chemostat culture in response to nutrient deficiencies (Mandelstam, 1969; Dawes & Mandelstam, 1970).

In the study by Ng *et al.* (1973) an interesting feature of the conidiation of *Aspergillus niger* in chemostat culture was the occurrence of a considerable reduction in the complexity of the conidial apparatus. Previous observations have shown that the conidial apparatus of *A. niger* produced in submerged culture under a variety of conditions is essentially similar, although smaller than, the normal subaerial structure (Galbraith & Smith, 1969*b*; Anderson & Smith, 1971*a*, *b*). In chemostat culture they were characterized by possessing small vesicles with few phialides and occasionally conidia were observed to develop from modified hyphal tips. Carter & Bull (1971) have reported the formation of free conidia in the absence of conidiophores in chemostat culture of *A. nidulans*. The production of conidia from hyphal tips in submerged culture was also observed in *A. oryzae* by El Kotry (1970) while also in this fungus the direct production of phialides on the tips of germ tubes has been observed (J. G. Anderson & J. E. Smith, unpublished results). In some species of *Aspergillus*, for example *A. asperescens*, various degrees of reduction in conidiophore complexity are common even under subaerial conditions (Raper & Fennell, 1965).

These reductions in conidiophore complexity may represent only a partial switch on of the conidiation mechanism. It does indicate that under certain conditions the morphological and biochemical events of conidiophore development which prelude conidiospore formation can be by-passed. This idea is substantiated by observations by Clutterbuck (1969*b*) of morphological mutants of *A. nidulans* which produce normal conidia on abnormal simplified conidiophores.

Microcycle conidiation

The various culture techniques already described have each been used, with at least some measure of success, to provide fundamental information on the mechanism of conidiation in the aspergilli. However, it is clear that the nature and extent of this information has been largely determined by the inherent limitations in the culture system itself. The major limitation with these conventional culture techniques is that conidiation is preceded by a period of vegetative filamentous growth which inevitably creates a heterogeneous cell population. Recently, Anderson & Smith (1971*b*, 1972) have described a unique microcycle conidiation technique (Fig. 3) which, by eliminating the normal hyphal vegetative growth phase of the fungus, promises a novel approach to studies on the mechanism of conidiation in the aspergilli.

Striking morphological changes occur when the conidia of *A. niger* are cultured under submerged conditions at elevated temperatures

(38–44°) in a basal growth medium supplemented with glutamate (Anderson & Smith, 1972). Whereas all conidia produce germ tubes at 30° (Plate 1, fig. 1) at temperatures from 38° to 43° the proportion of conidia which produce germ tubes progressively decreases and at 44° germ tube formation is completely inhibited. However, at this temperature swelling of the conidia continues to occur over a prolonged period to produce large spherical cells (20 μm mean diameter) (Plate 1, fig. 4). These cultural conditions also induce morphogenetic changes in the spores which lead to the direct outgrowth of conidiophores (Anderson & Smith, 1971*b*). Although conidiophore initiation is stimulated at temperatures between 35° and 43° maturation is poor and optimum development can be obtained by incubation at these temperatures followed by 30°. An even more remarkable sequence of morphological events occurs after incubation of the conidia at 44°. A prolonged (48 h) period of incubation at this temperature followed by incubation at 30° results in the direct outgrowth of a conidiophore from the enlarged conidium in the complete absence of normal vegetative growth (Plate 2, fig. 3). Initially a single conidiophore develops and this is then frequently followed by a second and occasionally up to five are produced by the enlarged conidium (Plate 2, fig. 4). The conidiophores produced are similar, but smaller than, normal subaerial conidiophores and viable conidia are produced.

In reports of secondary spore formation in various other fungi (see Anderson & Smith, 1971*b*) as well as 'microcycle sporogenesis' in bacterial spores (Vinter & Slepecky, 1965) induction of sporulation can be attributed to factors which interfere with the metabolic condition characteristic of the somatic growth phase. Microcycle conidiation in *Aspergillus niger* is stimulated by factors which inhibit the apical growth process associated with rapid vegetative growth. These conditions impose a minimal growth rate and induce enlargement of the conidia and abnormal swelling and branching of the germ tubes. While these morphological changes occur in a basal growth medium at elevated temperatures an additional factor, glutamate, is required for the maximum expression of conidiation. Results with *A. niger* using the chemostat culture technique demonstrate that while conidiation can be controlled by growth rate the composition of the medium must also be considered. The phenomenon of microcycle conidiation is consistent with these findings since both apical growth inhibition and medium composition interact to induce conidiophore development.

The complete loss of the ability of the conidia to produce vegetative growth in a complete growth medium at 30° after a prolonged period of

enlargement at 44° while retaining the ability to form a complex reproductive structure and viable conidia has not yet been explained. However, studies on this phenomenon should provide valuable information
relating to the control of both the vegetative and asexual reproductive
phases of this fungus. As a culture technique for studies on conidiation
it should simplify the studies of regulatory mechanisms preceding
conidiation and of the processes accompanying conidiophore development and conidia production.

Biochemistry of conidiation

Filamentous multicellular micro-organisms such as *Aspergillus* are
inherently difficult systems from which to obtain meaningful results on
specific biochemical changes that may be associated with the process of
differentiation. As with other microbial systems most physiological and
biochemical studies have used crude homogenates of the entire culture.
With synchronously growing unicellular cultures such methods may
closely indicate the true changes occurring at specific phases of development. However, with mycelial cultures, and in particular because of the
phenomenon of apical growth there exists within each mycelium, and
indeed each hypha, a spatial distribution of differing biochemical
activities. As a result, particular aspects of fungal differentiation, e.g.
conidiation may not necessarily involve the entire thallus and since
normally the vegetative cells far outnumber the cells actively involved in
conidiation important specific and highly characteristic biochemical
changes may be masked by the vegetative physiology. The problems
inherent in the analytical methods are further multiplied in static
mycelial cultures where variation of the physiological and biochemical
status of the mycelium is increased by the additional variants of oxygen
tension and nutrient concentration. Submerged batch or continuous
cultivation reduce these complications only if the mycelium grows in a
filamentous form rather than in the pellet form, the more usual condition
of growth in submerged culture (Galbraith & Smith, 1969a). With
pellet growth, nutrients and oxygen are only readily available to the
peripheral hyphae and autolysis rapidly occurs at the centre while
dense growth continues at the edge. Although submerged growth is
more desirable for biochemical studies of conidiation it is often beset
with the problem that conidiation is generally suppressed in submerged
culture even in those species which conidiate freely in static surface
culture. However, with the aspergilli conidiation can be induced in
submerged culture by the manipulation of the carbon and nitrogen
sources and by the presence of various trace elements (see p. 314).

Changes in enzyme activity

A major criticism of the widespread implication of enzymes as the causal agents of developmental processes has been the question of the reliability and reproducibility of in-vitro enzyme determinations. *In vivo*, the activity of a given enzyme concentration is dependent in part on substrate and cofactor availability, presence or absence of inhibitors or activators and the ionic environment. Furthermore, the specific activity of an enzyme under conditions of optimum substrate concentration, as will occur in in-vitro assays, only measures change in the enzyme concentration. In in-vitro studies the physiological system is massively disrupted by the extraction procedures and in particular damage to organelles such as nuclei, mitochondria and lysosomes may cause gross artifacts by allowing enzymes and metabolites to meet which are normally compartmentalized and kept apart. However, numerous studies have dealt specifically with quantitative and qualitative changes in enzyme levels during differentiation and suggestive correlations have been described between increasing levels of specific enzymes and the changing requirements of differentiation in *Aspergillus* spp.

The changing patterns of enzymes of phosphorus metabolism in developing mycelium of *Aspergillus niger* have been studied by Nagasaki (1968b). Ribonuclease, deoxyribonuclease, phosphomonoesterase and pyrophosphatase showed two pH optima at acid and neutral values and in general the enzymes showing the high pH optima were active in younger mycelia, the others in older cultures. Major changes in enzyme activity occurred during and after conidia formation. Cytochemical methods have also been used to study the intracellular localization of phosphatases in the hyphae and conidia of *A. niger* (Nagasaki, 1968a). In aerial and substrate hyphae the activity of alkaline phosphatases exceeded that of acid phosphatases in young hyphae while in older hyphae the situation was reversed. The activities of both enzymes were much higher in conidiophores than in vegetative hyphae. In the growing conidiophore the apical regions were slightly less active but activity became highly concentrated in the vesicles and phialides as the conidiophores developed. As conidiation progressed cytoplasmic materials containing phosphatases moved up the conidiophore towards the vesicle.

In *Aspergillus niger* grown in flask culture lipolytic esterases were always present in mycelial extracts during conidiation irrespective of the mode of induction of conidiation but could not be detected by electrophoresis in the vegetative mycelium of cultures which would ultimately

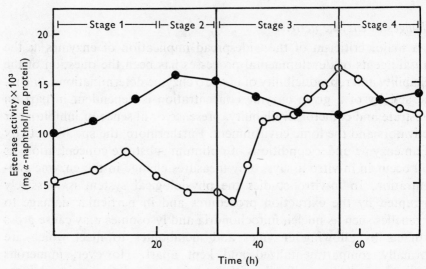

Fig. 4. Comparison of esterase activity and lipid content of *Aspergillus niger* during growth and morphogenesis in replacement fermenter culture. Vertical lines at 32 h and 56 h indicate times of first and second nutrient replacements. ○—○, Esterase activity of the extracts; ●—●, lipid content of the mycelium. (Lloyd *et al.* 1972.)

conidiate or in the mycelium of sterile cultures (Lloyd, Morris & Smith, 1970). However, quantitative methods did show that even in vegetative cultures a low basal level of esterase activity was always present and at conidiation this level greatly increased (Lloyd *et al.* 1972). These results would suggest that esterases are only detectable by electrophoresis above a certain enzyme concentration.

In the replacement fermenter technique quantitative esterase determinations showed that a basal level of esterase activity was present during vegetative development (Fig. 4). Esterase activity increased greatly immediately prior to vesicle and phialide formation and persisted in these structures once formed (Lloyd *et al.* 1972). The increase in esterase activity during conidiation was observed cytochemically to occur in the conidiophore tip prior to the formation of the vesicle and phialides and in the latter structures after their formation (Lloyd *et al.* 1972). These results may infer that lipids function as a source of carbon and energy during conidiation. Esterase production is also associated with conidiation in continuous culture (A. Ng, personal communication) and in microcycle conidiation (G. I. Lloyd, personal communication).

Thus a causal relationship between a biochemical event and a morphological character in a fungus has been firmly established. Esterase production is always associated with conidiation in *Aspergillus niger* while vegetative cultures have been characterized by a low basal level of esterase activity. Thus, according to the criteria of White & Sussman (1961) that a causal relationship can only be proven if the biochemical event always accompanies differentiation, no matter what the manner of induction of differentiation, and if failure of differentiation of the said structure is always accompanied by loss of that particular event, esterase production can be seen to be an important event in conidiation in *A. niger*.

Clutterbuck (1972) has recently shown that the enzyme laccase or *p*-diphenyl oxidase is associated with conidiation in green-spored strains of *Aspergillus nidulans* and not in yellow-spored mutants. The enzyme is obviously involved in the synthesis of the green spore pigment.

Studies of the tricarboxylic acid (TCA) and glyoxylic acid (GLC) cycles of *Aspergillus niger* during the onset of conidiation (Galbraith & Smith, 1969c), gave results strongly reminiscent of the enzyme activities associated with differentiation in *Blastocladiella* and *Neurospora* (for reference see Smith & Galbraith, 1971). NADPH-dependent isocitrate dehydrogenase and isocitrate lyase showed much higher specific activities at the period preceding conidiophore development than in vegetative mycelium of the same physiological age. Malate dehydrogenase, NADP-dependent isocitrate dehydrogenase and malate synthetase were relatively similar in pre-conidiating and vegetative mycelium whereas glycine–alanine transferase was detectable qualitatively only in pre-conidiating mycelium. α-Oxo-glutarate dehydrogenase was not detectable at any stage in either type of mycelium.

These studies were insufficient to correlate with the conclusion of Behal & Eakin (1959a) that conidia formation in *Aspergillus niger* is dependent on a functionally complete TCA cycle. They interpreted inhibition of conidia development but not growth by 6-ethylthiopurine as an interference with methionine metabolism, causing a decrease in the activity of the condensing enzyme. The increase in glucose uptake and carbon dioxide evolved after a period of inhibition by 6-ethylthiopurine, together with the activity of TCA cycle intermediates and saccharides as counteractants of 6-ethylthiopurine were regarded as an indication of a requirement for a respiratory process providing energy for spore formation (Behal & Eakin, 1959b).

Recently there has been extensive examination of carbon catabolism during differentiation of *Aspergillus niger* (Valenzuela-Perez & Smith, 1971; Smith *et al.* 1971; Ng *et al.* 1972). In these experiments *in vitro* enzyme determinations were coupled with the radiorespirometric analysis of glucose dissimilation *in vivo* in order to give a more reliable estimation of the in-vivo changes occurring in glucose catabolism.

The first series of experiments were carried out on mycelium grown in flask culture, using glutamate as the conidiophore inducing factor, and the results indicated that the contribution of the EMP pathway was highest in conidiating mycelium whereas the contribution of the PP pathway was highest in young conidiating mycelium growing under conditions that would not induce conidiation (Smith *et al.* 1971). These results were in direct opposition to other studies of carbon dissimilation during spore formation in other filamentous fungi. Radiorespirometric studies with *Endothia parasitica* (McDowell & DeHertogh, 1968) and *Aspergillus nidulans* (Carter & Bull, 1969) have indicated an enhanced contribution of the PP pathway prior to and during conidiation. Furthermore, an active PP pathway has been shown to operate throughout conidiation in *Neurospora crassa* (Turian, 1962) while Daly, Sayre & Pazur (1957) have demonstrated that the PP pathway is the major respiratory pathway during conidiation of the obligate rust fungus *Puccinia carthami* growing on safflower.

When carbon catabolism was investigated using the replacement fermenter technique it was found that in general during conidiophore development the PP pathway enzymes were higher in activity than the EMP pathway enzymes and when taken together with the radiorespirometric analysis strongly implied that the direct oxidation of glucose through the PP pathway may be of major importance during conidiophore development of *Aspergillus niger* (Ng *et al.* 1972). High EMP activity was obtained in LNO medium which did not support conidiophore initiation but did allow vegetative growth to occur.

The differences in results between the flask and the replacement fermenter experiments are probably related to the presence of glutamate in the former. One of the main functions of the PP pathway in cellular metabolism is to produce NADPH essential for reductive biosynthesis, in particular amino acid synthesis from inorganic nitrogen sources. The ability of the mycelium to incorporate glutamic acid from the medium directly into the amino acid pool could greatly reduce the involvement of the PP pathway for this particular aspect of cellular biosynthesis. Thus the presence of glutamate in the growth medium can mask the high biosynthetic needs of a differentiating system and in turn lead to a reduced contribution of specific enzymes involved in its formation. However, the present studies would imply that the morphogenetic process of conidiation can be achieved in *Aspergillus niger* with a variety of enzyme assemblies.

Intermediate flux and conidiation

It has been considered that the properties of a cell or micro-organism can be largely determined by the rhythmic pattern of variation in constituent levels of metabolites resulting from the operation of numerous metabolic control circuits. Thus the process of cellular differentiation may result from alterations in these periodicities in so far that changes in intermediate levels may control enzyme activity as well as the sequential release of nuclear information required for progressive differentiation. Wright (1968) considers that changes in enzyme activity may be irrelevant to the control of differentiation and that changes in the concentration of key metabolites during differentiation are in themselves sufficient to cause the qualitative and quantitative changes in metabolic flux that occur during morphogenesis. Furthermore, changes in metabolic fluxes can be achieved by changes in substrate levels without altering the amount of enzyme present (Wright, 1966).

Studies with *Aspergillus niger* using mycelium from flask culture (Smith & Valenzuela-Perez, 1971) and replacement fermenter culture (Smith & Ng, 1972) have shown that major changes occur in the levels of certain intermediates of glycolysis during differentiation in each test system. The fact that there was little apparent correlation between intermediate levels in the two systems would substantiate the results of Wright (1968) that the intracellular concentration of certain metabolites essential to differentiation can vary from one study to another and yet normal morphogenesis can occur. This would imply that critical changes in the concentration of metabolites essential to differentiation can be met and balanced by compensatory mechanisms within the cell.

Figs. 5 and 6 show the levels of glycolytic intermediates from mycelium grown in the complete replacement fermenter technique and also in LNO medium. It is interesting to compare the levels of the intermediates in the initial LN medium (where conidiophores are initiated) and LNO medium (in which conidiophores are not initiated). The levels of glucose 6-phosphate and fructose 6-phosphate were in general higher in LNO medium. Furthermore, the pattern of levels was different; in LN medium the levels were initially low and increased with time whereas in LNO medium the maximum levels occurred initially and subsequently decreased with time. In each case the changes in the levels of the hexose phosphates were generally in phase but out of phase with fructose diphosphate. In LN medium the fructose diphosphate values were initially tenfold greater than in LNO medium but subsequently

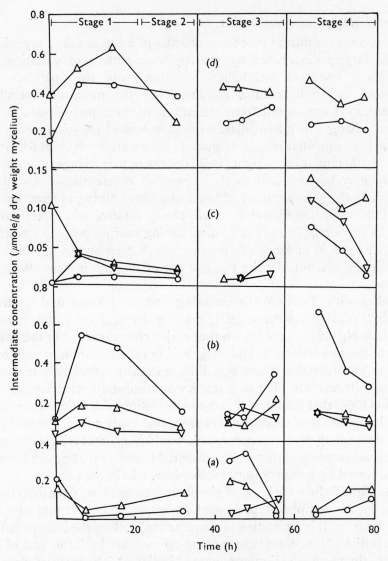

Fig. 5. Changes in the concentrations of glycolytic intermediates during growth
of *Aspergillus niger* in replacement fermenter culture.

	O—O	△—△	▽—▽
(a)	G6P	F6P	G1P
(b)	FDP	DHAP	GA3P
(c)	1:3 DPG	3 PGA	2 PGA
(d)	PEP	PYR	

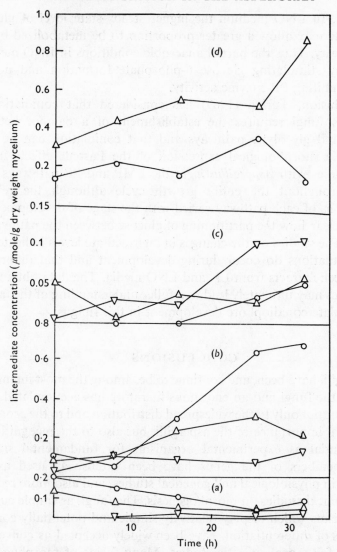

Fig. 6. Changes in the concentrations of glycolytic intermediates during growth of
Aspergillus niger in LNO medium. Symbols as in Fig. 5.

decreased whereas in LNO medium the levels increased. The build-up of
pyruvate in LNO medium may indicate a partial blockage in its further
metabolism. The decrease in pyruvate in the aerobic conditions of LN
medium may indicate the involvement of the TCA cycle.

Thus the low steady-state level of glucose 6-phosphate in LN medium
coupled with a possible high affinity for glucose 6-phosphate by glucose
6-phosphate dehydrogenase may regulate the pathway of glucose

katabolism. In LNO medium the higher steady-state level of glucose 6-phosphate may allow a greater proportion to be metabolized by the EMP pathway. Thus the partial anaerobic conditions in LNO medium may act by stimulating glucose 6-phosphate formation and not by directly inhibiting PP enzyme activity.

In conclusion, Turian (1969) has considered that conidiation in filamentous fungi requires the establishment of a balance between oxidative and glycolytic pathways and that conidiation can be considered as a morphological expression of the Pasteur effect. In the present studies with *Aspergillus niger* the EMP and PP pathways were present throughout the entire growth cycle although the relative contributions of each pathway varied with the stage of development. It is not yet clear how the partitioning of glucose between the pathways is achieved. The studies on the changes in intermediate levels indicate that major fluctuations do occur during development and that differences exist between extracts from LN and LNO media. The data obtained in these studies may ultimately lead to a fuller understanding of the factors which regulate conidiophore development in this fungus.

CONCLUSIONS

The aspergilli have been, and continue to be, among the most intensively studied of the fungi and an enormous literature has accumulated. This has been due not only to the widespread distribution and to the economic and medical importance of the aspergilli, but also to the recognition of their potential as experimental organisms for fundamental studies. Various members of the genus have been extensively used as test organisms in physiological and genetical studies, and also more recently for biochemical studies on morphogenesis. The fungi, as simple eukaryotic organisms which display relatively simple and potentially controllable forms of differentiation, have been widely accepted as convenient organisms for morphogenetic studies. Many forms of morphogenesis occur in the fungi and those encountered in the aspergilli are among the most common. Like the vast majority of fungi the aspergilli possess a filamentous somatic growth phase which on account of its coenocytic nature gives these organisms properties which are unique among the eukaryotes. The formation of conidia on specialized aerial structures, displayed in a most advanced form in the aspergilli, represents the most common and evidently the most successful form of asexual reproduction in the fungi. Therefore, it can be anticipated that the results of studies on the metabolic processes which are involved in hyphal morphogenesis in the aspergilli and in the formation and germination of the conidia of these organisms may well be applicable to the numerous other fungi which exhibit similar forms of morphogenesis. It is also now apparent that certain correlations exist between the changing patterns of metabolism which lead to the formation of differential structures in microorganisms and those which lead to the elaboration of secondary metabolites, a phenomenon which occurs with particular frequency in the aspergilli. Consequently, fundamental studies on the differentiation of the aspergilli may well lead to a better understanding of the precise conditions which result in the accumulation of both the beneficial and detrimental products which are associated with these organisms.

REFERENCES

ADEBAYO, A. A., HARRIS, R. F. & GARDNER, W. R. (1971). Turgor pressure of fungal mycelia. *Transactions of the British Mycological Society*, **57**, 145–51.

AITKEN, W. B. & NIEDERPRUEM, D. J. (1970). Ultrastructural changes and biochemical events in basidiospore germination of *Schizophyllum commune*. *Journal of Bacteriology*, **104**, 981–8.

ALEXOPOULOS, C. J. (1962). *Introductory mycology*. New York: John Wiley and Sons.

ALLEN, P. J. (1965). Metabolic aspects of spore germination in fungi. *Annual Review of Phytopathology*, **3**, 313–42.

ANDERSON, J. G. & SMITH, J. E. (1971a). Synchronous initiation and maturation of *Aspergillus niger* conidiophores in culture. *Transactions of the British Mycological Society*, **56**, 9–29.

ANDERSON, J. G. & SMITH, J. E. (1971b). The production of conidiophores and conidia by newly germinated conidia of *Aspergillus niger* (microcycle conidiation). *Journal of General Microbiology*, **69**, 185–97.

ANDERSON, J. G. & SMITH, J. E. (1972). The effects of elevated temperature on spore swelling and germination in *Aspergillus niger*. *Canadian Journal of Microbiology*, **18**, 289–97.

ARLETT, C. F., GRINDLE, M. & JINKS, J. L. (1962). The red cytoplasmic variant of *Aspergillus nidulans*. *Heredity*, **17**, 197–209.

AUSTWICK, P. K. C. (1965). Pathogenicity. In *The Genus Aspergillus*, ed. K. B. Raper & Dorothy I. Fennell, pp. 82–126. Baltimore: Williams & Wilkins.

AUSTWICK, P. K. C. (1966). The role of spores in the allergies and mycoses of man and animals. In *The Fungus Spore*, ed. M. F. Madelin, pp. 321–37. London: Butterworths.

BAINBRIDGE, B. W. (1971). Macromolecular composition and nuclear division during spore germination in *Aspergillus nidulans*. *Journal of General Microbiology*, **66**, 319–25.

BAINBRIDGE, B. W., BULL, A. T., PIRT, S. J., ROWLEY, B. I. & TRINCI, A. P. J. (1971). Biochemical and structural changes in non-growing maintained and autolysing cultures of *Aspergillus nidulans*. *Transactions of the British Mycological Society*, **56**, 371–85.

BAKER, G. E. (1945). Conidium formation in species of *Aspergillus*. *Mycologia*, **37**, 582–600.

BARTNICKI-GARCIA, S. (1968). Cell wall chemistry, morphogenesis, and taxonomy of fungi. *Annual Review of Microbiology*, **22**, 87–108.

BARTNICKI-GARCIA, S., NELSON, N. & COTA-ROBLES, E. (1968). A novel apical corpuscle in hyphae of *Mucor rouxii*. *Journal of Bacteriology*, **95**, 2399–402.

BEHAL, F. J. & EAKIN, R. F. (1959a). Inhibition of mold development by purine and pyrimidine analogues. *Archives of Biochemistry and Biophysics*, **82**, 439–47.

BEHAL, F. J. & EAKIN, R. E. (1959b). Metabolic changes accompanying the inhibition of spore formation in *Aspergillus niger*. *Archives of Biochemistry and Biophysics*, **82**, 448–54.

BENT, K. J. & MORTON, A. G. (1963). Formation and nature of swollen hyphae in *Penicillium* and related fungi. *Transactions of the British Mycological Society*, **46**, 401–8.

BHATNAGAR, G. M. & KRISHNAN, P. S. (1960a). Enzymatic studies in spores of *Aspergillus niger*. II. Phosphatases. *Archiv für Mikrobiologie*, **36**, 169–74.

BHATNAGAR, G. M. & KRISHNAN, P. S. (1960b). Enzymatic studies in spores of *Aspergillus niger*. III. Enzymes of Embden–Meyerhof–Parnas pathway in germinating spores. *Archiv für Mikrobiologie*, **37**, 211–14.

BLEUL, J. (1962). Eine Sektorenvariante mit proliferierten Sterigmen von *Aspergillus amstelodami* (Mangin) Thom and Church. *Archiv für Mikrobiologie*, **44**, 105–12.

BLOOMFIELD, B. & ALEXANDER, M. (1967). Melanins and resistance of fungi to lysis. *Journal of Bacteriology*, **93**, 1276–80.

BORDER, D. J. & TRINCI, A. P. J. (1970). Fine structure of the germination of *Aspergillus nidulans* conidia. *Transactions of the British Mycological Society*, **54**, 143–52.

BULL, A. T. (1970). Chemical composition of wild-type and mutant *Aspergillus nidulans* cell walls. The nature of polysaccharide and melanin constituents. *Journal of General Microbiology*, **63**, 75–94.

CAMPBELL, C. K. (1970). Fine structure of vegetative hyphae of *Aspergillus fumigatus*. *Journal of General Microbiology*, **64**, 373–6.

CAMPBELL, C. K. (1971). Fine structure and physiology of conidial germination in *Aspergillus fumigatus*. *Transactions of the British Mycological Society*, **57**, 393–402.

CARTER, B. L. A. & BULL, A. T. (1969). Studies of fungal growth and intermediary carbon metabolism under steady and non-steady state conditions. *Biotechnology and Bioengineering*, **11**, 785–804.

CARTER, B. L. A. & BULL, A. T. (1971). The effect of oxygen tension in the medium on the morphology and growth kinetics of *Aspergillus nidulans*. *Journal of General Microbiology*, **65**, 265–73.

CASTLE, E. S. (1945). The structure of the cell walls of *Aspergillus* and the theory of cellulose particles. *American Journal of Botany*, **32**, 148–51.

CHILDS, E. A., AYRES, J. C. & KOEHLER, P. E. (1971). Differentiation in *Aspergillus flavus* as influenced by L-canavanine. *Mycologia*, **63**, 181–4.

CLARK, D. S. (1962). Submerged citric acid fermentation of ferrocyanide-treated beet molasses: morphology of pellets of *Aspergillus niger*. *Canadian Journal of Microbiology*, **8**, 133–6.

CLUTTERBUCK, A. J. (1969a). Cell volume per nucleus in haploid and diploid strains of *Aspergillus nidulans*. *Journal of General Microbiology*, **55**, 291–9.

CLUTTERBUCK, A. J. (1969b). A mutational analysis of conidial development in *Aspergillus nidulans*. *Genetics*, **63**, 317–27.

CLUTTERBUCK, A. J. (1970). Synchronous nuclear division and septation in *Aspergillus nidulans*. *Journal of General Microbiology*, **60**, 133–5.

CLUTTERBUCK, A. J. (1972). Absence of laccase from yellow-spored mutants of *Aspergillus nidulans*. *Journal of General Microbiology*, **70**, 423–35.

COHEN, J., KATZ, D. & ROSENBERGER, R. F. (1969). Temperature-sensitive mutant of *Aspergillus nidulans* lacking amino sugars in its cell wall. *Nature, London,* **224**, 713–15.

CURRAN, P. M. T. (1971). Sporulation in some members of the *Aspergillus glaucus* group in response to osmotic pressure, illumination and temperature. *Transactions of the British Mycological Society,* **57**, 201–11.

DALY, J. M., SAYRE, R. M. & PAZUR, J. H. (1957). The hexosemonophosphate shunt as the major respiratory pathway during sporulation of rust of safflower. *Plant Physiology,* **32**, 44–8.

DAWES, I. W. & MANDELSTAM, J. (1970). Sporulation of *Bacillus subtilis* in continuous culture. *Journal of Bacteriology,* **103**, 529–35.

DETROY, R. W. & CIEGLER, A. (1971). Induction of yeastlike development in *Aspergillus parasiticus*. *Journal of General Microbiology,* **65**, 259–64.

DORN, G. L. (1970). Genetic and morphological properties of undifferentiated and invasive variants of *Aspergillus nidulans*. *Genetics,* **66**, 267–79.

DORN, G. & RIVERA, W. (1966). Kinetics of fungal growth and phosphatase formation in *Aspergillus nidulans*. *Journal of Bacteriology,* **92**, 1618–22.

DORN, G. L., MARTIN, G. M. & PURNELL, D. M. (1967). Genetical and cytoplasmic control of undifferentiated growth in *Aspergillus nidulans*. *Life Sciences,* **6**, 629–33.

EL KOTRY, R. A. R. (1970). Physiological studies on the production of amylase by *Aspergillus oryzae* in batch and continuous cultures. Ph.D. Thesis, Strathclyde University, Glasgow.

FARR, W. K. (1954). Structure and composition of the walls of the conidiophores of *Aspergillus niger* and *A. carbonarius*. *Transactions of the New York Academy of Science,* **16**, 209–14.

FENCL, Z. (1970). Comments on differentiation and product formation in molds. *Biotechnology and Bioengineering,* **12**, 845–7.

FINCHAM, J. R. S. & DAY, P. R. (1971). *Fungal Genetics*. Oxford: Blackwell.

FLETCHER, J. (1969). Morphology and nuclear behavior of germinating conidia of *Penicillium griseofulvum*. *Transactions of the British Mycological Society,* **53**, 425–32.

FLORANCE, E. R., DENISON, W. C. & ALLEN, T. C. (1972). Ultrastructure of dormant and germinating conidia of *Aspergillus nidulans*. *Mycologia,* **69**, 115–23.

GALBRAITH, J. C. & SMITH, J. E. (1969a). Filamentous growth of *Aspergillus niger* in submerged shake culture. *Transactions of the British Mycological Society,* **52**, 237–46.

GALBRAITH, J. C. & SMITH, J. E. (1969b). Sporulation of *Aspergillus niger* in submerged liquid culture. *Journal of General Microbiology,* **59**, 31–45.

GALBRAITH, J. C. & SMITH, J. E. (1969c). Changes in activity of certain enzymes of the tricarboxylic acid cycle and the glyoxylate cycle during the initiation of conidiation of *Aspergillus niger*. *Canadian Journal of Microbiology,* **15**, 1207–12.

GOTTLIEB, D. (1966). Biosynthetic processes in germinating spores. In *The Fungus Spore*, ed. M. F. Madelin, pp. 217–33. London: Butterworths.

GREGORY, P. H. (1966). The fungus spore: what it is and what it does. In *The Fungus Spore*, ed. M. F. Madelin, pp. 1–13. London: Butterworths.

GROVE, S. N. (1972). Apical vesicles in germinating conidia of *Aspergillus parasiticus*. *Mycologia,* **64**, 638–41.

GROVE, S. N. & BRACKER, C. E. (1970). Protoplasmic organization of hyphal tips among fungi: vesicles and Spitzenkörper. *Journal of Bacteriology,* **104**, 989–1009.

HAWKER, L. E. (1966). Germination: morphological and anatomical changes. In *The Fungus Spore*, ed. M. F. Madelin, pp. 151–62. London: Butterworths.

HEATH, L. A. F. & EGGINS, H. O. W. (1965). Effects of light, temperature and nutrients on the production of conidia and sclerotia by forms of *Aspergillus japonicus. Experientia*, **21**, 385–6.

HOLLOMON, D. W. (1969). Biochemistry of germination in *Peronospora tabacina* conidia: evidence for the existence of stable messenger RNA. *Journal of General Microbiology*, **55**, 267–74.

HORIKOSHI, K., IIDA, S. & IKEDA, Y. (1965). Mannitol and mannitol dehydrogenases in conidia of *Aspergillus oryzae. Journal of Bacteriology*, **89**, 326–30.

HORIKOSHI, K. & IKEDA, Y. (1966). Trehalose in conidia of *Aspergillus oryzae. Journal of Bacteriology*, **91**, 1883–7.

HORIKOSHI, K. & IKEDA, Y. (1968). Studies on the conidia of *Aspergillus oryzae.* VII. Development of protein synthesizing activity during germination. *Biochimica et Biophysica Acta*, **166**, 505–11.

HORIKOSHI, K. & IKEDA, Y. (1969). Studies on the conidia of *Aspergillus oryzae.* IX. Protein synthesizing activity of dormant conidia. *Biochimica et Biophysica Acta*, **190**, 187–92.

HUGHES, S. J. (1953). Conidiophores, conidia and classification. *Canadian Journal of Botany*, **31**, 577–659.

ISHITANI, C., IKEDA, Y. & SAKAGUCHI, K. (1956). Hereditary variation and genetic recombination in koji-molds (*Aspergillus oryzae* and *A. sojae*). VI. Genetic recombination in heterozygous diploids. *Journal of General and Applied Microbiology*, **2**, 401–9.

JINKS, J. L. (1957). Selection for cytoplasmic differences. *Proceedings of the Royal Society*, B, **146**, 527–40.

JINKS, J. L. (1958). Cytoplasmic differentiation in fungi. *Proceedings of the Royal Society*, B, **148**, 314–21.

JINKS, J. L. (1959a). Selection for adaptability to new environments in *Aspergillus glaucus. Journal of General Microbiology*, **20**, 223–36.

JINKS, J. L. (1959b). The genetic basis of 'duality' in imperfect fungi. *Heredity*, **15**, 525–8.

JINKS, J. L. (1966). Extranuclear inheritance. In *The Fungi*, vol. II, ed. G. C. Ainsworth & A. S. Sussman, pp. 619–60. New York and London: Academic Press.

JOHNSON, I. R. (1965). The composition of the cell wall of *Aspergillus niger. Biochemical Journal*, **96**, 651–8.

KATZ, D. & ROSENBERGER, R. F. (1970). A mutation in *Aspergillus nidulans* producing hyphal walls which lack chitin. *Biochimica et Biophysica Acta*, **208**, 452–60.

KATZ, D. & ROSENBERGER, R. F. (1971). Hyphal wall synthesis in *Aspergillus nidulans*: effect of protein synthesis inhibition and osmotic shock on chitin insertion and morphogenesis. *Journal of Bacteriology*, **108**, 184–90.

KESSEL, M. & ROSENBERGER, R. F. (1968). Regulation and timing of deoxyribonucleic acid synthesis in hyphae of *Aspergillus nidulans. Journal of Bacteriology*, **95**, 2275–81.

KLEIN, R. (1944). Development studies in the fungi. I. The foot-cell in *Aspergillus clavatus* Desm. *Transactions of the British Mycological Society*, **27**, 121–30.

KOGANE, F. & YANAGITA, T. (1964). Isolation and purification of DNA from *A. oryzae* conidia. *Journal of General and Applied Microbiology*, **10**, 61–8.

LEOPOLD, H. & SEICHERTOVA, O. (1959). The influence of some compounds on the formation of large thick-walled cells in the fungus *Aspergillus niger. Folia Microbiologica, Praha*, **4**, 202–5.

LEWIS, L. A. (1969). Correlated meiotic and mitotic maps in *Aspergillus amstelodami. Genetical Research, Cambridge*, **14**, 185–93.

LLOYD, G. I., ANDERSON, J. G., SMITH, J. E. & MORRIS, E. O. (1972). Conidiation and esterase synthesis in *Aspergillus niger*. *Transactions of the British Mycological Society*, **59**, in press.

LLOYD, G. I., MORRIS, E. O. & SMITH, J. E. (1970). A study of the esterases and their function in *Candida lipolytica*, *Aspergillus niger* and a yeast-like fungus. *Journal of General Microbiology*, **63**, 141–50.

McCLURE, W. K., PARK, D. & ROBINSON, P. M. (1968). Apical organization in the somatic hyphae of fungi. *Journal of General Microbiology*, **50**, 177–82.

McDOWELL, L. L. & DEHERTOGH, A. A. (1968). Metabolism of sporulation in filamentous fungi. 1. Glucose and acetate oxidation in sporulating and non-sporulating cultures of *Endothia parasitica*. *Canadian Journal of Botany*, **46**, 449–51.

MACKINTOSH, M. E. & PRITCHARD, R. H. (1963). The production and replica plating of micro-colonies of *Aspergillus nidulans*. *Genetical Research, Cambridge*, **4**, 320–2.

MANDELS, G. R. (1963). Endogenous respiration of fungus spores in relation to dormancy and germination. *Annals of the New York Academy of Sciences*, **102**, 724–39.

MANDELSTAM, J. (1969). Regulation of bacterial spore formation. In *Microbial Growth, Symposia of the Society for General Microbiology*, **19**, 377–402.

MANNERS, J. G. (1966). Assessment of germination. In *The Fungus Spore*, ed. M. F. Madelin, pp. 165–74. London: Butterworths.

MARTINELLI, S. D. & CLUTTERBUCK, A. J. (1971). A quantitative survey of conidiation mutants in *Aspergillus nidulans*. *Journal of General Microbiology*, **69**, 261–8.

MATHER, K. & JINKS, J. L. (1958). Cytoplasm in sexual reproduction. *Nature, London*, **182**, 1188–90.

MEGEE, R. D., KINOSHITA, S., FREDRICKSON, A. G. & TSUCHIYA, H. M. (1970). Differentiation and product formation in molds. *Biotechnology and Bioengineering*, **12**, 771–801.

MICHELI, P. A. (1729). *Nova plantarum genera juxta Tournefortii methodium disposita Illus Florence.*

MILLER, C. W. & ANDERSON, N. A. (1961). Proliferations of conidiophores and intrahyphal hyphae in *Aspergillus niger*. *Mycologia*, **53**, 433–6.

NAGASAKI, S. (1968*a*). Cytological and physiological studies on phosphatases in developing cultures of *Aspergillus niger*. *Journal of General and Applied Microbiology*, **14**, 263–77.

NAGASAKI, S. (1968*b*). Physiological aspects of various enzyme activities in relation to the culture age of *Aspergillus niger* mycelia. *Journal of General and Applied Microbiology*, **14**, 147–61.

NG, A., SMITH, J. E. & McINTOSH, A. F. (1973). Conidiation of *Aspergillus niger* in continuous culture. *Archiv für Mikrobiologie*, in press.

NG, W. S., SMITH, J. E. & ANDERSON, J. G. (1972). Changes in carbon catabolic pathways during synchronous development of *Aspergillus niger*. *Journal of General Microbiology*, **71**, 495–504.

NISHI, A. (1961). Role of polyphosphate and phospholipid in germinating spores of *Aspergillus niger*. *Journal of Bacteriology*, **81**, 10–19.

NISHI, A., YANAGITA, T. & MARUYAMA, Y. (1968). Cellular events occurring in growing hyphae of *Aspergillus oryzae* as studied by autoradiography. *Journal of General and Applied Microbiology*, **14**, 171–82.

PEBERDY, J. F. (1972). Protoplasts from fungi. *Science Progress*, **60**, 73–86.

PEBERDY, J. F. & GIBSON, R. K. (1971). Regeneration of *Aspergillus nidulans* protoplasts. *Journal of General Microbiology*, **69**, 325–30.

PONTECORVO, G., ROPER, J. A. & FORBES, E. (1953). Genetic recombination without sexual reproduction in *Aspergillus niger*. *Journal of General Microbiology*, **8**, 198–210.

PONTECORVO, G., ROPER, J. A., HEMMONS, L. M., MACDONALD, K. D. & BUFTON, A. W. J. (1953). The genetics of *Aspergillus nidulans*. *Advances in Genetics*, **5**, 141–238.

RAPER, K. B. & FENNELL, D. I. (1965). *The genus Aspergillus*. Baltimore: Williams and Wilkins Co.

RAPER, K. B., FENNELL, D. I. & TRESNER, H. D. (1953). The ascosporic stage of *Aspergillus citrisporus* and related forms. *Mycologia*, **45**, 671–92.

RAPER, K. B. & THOM, C. (1944). New *Aspergilli* from soil. *Mycologia*, **36**, 555–75.

RIGHELATO, R. C., TRINCI, A. P. J., PIRT, S. J. & PEAT, A. (1968). The influence of maintenance energy and growth rate on the metabolic activity, morphology and conidiation of *Penicillium chrysogenum*. *Journal of General Microbiology*, **50**, 399–412.

RIPPEL, A. & BORTELS, H. (1927). Vorlänfige Versuche über die allgemeine Bedeutung der Kohlensäure für die Pflanzenzelle (Versuche an *Aspergillus niger*). *Biochemische Zeitschrift*, **184**, 237–44.

ROBINOW, C. F. & CATEN, C. E. (1969). Mitosis in *Aspergillus nidulans*. *Journal of Cell Science*, **5**, 403–31.

ROPER, J. A. (1958). Nucleo-cytoplasmic interactions in *Aspergillus nidulans*. *Cold Spring Harbor Symposia on Quantitative Biology*, **23**, 141–54.

ROSENBERGER, R. F. & KESSEL, M. (1967). Synchrony in nuclear replication in individual hyphae of *Aspergillus nidulans*. *Journal of Bacteriology*, **94**, 1464–9.

ROSENBERGER, R. F. & KESSEL, M. (1968). Non-random sister chromatid segregation and nuclear migration in hyphae of *Aspergillus niger*. *Journal of Bacteriology*, **96**, 1208–13.

RUIZ-HERRERA, J. (1967). Chemical composition of the cell wall of an *Aspergillus* species. *Archives of Biochemistry and Biophysics*, **122**, 118–25.

SAKAGUCHI, K. & ISHITANI, C. (1952). Studies on the natural variation of Koji-molds (*Aspergillus oryzae* and *Aspergillus sojae*). 2. Cytological studies on the natural variation of Koji-molds. *Journal of the Agricultural Chemical Society of Japan*, **26**, 85–90.

SEMENIUK, G., HARSHFIELD, G. S., CARLSON, C. W., HESSELTINE, C. W. & KWOLEK, W. F. (1971). Mycotoxins in *Aspergillus*. *Mycopathologia et Mycologia Applicata*, **43**, 137–52.

SHEPHERD, C. J. (1957). Changes occurring in the composition of *Aspergillus nidulans* conidia during germination. *Journal of General Microbiology*, **16**, 775.

SMITH, G. (1969). *An Introduction to Industrial Mycology*. London: Edward Arnold.

SMITH, J. E. & GALBRAITH, J. C. (1971). Biochemical and physiological aspects of differentiation in the fungi. *Advances in Microbial Physiology*, **5**, 45–134.

SMITH, J. E. & NG, W. S. (1972). Fluorometric determination of glycolytic intermediates and adenylates during sequential changes in replacement culture of *Aspergillus niger*. *Canadian Journal of Microbiology*, in press.

SMITH, J. E. & VALENZUELA-PEREZ, J. (1971). Changes in intracellular concentrations of glycolytic intermediates and adenosine phosphates during growth cycle of *Aspergillus niger*. *Transactions of the British Mycological Society*, **57**, 103–10.

SMITH, J. E., VALENZUELA-PEREZ, J. & NG, W. S. (1971). Changes in activities of the Embden–Meyerhoff–Parnas and pentose phosphate pathways during the growth cycle of *Aspergillus niger*. *Transactions of the British Mycological Society*, **57**, 93–101.

STAPLES, T., SYAMANANDA, R., KAO, V. & BLOCK, R. J. (1962). Comparative biochemistry of obligately parasitic and saprophytic fungi. II. Assimilation of

C^{14}-labeled substrates by germinating spores. *Contributions Boyce Thompson Institute of Plant Research*, **21**, 345–62.

STROMNAES, Ö. & GARBER, E. D. (1963). Heterocaryosis and the parasexual cycle in *Aspergillus fumigatus*. *Genetics*, **48**, 653–62.

SUMI, M. (1928). Über die chemischen Bastandteile der Sporen von *Aspergillus oryzae*. *Biochemische Zeitschrift*, **195**, 161–74.

SUSSMAN, A. S. (1966). Dormancy and spore germination. In *The Fungi*, vol. II, ed. G. C. Ainsworth & A. S. Sussman, pp. 733–64. London: Academic Press.

SUSSMAN, A. S. & HALVORSON, H. O. (1966). *Spores, Their Dormancy and Germination*. New York and London: Harper and Row.

TAKEBE, I. (1960). Choline sulfate as a major soluble sulfur component of conidiospores of *Aspergillus niger*. *Journal of General and Applied Microbiology*, **6**, 83–9.

TANAKA, K. (1966). Changes in ultrastructure of *Aspergillus oryzae* conidia during germination. *Journal of General and Applied Microbiology*, **12**, 239–46.

TANAKA, K., KOGANE, F. & YANAGITA, T. (1965). Is deoxyribonucleic acid of *Aspergillus oryzae* conidia modified chemically in the early period of germination? *Journal of General and Applied Microbiology*, **11**, 85–90.

TANAKA, K. & YANAGITA, T. (1963a). Electron microscopy of ultrathin sections of *Aspergillus niger*. I. The fine structure of hyphal cells. *Journal of General and Applied Microbiology*, **9**, 101–18.

TANAKA, K. & YANAGITA, T. (1963b). Electron microscopy of ultrathin sections of *Aspergillus niger*. II. Fine structure of conidia-bearing apparatus. *Journal of General and Applied Microbiology*, **9**, 189–203.

TERUI, G. & MOCHIZUKI, T. (1955). Studies on the metabolism of mold spores in relation to germination. *Technology Reports of Osaka University*, **5**, 219–27.

TESTI-CAMPOSANO, A. (1959). Morphological observations on the growth of *Aspergillus flavus* (Link) in shake flasks. *Selected Scientific Papers from the Instituto Superiore di Sanita*, **2**, 448–53.

THIELKE, C. (1958). Studien zur Entwicklungsphysiologie von *Aspergillus*. I. Sterigmenproliferation bei *Aspergillus repens*. *Planta, Berlin*, **51**, 308–20.

THIELKE, C. & PARAVICINI, R. (1962). Studien zur Entwicklungsphysiologie von *Aspergillus*. III. Die Entwicklung mehrkerniger Ascosporen und die Modifizierbarkeit von Conidien und Ascosporen bei *Aspergillus echinulatus*. *Archiv für Mikrobiologie*, **44**, 75–86.

THOM, C. & RAPER, K. B. (1945). *A Manual of the Aspergilli*. Baltimore: Williams and Wilkins.

TRINCI, A. P. J. (1969). A kinetic study of the growth of *Aspergillus nidulans* and other fungi. *Journal of General Microbiology*, **57**, 11–24.

TRINCI, A. P. J. (1970a). Kinetics of the growth of mycelial pellets of *Aspergillus nidulans*. *Archiv für Mikrobiologie*, **73**, 353–67.

TRINCI, A. P. J. (1970b). Kinetics of apical and lateral branching in *Aspergillus nidulans* and *Geotrichum lactis*. *Transactions of the British Mycological Society*, **55**, 17–28.

TRINCI, A. P. J. (1971a). Exponential growth of the germ-tubes of fungal spores. *Journal of General Microbiology*, **67**, 345–8.

TRINCI, A. P. J. (1971b). Influence of the width of the peripheral growth zone on the radial growth rate of fungal colonies on solid media. *Journal of General Microbiology*, **67**, 325–44.

TRINCI, A. P. J. & BANBURY, G. H. (1967). A study of the growth of tall conidiophores of *Aspergillus giganteus* Wehmer. *Transactions of the British Mycological Society*, **50**, 525–38.

TRINCI, A. P. J. & BANBURY, G. H. (1968). Phototropism and light-growth responses of the tall conidiophores of *Aspergillus giganteus*. *Journal of General Microbiology*, **54**, 427–38.

TRINCI, A. P. J. & BANBURY, G. H. (1969). Effect of light on growth and carotenogenesis of the tall conidiophores of *Aspergillus giganteus*. *Transactions of the British Mycological Society*, **52**, 73–86.

TRINCI, A. P. J. & GULL, K. (1970). Effect of actidione, griseofulvin and triphenyltin acetate on the kinetics of fungal growth. *Journal of General Microbiology*, **60**, 287–92.

TRINCI, A. P. J., PEAT, A. & BANBURY, G. H. (1968). Fine structure of phialide and conidiospore development in *Aspergillus giganteus* 'Wehmer'. *Annals of Botany*, **32**, 241–9.

TSUKAHARA, T. (1968). Electron microscopy of germinating conidiospores of *Aspergillus niger*. *Sabourandia*, **6**, 185–91.

TSUKAHARA, T. & YAMADA, M. (1965). Cytological structure of *Aspergillus niger* by electron microscopy. *Japanese Journal of Microbiology*, **9**, 35–48.

TUBAKI, K. (1958). Studies on the Japanese Hyphomycetes. V. Leaf and stem group with discussion of the classification of Hyphomycetes and their perfect stages. *Journal Hattori Botanical Laboratory*, **31**, 142–244.

TURIAN, G. (1962). The hexose monophosphate shunt as an alternate metabolic pathway for conidial differentiation in *Neurospora*. *Neurospora Newsletter*, **2**, 15.

TURIAN, G. (1969). *Différenciation Fongique*. Paris: Masson.

TYLUTKI, E. E. (1952). The production of different types of conidial heads by *Aspergillus janus*. M.S. Thesis, University of Illinois, Urbana, Illinois.

VAKIL, J. R., RAO, M. R. R. & BHATTACHARYYA, P. K. (1961). Effect of CO_2 on the germination of conidiospores of *Aspergillus niger*. *Archiv für Mikrobiologie*, **39**, 53–7.

VALENZUELA-PEREZ, J. & SMITH, J. E. (1971). Role of glycolysis in sporulation of *Aspergillus niger* in submerged culture. *Transactions of the British Mycological Society*, **57**, 111–19.

VAN ETTEN, J. L. (1969). Protein synthesis during fungal spore germination. *Phytopathology*, **59**, 1060–4.

VEZINA, C., SINGH, K. & SEHGAL, S. N. (1965). Sporulation of filamentous fungi in submerged culture. *Mycologia*, **62**, 722–36.

VINTER, V. & SLEPECKY, R. A. (1965). Direct transition of outgrowing bacterial spores to new sporangia without intermediate cell division. *Journal of Bacteriology*, **90**, 803–7.

WATSON, K., PATON, W. & SMITH, J. E. (1969). Oxidative phosphorylation and respiratory control in mitochondria from *Aspergillus oryzae*. *Canadian Journal of Microbiology*, **15**, 975–7.

WATSON, K. & SMITH, J. E. (1967). Oxidative phosphorylation and respiratory control in mitochondria from *Aspergillus niger*. *Biochemical Journal*, **104**, 332–9.

WEIJER, J. & WEISBERG, S. H. (1966). Karyokinesis of the somatic nuclei of *Aspergillus nidulans*. I. The juvenile chromosome cycle (Feulgen staining). *Canadian Journal of Genetics and Cytology*, **8**, 361–74.

WHITE, G. J. & SUSSMAN, M. (1961). Metabolism of major cell components during slime mould morphogenesis. *Biochimica et Biophysica Acta*, **53**, 285–93.

WILDMAN, J. D. (1966). Note on occurrence of giant cells in *Aspergillus flavus* Link. *Journal of the Association of Official Analytical Chemists*, **49**, 563–4.

WOGAN, G. N. & PONG, R. S. (1970). Aflatoxins. *Annals of the New York Academy of Sciences*, **174**, 623–35.

PLATE 1

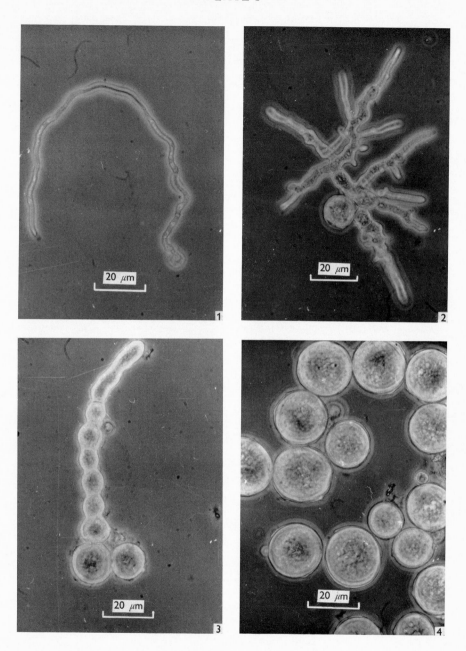

PLATE 2

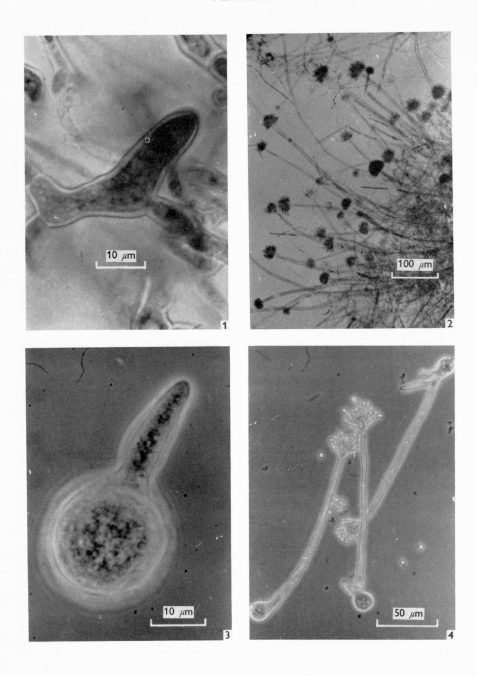

WRIGHT, B. E. (1966). Multiple causes and controls in differentiation. *Science, New York*, **153**, 830–7.

WRIGHT, B. E. (1968). An analysis of metabolism underlying differentiation in *Dictyostelium discoideum*. *Journal of Cellular Physiology*, **72**, 145–60.

YANAGITA, T. (1957). Biochemical aspects of the germination of conidiospores of *Aspergillus niger*. *Archiv für Mikrobiologie*, **26**, 329–44.

YANAGITA, T. (1963). Carbon dioxide fixation in germinating conidiospores of *Aspergillus niger*. *Journal of General Microbiology*, **9**, 343–51.

YANAGITA, T. (1964). Germinating conidiospores of *Aspergillus niger*. In *Synchrony in Cell Division and Growth*, ed. E. Zeuthen, pp. 391–420. New York: John Wiley and Sons.

YANAGITA, T. & KOGANE, F. (1962). Growth and cytochemical differentiation of mold colonies. *Journal of General and Applied Microbiology*, **8**, 201–13.

YANAGITA, T. & KOGANE, F. (1963a). Cellular differentiation of growing mold colonies with special reference to phosphorus metabolism. *Journal of General and Applied Microbiology*, **9**, 313–30.

YANAGITA, T. & KOGANE, F. (1963b). Cytochemical and physiological differentiation of mold pellets. *Journal of General and Applied Microbiology*, **9**, 179–87.

YUILL, E. (1950). The numbers of nuclei in conidia of Aspergilli. *Transactions of the British Mycological Society*, **36**, 57–60.

ZONNEVELD, B. J. M. (1971). Biochemical analysis of the cell wall of *Aspergillus nidulans*. *Biochimica et Biophysica Acta*, **249**, 506–14.

EXPLANATION OF PLATES

PLATE 1

Changes in the morphology of germinating conidia and newly formed hyphae of *Aspergillus niger* associated with varying degrees of apical growth inhibition.

Fig. 1. Rapid hyphal elongation after spore germination at 30°. The conidium has remained small and the hypha is thin and unbranched (about 8 h after inoculation).

Fig. 2. Partially inhibited hyphal elongation after spore germination at 41°. The conidium is enlarged and the hyphae are thickened and excessively branched (about 15 h after inoculation).

Fig. 3. Severely inhibited hyphal elongation after spore germination at 43°. The conidium is enlarged and the 'hypha' assumes the form of a chain of spherical cells produced by regular septum formation (about 24 h after inoculation).

Fig. 4. Total inhibition of the hyphal form although enlargement (spherical growth) of the conidia continues (about 48 h after inoculation).

PLATE 2

Early and late stages in the development of conidiophores of *Aspergillus niger* from enlarged conidia (microcycle conidiation) and from mycelia under submerged agitated conditions.

Fig. 1. Foot cell and outgrowing conidiophore induced by nitrogen exhaustion in fermenter culture.

Fig. 2. Conidiophore morphology in fermenter culture after first medium replacement. Vesicle and phialide formation has occurred but conidiospore production is inhibited.

Fig. 3. Initiation of microcycle conidiation. Conidiophore outgrowth from the enlarged conidium in the absence of vegetative development was induced by incubation at 44° for 48 h followed by 30°.

Fig. 4. Completion of microcycle conidiation. The first formed conidiophores are fully developed and secondary conidiophores are being produced at their base.

DIFFERENTIATION IN TRYPANOSOMATIDAE

B. A. NEWTON, G. A. M. CROSS AND J. R. BAKER

Medical Research Council Biochemical Parasitology Unit,
The Molteno Institute, University of Cambridge

INTRODUCTION

The Trypanosomatidae are parasitic flagellate protozoa (Table 1); some (monogenetic species) have one host in their life cycle, others (digenetic species) have two. Hosts of the former are generally arthropods, usually insects, and the parasites most frequently inhabit their gut. The second hosts of digenetic species are vertebrates of all classes in which the parasites inhabit blood and other tissues.

Digenetic species of Trypanosomatidae may have evolved from monogenetic parasites of invertebrates and the acquisition of a second host can be regarded as an evolutionary advance resulting from the adoption of a vertebrate blood diet by some invertebrates (Baker, 1963; Hoare, 1972). Most species of *Trypanosoma* enter their vertebrate host through mucous membranes or skin lesions by contamination with or ingestion of infected faeces or intestinal contents of the invertebrate host, but some (the salivarian group of the genus *Trypanosoma* (Hoare, 1966) and members of the genus *Leishmania*) are injected into the vertebrate through the proboscis of the feeding insect vector. A few species (*T. vivax viennei, T. evansi, T. equinum* and *T. equiperdum*) do not develop in an invertebrate host and have become monogenetic parasites of vertebrates; the first three use blood sucking insects as transport hosts to carry them to another vertebrate while *T. equiperdum* is transmitted venereally.

A number of different morphological forms occur during the developmental cycles of Trypanosomatidae (Figs. 1 and 2) and in some species it is now known that these visible changes are associated with marked alterations in cell structure and metabolism: the aim of this paper is to review progress in this field.

MORPHOLOGICAL FORMS

For most of that part of their life cycle which is spent in the invertebrate hosts, trypanosomatids occur as flagellated organisms in the forms known as promastigote, epimastigote, choanomastigote or opistho-

Table 1. *Classification of parasites discussed in this paper*

See Hoare (1972) for a discussion of the validity of treating *T. rhodesiense* and *T. gambiense* as subspecies of *T. brucei*.

Family Trypanosomatidae

1. Monogenetic parasites of invertebrates: genera *Crithidia*, *Leptomonas*, *Blastocrithidia*, *Herpetomonas*

2. Digenetic parasites:
   (*a*) of invertebrates and plants: genus *Phytomonas*
   (*b*) of invertebrates and vertebrates
      (i) Vertebrate hosts fish, amphibia, reptiles or birds: genus *Trypanosoma* (e.g. *T. mega*, *T. avium*)
      (ii) Vertebrate hosts mammals: genus *Leishmania*
      Genus *Trypanosoma*

         section Stercoraria: subgenera *Megatrypanum* (e.g. *T. theileri*), *Herpetomonas* (e.g. *T. lewisi*, *T. conorrhini*), *Schizotrypanum* (e.g. *T. cruzi*)

         section Salivaria (development in invertebrate hosts occurs in *Glossina* spp. tsetse only, unless secondarily lost): subgenera *Duttonella* (e.g. *T. vivax vivax*, *T. vivax viennei**), *Nannomonas*, *Trypanozoon* (e.g. *T. brucei brucei*, *T. b. rhodesiense*, *T. b gambiense*, *T. evansi*,* *T. equinum*,* *T. equiperdum†*)

* No cyclical development in invertebrate; various Diptera serve as transport hosts.

† No invertebrate host; transmission venereal.

mastigote (Hoare & Wallace, 1966): these forms are illustrated in Fig. 2. In the external environment monogenetic species exist as encysted non-flagellated forms (amastigotes). Amastigotes also occur in the digenetic genus *Leishmania* but they are not encysted and occur within certain cells of the vertebrate host. In contrast, the other digenetic genus, *Trypanosoma*, has evolved a new form, the trypomastigote, as which it spends part or all of its life in the blood or tissue fluids of its vertebrate hosts. Most species of this genus, however, do not divide while in the trypomastigote form; their division stages, which are extracellular except in the subgenus *Schizotrypanum*, may be promastigotes, epimastigotes or amastigotes. Trypomastigotes are known to divide only in the three salivarian subgenera (*Duttonella*, *Nannomonas* and *Trypanozoon*). There are many reasons for regarding the Salivaria as the most recently evolved group of species of *Trypanosoma* (Woo, 1970) and it may well be that the acquisition of the ability to divide in the trypomastigote form represents a relatively late evolutionary development. The species referred to above as having become wholly or partly emancipated from their invertebrate hosts (*T. equiperdum*, *T. evansi*, *T. equinum* and *T. vivax viennei*) are believed to exist only as trypomastigotes.

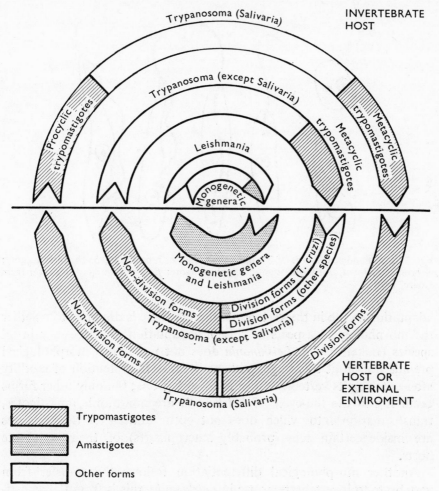

Fig. 1. Diagrammatic representation of the life cycles of Trypanosomatidae.

MORPHOLOGICAL CHANGES DURING
THE LIFE CYCLE

In the invertebrate host

The monogenetic genera differentiate into encysted amastigotes in the invertebrate's hind-gut before being expelled to face the harsher realities of the outside world. Those species of *Trypanosoma* which undergo cyclical development in an invertebrate host differentiate into trypomastigotes (called metacyclic trypomastigotes as they occur after the developmental cycle in the invertebrate) before entering the next (vertebrate) host; this change is thought to represent a pre-

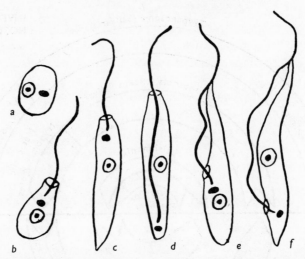

Fig. 2. Morphological forms of Trypanosomatidae. *a*, amastigote; *b*, choanomastigote; *c*, promastigote; *d*, opisthomastigote; *e*, epimastigote; *f*, trypomastigote (redrawn from Hoare, 1970).

adaptation to life in the vertebrate, though there is doubt as to whether the morphological aspects of this pre-adaptation are necessary in all species (Baker, 1966). *Leishmania* does not undergo a morphological pre-adaptation at this stage, perhaps because the retention of motility after entering its vertebrate host is advantageous; the only other forms occurring in the life-cycle of *Leishmania* are non-motile amastigotes, transformation into which does not occur until after the parasites are inside certain cells (probably macrophages) of their vertebrate hosts.

Another morphological differentiation follows the transfer from vertebrate to invertebrate hosts; in *Leishmania*, this is from the amastigote to promastigote form, while in *Trypanosoma* it is from trypomastigote to epimastigote. Usually this occurs almost immediately after transfer and before any cell-division has occurred (e.g. Muniz & de Freitas, 1945, 1946; Baker, 1956), but in the salivarian species of *Trypanosoma* the trypomastigote phase is carried over into the invertebrate gut where the parasites multiply and only subsequently differentiate into epimastigotes to continue the life cycle. These forms of the Salivaria which multiply in the gut of their invertebrate hosts (or *in vitro* in suitable growth media when incubated at 28°) differ morphologically and physiologically from those occurring in the vertebrate (see below), and they are commonly called 'mid-gut' or 'culture forms', although both are trypomastigotes by the criteria of Hoare & Wallace (1966). Mid-gut

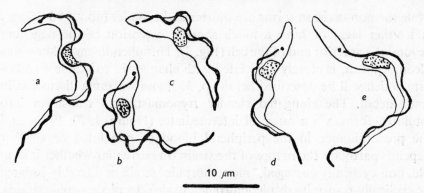

Fig. 3. *Trypanosoma brucei* sspp. *a–c*, haematozoic trypomastigotes: *a*, slender; *b*, intermediate; *c*, stumpy; *d, e*, procyclic trypomastigotes from midgut of *Glossina* (redrawn from Hoare, 1970).

forms, by analogy with the metacyclic trypomastigotes (the final developmental stage in the invertebrate host) could perhaps more conveniently be distinguished as procyclic trypomastigotes (Fig. 3).

In the vertebrate host

A further differentiation, which occurs during the life cycle of all species of *Trypanosoma*, is that from division forms to non-division forms in the vertebrate host (Deane, 1969). The presumed function of these non-dividing forms is to infect the next, invertebrate, host; an exactly analogous function to that of the metacyclic trypomastigotes in the invertebrate, whose only known function is to infect the vertebrate. It is possible that in some of the more 'primitive' species of *Trypanosoma* (e.g. some of those parasitising birds, reptiles, fishes or amphibia) division does not occur at all in the vertebrate host (Demaree & Marquardt, 1971), though this is uncertain. Perhaps evolution of the ability to divide in the vertebrate is relatively recent and only in the Salivaria has this function been acquired by the most recently developed morphological form, the trypomastigote. However, there is evidence that at least some strains of species belonging to the subgenus *Trypanozoon* may multiply in their vertebrate hosts as forms other than trypomastigotes (Ormerod & Venkatesan, 1971*a, b*).

In stercorarian species, division and non-division stages in the vertebrate are easily distinguished morphologically, since only the latter are trypomastigote, compared with the majority of salivarian species in which both are trypomastigotes. However, in the salivarian subgenus *Trypanozoon* there is a gross morphological distinction; the division forms are elongated slender organisms measuring about $30 \times 1.5 \ \mu m$

while the non-division forms are shorter and broader (about $20 \times 3 \mu m$) and either lack, or have a much shorter, extension of the flagellum beyond the anterior end of the cell (Fig. 3). This phenomenon, known as pleomorphism, is closely associated with changes in oxidative metabolism which will be described below (p. 352) and warrants discussion in some detail. The elongated slender trypomastigotes transform into 'stumpy' forms via a range of intermediates (Hoare, 1970, 1972), and the predominance in the peripheral blood of one form or another depends partly on the nature of the strain (in particular whether it is an old, non-cyclically passaged, 'monomorphic' strain or a freshly-isolated or cyclically-transmitted 'pleomorphic' strain). In pleomorphic strains, when parasitaemia is increasing, slender forms predominate; but at the onset of host antibody response, increasing proportions of intermediate and stumpy forms are found. Balber (1972) has suggested that this results from a diminution in the number of slender forms, due possibly to their being preferentially destroyed by antibody and/or sequestered in certain viscera rather than to an increase in the number of stumpy forms. Stumpy forms are believed (on inferential evidence only) to be those capable of continuing development either *in vitro* or in the tsetse's midgut as procyclic trypomastigotes. During subsequent waves of increasing parasitaemia, slender trypomastigotes reappear in the peripheral blood but their surface antigens are different from those of the preceding population (see p. 346). Pleomorphism occurs in populations (clones) arising from single trypanosomes and in the absence of demonstrable immune response by the host (Luckins, 1972; Balber, 1972). In the other salivarian subgenera (*Duttonella* and *Nannomonas*), there is no obvious morphological or detectable physiological distinction between division and non-division forms.

Biochemical and ultrastructural changes now known to be associated with these complex morphological changes will be discussed below. However, progress in these investigations has been, and still is, severely limited by the problems of obtaining sufficient material for metabolic studies and of inducing differentiation *in vitro* under defined conditions. An understanding of the limitations of existing experimental systems is therefore essential to a discussion of current work.

LIMITATIONS OF EXISTING EXPERIMENTAL SYSTEMS

Ideally, for experimental study of the differentiations referred to above the parasites' entire life cycle should be reproducible *in vitro* in a defined medium. This ideal is very far from realisation. That part of the cycle undergone by many of the monogenetic and digenetic species in their invertebrate hosts can be reproduced *in vitro* but, with few exceptions, not in defined media (see Taylor & Baker, 1968). The salivarian species of *Trypanosoma* are less amenable; only the initial phase (procyclic trypomastigotes) of their intra-invertebrate cycle has so far been grown *in vitro*, and this in media which are far from definition (e.g. Pittam, 1970). There have been reports of the infectivity of such cultures to mammals (Trager, 1959*a*, *b*; Amrein, Geigy & Kauffmann, 1965; G. A. M. Cross, unpublished results) suggesting that metacyclic trypomastigotes may occasionally develop *in vitro*, but no direct microscopical evidence of their presence has been produced. The intra-vertebrate part of the life cycle of digenetic species has proved even less tractable; that of *T. cruzi* and *Leishmania* spp. has been grown in tissue cultures (Pipkin, 1960; Zuckerman, 1966), but such systems are almost as complex as those using a living vertebrate. Very limited success has been achieved in growing normally intracellular forms (amastigotes of *Leishmania* and *T. cruzi*) in non-cellular media (Trager, 1953; Lemma & Schiller, 1964; Pan, 1971) and successful in-vitro maintenance of haematozoic forms of other species of *Trypanosoma* has been mainly restricted to species parasitising amphibia and to the stercorarian species infecting mammals (see Taylor & Baker, 1968, for references; also Sanchez & Dusanic, 1968). Haematozoic forms of *T. brucei* sspp. were first maintained *in vitro* by Yorke, Adams & Murgatroyd (1929), but not until 38 years later were conditions found under which increase in numbers would occur. These conditions involve the presence of living mammalian cells, and so are very complex; an even more severe limitation is that the maximum increase in cell numbers yet achieved *in vitro* is 8–10-fold (Le Page, 1967; Hawking, 1971; Chaloner, 1972).

Because of these difficulties, detailed studies of the physiological changes associated with differentiation in Trypanosomatidae have been restricted to relatively few species. The transformation of slender to stumpy haematozoic forms of *T. brucei* sspp. and of these forms into procyclic trypomastigotes (culture forms) has attracted most attention (reviewed by Vickerman, 1971) and much of this article will be concerned with these investigations. Other systems which have proved amenable to study include the transformation of haematozoic trypomastigotes of *T. cruzi* to epimastigotes (Muniz & de Freitas, 1945, 1946); epimastigotes of *T. cruzi* to metacyclic trypomastigotes (Fernandes, Castellani & Kimura, 1969); epimastigotes of *T. conorrhini* to haematozoic trypomastigotes (Deane & Kirchner, 1963) and amastigotes of *L. donovani* to promastigotes (Simpson, 1968).

DEVELOPMENTAL CHANGES IN THE CELL SURFACE

It is hoped that work in progress and the passage of time will justify discussion under this heading of some of the more interesting implications of the extensive immunological investigations carried out on trypanosomes. It is convenient to separate trypanosome antigens into two classes: common or invariant antigens and strain-specific or variable antigens. We believe that it is primarily the variable antigens that are relevant to a discussion of developmental changes in the trypanosome cell surface. For a more comprehensive treatment of trypanosome immunology the reader is referred to the reviews of Brown (1963), Weitz (1970) and Desowitz (1970).

Trypanosome infections may persist for long periods in the vascular system of the mammalian host, despite a strong immunological response. If antisera are prepared against populations of trypanosomes isolated on different occasions during a chronic infection, it can be seen from agglutination tests that the immunological properties of the cells are changing during the course of the infection. Trypanosomes isolated at any given stage in an infection may react only with homologous antisera, and not with antisera prepared against previous or subsequent populations. This phenomenon has been demonstrated most convincingly in the work of Gray (1965). Because of this capacity for antigenic variation, successful immunisation against trypanosomiasis has been impossible. Antigenic variation may also be partially responsible for persistence of reservoirs of infection in some wild animals.

The apparently unlimited capacity of *T. brucei* and possibly other species to change their antigenic character suggests the operation of an unusual system controlling the synthesis of a class of surface antigens. At the present time there is insufficient evidence to favour any particular hypothesis concerning the control of this phenomenon. For example, it is not known whether the generation of a new antigenic type occurs by selection of a minority population of variants pre-existent at the onset of antibody response against the prevailing majority variant or whether in response to antibody or other stimulus any given cell can modify or shed its existing antigen and replace it with a molecule having a different immunological specificity. Certainly there appears, in many trypanosome infections, to be an oscillation in the parasitaemia associated with the production of antibody. But the disappearance of trypanosomes from the bloodstream, however dramatic, is not proof of their destruction by the immune system; they may have migrated to the shelter of the tissues, where they are frequently found in large numbers, to restructure their vulnerable antigens. This question might be resolved by studies on total cell turnover rather than measurements on the level of the parasitaemia. Similar problems are posed at the genetic level regardless of whether the antigenic changes take place by antibody-induced changes in individual cells or by population selection processes. There is the possibility of mutation on the one hand or selective transcription from pre-existing alternative genes on the other. Gray's (1965) observations that different antigenic variants may revert to a 'basic strain antigen' following cyclical transmission through the tsetse fly, and that in subsequent infections the same antigenic variants might appear in a similar sequence, argue against a hypothesis based on mutation. The alternative hypothesis implies that gene duplication and mutation has occurred at some stage in the evolution of the system, and that we now have a stable population of closely-related genes which can be translated in a controlled manner. Extensive gene duplication might have resulted in the appearance of distinct 'satellite' DNA components (Fig. 5) such as appear to be particularly prominent in certain Salivaria (Newton & Burnett, 1972a, b). A further possibility, but one which would put definite limits on the capacity for variation, would be the existence of a group of protein subunits which, by being assembled in several possible configurations, could present a wide range of immunological identities. However, the most interesting hypothesis which we would like to suggest is that there exists in trypanosomes a system for antigen synthesis which is similar to that involved in immunoglobulin synthesis. Both of these systems have a requirement for the synthesis of molecules having similar gross structures, but having an infinite range of distinct immunological properties. To extend this analogy even further, the trypanosome antigens, assuming a definite cellular location, might require variable and common regions, the variable region being exposed to the environment and the common portion locating the antigen in the correct configuration at a particular cellular site.

There is good reason to believe that a major class of variable antigens exists and is located on the cell surface. All cell surfaces are immunogenic to some extent and generally contain a range of potential antigens. The most economical strategy for protecting functionally essential and varied surface components from attack by antibodies might therefore be to mask them with a single type of molecule which could be replaced as required. This may indeed be what the trypanosome does. Electron micrographs of bloodstream forms of *T. brucei* sspp., *T. vivax* and *T. congolense*, show the presence of a dense and regular surface coat extending 12–15 nm outwards from the plasma membrane (Plate 1)

(Kubo, 1968; Godfrey & Taylor, 1969; Vickerman, 1969a; Wright & Hales, 1970). This surface coat is absent from procyclic trypomastigotes found in the mid-gut of the invertebrate host (*Glossina* spp.) and in culture, but is re-acquired during transformation from epimastigote to infective metacyclic forms in the tsetse salivary gland (Vickerman, 1969a; Steiger, 1971). The presence or absence of the surface coat correlates with earlier studies of the surface charge on trypanosomes (Hollingshead, Pethica & Ryley, 1963). Evidence obtained by the use of ferritin-conjugated antibodies suggests that the surface coat does indeed contain variant-specific antigens (Vickerman & Luckins, 1969).

By labelling intact trypanosomes with a non-penetrating protein-reactive reagent, a glycoprotein which is believed to be the major component of the surface coat of *T. brucei* has been identified and isolated (Cross, 1972). This protein represents about 10 % of the total cell protein, as might be expected from calculations based on surface area and coat thickness. Small variations in electrophoretic mobility, amino acid and carbohydrate composition between coat proteins isolated from different cell clones provide preliminary indications of variations in structure which would be anticipated if the coat protein is the major variable antigen. As yet there is no direct evidence to equate this purified coat protein with the variant-specific antigen, but it is probably significant that the antigens studied by Allsopp, Njogu & Humphreys (1971) also appear to be the major soluble glycoproteins of the cell.

It has not yet been established whether the trypanosome surface coat is wholly of endogenous origin, or whether host proteins contribute to it by adsorption onto the plasma membrane. *Trypanosoma vivax* certainly adsorbs serum proteins to some extent (Ketteridge, 1970, 1971) and *Leishmania tarentolae* grown in the presence of homologous antiserum acquires a surface coat, due presumably to the adsorption of immunoglobulin molecules (Strauss, 1971). Desowitz (1970) has invoked host antibody in a theory concerning antigenic variation whereby an intrinsically invariant surface antigen is modified by interaction with antibody produced in response to the trypanosome. A new immunogenic identity would be produced and the process would be repeated as required by the appearance of subsequent waves of antibody.

In summary, the surface coat of trypanosomes appears to be an adaptation (or in the case of the metacyclic trypomastigotes, a pre-adaptation) to life in the mammalian bloodstream. Not all trypanosomes have such a distinct surface coat as *T. brucei* sspp. We know nothing at present of the mechanisms controlling the initiation of coat synthesis in metacyclic forms or variation of coat structure in the mammalian host. Detailed sequence studies on coat proteins isolated from clones of *T. b. brucei* may provide a stronger foundation on which to base hypotheses concerning the underlying control systems.

CHANGES IN PATHWAYS OF OXIDATIVE METABOLISM

Studies of the intermediary metabolism and ultrastructure of bloodstream and culture forms of *T. brucei* sspp. have shown that morphological changes occurring during their life cycle are accompanied by an alternating proliferation and regression of mitochondrial structures, Krebs cycle enzymes and cytochromes. The foundations of current ideas about the changes in energy metabolism which accompany differentiation in these organisms were laid by the early work of Von Brand (1951), Fulton & Spooner (1959), Ryley (1956, 1962) and Grant & Sargent (1960, 1961). These early metabolic studies (reviewed by Fulton, 1969) were later extended by application of cytochemical techniques to developmental stages occurring in the invertebrate host (which have never been obtained in sufficient quantities for biochemical investigation) and by a comparison of the ultrastructure of different developmental stages (Ryley, 1964; Vickerman, 1962, 1965, 1971). On the basis of these investigations Vickerman proposed an elegantly simple hypothesis relating the observed changes in morphology and metabolic activity to mitochondrial biogenesis and regression, but this hypothesis has not yet been fully substantiated and detailed exploration of some of the ideas contained in it is only just beginning.

The aerobic metabolism of haematozoic trypomastigotes of *T. brucei* sspp. is characterised by a voracious and incomplete metabolism of glucose, most of which is oxidised to pyruvate and excreted (Ryley, 1956). Motility and integrity of the cells is dependent on the presence of glucose, mannose, fructose or glycerol; no other substrate is known to be oxidised to any significant extent. Cytochrome pigments and several enzymes of the tricarboxylic acid cycle are absent, and respiration is unaffected by cyanide or carbon monoxide (Ryley, 1956, 1962; Fulton & Spooner, 1959). NADH formed during glycolysis is reoxidised by a cyclic system (Grant & Sargent, 1960, 1961; Grant, Sargent & Ryley, 1961) involving a soluble L-glycerol 3-phosphate:NAD oxidoreductase and a particulate L-glycerol 3-phosphate oxidase which is itself thought to consist of two flavoproteins, an L-glycerol 3-phosphate oxidase and a substrate-specific peroxidase (Grant & Bowman, 1963; Bide & Grant, 1964).

Cytochemical observations suggest that this oxidase system is localised in extramitochondrial cytoplasmic granules (Ryley, 1964; Vickerman, 1965). It is not known whether the operation of this system produces any ATP; if not, there would appear to be the possibility of a net formation of only two molecules of ATP by substrate-level phosphorylation during conversion of glucose to pyruvate. Anaerobic metabolism of glucose results in the production of equimolar amounts of pyruvate and glycerol (Ryley, 1956).

The oxidative pathways of culture forms of *T. brucei* sspp. are more complex than those of the bloodstream forms from which they are derived. Under aerobic conditions they can metabolise glucose completely to carbon dioxide and water and probably contain a full complement of tricarboxylic acid cycle enzymes coupled to a 'mammalian-type' cytochrome system with a cyanide-sensitive terminal oxidase (Ryley, 1956, 1962; Fulton & Spooner, 1959; Baernstein, 1963). Anaerobically, culture forms produce succinate by a CO_2-dependent pathway (Ryley, 1962) indicating a reversed flow through the tricarboxylic acid cycle, probably via phosphoenolpyruvate carboxy kinase (Bacchi, Ciaccio, Kaback & Hutner, 1970). The structure of the mitochondrion is more highly developed in culture forms than in bloodstream forms; it appears to be a single organelle having a highly branched structure extending along most of the length of the cell. A large mass of DNA is located in an area of the mitochrondion forming the kinetoplast: the structure and possible function of this DNA is discussed below. The most extensive mitochondrial network is seen in monogenetic trypanosomatids which are believed to be obligate aerobes (Plate 2); at the other extreme, salivarian trypanosomes which are no longer cyclically transmitted (e.g. *T. equiperdum*) or which no longer possess a normal kinetoplast (see below) may retain only a vestigial mitochondrial tubule.

Recent work on the electron transport systems of monogenetic trypanosomatids (Hill & Anderson, 1970; Hill, 1972; Toner & Weber, 1972; Kusel & Storey, 1972) seems likely to be relevant to the digenetic trypanosomes. The most interesting developments in this work have come from the realisation that the respiration of culture forms is not always cyanide-sensitive and that two alternative oxidative pathways may exist, one terminating in a mammalian-type, cyanide-sensitive cytochrome aa_3 oxidase, the other terminating in a cyanide-insensitive oxidase, which may be cytochrome o. The first intimation of this divergence from the accepted view came from the experiments of Evans & Brown (1971a, b) who studied culture forms which developed immediately following the transfer of bloodstream *T. brucei* sspp. into a

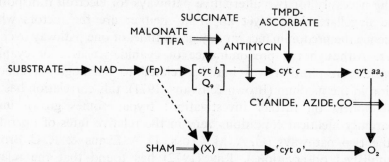

Fig. 4. Branched electron transport pathway proposed for *Trypanosoma mega* (Ray & Cross, 1972*a*) and *Crithidia fasciculata* (Hill, 1972). (TMPD, *N*,*N*,*N'*,*N'*-tetramethyl'-*p*-phenylenediamine; SHAM, salicyl hydroxamic acid; TTFA, thenoyl trifluoro-acetone.)

monophasic liquid medium. They found that these organisms developed a cyanide-insensitive pathway for succinate oxidation. The presence of this pathway was initially thought to be restricted to recently-transformed culture forms, but subsequent work (Ray & Cross, 1972*a*, *b*) in contrast to the earlier results of Ryley (1962), has shown that respiration in long-established culture forms can also proceed via a cyanide-insensitive pathway. It is probable that this discrepancy is due to differences in the culture media used in these studies.

The presence of a branched electron-transport pathway (Fig. 4) with cytochromes a_3 and o as alternative terminal oxidases, has been proposed to account for patterns of substrate oxidation and inhibitor activity observed in studies with *T. mega* (Ray & Cross, 1972*a*, *b*). Hill (1972) has proposed a similar pathway for the cyanide-insensitive respiration of *Crithidia fasciculata*. Determination of photochemical action spectra for light reversal of carbon monoxide-inhibited respiration has confirmed that the CO-binding pigment, cytochrome o, is a functional oxidase in several monogenetic trypanosomatids and in *T. mega* (Kusel & Kronick, 1972; P. Kronick & G. C. Hill, personal communication). There are indications that these pathways may also exist in *T. brucei* sspp. and detailed investigations are in progress. A *c*-type cytochrome has recently been purified from *T. b. rhodesiense*, and appears to have similar properties to *c*-type cytochromes isolated from other trypanosomatids, which show characteristic differences from cytochroms *c* from mammalian tissues (Hill, Gutteridge & Matthewson, 1971). Spectroscopic evidence has also been obtained for the presence of cytochromes *b*, *o* and aa_3 in culture forms of *T. brucei*, although the aa_3 component is not usually present in readily detectable amounts (G. A. M. Cross, unpublished observations).

The necessity for two alternative pathways for electron transport in these flagellates is not yet apparent, neither are the factors which influence the predominance, at any given time, of one pathway over the other. Although the predominance of cyanide-sensitive or cyanide-insensitive respiration was initially related to haemoglobin concentration in the medium (Brown & Evans, 1971), this correlation has not been upheld on further investigation: trypanosomes grown under apparently identical conditions vary in the relative rates of operation of the two pathways (G. A. M. Cross, D. A. Evans & R. C. Brown, unpublished observations). Ray (1972) has found that the relative proportions of cytochromes and predominance of cyanide-sensitive or cyanide-insensitive terminal oxidation in *T. mega* can be reproducibly influenced by the concentration of haemin in the growth medium. Studies on the insect trypanosomatid *Crithidia oncopelti* may also be relevant to this problem. Srivastava (1971) has published spectroscopic evidence that this flagellate contains only cytochrome *o* during the early logarithmic growth phase, but cytochrome aa_3 appears in stationary phase cultures; however, it must be stressed that this flagellate is atypical of trypanosomatids in general since it does not require an exogenous source of haemin for growth (Lwoff, 1937; Newton, 1956).

Before summarising present views on the development of various oxidative pathways at different stages in the life cycle of *T. brucei* sspp. it is necessary to return to the phenomenon of pleomorphism (p. 344). In their bloodstream phase *T. brucei* sspp. exist in a number of morphological forms ranging from slender through intermediate to stumpy forms. Vickerman (1965) obtained cytological evidence for the presence in stumpy forms, but not in slender forms, of NADH–nitrotetrazolium blue reductase activity associated with the mitochondrion. He also obtained evidence based on the retention of motility in the presence of glucose or α-oxoglutarate, that only stumpy forms metabolise the latter substrate. He therefore proposed that the first changes in metabolic activity occurred at the time of the morphological change from slender to stumpy forms. On the basis of more detailed biochemical studies using populations in which either slender or stumpy forms predominated, Bowman, Flynn & Fairlamb (1970) suggested that stumpy forms contain all the enzymes of the tricarboxylic acid cycle, with the possible exception of succinic dehydrogenase. However, it must be emphasised that not all of the enzymes of the tricarboxylic acid cycle have been unequivocally demonstrated in digenetic trypanosomes. Also, most investigations have concentrated on the energy-generating function of the cycle, its importance in biosynthetic pathways being largely ignored.

In haematozoic forms, particularly, we have little knowledge of the area of metabolism centred around acetyl-CoA: termination of carbohydrate metabolism at pyruvate and the absence of a functional tricarboxylic acid cycle must place severe limits on the biosynthetic versatility of these forms.

Thus the present picture of the switches in respiratory metabolism which occur during the life cycle of *T. brucei* sspp. may be summarised as follows. Activation of certain mitochondrial enzymes appears to take place during the slender–stumpy transition in the mammalian host. This is accompanied by some structural development of the mitochondrion, but complete development of the mitochondrion and appearance of cytochromes occurs only following transfer of bloodstream organisms to the insect host, or to culture medium. Regression of the mitochondrion, by morphological criteria, takes place in the tsetse salivary gland during the development of metacyclic forms capable of re-infecting the mammalian host.

Current research is largely centred around three main aspects of the problem.

1. Structure of the electron-transport pathways of culture and bloodstream forms, location of phosphorylation sites, and investigation of parameters affecting the operation of alternative pathways.

2. Exploration of the metabolic capabilities of bloodstream and culture forms.

3. Temporal relationships between activation of different pathways during differentiation, leading to considerations of control mechanisms especially in relation to the role of mitochondrial (kinetoplast) DNA.

This last aspect is currently being investigated in several laboratories, by transferring bloodstream forms of *T. brucei* sspp. into culture media and following the consequential metabolic changes. Under the conditions used by Evans & Brown (1971*a*, *b*), bloodstream forms transferred to Pittam's medium (Pittam, 1970) at 26° began to grow without any observable lag, with a doubling time of about 17 h. These organisms developed the ability to oxidise succinate via a cyanide-insensitive pathway over a period of 48 h; however, they could not be subcultured indefinitely, and after about 14 days, when the flagellates had developed a cyanide-sensitive oxidation, growth could be maintained only by modification of the culture medium. Evans & Brown (1972) also found that transformation to culture forms was accompanied by a switch from glucose to proline oxidation. The initial phase of transformation was characterised by a rapid decrease in glucose oxidation and pyruvate excretion, accompanied by the development of a highly active 'proline-oxidase' system. In a superficially similar study, Srivastava & Bowman

(1971, 1972) followed the appearance of ability to oxidise α-keto-glutarate, NADH, succinate and proline in bloodstream forms and during their development into culture forms; they found that in the case of succinate oxidation the change from partial to complete cyanide sensitivity occurred earlier than reported by Evans & Brown (1971*a*, *b*), but they do not state whether cell multiplication occurred during their experiments.

It is clear that present knowledge of oxidative metabolism of trypanosomatids is meagre: the performance of many simple and obvious experiments is limited by the difficulty of obtaining adequate material for biochemical studies and by the problems of variability resulting from the use of complex media containing blood. Conclusions about changes in metabolism based on observations of organisms in undefined and continuously varying environments are clearly highly conjectural. More meaningful investigations of the oxidative metabolism of *T. brucei* sspp. should be possible now that a synthetic medium has been devised which will support growth of procyclic forms of this flagellate (Cross, 1973).

CHANGES IN LIPID AND STEROL COMPOSITION DURING DIFFERENTIATION

Changes have been detected in the lipid composition of phytoflagellates, resulting from the development of proplastids into functional chloroplasts, when these organisms switch from a heterotrophic to a photoautotrophic existence (Hulanicka, Erwin & Bloch, 1964; Rosenberg & Pecker, 1964). As the changes in oxidative metabolism described above for *T. brucei* sspp. are accompanied by alternate development and regression of their mitochondrial network, and as, in other cell types, the lipid composition of mitochondrial membranes is known to differ significantly from that of plasma membranes (Korn, 1966), it might be expected that the metabolic changes in trypanosomes would be accompanied by variations in their overall lipid and sterol composition. Williamson & Ginger (1965) examined this possibility by analysing lipids extracted from culture and bloodstream forms of *T. b. rhodesiense*; culture forms were found to contain about half the amount of phospholipid present in bloodstream forms and the lipoprotein/lipopolysaccharide ratio was much lower in bloodstream than in culture forms. Further work (Meyer & Holtz, 1966; Godfrey, 1967; Dixon & Williamson, 1970) has demonstrated that the phospholipids of trypanosomes are, in general, similar to those found in mammalian cells, consisting largely of phosphatidyl choline, phosphatidyl ethanolamine and some phosphatidyl inositol. However, some interesting differences between species and between developmental stages of the same species are now coming to light as more detailed analyses are carried out. Dixon & Williamson (1970) found sphingomyelin accounted for about 23 % of the total phospholipid in the bloodstream form of *T. b. rhodesiense*, compared with only 5 % in culture forms. There is also evidence that the sterol content of this organism changes; cholesterol is the principal sterol in bloodstream forms and ergosterol predominates in culture forms (Threlfall, Williams & Goodwin, 1965; Williams, Goodwin & Ryley, 1966). It seems that cholesterol can be readily absorbed by both bloodstream and culture forms of *T. b. rhodesiense* and esterified to a limited extent, but sterol biosynthesis from acetate and methionine could be demonstrated in culture forms only (Dixon, Ginger & Williamson, 1972). It is interesting to note that in yeast, ergosterol is involved in the development and regression of mitochondria in response to aerobiosis (Parks & Starr, 1963; Luckins, Tham, Wallace & Linnane, 1966): perhaps acquisition of ability to synthesise sterols in the culture forms of *T. b. rhodesiense* is related to the requirement of these forms for a functional mitochondrion. In contrast to these observations on *T. b. rhodesiense*, the stercorarian trypanosome *T. lewisi*, which has a fully developed mitochondrion in both bloodstream and culture forms, does not lose the ability to synthesise sterols during development in the bloodstream (Dixon, Ginger & Williamson, 1971).

While all these findings suggest that differentiation in certain species of Trypanosomatidae may involve major changes in the pathways of lipid and sterol biosynthesis, there is need for caution in their interpretation. All the data published can be criticised to some extent as the techniques used for extraction, separation and analysis do not permit unambiguous identification of lipid components. A second and more serious criticism is that only haematozoic forms of monomorphic strains have been studied and the haematozoic and culture forms used in comparative studies have been derived from different strains. These investigations should be repeated with recently isolated pleomorphic strains of *T. brucei* sspp.

INITIATION AND CONTROL OF DIFFERENTIATION

It will now be clear to the reader that knowledge of the biochemical changes occurring during the developmental cycles of Trypanosomatidae is still far from complete. While we know even less about the factors which initiate and control these changes, some recent ideas and hypotheses are providing a useful basis for further investigations and warrant discussion. We will consider some of these ideas in relation to three fundamental questions posed recently by Trager (1970). What environmental changes initiate the development of one morphological form into another? Must flagellates undergo some intrinsic change before they can respond to environmental changes? If so, what are the molecular bases of these changes?

Extrinsic factors

Once comparative studies revealed differences in the oxidative metabolism of bloodstream and culture forms of *T. brucei* sspp. it was an obvious step to consider whether these could have developed in response to a change in environment. For example, when trypanosomes move from a vertebrate host to an insect vector they experience among other changes a temperature drop, a decrease in concentration of exogenous nutrients (particularly glucose), a lowering of oxygen tension and a change in osmotic conditions. Many workers have speculated that synthesis of mitochondrial enzymes might be switched on in response to these and/or other changes.

Temperature

There is now considerable evidence from experiments *in vitro* that temperature can influence transformation in a number of species of Trypanosomatidae. When trypomastigotes of *T. brucei* sspp. are transferred from the mammalian bloodstream to a blood-containing nutrient medium of the type described by Weinman (1953) and incubated at 37° they die rapidly, but incubation at 28° in the same medium permits transformation to culture forms and subsequent multiplication. This transformation generally results in a loss of infectivity for both the insect vector and mammalian host, but occasionally infective organisms have been detected in cultures and it has been suggested that an increase in temperature may initiate the production of infective 'metacyclic' forms (Trager, 1959*a*, *b*). Similarly, non-motile amastigote forms of *Leishmania donovani* may transform into motile promastigotes in a suitable medium at 28° but not at 37°; conversely *T. cruzi* cultured with L cells in Eagles' medium will grow extracellularly in the epimastigote form but if the temperature is raised to 38° these forms invade the L cells, develop as amastigotes and later transform into trypomastigotes (Trejos, Godoy, Greenblatt & Cedillos, 1963).

While these findings clearly indicate that temperature can influence differentiation *in vitro* there is good evidence that it is not the sole initiating or controlling factor. The results of cytochemical studies and cell motility experiments (Ryley, 1964; Vickerman, 1965) already referred to indicate that some increase in mitochondrial activity occurs in trypomastigotes of *T. brucei* sspp. while they are developing in the blood-stream of the vertebrate host and before they are subjected to a temperature change. It has been proposed that the immune response of the host may be the initiating factor (Ashcroft, 1957) but recent work does not support this view (Balber, 1972; Luckins, 1972) and at the time of writing there is no indication of what triggers the respiratory switch *in vivo*.

Chemical factors

Studies of transformation *in vitro* have now led to the identification of some chemical factors which can influence this process in *Leishmania donovani* and *Trypanosoma mega* (Steinert, 1958; Simpson, 1968) but unfortunately transformation of *T. brucei* sspp. and *T. cruzi* has not yet been achieved in a defined medium which makes the task of identifying chemical 'initiators' more difficult. The addition of trehalose to complex media has been reported to produce infective metacyclic trypomastigotes in cultures of *T. b. rhodesiense* (Weinman, 1957; Geigy & Kauffmann, 1964); however, other workers (Lehman, 1961; Bowman, von Brand & Tobie, 1960) failed to substantiate this observation. Further study of the epimastigote to amastigote transformation in cultures of *T. cruzi* (Pan, 1968) has led to the development of media in which transformation will occur in the absence of L cells and Pan (1971) concludes that in certain media the predominance of amastigotes is related to the presence of chicken plasma and chick embryo extract rather than to temperature. In cultures of *T. cruzi* maintained at 28° epimastigote forms are generally non-infective to mammals, but some infective metacyclic forms may develop at the end of the exponential growth phase (Camargo, 1964) and their appearance has been correlated with a fall in pH of the medium and an accumulation of organic acids.

Leishmania donovani is the only species of trypanosomatid which has been observed to transform in a defined medium. Simpson (1968) obtained over 90 % transformation from amastigote to epimastigote forms in a buffered salts–glucose medium containing 17 amino acids and haemin. In the absence of amino acids only 14 % of the population transformed, and transformation in the presence of amino acids was inhibited by actinomycin D, puromycin and mitomycin C, indicating that the change involved RNA as well as protein synthesis. Under optimal conditions transformation was complete in 20–30 h and was accompanied by a 5–7-fold increase in Q_{O_2} and the appearance of new soluble antigens. The reverse change, from motile epimastigotes to aflagellate forms, occurred in *Leishmania tarentolae* when either the riboflavine or choline content of the growth medium was reduced to a low level (Trager, 1957).

The addition of urea (6×10^{-3} M) to cultures of *Trypanosoma mega*, a parasite of African toads, initiated transformation from epimastigote to trypomastigote forms (Steinert, 1958); however, the percentage of a given population transforming was variable and influenced by age of culture, anaerobiosis and pH. As with *Leishmania donovani*, transformation of *T. mega* can be inhibited by actinomycin D; Steinert (1965) has postulated that urea may initiate genetic transcriptions by reversibly and specifically modifying secondary and tertiary structures of certain macromolecules. Use of ^{14}C-urea has demonstrated that there is no incorporation of urea carbon into cells. Recent attempts to extend these studies (Ray, 1972) have failed because the strain of *T. mega* originally isolated by Boné & Steinert in 1956, which has been maintained in culture since that time, no longer responds adequately to urea treatment.

Intrinsic factors

Although certain environmental factors influence differentiation in trypanosomes not all of a given population necessarily respond to these factors. For example, in Steinert's (1965) experiments urea induced transformation in *T. mega* only at the end of exponential growth and, even then, only 10–50 % of the flagellates in a culture were competent to transform. Similarly there is evidence that when bloodstream forms of *T. brucei* sspp. enter the insect vector or are transferred to a culture medium only certain trypomastigotes, possibly the stumpy forms, can continue to develop (p. 344). These findings suggest that intrinsic factors may play an important part in the developmental cycle of some trypanosomatids but, as in other differentiating systems, we know nothing of the nature of these factors or of the ways in which they may lead to certain genes being repressed or derepressed to give rise to cells competent to respond to environmental stimuli.

Kinetoplast DNA and cell differentiation

The kinetoplast can be detected by light microscopy as a basophilic granule near the base of the flagellum of trypanosomatid flagellates; it was first reported to contain Feulgen-positive material by Bresslau & Scremin (1924). This structure is now recognised as a modified region of the single tubular mitochondrion (Plate 2) (Meyer, Musacchio &

Mendonça, 1958; Steinert, 1960; Clark & Wallace, 1960; Pitelka, 1961). In 1910 Werbitzki reported that acriflavine treatment of trypanosomes induced the loss of kinetoplasts, as judged by the loss of Feulgen-staining material; these forms have been referred to as akinetoplastic or dyskinetoplastic. Subsequent work (Guttman & Eisenman, 1965; Steinert & Van Assel, 1967) established that dyskinetoplasty is a lethal condition for trypanosomatids growing *in vitro*. In contrast to this finding, some dyskinetoplastic forms of *T. brucei* sspp. can multiply indefinitely as slender trypomastigotes in the bloodstream of the mammalian host; however, they cannot develop in the invertebrate host. Dyskinetoplastic organisms are also known to arise spontaneously in *T. brucei* sspp. (Hoare & Bennett, 1937; Hoare, 1954) and as stated earlier these forms can be transmitted by direct inoculation. These findings led to the hypothesis that kinetoplast DNA plays a vital role in the developmental cycle of *T. brucei* sspp. and that it may contain genetic information required for the synthesis of some or all of the oxidative enzymes present in culture forms but absent from bloodstream trypomastigotes of this group. Although this hypothesis is unsubstantiated at the time of writing, and all evidence which can be cited in its support is inferred rather than direct, it is clearly relevant to the present discussion and warrants consideration in some detail.

Characteristics of kinetoplast DNA

DNA extracted from trypanosomatids which contain a Feulgen-positive kinetoplast can be separated into a major component and at least one minor satellite by isopycnic centrifugation in caesium chloride (Schildkraut, Mandel, Levisohn, Smith-Sonneborn & Marmur, 1962) (Fig. 5). One of these minor satellites has been found to have unusual properties in that it forms ultra-sharp bands during the first two hours of centrifugation (DuBuy, Mattern & Riley, 1965). Kinetoplasts have now been isolated from a number of species of trypanosomatids (DuBuy, Mattern & Riley, 1965; Newton, 1967; Renger & Wolstenholme, 1970, 1971; Riou & Delain, 1969; Laurent, Van Assel & Steinert, 1971) and DNA extracted from these preparations has been shown to correspond to the rapidly banding component extracted from whole cells. What has not yet been established unequivocally is whether this satellite contains all the DNA of the kinetoplast or whether any of the rapidly banding DNA is present in the nucleus. DNA extracted from nuclear fractions of *T. equiperdum* and *T. cruzi* (Riou & Pautrizel, 1969) has been found to band as a single component in caesium chloride at the same density as the major component of whole cell DNA, and Simpson & de Silva

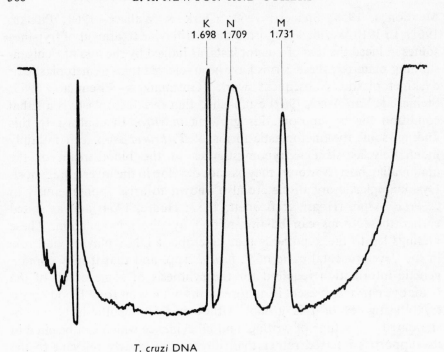

T. *cruzi* DNA

Fig. 5a. For legend see facing page.

(1971) failed to detect any hybridisation between nuclear and kineto-plast DNA: however, neither of these techniques is sufficiently sensitive to allow us to conclude from these results that no part of the kinetoplast DNA is represented in the nucleus.

The rapidly banding satellite, which will now be referred to as K-DNA, has in general a higher adenine + thymine content than nuclear DNA, and a survey of flagellates representing the major genera and subgenera of the order Kinetoplastida has shown that it may account for 6–28 % of the total cell DNA (Newton & Burnett, 1972b). Denatured K-DNA has been reported to re-anneal rapidly (DuBuy, Mattern & Riley, 1966; Riou & Paoletti, 1967; Simpson & de Silva, 1971) and its banding characteristics in the ultracentrifuge suggest that it may be extremely homogeneous and of high molecular weight. However, electron microscope observations (Riou & Delain, 1969; Simpson & da Silva, 1971) do not support the latter conclusion since isolated K-DNA has been seen as small circular molecules (Plate 3) ranging from 0.29 μm to 0.74 μm in contour length depending upon the species of flagellate from which it was derived. In this respect K-DNA resembles mito-chondrial DNA from other cell types with the exception that it is about

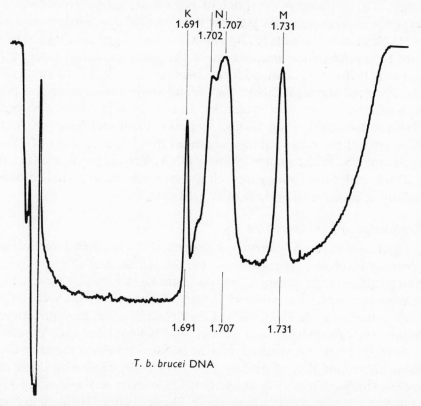

Fig. 5. Trypanosome DNA. Microdensitometer tracings of DNA bands obtained by analytical ultracentrifugation in caesium chloride (44000 rpm, 20 h at 20°). Buoyant densities are relative to marker DNA (M = 1.731). DNA from most Trypanosomatidae can be separated into a major nuclear component (N) and rapidly-banding kinetoplast DNA (K), e.g. *T. cruzi* DNA (5a). DNA from *T. brucei* sspp. (5b) characteristically separates into additional components (Newton & Burnett, 1972a, b).

one-tenth the size. K-DNA 'minicircles' have been seen in relaxed open form, figures of eight, supercoiled circles and catenates and it has been suggested that a parallel array of minicircles, each twisted to form a figure of eight, could form the basis of the fibrous structure observed in thin sections of *T. cruzi* kinetoplasts (Plate 4; Riou & Delain, 1969). Estimates based on the size of the circular molecules and the amount of DNA per kinetoplast suggest that there could be as many as 25–30 000 minicircles per cell (Simpson & da Silva, 1971; Laurent, Van Assel & Steinert, 1971). However, these ideas based on electron microscopy of purified K-DNA may be over-simplified and there is now reason to believe that not all K-DNA is in the form of minicircles: it seems that preparative procedures used in some of the earlier work selected these forms. The application of standard phenol extraction procedures to trypanosomes results in low yields of K-DNA (Newton, 1967; Riou *et al.* 1971; Newton & Burnett, 1972*b*) and it has been suggested that K-DNA may be bound to the membranes of the mitochondrion. Laurent & Steinert (1970) lysed isolated kinetoplasts on electron microscope grids and observed the total DNA: they found some minicircles but a predominance of long linear molecules. Linear molecules have also been observed by Ozeki, Ono, Okubo & Inoki (1970) and Simpson & da Silva (1971); the latter authors suggested that large numbers of minicircles may be held together by linear DNA. Very large associations of K-DNA molecules formed in such a way would be expected to band rapidly in caesium chloride density gradients.

Replication of kinetoplast DNA

Light and electron microscopy suggest that kinetoplast replication generally involves a lengthening of the fibrous band of DNA followed by a partition of the organelle into daughter kinetoplasts by constriction of the mitochondrial membranes (Burton & Dusanic, 1968; Vickerman, 1969*b*; Anderson & Hill, 1969). Kinetoplast division may precede or follow nuclear division and studies of ^{3}H-thymidine incorporation indicate that K-DNA synthesis may be initiated before, at the same time as, or after initiation of nuclear DNA synthesis, depending upon the species studied (Steinert & Steinert, 1962; Steinert & Van Assel, 1967; Hill & Anderson, 1969; Cosgrove & Skeen, 1970). ^{3}H-thymidine incorporation most frequently occurs at the periphery of the kinetoplast but nothing is known of the number or location of sites of DNA synthesis or of the mechanism of replication of K-DNA minicircles.

The role of kinetoplast DNA in the developmental cycle

Although there is evidence that kinetoplasts contain RNA and protein (Steinert & Van Assel, 1969; Kallinikova, 1969; Ozeki, Sooksri, Ono & Inoki, 1971) there is only indirect evidence, mainly from a comparison of normal and dyskinetoplastic cells, that kinetoplast DNA contains genetic information which is expressed during the developmental cycle. The general conclusion from these studies is that only flagellates which can respire by a non-mitochondrial L-α-glycerophosphate oxidase system, i.e. bloodstream trypomastigotes of *T. brucei* sspp., can survive the apparent loss of K-DNA (see below). Organisms lacking this oxidase system are dependent upon the mitochondrial respiratory chain to regenerate reduced NAD at all stages of their developmental cycle and it has been proposed that dyskinetoplasty is lethal in such flagellates because of an impairment of their respiratory chain. In support of this view, studies on the effect of acriflavine and other drugs known to inhibit selectively K-DNA replication have shown that the respiration of dyskinetoplastic cells is lower than that of normal cells (Simpson, 1968; Hill & Anderson, 1969). Acriflavine (5 μM) treatment of *Crithidia fasciculata* gave rise to a population which was 85 % dyskinetoplastic under the conditions used by Hill & Anderson (1969) and these workers attributed the reduced respiration of drug-treated cells to a decrease in cytochrome content and activity of mitochondrial enzymes. In the same system Bacchi & Hill (1972) detected an increase in the activity of a number of dehydrogenases, particularly α-glycerophosphate and glucose-6-phosphate dehydrogenases. However, it is not possible to deduce from any of these experiments that the changes in enzyme activity are a direct consequence of an inhibition of K-DNA replication: there is even some doubt as to whether drug-induced dyskinetoplastic organisms have entirely lost their K-DNA. We stated earlier that dyskinetoplastic organisms are, by definition, organisms which lack a Feulgen-positive kinetoplast. No fibrous DNA can be seen by electron microscopy in the kinetoplast of spontaneously occurring dyskinetoplastic forms of *T. brucei* sspp., but the kinetoplast envelope and a rudimentary mitochondrial tubule are present (Mühlpfordt, 1964). Drug-induced dyskinetoplastic forms also lack fibrous DNA but electron-dense material has been observed in the kinetoplast of acriflavine-treated cells (Trager & Rudzinska, 1964). Some workers (Stuart, 1971) claim that satellite DNA can be extracted from dyskinetoplastic forms in undiminished amounts but others have reported it to be greatly reduced or absent (Simpson, 1968; Renger & Wolstenholme, 1970; Newton & Burnett, 1971). This discrepancy may be a reflection of differences in the extraction procedures used. However, it seems to be generally agreed that if K-DNA still exists in drug-induced dyskinetoplastic strains it is in a modified form, as judged by electron microscopy and by the fact that it does not band rapidly in caesium chloride density gradients.

Inoki and his collaborators have proposed that retention of the kinetoplast envelope is another factor which determines whether dyskinetoplastic cells can survive and multiply. They reported that spontaneously occurring and p-rosaniline-induced dyskinetoplastic forms of *T. b. gambiense* were unable to divide while similar forms of *T. evansi* could (Inoki, Taniuchi, Matsushiro & Sakamoto, 1960). Electron microscopy showed that p-rosaniline treatment eliminated both K-DNA and the kinetoplast envelope from a strain of *T. b. gambiense* whereas

T. evansi after similar treatment lost the fibrous K-DNA but retained kinetoplast membranes (Inoki, Ozeki & Ono, 1969). Examination of DNA extracted from this dyskinetoplastic strain of *T. evansi* revealed some satellite DNA even though none was detected in the kinetoplast by electron microscopy; autoradiography of ^{3}H-thymidine labelled cells led Ono, Ozeki, Okubo & Inoki (1971) to the conclusion that the residual DNA was located in the kinetoplast envelope. These results suggest that structural differences may exist in the kinetoplast envelopes and mitochondrial membranes of different species of trypanosomes and that the relationship of K-DNA to these membranes may also vary.

In view of these findings it is interesting to speculate on the function of K-DNA in relation to recent work on its structure. Assuming K-DNA exists *in vivo* as a large number of small circular molecules each having a very similar base composition, an average contour length of 0.8 μm and a molecular weight of about 1.5×10^6 daltons, it can be calculated that there may be as many as 27000 copies. Such circular molecules could contain only sufficient information to determine the amino acid sequences of three or four relatively small polypeptides (of molecular weight 20000–30000 daltons), which would be a very minor contribution to the enzyme requirement for a functional citric acid cycle and respiratory chain. However, the possibility that at least some of the K-DNA is in the form of longer linear molecules (Laurent & Steinert, 1970; Simpson & da Silva, 1971; Ono *et al.* 1971; Riou *et al.* 1971) throws doubt on the significance of such calculations at the present time. The next important step in this work is to determine what proportion of K-DNA exists in a linear form and whether it is composed of short repetitive sequences of bases, homologous to the circular molecules, or whether it contains segments of high information capacity.

A final question which must be considered in any discussion of the role of K-DNA in the trypanosome developmental cycle concerns monomorphic strains of *T. brucei* sspp. It will be recalled that these strains result from prolonged non-cyclical transmission in laboratory animals; they develop in the bloodstream as long slender trypomastigotes but do not differentiate into stumpy forms, and do not have the ability to infect the insect vector or grow *in vitro*. In these respects monomorphic organisms resemble dyskinetoplastic organisms; however, they still retain an apparently normal complement of K-DNA. At the present time we have no idea why monomorphic strains fail to synthesise the additional oxidative enzymes which are a prerequisite for growth in the insect vector or *in vitro*. Mühlpfordt (1964) suggested that an interchange of genetic material between the nucleus and kinetoplast may be essential for the maintenance of K-DNA in a functional state. An opportunity for such an exchange occurs in the normal life cycle during the transition from trypomastigote forms to epimastigotes, which involves movement of the kinetoplast from a postnuclear to a juxtanuclear position; at this time kinetoplast and nucleus are frequently observed to be in close contact and fusion could occur. Electron micrographs which appear to show nuclear–kinetoplast fusion have been published (Mühlpfordt, 1964; Deane & Milder, 1972) but the possibility that apparent fusion is an artifact of sectioning cannot be excluded. In non-cyclically transmitted strains the normal life cycle is curtailed, epimastigote forms never develop and the kinetoplast is never in a juxtanuclear position; thus there would be no opportunity for exchange of genetic material. This hypothesis poses many interesting questions concerning possible relationships between nuclear and kinetoplast DNA and their importance to differentiation in Trypanosomatidae. The whole question of K-DNA function is being actively investigated at the time of writing and it seems likely that major advances will be made before this contribution is discussed at the symposium.

The authors wish to thank Dr B. E. Brooker, Dr K. A. Wright and Dr G. Riou for photographs reproduced in Plates 1 and 2. They are also indebted to the editors of *The Journal of Parasitology* (Plate 1) and the Crown Copyright holders (Figs. 2 and 3) for permission to reproduce plates and diagrams.

REFERENCES

ALLSOPP, B. A., NJOGU, A. R. & HUMPHREYS, K. C. (1971). Nature and location of *Trypanosoma brucei* subgroup exoantigen and its relationship to 4S antigen. *Experimental Parasitology*, **29**, 271–84.

AMREIN, Y. U., GEIGY, R. & KAUFFMANN, M. (1965). On the reacquisition of virulence in trypanosomes of the *brucei*-group. *Acta Tropica*, **22**, 193–203.

ANDERSON, W. & HILL, G. C. (1969). Division and DNA synthesis in the kinetoplast of *Crithidia fasciculata. Journal of Cell Science*, **4**, 611–20.

ASHCROFT, M. T. (1957). The polymorphism of *Trypanosoma brucei* and *T. rhodesiense*, its relation to relapses and remissions of infections in white rats, and the effect of cortisone. *Annals of Tropical Medicine and Parasitology*, **51**, 301–12.

BACCHI, C. J., CIACCIO, E. I., KABACK, D. B. & HUTNER, S. H. (1970). Oxaloacetate production via carboxylations in *Crithidia fasciculata* preparations. *Journal of Protozoology*, **17**, 305–11.

BACCHI, C. J. & HILL, G. C. (1972). *Crithidia fasciculata*: acriflavine-induced changes in soluble enzyme levels. *Experimental Parasitology*, **31**, 290–8.

BAERNSTEIN, H. D. (1963). A review of electron transport mechanisms in parasitic protozoa. *Journal of Parasitology*, **49**, 12–21.

BAKER, J. R. (1956). Studies on *Trypanosoma avium* Danilewsky 1885. III. Life-cycle in vertebrate and invertebrate hosts. *Parasitology*, **46**, 335–52.

BAKER, J. R. (1963). Speculations on the evolution of the family Trypanosomatidae Doflein, 1901. *Experimental Parasitology*, **13**, 219–33.

BAKER, J. R. (1966). Studies on *Trypanosoma avium*. IV. The development of infective metacyclic trypanosomes in cultures grown *in vitro*. *Parasitology*, **56**, 15–19.

BALBER, A. E. (1972). *Trypanosoma brucei*: fluxes of the morphological variants in intact and irradiated mice. *Experimental Parasitology*, **31**, 307–10.

BIDE, R. W. & GRANT, P. T. (1964). L-α-glycerophosphate oxidase of *Trypanosoma rhodesiense*. *Abstracts 1st Meeting Federation of European Biochemical Societies*, London, p. 72.

BONÉ, G. J. & STEINERT, M. (1956). Isotopes incorporated into the nucleic acids of *Trypanosoma mega*. *Nature, London*, **178**, 308–9.

BOWMAN, I. B. R., FLYNN, I. W. & FAIRLAMB, A. H. (1970). Carbohydrate metabolism of pleomorphic strains of *Trypanosoma rhodesiense* and sites of action of arsenical drugs. *Journal of Parasitology*, **56**, 402.

BOWMAN, I. B. R., VON BRAND, T. & TOBIE, E. J. (1960). The cultivation and metabolism of trypanosomes in the presence of trehalose with observations on trehalose in blood serum. *Experimental Parasitology*, **10**, 274–83.

BRESSLAU, E. & SCREMIN, L. (1924). Die Kerne der Trypanosomen und ihr Verhatten zur Nuklealreaktion. *Archiv für Protistenkunde*, **48**, 509–15.

BROWN, K. N. (1963). The antigenic character of the '*Brucei*' trypanosomes. In *Immunity to Protozoa*, a Symposium of the British Society for Immunology, ed. P. C. C. Garnham, A. E. Pierce & I. Roitt, pp. 204–12. Oxford: Blackwell.

BROWN, R. C. & EVANS, D. A. (1971). Factors affecting the establishment of *Trypanosoma brucei* in culture. *Transactions of the Royal Society of Tropical Medicine and Hygiene*, **65**, 256–7.

BURTON, P. R. & DUSANIC, D. G. (1968). Fine structure and replication of the kinetoplast of *Trypanosoma lewisi*. *Journal of Cell Biology*, **39**, 318–31.

CAMARGO, E. P. (1964). Growth and differentiation in *Trypanosoma cruzi*. I. Origin of metacyclic trypanosomes in liquid medium. *Revista do Instituto de Medicina Tropical de São Paulo*, **3**, 93–100.

CHALONER, L. A. (1972). Multiplication of bloodstream forms of *Trypanosoma brucei in vitro* at 37 °C in the presence of mouse blood. *Transactions of the Royal Society of Tropical Medicine and Hygiene*, **66**, 527.

CLARK, T. B. & WALLACE, F. G. (1960). A comparative study of kinetoplast ultrastructure in trypanosomatids. *Journal of Protozoology*, **7**, 115–24.

COSGROVE, W. & SKEEN, M. (1970). The cell cycle in *Crithidia fasciculata*. Temporal relationships between synthesis of deoxyribonucleic acid in the nucleus and in the kinetoplast. *Journal of Protozoology*, **17**, 172–7.

CROSS, G. A. M. (1972). Identification and isolation of a surface coat glycoprotein from *Trypanosoma brucei*. *Abstracts 8th Meeting Federation of European Biochemical Societies*, Amsterdam. Abs. No. 63.

CROSS, G. A. M. (1973). Development of a defined medium for growth of *Trypanosoma brucei sspp*. *Transactions of the Royal Society of Tropical Medicine and Hygiene* (in press).

DEANE, M. P. (1969). On the life cycle of trypanosomes of the *lewisi* group and their relationships to other mammalian trypanosomes. *Revista do Instituto de Medicina Tropica de São Paulo*, **11**, 34–43.

DEANE, M. P. & KIRCHNER, E. (1963). Life-cycle of *Trypanosoma conorhini*. Influence of temperature and other factors on growth and morphogenesis. *Journal of Protozoology*, **10**, 391–400.

DEANE, M. P. & MILDER, R. (1972). Ultrastructure of the 'cyst-like bodies' of *Trypanosoma conorhini*. *Journal of Protozoology*, **19**, 28–42.

DEMAREE, R. S., JR & MARQUARDT, W. C. (1971). Avian trypanosome division: a light and electron microscope study. *Journal of Protozoology*, **18**, 388–91.

DESOWITZ, R. S. (1970). African trypanosomes. In *Immunity to Parasitic Animals*, ed. G. J. Jackson, R. Herman & I. Singer. New York: Appleton-Century-Crofts, Meredith Corporation.

DIXON, H. & WILLIAMSON, J. (1970). The lipid composition of blood and culture forms of *Trypanosoma lewisi* and *Trypanosoma rhodesiense* compared with that of their environment. *Comparative Biochemistry and Physiology*, 33, 111–28.

DIXON, H., GINGER, C. D. & WILLIAMSON, J. (1971). The lipid metabolism of blood and culture forms of *Trypanosoma lewisi* and *Trypanosoma rhodesiense*. *Comparative Biochemistry and Physiology*, 39B, 247–66.

DIXON, H., GINGER, C. D. & WILLIAMSON, J. (1972). Trypanosome sterols and their metabolic origins. *Comparative Biochemistry and Physiology*, 41B, 1–18.

DuBUY, H. G., MATTERN, C. F. T. & RILEY, F. L. (1965). Isolation and characterization of DNA from kinetoplasts of *Leishmania enriettii*. *Science, New York*, 147, 754–6.

DuBUY, H. G., MATTERN, C. F. T. & RILEY, F. L. (1966). Comparison of the DNA's obtained from brain nuclei and mitochondria of mice and from the nuclei and kinetoplasts of *Leishmania enriettii*. *Biochimica et Biophysica Acta*, 123, 298–305.

EVANS, D. A. & BROWN, R. C. (1971a). The appearance of a novel respiratory pathway during the *in vitro* transformation of *Trypanosoma brucei*. *Transactions of the Royal Society of Tropical Medicine and Hygiene*, 65, 256.

EVANS, D. A. & BROWN, R. C. (1971b). Cyanide-insensitive culture form of *Trypanosoma brucei*. *Nature, London*, 230, 251–2.

EVANS, D. A. & BROWN, R. C. (1972). Utilization of glucose and proline by culture forms of *Trypanosoma brucei*. *Journal of Protozoology*, 19, Supplement, 47.

FERNANDES, J. F., CASTELLANI, O. & KIMURA, E. (1969). Physiological events in the course of the growth and differentiation of *Trypanosoma cruzi*. *Genetics*, 61, supplement, 214–26.

FULTON, J. D. (1969). Metabolism and pathogenic mechanisms of parasitic protozoa. In *Research in Protozoology*, vol. 3, ed. T.-T. Chen, pp. 389–504. Oxford: Pergamon Press.

FULTON, J. D. & SPOONER, D. F. (1959). Terminal respiration in certain mammalian trypanosomes. *Experimental Parasitology*, 8, 137–62.

GEIGY, R. & KAUFFMANN, M. (1964). On the effect of substances found in *Glossina* tissues on culture trypanosomes of the *brucei*-subgroup. *Acta Tropica*, 21, 169–73.

GODFREY, D. G. (1967). Phospholipids of *Trypanosoma lewisi, T. vivax, T. congolense* and *T. brucei*. *Experimental Parasitology*, 20, 106–18.

GODFREY, D. G. & TAYLOR, A. E. R. (1969). Studies of the surface of trypanosomes. *Transactions of the Royal Society of Tropical Medicine and Hygiene*, 63, 115–16.

GRANT, P. T. & BOWMAN, I. B. R. (1963). Respiratory systems in protozoa with special reference to the Trypanosomidae. *Biochemical Journal*, 89, 89P.

GRANT, P. T. & SARGENT, J. R. (1960). Properties of L-α-glycerophosphate oxidase and its role in the respiration of *Trypanosoma rhodesiense*. *Biochemical Journal*, 76, 229–37.

GRANT, P. T. & SARGENT, J. R. (1961). L-α-glycerophosphate dehydrogenase, a component of an oxidase system in *Trypanosoma rhodesiense*. *Biochemical Journal*, 81, 206–14.

GRANT, P. T., SARGENT, J. R. & RYLEY, J. F. (1961). Respiratory systems in the Trypanosomidae. *Biochemical Journal*, 81, 200–6.

GRAY, A. R. (1965). Antigenic variation in a strain of *Trypanosoma brucei* transmitted by *Glossina morsitans* and *G. palpalis*. *Journal of General Microbiology*, 41, 195–214.

GUTTMAN, H. N. & EISENMAN, R. N. (1965). Acriflavin-induced loss of kinetoplast deoxyribonucleic acid in *Crithidia fasciculata* (*Culex pipiens* strain). *Nature, London*, **207**, 1280–1.

HAWKING, F. (1971). The propagation and survival of *Trypanosoma brucei in vitro* at 37 °C. *Transactions of the Royal Society of Tropical Medicine and Hygiene*, **65**, 672–5.

HILL, G. C. (1972). Recent studies on the characterization of the cytochrome system in Kinetoplastidae. In *Comparative Biochemistry of Parasites*, ed. H. Van den Bossche, pp. 395–415. New York and London: Academic Press.

HILL, G. C. & ANDERSON, W. A. (1969). Effects of acriflavine on the mitochondria and kinetoplast of *Crithidia fasciculata. Journal of Cell Biology*, **41**, 547–61.

HILL, G. C. & ANDERSON, W. A. (1970). Electron transport systems and mito-chondrial DNA in Trypanosomatidae: a review. *Experimental Parasitology*, **28**, 356–80.

HILL, G. C., GUTTERIDGE, W. E. & MATTHEWSON, N. W. (1971). Purification and properties of cytochromes *c* from trypanosomatids. *Biochimica et Biophysica Acta*, **243**, 225–9.

HOARE, C. A. (1954). The loss of the kinetoplast in trypanosomes, with special reference to *Trypanosoma evansi. Journal of Protozoology*, **1**, 28–33.

HOARE, C. A. (1966). The classification of mammalian trypanosomes. *Ergebnisse der Mikrobiologie Immunitätsforschung und experimentellen Therapie*, **39**, 43–57.

HOARE, C. A. (1970). The mammalian trypanosomes of Africa. In *The African Trypanosomiases*, ed. H. W. Mulligan & W. H. Potts, pp. 3–50. London: Allen & Unwin.

HOARE, C. A. (1972). *The Trypanosomes of Mammals*. Oxford and Edinburgh: Blackwell.

HOARE, C. A. & BENNETT, S. C. J. (1937). Morphological and taxonomic studies on mammalian trypanosomes. III. Spontaneous occurrence of strains of *Trypano-soma evansi* devoid of the kinetoplast. *Parasitology*, **29**, 43–56.

HOARE, C. A. & WALLACE, F. G. (1966). Developmental stages of trypanosomatid flagellates: a new terminology. *Nature, London*, **212**, 1385–6.

HOLLINGSHEAD, S., PETHICA, B. A. & RYLEY, J. F. (1963). The electrophoretic behaviour of some trypanosomes. *Biochemical Journal*, **89**, 123–7.

HULANICKA, D., ERWIN, J. & BLOCH, K. (1964). Lipid metabolism of *Euglena gracilis. Journal of Biological Chemistry*, **239**, 2778–87.

INOKI, S., OZEKI, Y. & ONO, T. (1969). Effects of *p*-rosaniline on the ultrastructure of the kinetoplast in *Trypanosoma gambiense* and *Trypanosoma evansi. Biken Journal*, **12**, 187–99.

INOKI, S., TANIUCHI, Y., MATSUSHIRO, S. & SAKAMOTO, H. (1960). Multiplication ability of the akinetoplastic forms of *Trypanosoma evansi. Biken Journal*, **3**, 123–9.

KALLINIKOVA, V. D. (1969). Functional value of trypanosomatid kinetoplast in the light of cytochemical investigations. *Progress in Protozoology: IIIrd International Congress on Protozoology*, pp. 31–2.

KETTERIDGE, D. (1970). The presence of host serum components on the surface of rodent-adapted *Trypanosoma vivax. Journal of Protozoology*, **17**, Supplement, 24.

KETTERIDGE, D. (1971). Host antigens adsorbed by rodent-adapted *Trypanosoma vivax. Transactions of the Royal Society of Tropical Medicine and Hygiene*, **65**, 260.

KORN, E. D. (1966). Structure of biological membranes. *Science, New York*, **153**, 1491–8.

KUBO, R. (1968). Fine structures of haemoflagellates *Trypanosoma gambiense* and *T. lewisi. Journal of Nara Medical Association*, **19**, 309–24.

PLATE 1

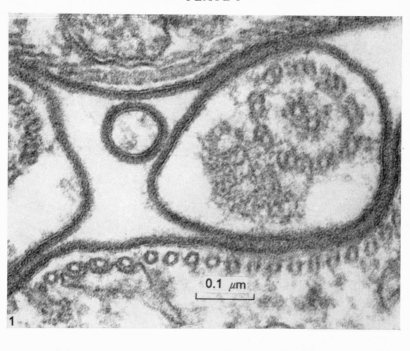

0.1 μm

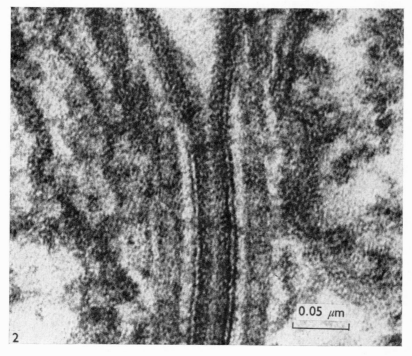

0.05 μm

PLATE 2

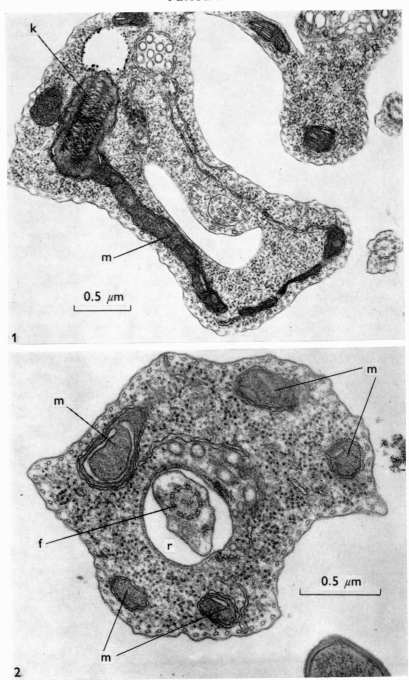

PLATE 3

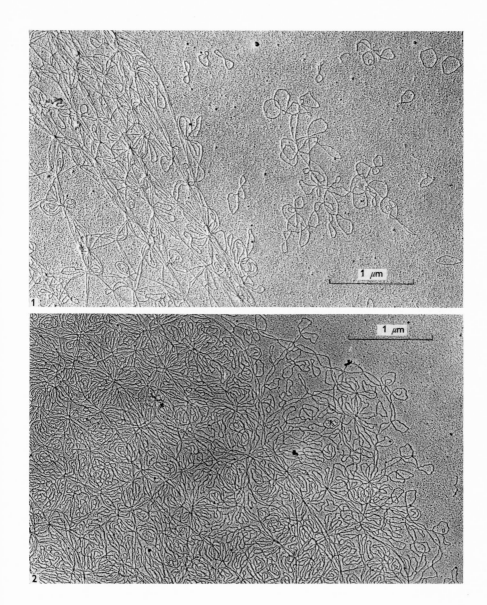

PLATE 4

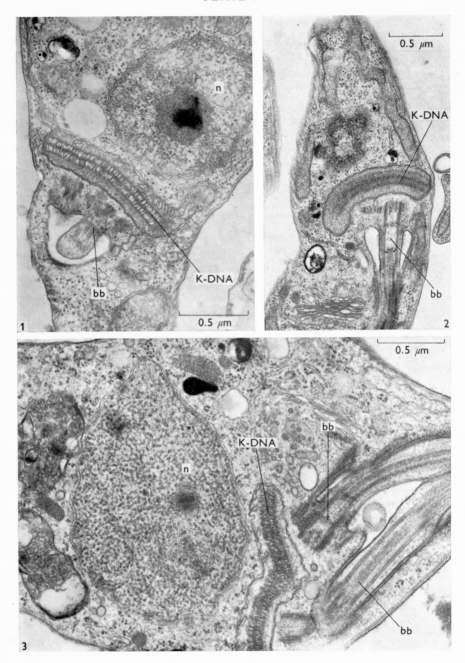

KUSEL, J. P. & KRONICK, P. (1972). The function of cytochrome o in the trypanosomatid haemoflagellate *Crithidia fasciculata*. *Federation Proceedings*, **31**, 878.

KUSEL, J. P. & STOREY, B. T. (1972). Evidence for the presence of two phosphorylation sites in mitochondria isolated from the trypanosomatid hemoflagellate *Crithidia fasciculata*. *Biochemical and Biophysical Research Communications*, **46**, 501–7.

LAURENT, M. & STEINERT, M. (1970). Electron microscopy of kinetoplastic DNA from *Trypanosoma mega*. *Proceedings of the National Academy of Sciences, U.S.A.* **66**, 419–24.

LAURENT, M., VAN ASSEL, S. & STEINERT, M. (1971). Kinetoplast DNA. A unique macromolecular structure of considerable size and mechanical resistance. *Biochemical and Biophysical Research Communications*, **43**, 278–84.

LEHMAN, D. L. (1961). Investigations on the infectivity of early cultural forms of Rhodesian trypanosomiasis. *Annals of Tropical Medicine and Parasitology*, **55**, 151–3.

LEMMA, A. & SCHILLER, E. L. (1964). Extracellular cultivation of the leishmanial bodies of species belonging to the protozoan genus *Leishmania*. *Experimental Parasitology*, **15**, 503–13.

LE PAGE, R. W. F. (1967). Short term cultivation of *Trypanosoma brucei in vitro* at 37 °C. *Nature, London*, **216**, 1141–2.

LUCKINS, A. G. (1972). Effects of X-irradiation and cortisone treatment of albino rats on infections with *brucei*-complex trypanosomes. *Transactions of the Royal Society of Tropical Medicine and Hygiene*, **66**, 130–9.

LUCKINS, H. B., THAM, S. H., WALLACE, P. G. & LINNANE, A. W. (1966). Correlation of membrane bound succinate dehydrogenase with the occurrence of mitochondrial profiles in *Saccharomyces cerevisiae*. *Biochemical and Biophysical Research Communications*, **23**, 363–7.

LWOFF, M. (1937). Aneurin as a growth factor for the trypanosomatid flagellate *Strigomonas oncopelti*. *Comptes Rendus des Séances de la Société de Biologie*, **126**, 771–3.

MEYER, F. & HOLTZ, G. G. (1966). Biosynthesis of lipid by kinetoplastid flagellates. *Journal of Biological Chemistry*, **241**, 5001–7.

MEYER, H. M., MUSACCHIO, M. DE O. & MENDONÇA, I. DE A. (1958). Electron microscope study of *Trypanosoma cruzi* in thin sections of infected tissue cultures and of blood-agar forms. *Parasitology*, **48**, 1–8.

MÜHLPFORDT, H. (1964). Über die Bedeutung und Feinstruktur des Blepharoplasten bei parasitischen Flagellaten. *Zeitschrift für Tropenmedizin und Parasitologie*, **14**, 475–501.

MUNIZ, J. & DE FREITAS, G. (1945). Estudo sôbre o determinismo da transformação das formas sanguìcolas do *Schizotrypanum cruzi* em critídias. *Revista brasileira de Medicina*, **2**, 995–9.

MUNIZ, J. & DE FREITAS, G. (1946). Realização *in vitro* do ciclo do *S. cruzi* no vertebrado, em meios de caldoliquido peritoneal. *Revista brasileira de Biologia*, **6**, 467–84.

NEWTON, B. A. (1956). A synthetic growth medium for the trypanosomid flagellate *Strigomonas (Herpetomonas) oncopelti*. *Nature, London*, **177**, 279–80.

NEWTON, B. A. (1967). Isolation of DNA from kinetoplasts of *Crithidia fasciculata*. *Journal of General Microbiology*, **48**, iv.

NEWTON, B. A. & BURNETT, J. K. (1971). The heterogeneity of trypanosome DNA. *Transactions of the Royal Society of Tropical Medicine and Hygiene*, **65**, 243–4.

NEWTON, B. A. & BURNETT, J. K. (1972a). The satellite DNAs of salivarian trypanosomes. *Transactions of the Royal Society of Tropical Medicine and Hygiene*, **66**, 353–4.

NEWTON, B. A. & BURNETT, J. K. (1972b). DNA of Kinetoplastidae: a Comparative Study. In *Comparative Biochemistry of Parasites*, ed. H. Van den Bossche, pp. 127–38. New York and London: Academic Press.

ONO, T., OZEKI, Y., OKUBO, S. & INOKI, S. (1971). Characterization of nuclear and satellite DNA from trypanosomes. *Biken Journal*, **14**, 203–15.

ORMEROD, W. E. & VENKATESAN, S. (1971a). The occult visceral phase of mammalian trypanosomes with special reference to the life cycle of *Trypanosoma* (*Trypanozoon*) *brucei*. *Transactions of the Royal Society of Tropical Medicine and Hygiene*, **65**, 722–35.

ORMEROD, W. E. & VENKATESAN, S. (1971b). An amastigote phase of the sleeping sickness trypanosome. *Transactions of the Royal Society of Tropical Medicine and Hygiene*, **65**, 736–41.

OZEKI, Y., ONO, T., OKUBO, S. & INOKI, S. (1970). Electron microscopy of DNA released from ruptured kinetoplasts of *Trypanosoma gambiense*. *Biken Journal*, **13**, 387–93.

OZEKI, Y., SOOKSRI, V., ONO, T. & INOKI, S. (1971). Studies on the ultrastructure of kinetoplasts of *Trypanosoma cruzi* and *Trypanosoma gambiense* by autoradiography and enzymatic digestion. *Biken Journal*, **14**, 97–118.

PAN, C.-T. (1968). Cultivation of the leishmaniform stage of *Trypanosoma cruzi* in cell-free media at different temperatures. *American Journal of Tropical Medicine and Hygiene*, **17**, 823–32.

PAN, C.-T. (1971). Cultivation and morphogenesis of *Trypanosoma cruzi* in improved liquid media. *Journal of Protozoology*, **18**, 556–60.

PARKS, L. W. & STARR, P. R. (1963). A relationship between ergosterol and respiratory competency in yeast. *Journal of Cellular and Comparative Physiology*, **61**, 61–5.

PIPKIN, A. C. (1960). Avian embryos and tissue culture in the study of parasitic Protozoa. II. Protozoa other than *Plasmodium*. *Experimental Parasitology*, **9**, 167–203.

PITELKA, D. R. (1961). Observations on the kinetoplast mitochondrion and cytostome of *Bodo*. *Experimental Cell Research*, **25**, 87–93.

PITTAM, M. D. (1970). Medium for *in vitro* culture of *Trypanosoma rhodesiense* and *T. brucei*. [Appendix to paper by Dixon, H. & Williamson, J.] *Comparative Biochemistry and Physiology*, **33**, 111–28.

RAY, S. K. (1972). Thesis submitted for the degree of Doctor of Philosophy, University of Cambridge.

RAY, S. K. & CROSS, G. A. M. (1972a). Branched electron transport chain in *Trypanosoma mega*. *Nature, New Biology, London*, **237**, 174–5.

RAY, S. K. & CROSS, G. A. M. (1972b). Terminal respiration in *Trypanosoma mega*. *Journal of Protozoology*, **19**, Supplement, 51.

RENGER, H. C. & WOLSTENHOLME, D. R. (1970). Kinetoplast deoxyribonucleic acid of the haemoflagellate *Trypanosoma lewisi*. *Journal of Cell Biology*, **47**, 689–702.

RENGER, H. C. & WOLSTENHOLME, D. R. (1971). Kinetoplast and other satellite DNAs of kinetoplastic and dyskinetoplastic strains of *Trypanosoma*. *Journal of Cell Biology*, **50**, 533–40.

RIOU, G. & DELAIN, E. (1969). Electron microscopy of the circular kinetoplastic DNA from *Trypanosoma cruzi*: occurrence of catenated forms. *Proceedings of the National Academy of Sciences, U.S.A.* **62**, 210–17.

RIOU, G., LÂCOME, A., BRACK, C., DELAIN, E. & PAUTRIZEL, R. (1971). Importance de la méthode d'extraction dans l'isolement de l'ADN de kinétoplaste de trypanosomes. *Comptes Rendus Hebdomadaire des Séances de l'Académie des Sciences*, **273**, 2150–3.

RIOU, G. & PAOLETTI, C. (1967). Preparation and properties of nuclear and satellite deoxyribonucleic acid of *Trypanosoma cruzi*. *Journal of Molecular Biology*, **28**, 377–82.

RIOU, G. & PAUTRIZEL, R. (1969). Nuclear and kinetoplastic DNA from trypanosomes. *Journal of Protozoology*, **16**, 509–13.

ROSENBERG, A. & PECKER, M. (1964). Lipid alterations in *Euglena gracilis* during light induced greening. *Biochemistry*, **3**, 254–8.

RYLEY, J. F. (1956). Studies on the metabolism of the protozoa. 7. Comparative carbohydrate metabolism of eleven species of trypanosomes. *Biochemical Journal*, **62**, 215–22.

RYLEY, J. F. (1962). Studies on the metabolism of the protozoa. 9. Comparative metabolism of bloodstream and culture forms of *Trypanosoma rhodesiense*. *Biochemical Journal*, **85**, 211–23.

RYLEY, J. F. (1964). Histochemical studies on the blood and culture forms of *Trypanosoma rhodesiense*. *Proceedings of 1st International Congress of Parasitology*, vol. 1, ed. A. Corradetti, pp. 41–2. London: Pergamon Press.

SANCHEZ, G. & DUSANIC, D. G. (1968). Growth of the bloodstream form of *Trypanosoma lewisi in vitro*. *Journal of Parasitology*, **54**, 601–5.

SCHILDKRAUT, C. L., MANDEL, M., LEVISOHN, S., SMITH-SONNEBORN, J. E. & MARMUR, J. (1962). Deoxyribonucleic acid base composition and taxonomy of some protozoa. *Nature, London*, **196**, 795–6.

SIMPSON, L. (1968). The leishmania–leptomonas transformation of *Leishmania donovani*: nutritional requirements, respiration changes and antigenic changes. *Journal of Protozoology*, **15**, 201–7.

SIMPSON, L. & DA SILVA, A. (1971). Isolation and characterization of kinetoplast DNA from *Leishmania tarentolae*. *Journal of Molecular Biology*, **56**, 443–73.

SRIVASTAVA, H. K. (1971). Carbon monoxide-reactive haemoproteins in parasitic flagellate *Crithidia oncopelti*. *FEBS Letters*, **16**, 189–91.

SRIVASTAVA, H. K. & BOWMAN, I. B. R. (1971). Adaptation in oxidative metabolism of *Trypanosoma rhodesiense* during transformation in culture. *Comparative Biochemistry and Physiology*, **40B**, 973–81.

SRIVASTAVA, H. K. & BOWMAN, I. B. R. (1972). Metabolic transformation of *Trypanosoma rhodesiense* in culture. *Nature, New Biology, London*, **235**, 152–3.

STEIGER, R. (1971). Some aspects of surface coat formation in *Trypanosoma brucei*. *Acta Tropica*, **28**, 341–6.

STEINERT, M. (1958). Études sur le déterminisme de la morphogénèse d'un trypanosome. *Experimental Cell Research*, **15**, 560–9.

STEINERT, M. (1960). Mitochondria associated with the kinetonucleus of *Trypanosoma mega*. *Journal of Biophysical and Biochemical Cytology*, **8**, 542–6.

STEINERT, M. (1965). The morphogenesis of trypanosomes. *Progress in Protozoology: 2nd International Congress of Protozoology, London*, Excerpta Medica Series 91, 40–1.

STEINERT, M. & STEINERT, G. (1962). La synthèse de l'acide désoxyribonucléique au cours du cycle de division de *Trypanosoma mega*. *Journal of Protozoology*, **9**, 203–11.

STEINERT, M. & VAN ASSEL, S. (1967). Réplications coordonnées des acides désoxyribonucléiques nucléaires et mitochondriales chez *Crithidia luciliae*. *Archives Internationales de Physiologie et de Biochimie*, **75**, 370–1.

STEINERT, M. & VAN ASSEL, S. (1969). Étude par autoradiographie, des effets du bromure d'ethidium sur la synthèse des acides nucléiques de *Crithidia luciliae*. *Experimental Cell Research*, **56**, 69–74.

STUART, K. D. (1971). Evidence for the retention of kinetoplast DNA in an acriflavine-induced dyskinetoplastic strain of *Trypanosoma brucei* which replicates the altered central element of the kinetoplast. *Journal of Cell Biology*, **49**, 189–95.

STRAUSS, P. R. (1971). The effect of homologous rabbit antiserum on the growth of *Leishmania tarentolae* – a fine structure study. *Journal of Protozoology*, **18**, 147–56.

TAYLOR, A. E. R. & BAKER, J. R. (1968). *The Cultivation of Parasites in vitro.* Oxford and Edinburgh: Blackwell.

THRELFALL, D. R., WILLIAMS, B. L. & GOODWIN, T. W. (1965). Terpenoid quinones and sterols in the parasitic and culture forms of *Trypanosoma rhodesiense.* *Progress in Protozoology: 2nd International Congress of Protozoology, London,* Excerpta Medica Series 91, p. 141.

TONER, J. J. & WEBER, M. M. (1972). Respiratory control in mitochondria from *Crithidia fasciculata. Biochemical and Biophysical Research Communications,* **46**, 652–60.

TRAGER, W. (1953). The development of *Leishmania donovani in vitro* at 37 °C. Effects of the kind of serum. *Journal of Experimental Medicine*, **97**, 177–88.

TRAGER, W. (1957). Nutrition of a haemoflagellate (*Leishmania tarentolae*) having an interchangeable requirement for choline and pyridoxal. *Journal of Protozoology*, **4**, 269–76.

TRAGER, W. (1959a). Tsetse-fly tissue culture and the development of trypanosomes to the infective stage. *Annals of Tropical Medicine and Parasitology*, **53**, 473–91.

TRAGER, W. (1959b). Development of *Trypanosoma vivax* to infective stage in tsetse fly tissue culture. *Nature, London*, **184**, B.A. 30–B.A. 31.

TRAGER, W. (1970). Recent progress in some aspects of the physiology of parasitic protozoa. *Journal of Parasitology*, **56**, 627–33.

TRAGER, W. & RUDZINSKA, M. (1964). The riboflavin requirements and effects of acriflavin on the fine structure of the kinetoplast of *Leishmania tarentolae. Journal of Protozoology*, **11**, 133–45.

TREJOS, A., GODOY, G. A., GREENBLATT, G. & CEDILLOS, R. (1963). Effects of temperature on morphologic variation of *Schizotrypanum cruzi* in tissue culture. *Experimental Parasitology*, **13**, 211–18.

VICKERMAN, K. (1962). The mechanism of cyclical development in trypanosomes of the *Trypanosoma brucei* subgroup: an hypothesis based on ultrastructural observations. *Transactions of the Royal Society of Tropical Medicine and Hygiene,* **56**, 487–95.

VICKERMAN, K. (1965). Polymorphism and mitochondrial activity in sleeping sickness trypanosomes. *Nature, London*, **208**, 762–6.

VICKERMAN, K. (1969a). On the surface coat and flagellar adhesion in trypanosomes. *Journal of Cell Science*, **5**, 163–93.

VICKERMAN, K. (1969b). The fine structure of *Trypanosoma congolense* in its bloodstream phase. *Journal of Protozoology*, **16**, 54–69.

VICKERMAN, K. (1971). Morphological and physiological considerations of extracellular blood protozoa. In *Ecology and Physiology of Parasites: a Symposium,* ed. A. M. Fallis, pp. 58–91. Toronto: University Press.

VICKERMAN, K. & LUCKINS, A. G. (1969). Localisation of variable antigens in the surface coat of *Trypanosoma brucei* using ferritin conjugated antibody. *Nature, London*, **224**, 1125–7.

VON BRAND, T. (1951). Metabolism of Trypanosomidae and Bodonidae. In *Biochemistry and Physiology of Protozoa,* vol. 1, ed. A. Lwoff, pp. 177–250. New York and London: Academic Press.

WEINMAN, D. (1953). African sleeping sickness trypanosomes: cultivation and properties of the culture forms. *Annals of the New York Academy of Sciences,* **56**, 995–1003.

WEINMAN, D. (1957). Cultivation of trypanosomes. *Transactions of the Royal Society of Tropical Medicine and Hygiene*, **51**, 560–1.

WEITZ, B. G. F. (1970). Infection and resistance. In *The African Trypanosomiases*, ed. H. W. Mulligan & W. H. Potts, pp. 97–124. London: Allen & Unwin.

WERBITZKI, F. W. (1910). Über blepharoplastlose Trypanosomen. *Zentralblatt für Bakteriologie, Parasitenkunde, Infecktions-Krankheiten und Hygiene* (Abteilung I) Originale, **53**, 303–15.

WILLIAMS, B. L., GOODWIN, T. W. & RYLEY, J. F. (1966). The sterol content of some protozoa. *Journal of Protozoology*, **13**, 227–30.

WILLIAMSON, J. & GINGER, C. D. (1965). Lipid constitution of some protozoa, spirochaetes and bacteria. *Transactions of the Royal Society of Tropical Medicine and Hygiene*, **59**, 366–7.

WOO, P. T. K. (1970). Origin of mammalian trypanosomes which develop in the anterior station of blood-sucking arthropods. *Nature, London*, **228**, 1059–62.

WRIGHT, K. A. & HALES, H. (1970). Cytochemistry of the pellicle of blood-stream forms of *Trypanosoma (Trypanozoon) brucei*. *Journal of Parasitology*, **56**, 671–83.

YORKE, W., ADAMS, A. R. D. & MURGATROYD, F. (1929). Studies on chemotherapy. I. A method for maintaining pathogenic trypanosomes alive *in vitro* at 37 °C for 24 hours. *Annals of Tropical Medicine and Parasitology*, **23**, 501–18.

ZUCKERMAN, A. (1966). Propagation of parasitic protozoa in tissue culture and avian embryos. *Annals of the New York Academy of Science*, **139**, 24–38.

EXPLANATION OF PLATES

PLATE 1

Electron micrograph of *Trypanosoma b. brucei* (Wright & Hales, 1970). Preparations fixed in 5 % glutaraldehyde–cacodylate and stained with osmium, uranyl acetate and lead citrate showing pellicle structure consisting of trilaminate cell membrane plus external cell coat.

Fig. 1. Section through two flagella and trypanosome bodies.

Fig. 2. The pellicle of two adjacent trypanosomes.

PLATE 2

Electron micrographs of *Crithidia fasciculata*. (Reproduced by kind permission of Dr B. E. Brooker.) Preparations fixed in buffered glutaraldehyde and stained with uranyl acetate and lead citrate.

Fig. 1. Longitudinal section through the anterior half of the cell to show the relationship between the kinetoplast (k) and mitochondrial tubule (m).

Fig. 2. Section in the area of the flagellum (f) and reservoir (r) showing transverse sections of five branches of the single mitochondrial tubule (m).

PLATE 3

Portions of a large association of kinetoplast DNA molecules extracted from *Trypanosoma cruzi*. (Reproduced by kind permission of Dr G. Riou and Dr E. Delain.) Figs. 1 and 2 show densely interlocked circular and linear molecules. Some circular molecules (contour length 0.5 μm) are free and others apparently catenated.

PLATE 4

Electron micrographs of *Trypanosoma cruzi* (Reproduced by kind permission of Dr G. Riou and Dr E. Delain.) Preparations fixed in glutaraldehyde–osmium and stained with uranyl acetate. Figs. 1–3 show the ultrastructure of the kinetoplast situated between the basal body of the flagellum (bb) and the nucleus (n). The kinetoplast DNA (K–DNA) can be seen as an organised double layered structure.

DIFFERENTIATION IN
PHYSARUM POLYCEPHALUM

H. W. SAUER

*Zoologisches Institut der Universität Heidelberg,
69 Heidelberg 1, Germany*

INTRODUCTION

Physarum polycephalum has attracted a number of students since it has been possible to grow this myxomycete axenically in liquid medium (Daniel & Rusch, 1961) and to obtain large plasmodia suitable for biochemical analysis. Synchronous mitosis and the distinct stages of the life cycle have made this organism a candidate for analysing major questions of growth and differentiation. There are several reviews covering the biology of myxomycetes (Alexopoulos, 1966; Gray & Alexopoulos, 1968), the physiology of growth and development of myxomycetes (von Stosch, 1965), the life cycle of *Physarum polycephalum* as a tool for cell science (Cummins & Rusch, 1968; Kleinig, 1972) and the biochemistry of the life cycle of *Physarum* (Rusch, 1969; Rusch, 1970; Cummins, 1968). This paper will review some results obtained on *Physarum* which are believed to be relevant to the question of differentiation.

THE LIFE CYCLE OF *PHYSARUM POLYCEPHALUM*

The life cycle of *Physarum* consists of alternating haploid and diploid generations which can be studied in the laboratory. Haploid amoebae germinate from spores; amoebae of a matching mating type can fuse in pairs (syngamy) and from zygotes after nuclear fusion (karyogamy) a diploid plasmodium can arise by nuclear divisions which are not accompanied by cell division (cytokinesis). The life cycle becomes completed when a vegetative plasmodium is transformed into sporangia and after meiosis haploid spores are formed. A schematic presentation is given in Fig. 1. Both the haploid and diploid vegetative stages of the life cycle can be extended indefinitely through mitotic cycles. Mitosis in the plasmodia has been extensively studied in *Physarum polycephalum*.

Plasmodia can be grown as a shaken suspension of small cells, microplasmodia, which display mitotic synchrony only within an individual plasmodium. Microplasmodia will fuse with one another if placed on a membrane and will grow as a macroplasmodium, containing 10^8 nuclei

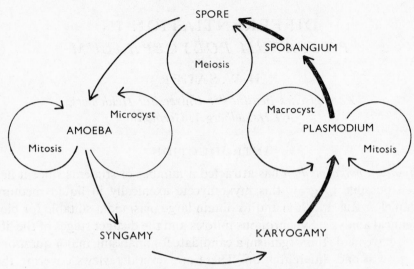

Fig. 1. A schematic representation of the life cycle of *Physarum polycephalum* (see text for details).

in which synchronous mitoses occur every 8–10 h. Certain cultural conditions can interfere with mitotic growth and lead to an encystment of amoebae (microcysts) and plasmodia (macrocysts); microplasmodia turn into spherules while macroplasmodia are transformed into sclerotia. Detailed culture methods have been published for stock cultures (Daniel & Baldwin, 1964), macroplasmodia (Guttes & Guttes, 1964) and for sporulation (Daniel, 1966).

In the following discussion we have applied a rather wide definition of differentiation to include studies of differential changes in the mitotic cycle, transformations from the growing to encysted stages, and sporulation.

THE MITOTIC CYCLE

There are at least two differential events in the growth cycle of *Physarum*, mitosis and S phase, which might be considered as models of differential activities in a cell under constant external conditions, since they occur with natural synchrony. Since cell mass increases continually during growth and the size of nuclear components remains rather constant during interphase, a specific relation (Kern–Plasma–Relation) of the volumes occupied by cytoplasm and nucleus has long been thought to determine mitosis (Hertwig, 1903). In *Physarum*, bulk protein and RNA increase steadily during growth cycles while DNA duplicates in steps after each mitosis (Mohberg & Rusch, 1969a). This indicates that many syntheses can go on at all times of the cell cycle. Experiments with actinomycin D have indicated that some RNA made during one cell cycle could still be translated in the following cycle (Mittermayer, Braun & Rusch, 1965; Sachsenmaier & Becker, 1965). We have shown (unpublished results) that actinomycin D present for longer than one generation time has only a limited effect as judged from amino acid incorporation and polysome profiles. Therefore, a large proportion of protein synthesis seems to go on without concomitant RNA synthesis. Other experiments, performed with actinomycin D, have demonstrated that some RNA must be made up to 35 min before mitosis (Sachsenmaier, von Fournier & Gürtler, 1967). Differential events may occur during a cell cycle at the transcription level which might determine mitosis.

Mitosis

Fusion of a population of microplasmodia is followed by a synchronous mitosis of most nuclei after one-half a generation time (Guttes & Guttes, 1964). From this observation it is certain that nuclei from individual plasmodia divide sooner or later than would occur in a random population of microplasmodia. The synchronizing effect could have been mediated by starvation, which facilitates fusion, or by the production, distribution and accumulation of cytoplasmic factors. The latter argument was strengthened by controlled fusion experiments (Rusch, Sachsenmaier, Behrens & Gruter, 1966). Control plasmodia were set up to divide at 1 p.m. or 3 p.m. (A and B, respectively) while nuclei of a mixed plasmodium, a sandwich of equal portions of the control plasmodia, divided at 2 p.m. If the proportion contributed from plasmodium A was larger, mitosis in the mixed plasmodium occurred before 2 p.m. The latter experiment could indicate that, during interphase, the concentration of certain cytoplasmic factors increased to a maximum value at which time mitosis was triggered.

Further support for a continuous accumulation of cytoplasmic components in G2 phase came from experiments with ultraviolet light (Sachsenmaier, Dönges & Rupff, 1970). Irradiation in S phase prolonged the cell cycle and caused a reduction of the DNA content by eliminating damaged nuclei during the following S phase. This and subsequent cycles, however, were shorter than those of unirradiated control plasmodia. Protein content was not altered significantly by irradiation leaving unbalanced quantities of nuclei and cytoplasm. The shortening of the cell cycle has been explained as a consequence of a smaller number of nuclei which could be saturated by a low concentration of the hypothetical cytoplasmic mitotic factors which, in turn, would be synthesized in a shorter period of time. Devi, Guttes & Guttes (1968) have not excluded the possibility that the remaining nuclei, after ultraviolet irradiation, were provided with higher concentrations of precursors and replicated DNA more rapidly; speed of replication could indirectly control the duration of G2 phase.

The importance of the G2 phase for the preparation of mitosis was stressed by heat shock experiments (Brewer & Rusch, 1968). Plasmodia were shifted from 26 to 37 °C for 30 min. There was no delay when S phase plasmodia were heated. A maximum delay of 2 h was observed when plasmodia were heated 2 h before mitosis. Repeated heat shocks during one G2 phase prolonged that cycle by 3 h but the subsequent cycle was shortened by about the same length of time. Therefore, only some events necessary for mitosis were affected by heat and might have resulted in an imbalance of cytoplasm and nuclei.

It had been demonstrated in *Physarum* that one effect of a heat shock is a dissociation of polysomes and a significant inhibition of protein synthesis (Schiebel, Chayka, DeVries & Rusch, 1969). Experiments with cycloheximide, a potent inhibitor of protein synthesis in *Physarum*, have clearly demonstrated that protein must be made up to 15 min before metaphase to allow completion of mitosis (Cummins & Rusch, 1966). This line of argument was consistent with a hypothesis of selective protein synthesis during G2 phase and an interaction of protein with nuclei before mitosis.

Synthesis or selective accumulation of a nuclear protein in G2 phase has been described (Jockusch, Brown & Rusch, 1970). This protein has been localized in the nucleolus and identified as actin (Jockusch, Brown & Rusch, 1971). It could be that this protein, in the absence of the proper spindle protein tubulin, became a structural part of the intranuclear mitotic spindle of *Physarum*. Electron microscopic studies have demonstrated fibrils and microtubuli, which make up spindle fibres, in the proximity of the nucleolus when it started to dissolve during prophase (Guttes, Guttes & Ellis, 1968).

If the degradation of the nucleolus was prevented by the application of certain amounts of actinomycin D (50–100 μg/ml) no spindle was formed (Guttes, Guttes & Devi, 1969). Mitosis cannot proceed although a stage, comparable to late telophase with prenucleolar bodies and DNA synthesis, was observed. However, dissolution of a nucleolus alone, which could be induced by the application of cyclic AMP, was not sufficient to trigger mitosis (Murray & Bigler, 1970).

Two classes of nuclear protein which are synthesized in S phase, histone (Mohberg & Rusch, 1969b) and acidic protein (LeStourgeon &

Rusch, 1971), have been analysed by gel electrophoresis during the cell cycle. No selective changes have been observed, although selective phosphorylation of some acidic proteins could have occurred during G2 phase.

From the arguments presented it is easy, theoretically, to set up conditions for a bio-assay: cytoplasmic material prepared in late G2 phase (prophase) should stimulate nuclear division in a plasmodium that has just entered G2 phase. In-vivo experiments have shown that nuclei can divide after completion of S phase in a mixed plasmodium (Guttes & Guttes, 1968). Practically, adequate controls have not been devised to identify a general growth-promoting factor from a mitotic stimulator. However, it has been claimed that extracts from late G2 plasmodia contain a substance, possibly a protein, that shortens the generation time of the early G2 phase recipient plasmodium by 1 h (Oppenheim & Katzir, 1971). A characterization of this 'stimulator' or its mode of action from an extracellular position on the nuclei cannot be given. Injection of cytoplasm from various stages into recipient G2 phase plasmodia has failed to demonstrate significant mitotic stimulation, but has hinted at a negative control mechanism via 'inhibitors' (E. M. Goodman & W. Schiebel, personal communication).

Experiments described so far, and conclusions from a review of the biochemistry of mitosis (Duspiva, 1971), are consistent with the old hypothesis of quantitative nuclear cytoplasmic relations. However, refined fusion experiments and a more complete analysis of ultraviolet irradiation effects argue for selective events, not continuous or linear increase of factors during all of G2 phase, which trigger mitosis. In these fusion experiments (Guttes, Devi & Guttes, 1969) a dominant effect of prophase over early G2 phase was demonstrated. Macroplasmodia of these two stages were first fragmented by vigorous shaking and then allowed to fuse again in equal quantitative combinations of G2 phase and prophase nuclei. In this combination the nuclei of the composite plasmodium did not divide at a time half-way between the times of the controls but significantly sooner. The following interpretation could be given: if prophase nuclei contained a trigger for mitosis they would have divided immediately and would not stimulate G2 nuclei. Since the prophase component of the mixed plasmodium stimulated G2 nuclei to divide sooner than at the intermediate time, a diffusable factor that stimulated mitosis might only be synthesized late in G2 phase.

Experiments with ultraviolet light (Guttes, communicated at the *Physarum* Conference, Madison, 1968) allowed two further conclusions. First, plasmodia became insensitive to starvation. In control plasmodia, in the absence of nutrients, the cell cycle had become prolonged from 8–10 h to 24–48 h. Under these experimental conditions mitotic stimulation had become dissociated from a general increase of cytoplasm during growth. Secondly, plasmodia became insensitive to actinomycin D, although they continued to have synchronous mitotic cycles. This result indicated that, under certain conditions and provided that actinomycin D inhibited RNA synthesis as in unirradiated plasmodia, cyclic stimulation of mitosis in *Physarum* might be controlled by mechanisms other than transcription. It was not improbable, therefore, that a cell could store factors that regulate mitosis for several generations.

Thus, in *Physarum* there is indirect evidence for mitotic stimulation by the production of specific cytoplasmic, diffusible protein factors late in G2 phase (prophase). For their isolation, characterization and identification we need a bio-assay. At this point it is not yet possible to decide whether transcription during G2 phase plays an essential role in regulating the structural changes associated with mitosis.

S phase

It can be considered that, in order to drive a cell through consecutive growth cycles, some differential events like mitosis and duplication of DNA (S phase) must take place at the right time. One possible mechanism to determine a cell cycle would be to provide for a starting point and a programmed sequence at the molecular level, possibly in the chromatin.

Synchronized mitoses, or metaphase, the only time when RNA is not made in the cell cycle of *Physarum* (Kessler, 1967), could be considered a starting point. Replication of DNA during S phase could determine a programme for the following cell cycle. While mitosis is triggered by cytoplasmic factors, S phase, once it has been initiated in late telophase, is a local nuclear affair that is little affected by the cytoplasmic environment. This conclusion has been drawn from more fusion experiments (Guttes & Guttes, 1968) and from experimentally desynchronized plasmodia (Guttes & Telatnyk, 1971). In the first experiment G2 phase nuclei were not turned on to make DNA in an S phase cytoplasm and S phase nuclei were not turned off in a G2 phase cytoplasm. Since the evidence came from autoradiography it was not clear whether full replication of the DNA was achieved. The second experiment made use of the shuttle streaming in plasmodia and the observation that, when a plasmodium was covered with a membrane in prophase, only the nuclei in the periphery divided on time while mitosis was delayed in the central part of the plasmodium. Prophase nuclei from the retarded area, clearly identified by cytological criteria, were transported into the periphery by cytoplasmic streaming where postmitotic nuclei had already entered S phase. In these translocated prophase nuclei no chromosomal DNA synthesis was detected although nucleolar DNA synthesis, possibly an amplification of rDNA, was demonstrated by autoradiography as in G2 phase of normal plasmodia. This experiment stressed the necessity of mitosis for bulk DNA synthesis and ruled out the possibility that a sudden burst of DNA polymerase activity alone defined S phase. Although isolated nuclei incorporated [³H]dATP only in S phase (Brewer & Rusch, 1965), it has been claimed that template structure and not DNA polymerase was limiting (Brewer & Rusch, 1966). This conclusion was strengthened by the recent finding that isolated S phase and G2 phase nuclei yielded equal activities of a soluble DNA polymerase (Schiebel & Bamberg, 1972). It has also been demonstrated that, like nucleolar DNA, mitochondrial DNA is made throughout the cell cycle (Guttes, Hanawalt & Guttes, 1967).

Once chromosomal DNA synthesis has started there is good evidence for sequential replication during S phase (Braun, Mittermayer & Rusch, 1965). In these experiments DNA was labelled with [³H]-thymidine during the second half of one S phase. ¹⁴C-labelled BUDR (bromodeoxyuridine) was added before the next S phase. DNA was extracted at various points of that S phase and characterized by CsCl-gradient centrifugation. Newly synthesized DNA became more dense and labelled with ¹⁴C and contained no ³H label if the sample was prepared before the second half of S phase. These results were confirmed for shorter segments of the S phase (Braun & Wili, 1969).

There is evidence that sequential DNA synthesis required concomitant protein synthesis, since, in the presence of cycloheximide added during mitosis, only a limited amount (20 %) of the DNA was replicated (Cummins & Rusch, 1966). There is also an indication that the necessary protein is only made during DNA synthesis. This had been concluded from an experiment where DNA synthesis was blocked by FUDR (fluorodeoxyuridine) (Cummins & Rusch, 1966). When the blockage was reversed by thymidine after 2 h, DNA synthesis occurred again but was still significantly inhibited by cycloheximide added at the time of reversal. A similar experiment was performed with actinomycin D to demonstrate that RNA synthesis was necessary to complete an artificially prolonged S phase (Cummins, communicated in the *Physarum* Newsletter, 1968). These results indicated that during S phase concomitant protein and RNA synthesis were essential and that the resulting products were not accumulated in the absence of DNA synthesis.

It was possible therefore that proteins, other than polymerase, had to be made on RNA transcribed from 'early genes' in order to complete DNA synthesis in the right sequence. This hypothesis was supported by careful measurements of DNA synthesis in the presence of cyclo-heximide added at many points during mitosis and S phase, and by density shift experiments with BUDR. It was shown that DNA synthesis occurred in at least ten consecutive discrete steps and that protein synthesis was required for each 'round of replication' (Muldoon, Evans, Nygaard & Evans, 1971).

There were two classes of nuclear proteins that were synthesized during S phase: histone and acidic protein. No changes in the composition of a phenol-soluble fraction of acidic protein were observed during S phase (LeStourgeon & Rusch, 1971). It could be argued that histone synthesis might be a general requirement for the reconstruction of a newly replicated chromatin segment. However, cycloheximide affected DNA synthesis prior to histone synthesis (Mohberg & Rusch, 1969b). A prolonged inhibition of DNA synthesis led to a threefold increase in histone (histone/DNA ratio), but as concluded above, 'replication protein' did not accumulate in the absence of DNA synthesis. Therefore the synthesis of the main histone fractions, which did not change in the cell cycle, did not control the sequential replication. A minor fraction in the histone preparation, band number 1, showed distinct alterations early in S phase and could be involved in the regulation of DNA synthesis since this protein bound very strongly *in vitro* to DNA (Mohberg & Rusch, 1969b).

Although the evidence is merely circumstantial it can be argued that the synthesis of short-lived specific protein factors regulates the sequential DNA synthesis and restriction of the chromosomal DNA duplication to S phase. If we assume that a sequential transcription guarantees an ordered replication, this hypothesis might be extended to include sequential transcription for other events in the cell cycle, including mitosis. There is evidence for selective RNA synthesis from newly synthesized DNA as concluded from an inhibitory effect of FUDR on RNA synthesis only in S phase (Rao & Gontcharoff, 1969). Changes in base composition of RNA from different points of the cell cycle have also been claimed (Cummins, Weisfeld & Rusch, 1966; Cummins, Rusch & Evans, 1967). However, our results from DNA–RNA hybridization experiments (unpublished) have not shown any changes in the composition of RNA extracted from S phase or G2 phase plasmodia, nor has a careful analysis of RNA utilization improved DNA–RNA hybridization techniques (Zellweger & Braun, 1971).

Although direct data on sequential transcription are not available, it has been suggested that events which might be functionally related might also be regulated by related controls at the molecular level. Therefore, the well established correlation of thymidine kinase activity with S phase (Sachsenmaier & Ives, 1965) might be more than a temporal correlation and indicate a control mechanism not unlike an operon (Sachsenmaier, von Fournier & Gürtler, 1967). We have confirmed cyclic changes of thymidine kinase activity in consecutive cell cycles (A. Hildebrandt & H. W. Sauer, unpublished results), but we have also noticed a high level of enzyme activity when DNA synthesis was blocked with FUDR or hydroxyurea (HU). When the blockage of DNA synthesis was reversed DNA could be synthesized at a low level of thymidine kinase; after FUDR treatment and reversal, enzyme activity was even higher than in controls. Since both inhibitors affected DNA synthesis indirectly and could have interfered with nucleotide pools it could be that temporal correlation of increased enzyme activity with prophase and S phase was a coincidence. We have further shown that thymidylate kinase, an enzyme in the direct pathway of thymidine triphosphate (TTP) formation, did not show significant fluctuations in the cell cycle. More information will be expected from another peak enzyme, NAD pyrophosphorylase, which has been located in the nucleus and is also correlated with S phase (Solao & Shall, 1971).

Mitosis and S phase of *Physarum* have been considered as differential events of the cell cycle distinguishable from the general physiology of growth. This was indicated by steady activities of several enzymes which, if they are not made up of many isoenzymes, seemed to be synthesized at all times during the cell cycle: glucose 6-phosphate dehydrogenase (Sachsenmaier & Ives, 1965), glucose 6-phosphate dehydrogenase, isocitrate dehydrogenase, acid phosphatase, phosphodiesterase, β-glucosidase and histidase (Hüttermann, Porter & Rusch, 1970*a*) as well as hexokinase, phosphoglucomutase, glucose 6-phosphate dehydrogenase (pentose phosphate pathway); glyceraldehyde 3-phosphate dehydrogenase, pyruvate kinase and lactate dehydrogenase (glycolysis); citrate synthase (citric acid cycle); and 3-hydroxy-acyl CoA dehydrogenase (β-oxidation of fatty acids) (G. Wegener & H. W. Sauer, unpublished results).

To summarize, while there is good though indirect evidence for differential synthesis of 'initiator proteins' which are required for sequential replication of chromosomal DNA, other correlations, like the activity of thymidine kinase with S phase, might be temporal and of little significance for the cell cycle and could result from oscillations in metabolite pools.

SPHERULATION

As a response to unfavourable conditions, vegetative plasmodia of *Physarum* can acquire a dormant state in which the organism survives for years. In a process called sclerotization a plasmodium turns into a crust. In nature, both desiccation and low temperature can induce formation of sclerotia. Other factors like low pH (pH 2), high osmotic pressure (0.5 M sucrose) and sublethal concentrations of heavy metal ions also induce sclerotization (Jump, 1954). In this process much slime is produced, the cytoplasmic shuttle streaming stops and walls are formed, breaking up the plasmodium into multinucleate cells (macrocysts). In a mature sclerotium the nuclear diameter is reduced by one half. Upon rehydration the process is reversed: nuclei swell, macrocysts fuse and cytoplasmic streaming is resumed. In the laboratory, plasmodia of *Physarum* can be transformed into clusters of macrocysts after reaching the stationary growth phase (Stewart & Stewart, 1961).

Since glucose is depleted from the growth medium before spherules are observed in a shaken culture of microplasmodia, a non-nutrient salts medium has been devised for the production of spherules under controlled conditions (Daniel & Rusch, 1961). As a routine procedure, microplasmodia in the exponential growth phase are transferred to a salts medium and spherules are formed with relatively good synchrony after 24 to 36 h (Goodman, Sauer, Sauer & Rusch, 1969). The term 'spherulation' should be used to distinguish this kind of sclerotization, which occurs in liquid medium, from the sclerotization induced by desiccation. For practical purposes spherules can be dried on sterile millipore membranes and stored for years at 4°.

While in the growing plasmodium nuclear divisions occur in the absence of cell divisions, during spherulation the formation of cell walls can be studied in the absence of nuclear divisions. Therefore, cytokinesis and wall formation could be considered structural markers of a differentiation process in the life cycle of *Physarum*. It should be noted, however, that this kind of spherulation occurs during starvation and the results discussed below may be attributed to both or any one of the two processes.

Spherulation during starvation

Several ultrastructural changes have been observed during spherulation. The mechanism of cleavage of a plasmodium is thought to result from fusion of vesicles (Stewart & Stewart, 1961). The alignment of vesicles of unknown origin, their swelling and fusion during cleavage of a plasmodium after 24 h of starvation, has been confirmed (Goodman &

Rusch, 1970), and a similarity to cell plate formation in plant cell division has been suggested. After cleavage a thick wall is formed around a macrocyst which contains fibrils, 3–4 nm in diameter. At that time, after approximately 25 h of starvation, an elaboration of rough endoplasmic reticulum can be demonstrated and heterochromatic areas in the nuclei become less electron dense. At first sight these ultra-structural changes could indicate gene activation along with a differentiation step which might be suitable for biochemical analysis.

Two other morphological changes can be detected before cleavage. The number of glycogen granules decreases significantly and it has been confirmed that only about 30 % of the amount of glycogen present at the onset of starvation is left at the time of cleavage. At about 18 h of starvation, after inoculating plasmodia into salts medium, Golgi apparatus can be detected in the cytoplasm. Such structures cannot be clearly demonstrated in plasmodia in exponential growth phase. The appearance of Golgi apparatus coincides with the maximum production of slime, although some of this extracellular polysaccharide is also produced during growth. The slime has been identified as a sulphated galactose polymer (McCormick, Blomquist & Rusch, 1970a) and is identical in composition during growth and spherulation. It has also been demonstrated that ^{14}C-labelled slime cannot be metabolized by plasmodia, nor is it a precursor to the spherule wall material, a galactosamine polymer (McCormick, Blomquist & Rusch, 1970b).

We have observed a significant decrease in the protein content during spherulation and have presented some evidence from isotope dilution data that protein made during growth is degraded more rapidly than newly made protein (Sauer, Babcock & Rusch, 1970). We have also speculated on the induction of a protease early in starvation from the stimulatory effect of actinomycin D on amino acid incorporation. Further evidence for a selective increase in protein turnover during starvation has come from the effect of cycloheximide on some enzymes studied under conditions of growth and starvation (Hüttermann, Porter & Rusch, 1970a, b). Enzyme activity decreased with a half-life of approximately 6 h during growth and about twice as fast during starvation.

It is not known which proportion of the decrease in glycogen or protein is converted into increasing amounts of slime or into new molecules characteristic for spherules (galactosamine–wall) or is utilized in energy metabolism.

Experiments with cycloheximide demonstrated that protein synthesis was necessary for plasmodia to be transformed into spherules. Massive production of slime, however, could occur even when the inhibitor was added as early as the onset of starvation, therefore showing that slime production is not dependent on new enzyme synthesis (McCormick, Blomquist & Rusch, 1970a). Cycloheximide added just prior to the formation of spherules, at 20 h of starvation, still inhibited cleavage of the plasmodium and wall formation, both specific markers of spherulation (Goodman & Rusch, 1970).

Increase in activity of two enzymes, glutamate dehydrogenase (GLDH) and a phosphodiesterase, was also sensitive to cycloheximide (Hüttermann, Porter & Rusch, 1970b). These enzymes increased up to tenfold prior to spherule formation and could be products of selective gene activation as part of a spherulation programme controlled at the level of transcription. Although the functions of these enzymes during spherulation are not known it could be argued that GLDH might stimulate gluconeogenesis from amino acids which arise from the degradation of protein and might end up in new polysaccharides, while phosphodiesterase could degrade nucleic acids and provide new precursor molecules for transcription during starvation.

De novo synthesis of GLDH during spherulation has been clearly demonstrated by density shift experiments (Hüttermann, Elsevier & Eschrich, 1971). After incubation in deuterated amino acids, a new heavy band of enzyme activity was detected, following CsCl gradient centrifugation, which was sensitive to cycloheximide. Gradient centrifugation indicated two isoenzymes of phosphodiesterase during growth, one of which disappeared during spherulation (Hüttermann, 1972). Therefore, the increase in enzyme activity during spherulation reflects a very strong stimulation of one of the two isoenzymes. De novo synthesis was again proved by density shift.

Experiments to induce these enzymes by increasing the concentration of glutamic acid or of phosphate lead to negative results (Hüttermann, Porter & Rusch, 1970a). Therefore, the mechanism for the stimulation of enzyme synthesis in Physarum is probably not mediated by substrate induction via derepression.

An essential role for transcription during the spherulation process has not been established. Experiments with actinomycin D showed that there was no immediate inhibitory effect on protein synthesis during 20 h of starvation. However, shortly before spherule formation, amino acid incorporation was strongly inhibited (Sauer, Babcock & Rusch, 1970). These results have been confirmed for GLDH synthesis which was only affected by actinomycin D after approximately 20 h of starvation (Hüttermann, Elsevier & Eschrich, 1971). The increase in enzyme level, having occurred 10 h sooner, might be explained by increased translation or decreased degradation rates.

As the time of effectiveness of actinomycin D coincided with cleavage and wall production of the plasmodia, which in turn correlated with new rough endoplasmic reticulum, the effect of the inhibitor on the structural markers of spherulation were directly followed with the electron microscope (Goodman & Rusch, 1970). Cleavage and wall formation were not affected, although no rough endoplasmic reticulum was formed. Since spherules that were produced in the presence of actinomycin D had a low viability, these results indicated that under normal conditions the endoplasmic reticulum contained some RNA that was made during spherulation, but was not utilized before germination.

Actinomycin D did not block more than 80 % of RNA synthesis in any stage of the life cycle of *Physarum*. Therefore we investigated RNA after its extraction from plasmodia (Sauer, Babcock & Rusch, 1970). We observed with RNA eluted over a wide range of NaCl-concentration from MAK columns, that cytoplasmic RNA from starved plasmodia, after 30 min of labelling with [^{3}H]-uridine, contained a larger proportion of smaller molecules in the 10_ΔS region of sucrose gradients. These results could have been extraction artifacts due to an increase in degrading enzymes. Preliminary DNA–RNA hybridization competition experiments have shown that homologous unlabelled RNA in a 25-fold excess over DNA did not compete for more than 50–60 % of the labelled RNA. Under these conditions competition experiments with RNA from various points during a 24 h starvation period revealed no significant changes in the RNA composition.

Spherulation in growth medium

All data presented above were obtained when spherulation had been induced by inoculating growing plasmodia into a non-nutrient salts medium. Good biochemical evidence for *de novo* synthesis of GLDH and phosphodiesterase in response to spherulation conditions has been presented. The significance of these selective syntheses has been further investigated.

As we have reported above, many stimuli including osmotic pressure can induce sclerotization. Recently, controlled conditions have been worked out to induce spherulation in liquid defined growth medium by the addition of 0.5 M mannitol (Chet & Rusch, 1969). Under these conditions all plasmodia are transformed into spherules within 35–65 h in the growth medium. Spherules appear as single units, macrocysts, and not in clusters, which were common during induction by starvation. An explanation might be the low production of slime, which correlated with a less rapid decrease of glycogen content in the presence of mannitol. Mannitol did not block the uptake of glucose from the medium, since [^{14}C]glucose label could be traced in glycogen. Nor did the polyol induce spherulation by an osmotic pressure change alone, since mannitol added to plasmodia in salts medium did not lead to spherulation. However, as in starvation, a drastic decrease in protein and RNA concentration occurred before spherules were formed.

Therefore, there exist three distinct conditions under which the two enzyme markers could be tested: (1) starvation and spherulation (Hüttermann, Porter & Rusch, 1970b), (2) growth and spherulation, and (3) starvation and no spherulation (Hüttermann & Chet, 1971). In the first case, as has been discussed above, GLDH increased significantly and while one of the isoenzymes of phosphodiesterase was lost, the other increased. In the second case no significant increase in GLDH was observed and both isoenzymes of phosphodiesterase were present (Hüttermann, 1972). In the third case GLDH increased twofold and one isoenzyme was lost. From these results it seems clear that the biochemical markers observed during spherulation in salts medium are associated with events of starvation.

However, GLDH as such, no matter at what level, might still be involved in spherulation as could be concluded from the effect of glutaric acid. This substance was a potent but reversible inhibitor of GLDH and also inhibited spherulation reversibly (Hüttermann & Chet, 1971). Therefore, it could be argued that not the amount of enzyme, but the substrate availability, played a role in reversible differentiations of *Physarum*.

When spherulation was induced by mannitol, the production of spherules, as in starvation, could not be prevented by actinomycin D. However, only a few spherules germinated. Since germination required protein synthesis and occurred in the presence of actinomycin D (Chet & Rusch, 1970c), it could be speculated that some RNA made during spherulation was stored for long periods of time, perhaps for years.

It was concluded from an early inhibitory effect of actinomycin D on amino acid incorporation that some RNA was translated quickly after transcription (Chet & Rusch, 1969). Such was not observed when spherulation was studied during starvation. Analysis of extracted RNA confirmed the prevalence of smaller molecules as described above, but DNA–RNA hybridization data suggested differences in the composition of RNA made in the presence of mannitol (Chet & Rusch, 1970a).

Although the concentrations of RNA used were not at saturation level and only repetitive base sequences of the DNA could have reacted, large differences were detected between RNA from growing and spherulating plasmodia. The degree of hybridization of RNA fractions isolated after gel electrophoresis also revealed a different pattern of RNA during spherulation (Chet & Rusch, 1970*b*). These results could indicate a decrease of some RNA species and an addition of new species in the process of spherulation, and were consistent with a hypothesis of gene activation in differentiation. However, the possibility of only a partial extraction of RNA from *Physarum* by conventional methods (Jacobson & Holt, 1969) must be kept in mind. Further evidence for the significance of possible changes in RNA composition would come from a characterization of RNA from the three types of culture conditions mentioned above.

It must be emphasized again that many different stimuli result in a rather uniform structural process of sclerotization and it is reasonable to assume that a plasmodium can be transformed into cysts by one specific effect of heterogeneous stimuli, which interferes with the growth cycle. As was mentioned above, there is no G1 phase in the cell cycle of *Physarum*. However, spherule nuclei contain 0.6 pg of DNA, which is half the 4C amount found in vegetative plasmodia (Mohberg & Rusch, 1971). This finding correlates with the small nuclear volume of macrocyst nuclei.

One way to explain the low DNA content is to assume the introduction of a G1 phase which, once established, might determine spherule formation. It would be important to learn at which time a G1 phase appears since it could be a cause or a consequence of the metabolic changes (i.e. breakdown of protein and RNA) that occur before any kind of sclerotization.

One mitosis has been described at 6 h after a transfer of plasmodia to salts medium (Hemphill, 1962). We observed an increase in thymidine incorporation at that time and no significant DNA synthesis after 12 h of starvation (Sauer, Goodman, Babcock & Rusch, 1969). It seems, therefore, that a second mitosis followed by a G1 phase must be postulated, along with DNA duplication prior to a mitosis after germination. It can also be suggested that a G1 phase is not a general consequence of starvation since it did not occur during a 4 day starvation period which is required for sporulation (Mohberg & Rusch, 1971). It is not uncommon to assume a link between G1 phase and differentiation, and, if a G1 phase could be confirmed, we would no longer have to classify spherulation as a completely reversible differentiation process.

To summarize, spherulation describes a differentiation of micro-plasmodia of *Physarum* in liquid medium which can be induced, under controlled laboratory conditions, by osmotic shock during growth or by starvation. Cleavage of the acellular plasmodium and wall formation are selective structural markers and depend on protein synthesis. There is no evidence for selective transcription in the second kind of spherulation and the significance of changes in RNA after osmotic shock is not yet clear, although some RNA in the mature spherule seems to be very stable.

However, it can be argued that a programme for encystment exists at the molecular level of chromatin, which can be realized, once the tight correlation of mitosis and S phase is broken by sudden metabolic changes.

SPORULATION

Sporulation is an essential part of the life cycle of *Physarum* as it leads to genetic recombination during meiosis and to propagation of the organism by spores. The formation of fruiting bodies is a remarkable example of morphogenesis by cytoplasmic streaming in a uniform protoplast. Some work on the physiology of this differentiation process has been cited in the reviews on *Physarum* mentioned previously. Results obtained in axenic culture have been described in detail by Daniel (1966).

Although available information is not sufficient to propose a regu-lative mechanism of sporulation, necessary conditions for this process have been determined. It has become clear that two sets of factors are required to make a plasmodium sporulate, one coming from the substrate and the other through illumination. A delicate balance between these factors can be concluded from the influence of variable lengths of incubation with or without nutrients and the beginning and duration of illumination. Therefore, it is imperative to work under controlled conditions to achieve homogeneous material and a high incidence of sporulation. The effects of substrate and light are applied consecutively. Under the conditions specified by Daniel (1966) – i.e. inoculation of microplasmodia at the end of the exponential growth phase onto filter paper, a time of fusion of 6 h in the absence of medium, a 4 day period of starvation in sporulation medium in the dark, followed by a 4 h illumination period – sporulation resulted in over 90 % of the plasmodia with a high degree of synchrony.

To classify the events connected with sporulation one can distin-guish: (*a*) a period of starvation (in the presence of niacin) when a plasmodium becomes competent to light stimulation, (*b*) illumination

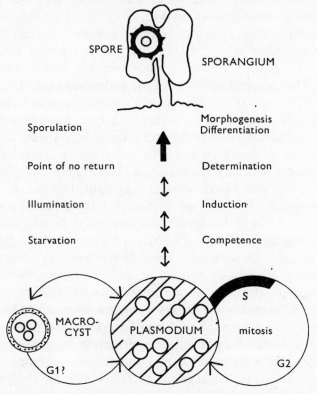

Fig. 2. Differentiations of a plasmodium of *Physarum polycephalum* (see text for details).

(induction of sporulation), (*c*) determination, the time at which a plasmodium had become committed to sporulate, (*d*) differentiation and (*e*) morphogenesis. The latter period was composed of various structural changes that have been described in detail (Guttes, Guttes & Rusch, 1961): transformation of plasmodial strands into droplets, their segregation into stalk and head, a synchronous presporangial mitosis, cleavage of the protoplast, formation of spores, spore wall, capillitium (a structure effective in spore dispersal) and melanization of the sporangia. A schematic presentation is given in Fig. 2.

Experiments during starvation have shown that nicotinic acid or related metabolites must be present in the non-nutrient medium for several days to ensure sporulation (Daniel, 1966). It could be argued that this substance, which is not required in the growth medium, channels metabolic pathways, possibly by the production of pyrimidine nucleotides, so that illumination becomes effective. Glucose overrides the effect of nicotinic acid, which may imply that sporulation is repressed under conditions of growth.

Since unpigmented *Physarum compressum* did not require illumination to sporulate (Gray, 1938), the yellow pigment of *Physarum polycephalum* should contain a photoreceptor system, as yet unidentified. The chemical structure of the yellow pigment, described as a conjugated polyene, is not known and analysis is complicated by the finding that the pigment is a mixture of several components. Careful analysis of the effectiveness of monochromatic light on the incidence of sporulation in a related species, *Physarum nudum*, revealed a broad action spectrum with two main ranges (330–540 nm and 630–713 nm) (Rakoczy, 1965) and could indicate that light does not react with a single component. However, sporulation was not induced by green light (540–620 nm). Light of that quality even inhibited the stimulation of the effective wavelengths (Rakoczy, 1967). Changes of the pigment from yellow to an olive brown colour were observed during intensive illumination during growth or starvation and have been interpreted as protective pigments for the photoreceptor system (Rakoczy, 1965).

The effect of light was not preserved when a plasmodium was kept for 10–20 h at 4° after illumination. Conditioned medium or extracts from illuminated cultures did not induce sporulation in starved plasmodia. When one half of a plasmodium was illuminated, only that half sporulated (unpublished results). It seems, therefore, that the significant effect of light is a local one that is not transported by cytoplasmic streaming, nor is it diffusible.

It is also suggestive that the state of 'no sporulation' is dominant over 'sporulation'. Some strains of *Physarum* after a long period of vegetative growth by serial transfers lose the capacity to sporulate. Recombination of these plasmodia by fusion with other plasmodia capable of sporulation did not result in sporulation (Daniel, 1966). Direct changes occurred during starvation followed by illumination in the presence of niacin (Daniel, 1966) and included a transient increase in ATP and a less significant decrease of RNA during illumination, a temporal increase of polysaccharide shortly after the light, and a doubling of the DNA content during sporulation. The pH value increased during starvation and again after illumination (pH 4 to pH 6). From these observations, changes in the metabolism of plasmodia were suggested.

Illumination inhibited respiration within 15 sec. This inhibition was reversible and could also be demonstrated in isolated mitochondria (Daniel, 1966). However, these effects were seen in growth and in starvation. Light affected the distribution of calcium in the plasmodium as was indicated by a rapid loss of mitochondrial calcium deposits. It was suggested that calcium was utilized in capillitium formation late in sporulation (Nicholls, 1972).

Availability of sulphydryl (SH) groups has been associated with preparations of the sporulation process. In the presence of NEM (*N*-ethyl-maleimide), a SH group reagent, the time required before sporulation could occur was reduced by half in *Physarum* not grown in axenic culture (Ward, 1959). In axenic culture, changes in pigmentation that occur during sporulation (photobleaching) could be mimicked by NEM in unilluminated plasmodia, growing or starved (Daniel, 1966). Pigment was confined to granules, and illumination, as well as NEM, resulted in bleaching of plasmodia as the pigment granules dissolved. In the process induced by NEM, and only in the plasmodia starved for sporulation, polyphenol oxidase became activated. This enzyme was normally found during melanization of the sporangia, a very late process during morphogenesis. However, morphogenesis or spore formation were not induced by NEM, and melanization, like changes in respiration, may not be a reliable marker of sporulation.

It could be argued that shifts in metabolic pathways, mediated by starvation, niacin and illumination, alone do not determine sporulation since there is evidence that transcription of the genome and differential protein synthesis are also part of the sporulation programme. Contrary to spherulation, sporulation is a differentiation process that is sensitive to actinomycin D.

During starvation mitoses occurred in intervals of 24–48 h. Therefore, at least one mitosis had taken place before light became effective. We observed a mitosis after 4 days of starvation, shortly before plasmodia were illuminated during routine procedure to ensure high incidence of sporulation (Sauer, Babcock & Rusch, 1969a). Thymidine incorporation was observed thereafter, and DNA synthesis seemed necessary since its inhibition by FUDR or HU inhibited sporulation. The number of nuclei did not increase during starvation and the DNA content decreased substantially over that period along with RNA and protein (Daniel, 1966; Sauer, Babcock & Rusch, 1969a). A high turnover of nuclei must have occurred, as indicated from an increase in pycnotic nuclei (Guttes, Guttes & Rusch, 1961), possibly due to a selective elimination process. The remaining nuclei contained approximately 1 pg DNA which equalled the 4C content of vegetative plasmodia (Mohberg & Rusch, 1971). From these data it could be concluded that no G1 period existed during starvation, and that light was very effective in the induction of sporulation during S phase.

The histone content of nuclei was somewhat smaller during starvation (DNA/histone ratio was 0.8 compared with 1.1 during growth) but the composition of histone fractions as analysed on acrylamide gels was unchanged (Mohberg & Rusch, 1970).

Significant differences could be detected in the acidic nuclear protein fraction (LeStourgeon & Rusch, 1971). There were at least four additional bands in preparations from starved cultures while one fraction of the growth stage was missing. Arguments have been presented which suggest that these changes may have caused an alteration of the gene activity in photosensitive plasmodia. These changes in nuclear protein composition became reversed under growth conditions within 24 h.

The RNA composition of starved plasmodia also changes and preliminary DNA–RNA hybridization experiments indicate additional RNA species as concluded from saturation and competition data (Sauer, Babcock & Rusch, 1969b). Only repetitive base sequences could have reacted under these conditions. RNA made during starvation was not immediately translated since, after two days of starvation, protein synthesis did not depend on concomitant RNA synthesis, as judged from amino acid incorporation and polysome profiles in the presence of actinomycin D (unpublished results). Whatever changes had occurred in the nuclear transcription programme during starvation, they were insufficient to induce sporulation, as further RNA synthesis must be allowed until after illumination.

In addition to the effects of light discussed above, the activity of the genome could also have been altered. There are several lines of indirect evidence for selective RNA synthesis during illumination and 2–3 h thereafter (Sauer, Babcock & Rusch, 1969a, b; Sauer & Rusch, 1969):

(a) Actinomycin D added within that period inhibited sporulation.

(b) The incorporation rate of [³H]uridine into RNA was somewhat higher than in unilluminated plasmodia. [³²P]incorporation into RNA indicated a five-fold increase during illumination; incorporation into a polyphosphate-like material increased ten-fold during that time.

(c) Cytoplasmic radioactive RNA contained an additional peak in the 10 S region of sucrose gradients.

(d) Although most cytoplasmic labelled RNA was ribosomal RNA, there was relatively less rRNA in illuminated plasmodia. They also contained a higher proportion of larger molecules in the RNA fractions isolated from polyribosomes, as analysed after sucrose gradient centrifugation of double labelled extracts.

(e) Preliminary DNA–RNA hybridization experiments had the following results: total RNA, fractionated by gradient centrifugation, before hybridization, revealed different patterns of transcription in illumination and starvation. Hybridization competition experiments showed that the competition for labelled RNA from illuminated plasmodia with non-radioactive RNA from unilluminated cultures was less complete than competition with homologous RNA (33 % *v.* 47.5 %). Hybridization saturation values with RNA from sporulating plasmodia were higher, suggesting a more heterogeneous RNA composition after illumination of a plasmodium.

If we assumed these data to be indicative of a change in the transcription programme in response to light, possibly of a selective gene activation, an irreversible state of commitment to sporulation had not yet been achieved, since refeeding could still revert an illuminated plasmodium to growth.

Refeeding a starved and illuminated plasmodium with glucose was no longer effective after a 'point of no return' (2–4 h after illumination) had been reached. Although there was no significant change in [^{14}C]glucose uptake at that time, the transfer of glucose from pinocytosis vesicles into the cytoplasm could have become blocked by an inactivation of a glucose permease as proposed by Daniel (1966). The point of commitment coincided with the time after which actinomycin D no longer inhibited sporulation, although the inhibition of RNA synthesis was equally as effective as before.

Although, after commitment to sporulation, no major RNA synthesis seemed to be required for the completion of sporulation, protein synthesis must be allowed at all stages of sporulation including morphogenesis. This was concluded from the immediate inhibitory effect of cycloheximide (Sauer, Babcock & Rusch, 1969a). Protein synthesis was not dependent on concomitant RNA synthesis, when actinomycin D was applied before illumination or after the 'point of no return'. However, a significant inhibition of amino acid incorporation resulted when RNA synthesis was blocked during the first 2 h after illumination. It could be concluded that some of the RNA, made in response to the light, became translated rather rapidly. Since the rate of amino acid incorporation in untreated plasmodia was reduced after the illumination, contrary to the labelling rate of RNA, it could be assumed that, although new protein had to be made, other proteins were no longer synthesized. Such was reflected in the polysome patterns obtained during starvation, illumination and after determination of sporulation (Jockusch, Sauer, Brown, Babcock & Rusch, 1970). As compared with starved plasmodia, we observed a general decrease in the proportion of polysomes to monosomes after illumination. After commitment to sporulation, there were fewer small polysomes (di- and trimers) and more heavy polysomes than in starved plasmodia. Therefore it could be expected that labelled proteins, extracted from plasmodia after determination of sporulation, contained fewer small molecules and more large ones.

This could be confirmed for two classes of protein extracted from illuminated plasmodia after a 5–7 h labelling period, which started at the point of no return and was completed before any morphological changes were observed (Jockusch, Sauer, Brown, Babcock & Rusch, 1970). Buffer-soluble proteins contained relatively more heavy than light molecules after incubation in [¹⁴C]protein hydrolysate. Buffer insoluble proteins, extracted with 66 % acetic acid, allowed the characterization of several bands in the electropherogram after polyacrylamide gel electrophoresis in SDS and urea. Among the small molecules (15000–20000 MW), a smaller amount and less radioactivity of some protein fractions were detected. Since histone and some acidic protein fractions of the nuclei migrated to a similar position in SDS gels, it could be concluded that, following illumination, there was a decrease in some of these proteins. However, direct analysis of histone (Mohberg & Rusch, 1970) and acidic protein (LeStourgeon & Rusch, 1971) has not revealed any differences in the stained protein bands after gel electrophoresis. Among the large polypeptide molecules (40000–80000 MW) a higher rate of incorporation was observed in three distinct bands in plasmodia which were committed to sporulate.

A selective marker of sporangia, aside from their form, is spore walls, a unique structure in the life cycle of *Physarum*. Spore walls consist of galactosamine (as in spherule walls) and contain very little protein (2 % of dry weight) and melanin (McCormick, Blomquist & Rusch, 1970b). Upon coelectrophoresis of these proteins with buffer-insoluble proteins it was found that, judged from their mobility on SDS urea gels, two of three fractions detected in spore walls closely resembled two of the three additional radioactive peaks in acid-soluble extracts (Jockusch et al. 1970).

If we put together the circumstantial evidence, in addition to metabolic changes caused by starvation, niacin and illumination, it can be argued that: (a) a set of genes was somehow activated by illumination, (b) some RNA was preferentially translated into protein which later on was selectively involved in the construction of a spore wall, and (c) the synthesis of some proteins was stopped, possibly by a dissociation of polysomes. If the small protein molecules which decrease due to illumination turn out to be regulatory factors of the chromatin, it could be suggested that a negative control mechanism for sporulation operates at the gene level.

Even if these speculations were confirmed, we would need to identify numerous and specific markers during sporulation before we could ask whether a unique sequence of events or a net of unrelated reactions underlie morphogenesis of sporangia.

While transcription of the genome seems to be clearly involved in sporulation, much more drastic alterations occur in the nuclei themselves. Pycnosis of nuclei, followed by an elimination of their DNA, has been demonstrated cytologically just prior to the presporangial synchronous mitosis (Guttes, Guttes & Rusch, 1961). Since plasmodia of all laboratory strains are heterokaryotic, this phenomenon could be indicative of a segregation and elimination of nuclei which have become unable to divide. A low turnover of nuclei has been found characteristic for all Myxomycetes (von Stosch, 1965).

Before the degradation of nuclei in a sporulating plasmodium occurs, an incorporation of thymidine has been detected in the absence of mitosis (Sauer, Babcock & Rusch, 1969a). A characterization of the labelled product has not been made, but the possibility of an altered DNA composition in an irreversibly differentiating system has been considered in this context.

To summarize, although sporulation is inducible in axenic culture, occurs in synchrony in the absence of growth and can be readily observed, we need much more information on the conditioning effects of starvation, on the induction by light and on determination. We will have to find a sufficient number of specific biochemical markers for the morphogenetic process and therefore genetic studies in *Physarum* must be included in this analysis. The results analysed in this paper are consistent with a hypothesis of selective gene activation of differentiation, as has been concluded from several recent reviews on mitosis (Duspiva, 1971), the biochemistry of differentiation (Gross, 1968) and gene activity in early embryogenesis (Davidson, 1968). However, gene alterations cannot be excluded.

HAPLOID STAGES OF *PHYSARUM*

Almost no biochemical work has been done on the reversible transformations of myxamoebae into flagellated swarm cells or into microcysts. One reason has been that these cells have to be cultured in the presence of bacteria. It has now become possible to culture amoebae under defined conditions in a very complex liquid medium containing glucose, bovine serum albumin, embryo extracts, bactopeptone and liver infusion broth (E. H. Goodman, personal communication).

Most studies on amoebae have been concerned with genetic questions. Various mating types have been detected (Dee, 1966), and a procedure has been devised by Haugli (cited by Mohberg & Rusch, 1971) for effective mutagenesis of myxamoebae. A combination of caffeine and ultraviolet light was necessary to overcome a very active repair system in *Physarum*. There has been a report on the isolation of a homothallic strain of *Physarum polycephalum* (Wheals, 1970). If meiosis in this strain could be confirmed, it would be possible to obtain homozygous diploid macroplasmodia from a single haploid clone. It would then be possible to reinvestigate the significance of nuclear turnover in myxomycetes, to select mutants for all stages of the life cycle and to begin recombination experiments which were facilitated by natural fusion of plasmodia.

As for microcyst formation, a differentiation process similar to spherulation (macrocysts), one important piece of information has come from the determination of DNA content (Mohberg & Rusch, 1971). Amoebae grown exponentially for 48 h on a lawn of killed bacteria reached stationary growth phase after 3 days and encysted after 10 days. While nuclei from vegetative plasmodia contained 1–1.2 pg DNA (the 4C content), amoebae contained 0.6 pg as would be expected for a haploid cell after meiosis. However, encysted amoebae contained 0.3 pg (1C). This result could indicate the appearance of G1 phase when a stable transition, cyst formation, occurred. This argument was strengthened by the same result discussed above, which was obtained for macrocysts (spherulation).

The progress recently made in the analysis of haploid stages of *Physarum polycephalum* – axenic culture, mutagenesis, isogenetic diploid plasmodia – will make this organism a more attractive candidate for questions of growth and differentiation.

CONCLUSIONS

After a critical look at the differentiations in *Physarum polycephalum*, it is still not yet possible to devise a model for differentiation. However, we can observe some phenomena which have been discussed in other systems that seem more simple and more complicated than acellular slime moulds.

Starvation and an increase in protein turnover occur before bacterial sporulation (Mandelstam, 1969) and fruit-body formation of a cellular slime mould, *Dictyostelium* (Ashworth, 1971). In both systems there exists a distinct sequence of morphological and biochemical events under one set of conditions. However, such correlations could be different under another set of conditions. Therefore, the significance of correlations between enzyme or substrate levels and morphogenesis is not clear and could indicate general metabolic changes due to various growth conditions and not to differentiation.

Where genetic activity is involved in establishing a sequence of biochemical changes, it has been argued whether these are regulated by sequential transcription or sequential induction by variable concentrations of metabolites. As a unique group of 'differentiation' genes is not known, it cannot be tested experimentally as to which of the two mechanisms might regulate transcription of the genome. At any rate the time of transcription does not determine the time of translation, which can be delayed over long periods of time and therefore requires additional regulation.

Although changes in intermediary metabolism have also been observed in higher organisms, it has been argued that the primary cause for stable differentiations should be found at the gene level (Gross, 1968). At that site positive or negative control factors might act and their synthesis would also have to be regulated. Production of a specific stimulator could induce transcription of 'differentiation' genes or a variation in metabolites might overcome specific repression of such a set of genes. In any case, specific factors would have to be made either before (repressors) or at differentiation (stimulators). The latter case might involve changes of the RNA polymerase as has been speculated for bacterial sporulation (Losick, Shorenstein & Sonnenshein, 1970).

There are striking, though formal, analogies between sporulation of *Physarum* and embryogenesis (i.e. gastrulation) as indicated in the terminology of Fig. 2. One way to translate classical into molecular terms would be to substitute 'determination' for selective transcription and 'differentiation' for a selective protein pattern. In so doing we have once again stressed the general hypothesis of selective gene activation for differentiation (Gross, 1968; Davidson, 1968).

We have no idea how such selectivity is brought about and embryologists take refuge in a theory of cytoplasmic factors at specific locations in the embryo. Very complicated patterns of regulations must be postulated to allow for the fact that a variety of pathways result in one stable differentiation step. This is reflected in the many unrelated treatments which cause spherulation in *Physarum* and also in the 'regulation' processes of self-organization in embryonic 'fields'. One hypothesis would make use of a feed-back of information to the chromatin once the translation of an RNA into a specific 'luxury' protein molecule has been achieved. Since the flow of information from nucleic acids to protein is thought to be uni-directional, such feed-back of differential protein synthesis could involve a transcription from RNA into DNA. A stable alteration of a segment of the nuclear DNA could result. It has been argued that reverse transcriptase might explain the amplification of ribosomal DNA in amphibian oöcytes (Crippa & Tocchini-Valentini, 1971) and the high flexibility of the immune system (Opitz, Opitz, Koch & Jachertz, 1972).

As long as the mechanism of messenger RNA formation in eukaryotes is not clear, the function of most of the DNA present in the nuclei unknown and most of the nuclear RNA is not translated, one might speculate that, in addition to variable gene activity, alterations in the DNA composition during development might cause stable irreversible differentiation. This idea does not necessarily impose new regulatory mechanisms on the already highly complex system of a eukaryotic cell. It could be that a fraction of the short-lived nuclear RNA contained variable species of such RNA which did not match the conditions forced on any 'new' RNA by the cytoplasm before it would be accepted as a 'meaningful' factor that was worth preserving in the information reservoir of a somatic cell. Chromatin elimination and DNA puffs (reviewed in Davidson, 1968) are examples of changes in the nuclei. A turnover of nuclei in a multinucleate plasmodium might be an amplification of events happening in uninucleate cells. Therefore, careful analysis of DNA in *Physarum* might contribute to an evaluation of this unorthodox speculation.

In summary, we propose that the three modes of differentiation in *Physarum* – cyclic, transient and irreversible – reflect variable degrees of three regulative mechanisms in eukaryotic cells: programmed transcription, metabolic fluctuations and gene amplification or gene alteration, and we speculate that the mechanisms underlying evolution and development are not completely unrelated.

I thank Drs Goodman, Haugli, Hüttermann and Schiebel for permission to quote from their unpublished results, and Dr Harold P. Rusch and many of my colleagues for advice and criticism. Our investigations were in part supported by the Deutsche Forschungsgemeinschaft.

REFERENCES

ALEXOPOULOS, J. (1966). Morphogenesis in the Myxomycetes. In *The Fungi, An Advanced Treatise*, ed. G. C. Ainsworth, vol. II, pp. 211–34. New York and London: Academic Press.

ASHWORTH, J. M. (1971). Cell development in the cellular slime mould *Dictyostelium discoideum*. In *Control Mechanisms of Growth and Differentiation. Symposia of the Society for Experimental Biology*, 25, 27–49.

BRAUN, C., MITTERMAYER, C. & RUSCH, H. P. (1965). Sequential temporal replication of DNA in *Physarum polycephalum*. *Proceedings at the National Academy of Sciences, U.S.A.* 53, 924–31.

BRAUN, R. & WILI, H. (1969). Time sequence of DNA replication in *Physarum polycephalum*. *Biochimica et Biophysica Acta*, 174, 246–52.

BREWER, E. N. & RUSCH, H. P. (1965). DNA synthesis by isolated nuclei of *Physarum polycephalum*. *Biochemical and Biophysical Research Communications*, 21, 235–41.

BREWER, E. N. & RUSCH, H. P. (1966). Control of DNA replication: effect of spermine on DNA polymerase activity in nuclei isolated from *Physarum polycephalum*. *Biochemical and Biophysical Research Communications*, 25, 579–84.

BREWER, E. N. & RUSCH, H. P. (1968). Effect of elevated temperature shocks on mitosis and on the initiation of DNA replication in *Physarum polycephalum*. *Experimental Cell Research*, 49, 79–86.

CHET, I. & RUSCH, H. P. (1969). Induction of spherule formation in *Physarum polycephalum* by polyols. *Journal of Bacteriology*, 100, 673–8.

CHET, I. & RUSCH, H. P. (1970*a*). RNA differences between spherulating and growing microplasmodia of *Physarum polycephalum* as revealed by sedimentation pattern and DNA–RNA hybridization. *Biochemica et Biophysica Acta*, 209, 559–68.

CHET, I. & RUSCH, H. P. (1970*b*). Differences between hybridizable RNA during growth and differentiation of *Physarum polycephalum*. *Biochimica et Biophysica Acta*, 213, 478–83.

CHET, I. & RUSCH, H. P. (1970*c*). RNA and protein synthesis during germination of spherules of *Physarum polycephalum*. *Biochimica et Biophysica Acta*, 224, 620–2.

CRIPPA, M. & TOCCHINI-VALENTINI, G. P. (1971). Synthesis of amplified DNA that codes for ribosomal RNA. *Proceedings of the National Academy of Sciences*, 68, 2769–73.

CUMMINS, J. E. (1968). Nuclear DNA replication and transcription during the cell cycle of *Physarum*. In *The Cell Cycle: Gene–Enzyme Interactions*, ed. G. Padilla, G. Whitson & I. Cameron, pp. 141–58. New York and London: Academic Press.

CUMMINS, J. E. & RUSCH, H. P. (1966). Limited DNA synthesis in the absence of protein synthesis in *Physarum polycephalum*. *Journal of Cell Biology*, 31, 577–83.

CUMMINS, J. E. & RUSCH, H. P. (1968). Natural synchrony in the slime mould *Physarum polycephalum*. *Endeavour*, 27, 124–9.

CUMMINS, J. E., RUSCH, H. P. & EVANS, T. E. (1967). Nearest neighbour frequencies and the phylogenetic origin of mitochondrial DNA in *Physarum polycephalum*. *Journal of Molecular Biology*, 23, 281–4.

CUMMINS, J. E., WEISFELD, G. H. & RUSCH, H. P. (1966). Fluctuation of ^{32}P distribution in rapidly labelled RNA during the cell cycle of *Physarum polycephalum*. *Biochimica et Biophysica Acta*, 129, 240–8.

402   H. W. SAUER

Daniel, J. W. (1966). Light-induced synchronous sporulation of a Myxomycete – the relation of initial metabolic changes to the establishment of a new cell state. In *Cell Synchrony*, ed. I. Cameron & G. M. Padilla, pp. 117–52. New York and London: Academic Press.

Daniel, J. W. & Baldwin, H. (1964). Methods of culture for plasmodial Myxomycetes. In *Methods in Cell Physiology*, vol. i, ed. D. M. Prescott, pp. 9–41. New York and London: Academic Press.

Daniel, J. W. & Rusch, H. P. (1961). The pure culture of *Physarum polycephalum* on a partially defined soluble medium. *Journal of General Microbiology*, 25, 47–59.

Davidson, E. H. (1968). *Gene Activity in Early Development*. New York and London: Academic Press.

Dee, J. (1966). Multiple alleles and other factors affecting plasmodium formation in the true slime mold *Physarum polycephalum* Schw. *Journal of Protozoology*, 13, 610–16.

Devi, V. R., Guttes, E. & Guttes, S. (1968). Effects of ultraviolet light on mitosis of *Physarum polycephalum*. *Experimental Cell Research*, 50, 589–98.

Duspiva, F. (1971). Die Mitose. *Handbuch der Allgemeinen Pathologie*, ed. H. W. Altman *et al.*, vol. 2, part ii, pp. 480–568. Berlin, Heidelberg, New York: Springer.

Goodman, E. M. & Rusch, H. P. (1970). Ultrastructural changes during spherule formation in *Physarum polycephalum*. *Journal of Ultrastructure Research*, 30, 172–83.

Goodman, E. M., Sauer, H. W., Sauer, L. & Rusch, H. P. (1969). Polyphosphate and other phosphorus compounds during growth and differentiation of *Physarum polycephalum*. *Canadian Journal of Microbiology*, 15, 1325–31.

Gray, W. D. (1938). The effect of light on the fruiting of Myxomycetes. *American Journal of Botany*, 25, 511–22.

Gray, W. D. & Alexopoulos, C. J. (1968). *Biology of Myxomycetes*. New York: The Ronald Press Co.

Gross, P. R. (1968). Biochemistry of differentiation. *Annual Review of Biochemistry*, 37, 631–60.

Guttes, E. & Guttes, S. (1964). Mitotic synchrony in the plasmodia of *Physarum polycephalum* and mitotic synchronization by coalescence of microplasmodia. In *Methods in Cell Physiology*, vol. i, ed. D. M. Prescott, pp. 43–54. New York and London: Academic Press.

Guttes, E., Devi, V. R. & Guttes, S. (1969). Synchronization of mitosis in *Physarum polycephalum* by coalescence of postmitotic and premitotic plasmodial fragments. *Experientia*, 25, 615–16.

Guttes, E., Guttes, S. & Devi, V. R. (1969). Electron microscope study of mitosis in normal and actinomycin treated nuclei of *Physarum polycephalum*. *3rd International Congress on Protozoology. Leningrad.*

Guttes, E., Guttes, S. & Rusch, H. P. (1961). Morphological observations on growth and differentiation of *Physarum polycephalum* grown in pure culture. *Developmental Biology*, 3, 588–614.

Guttes, E., Hanawalt, P. C. & Guttes, S. (1967). Mitochondrial DNA synthesis and the mitotic cycle in *Physarum polycephalum*. *Biochimica et Biophysica Acta*, 142, 181–94.

Guttes, E. & Telatnyk, M. M. (1971). Continuous nucleolar DNA synthesis after inhibition of mitosis in *Physarum polycephalum*. *Experientia*, 27, 772–4.

Guttes, S. & Guttes, E. (1968). Regulation of DNA replication in the nuclei of the slime mold *Physarum polycephalum*. *Journal of Cell Biology*, 37, 761–72.

GUTTES, S., GUTTES, E. & ELLIS, R. (1968). Electron microscope study of mitosis in *Physarum polycephalum. Journal of Ultrastructure Research*, **22**, 508–29.

HEMPHILL, M. D. (1962). Studies on a resting phase of *Physarum polycephalum* in axenic liquid cultures. M.S. dissertation. University of Wisconsin, Madison.

HERTWIG, R. (1903). Über Korrelation von Zell- und Kerngrösse und ihre Bedeutung für die geschlechtliche Differenzierung und die Teilung der Zelle. *Biologisches Zentralblatt*, **23**, 49–62.

HÜTTERMANN, A. (1972). Isoenzyme pattern and de novo synthesis of phosphodiesterase during differentiation (spherulation) in *Physarum polycephalum. Archiv für Mikrobiologie*, in press.

HÜTTERMANN, A. & CHET, I. (1971). Activity of some enzymes in *Physarum polycephalum*. III. During sporulation (differentiation) induced by mannitol. *Archiv für Mikrobiologie*, **78**, 189–92.

HÜTTERMANN, A., ELSEVIER, S. M. & ESCHRICH, W. (1971). Evidence for the *de nova* synthesis of glutamate dehydrogenase during the spherulation of *Physarum polycephalum. Archiv für Mikrobiologie*, **77**, 74–85.

HÜTTERMANN, A., PORTER, M. T. & RUSCH, H. P. (1970a). Activity of some enzymes in *Physarum polycephalum*. I. In the growing plasmodium. *Archiv für Mikrobiologie*, **74**, 90–100.

HÜTTERMANN, A., PORTER, M. T. & RUSCH, H. P. (1970b). Activity of some enzymes in *Physarum polycephalum*. II. During spherulation (differentiation). *Archiv für Mikrobiologie*, **74**, 283–91.

JACOBSON, D. N. & HOLT, C. E. (1969). The ribosomal RNA synthesis in *Physarum. Journal of Cell Biology*, **43**, 57a.

JOCKUSCH, B., BROWN, D. F. & RUSCH, H. P. (1970). Synthesis of a nuclear protein in G_2-phase. *Biochemical and Biophysical Research Communications*, **38**, 279–83.

JOCKUSCH, B., BROWN, D. F. & RUSCH, H. P. (1971). Synthesis and some properties of an actin-like nuclear protein in the slime mold *Physarum polycephalum. Journal of Bacteriology*, **108**, 705–14.

JOCKUSCH, B., SAUER, H. W., BROWN, D. F., BABCOCK, K. L. & RUSCH, H. P. (1970). Differential protein synthesis during sporulation in the slime mold *Physarum polycephalum. Journal of Bacteriology*, **103**, 356–63.

JUMP, J. A. (1954). Studies on sclerotization in *Physarum polycephalum. American Journal of Botany*, **41**, 561–67.

KESSLER, D. (1967). Nucleic acid synthesis during and after mitosis in the slime mould, *Physarum polycephalum. Experimental Cell Research*, **45**, 676–80.

KLEINIG, H. (1972). Ein Schleimpilz als Objekt der Zellbiologie: *Physarum polycephalum. Biologie in unserer Zeit*, **2**, 18–25.

LESTOURGEON, W. M. & RUSCH, H. P. (1971). Nuclear acidic protein changes during differentiation in *Physarum polycephalum. Science, New York*, **174**, 1233–6.

LOSICK, R., SHORENSTEIN, R. G. & SONNENSHEIN, A. L. (1970). Structural alteration of RNA polymerase during sporulation. *Nature, London*, **227**, 910–13.

McCORMICK, J. J., BLOMQUIST, J. C. & RUSCH, H. P. (1970a). Isolation and characterization of an extracellular polysaccharide from *Physarum polycephalum. Journal of Bacteriology*, **104**, 1110–18.

McCORMICK, J. J., BLOMQUIST, J. C. & RUSCH, H. P. (1970b). Isolation and characterization of a galactosamine wall from spores and spherules of *Physarum polycephalum. Journal of Bacteriology*, **104**, 1119–25.

MANDELSTAM, J. (1969). Regulation of bacterial spore formation. In *Microbial Growth. Symposia of the Society for General Microbiology*, **19**, 377–402.

MITTERMAYER, C., BRAUN, R. & RUSCH, H. P. (1965). The effect of actinomycin D on the timing of mitosis in *Physarum polycephalum*. *Experimental Cell Research*, **38**, 33–41.

MOHBERG, J. & RUSCH, H. P. (1969a). Growth of large plasmodia of the Myxomycete *Physarum polycephalum*. *Journal of Bacteriology*, **97**, 1411–18.

MOHBERG, J. & RUSCH, H. P. (1969b). Isolation of the nuclear histone from the Myxomycete, *Physarum polycephalum*. *Archives of Biochemistry and Biophysics*, **134**, 577–89.

MOHBERG, J. & RUSCH, H. P. (1970). Nuclear histone in *Physarum polycephalum* during growth and differentiation. *Archives of Biochemistry and Biophysics*, **138**, 418–32.

MOHBERG, J. & RUSCH, H. P. (1971). Isolation and DNA content of nuclei of *Physarum polycephalum*. *Experimental Cell Research*, **66**, 305–16.

MULDOON, J. J., EVANS, T. E., NYGAARD, O. F. & EVANS, H. H. (1971). Control of DNA replication by protein synthesis at defined times during the S period in *Physarum polycephalum*. *Biochimica et Biophysica Acta*, **247**, 310–21.

MURRAY, A. W. & BIGLER, W. N. (1970). Variation in adenylate energy charge accompanying nuclear division in *Physarum polycephalum*. *Federation Proceedings*, **29**, 910 Abs.

NICHOLLS, T. J. (1972). The effects of starvation and light on intramitochondrial granules in *Physarum polycephalum*. *Journal of Cell Science*, **10**, 1–14.

OPITZ, H. G., OPITZ, U., KOCH, G. & JACHERTZ, D. (1972). Properties of an RNA-dependent DNA polymerase from spleen cells. *Hoppe-Seyler's Zeitschrift für physiologische Chemie*, **353**, 740–1.

OPPENHEIM, A. & KATZIR, N. (1971). Advancing the onset of mitosis by cell free preparations of *Physarum polycephalum*. *Experimental Cell Research*, **68**, 224–6.

RAKOCZY, L. (1965). Action spectrum in sporulation of slime mold *Physarum nudum*. *Acta Societatis Botanicorum Poloniae*, **34**, 97–112.

RAKOCZY, L. (1967). Antagonistic action of light in sporulation of the Myxomycete *Physarum nudum*. *Acta Societatis Botanicorum Poloniae*, **36**, 153–9.

RAO, B. & GONTCHAROFF, M. (1969). Functionality of newly synthesized DNA as related to RNA synthesis during mitotic cycle in *Physarum polycephalum*. *Experimental Cell Research*, **56**, 269–74.

RUSCH, H. P. (1969). Some biochemical events in the growth cycle of *Physarum polycephalum*. *Federation Proceedings*, **28**, 1761–70.

RUSCH, H. P. (1970). Some biochemical events in the life cycle of *Physarum polycephalum*. In *Advances in Cell Biology*, vol. I, ed. D. M. Prescott, L. Goldstein & E. McConkey, pp. 297–327. New York: Appleton-Century-Crofts, Meredith Corporation.

RUSCH, H. P., SACHSENMAIER, W., BEHRENS, K. & GRUTER, C. (1966). Synchronization of mitosis by the fusion of the plasmodia of *Physarum polycephalum*. *Journal of Cell Biology*, **31**, 204–9.

SACHSENMAIER, W. & BECKER, J. E. (1965). Wirkung von Actinomycin D auf die RNS-Synthese und die synchrone Mitosetätigkeit in *Physarum polycephalum*. *Monatshefte für Chemie*, **96**, 754–65.

SACHSENMAIER, W., DÖNGES, K. H. & RUPFF, H. (1970). Advanced initiation of synchronous mitoses in *Physarum polycephalum* during UV-irradiation. *Zeitschrift für Naturforschung*, **25**, 866–71.

SACHSENMAIER, W., VON FOURNIER, D. & GÜRTLER, K. F. (1967). Periodic thymidine kinase production in synchronous plasmodia of *Physarum polycephalum*: inhibition by actinomycin and actidione. *Biochemical and Biophysical Research Communications*, **27**, 655–60.

SACHSENMAIER, W. & IVES, D. H. (1965). Periodische Änderungen der Thymidin-kinase-Aktivität im synchronen Mitosecyclus von *Physarum polycephalum*. *Biochemische Zeitschrift*, **343**, 399–406.

SAUER, H. W., BABCOCK, K. L. & RUSCH, H. P. (1969*a*). Sporulation in *Physarum polycephalum*. A model system for studies on differentiation. *Experimental Cell Research*, **57**, 319–27.

SAUER, H. W., BABCOCK, K. L. & RUSCH, H. P. (1969*b*). Changes in RNA synthesis associated with differentiation (sporulation) in *Physarum polycephalum*. *Biochimica et Biophysica Acta*, **195**, 410–21.

SAUER, H. W., BABCOCK, K. L. & RUSCH, H. P. (1970). Changes in nucleic acid and protein synthesis during starvation and spherule formation in *Physarum poly-cephalum*. *Wilhelm Roux' Archiv*, **165**, 110–24.

SAUER, H. W., GOODMAN, E. M., BABCOCK, K. L. & RUSCH, H. P. (1969). Poly-phosphate in the life cycle of *Physarum polycephalum* and its relation to RNA synthesis. *Biochimica et Biophysica Acta*, **195**, 401–9.

SAUER, H. W. & RUSCH, H. P. (1969). Sporulation bei *Physarum polycephalum*; ein Modell gen-abhängiger Differenzierung. *Zoologischer Anzeiger*, Supplement Band, **33**, 350–3.

SCHIEBEL, W. & BAMBERG, U. (1972). Charakterisierung einer löslichen DNA-polymerase aus isolierten Kernen des synchron wachsenden Myxomyceten, *Physarum polycephalum*. *Hoppe-Seyler's Zeitschrift für physiologische Chemie*, **353**, 753.

SCHIEBEL, W., CHAYKA, T. G., DEVRIES, A. & RUSCH, H. P. (1969). Decrease of protein synthesis and breakdown of polyribosomes by elevated temperature in *Physarum polycephalum*. *Biochemical and Biophysical Research Communica-tions*, **35**, 338–45.

SOLAO, P. B. & SHALL, S. (1971). Control of DNA replication in *Physarum poly-cephalum*. 1. The specific activity of NAD pyrophosphorylase in isolated nuclei during the cell cycle. *Experimental Cell Research*, **69**, 295–300.

STEWART, P. A. & STEWART, B. T. (1961). Membrane formation during sclerotiza-tion of *Physarum polycephalum* plasmodia. *Experimental Cell Research*, **23**, 471–8.

VON STOSCH, H.-A. (1965). Wachstums- und Entwicklungsphysiologie der Myxo-myceten. In *Handbuch der Pflanzenphysiologie*, vol. 15, part I, ed. W. Ruhland, pp. 641–79. Berlin, Heidelberg, New York: Springer.

WARD, J. M. (1959). Biochemical systems involved in differentiation of the fungi. In *4th Internationaler Kongress für Biochemie* (Wien, 1958), *Symposium VI, Biochemie der Morphogenese*, ed. W. J. Nickerson, pp. 33–58. London: Pergamon Press.

WHEALS, A. E. (1970). A homothallic strain of the Myxomycete *Physarum poly-cephalum*. *Genetics*, **66**, 623–33.

ZELLWEGER, A. & BRAUN, R. (1971). RNA of *Physarum*; template replication and transcription in the mitotic cycle. *Experimental Cell Research*, **65**, 424–32.

DEVELOPMENT OF THE CELLULAR SLIME MOULD *DICTYOSTELIUM DISCOIDEUM*

D. GARROD AND J. M. ASHWORTH

Department of Biochemistry, School of Biological Sciences,
University of Leicester, Leicester LE1 7RH

INTRODUCTION

Development in multicellular organisms may be usefully, if somewhat arbitrarily, divided into three separate aspects. The first of these, differentiation, involves the structural and functional specialisation of individual cells from one of a number of common basic 'stem' cells which are usually competent to differentiate in several different ways. Differentiation may be regarded as essentially an intracellular process involving the appearance within cells of certain biochemically or cyto-logically recognisable characteristics. It occurs as a consequence of the differential activation of genes whose products directly or indirectly confer these characteristics on the cell.

The other two aspects of development are related to the multicellular level of organisation in that they involve cellular interactions as opposed to processes occurring within individual cells. The first, pattern formation, is concerned with the spatial organisation of differentiation: that is how cells of different types develop in the correct spatial, temporal and proportional relationships to each other. The second, morphogenesis, refers to the development of the shape and form of the organism and its individual parts and is largely a problem of the co-ordination of cellular movements (the term morphogenesis also has meaning at the intra-cellular level of organisation but there it is concerned with molecular, rather than cellular, interactions).

Dramatic advances in our understanding of the fundamental aspects of macromolecular syntheses have revolutionised our understanding of certain fields of biology. By their very nature such advances have contri-buted more to studies of differentiation than to pattern formation and morphogenesis. A central problem of developmental biology is thus the study of the relationship between intracellular and intercellular pro-cesses. Our interest in the cellular slime mould stems from the fact that it appears to be an organism ideally suited to such studies.

The life cycle of the cellular slime mould (Fig. 1) consists of two mutually exclusive phases. Growth, DNA synthesis, increase in cell

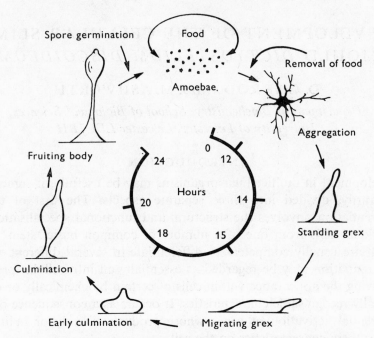

Fig. 1. Life cycle of the cellular slime mould *Dictyostelium discoideum*.
The times refer to development on Millipore filters (Sussman, 1966).

mass and virtually all cell division are restricted to the vegetative or
feeding phase whilst fruiting body construction and hence all differentiation, pattern formation and morphogenesis occur during the development phase. Development is initiated by removal of nutrients and may
be inhibited (up to the culmination stage) by the addition of nutrients.
In the vegetative phase the organism exists as unicellular amoebae which
feed on soil bacteria, but following the vegetative phase the cells aggregate forming a tissue-like assemblage of cells, the grex or pseudoplasmodium. The grex goes through a number of morphogenetic stages,
beginning with its vertical elongation into a slug shape. This slug lowers
itself on to the substratum and migrates, maintaining its elongate shape.
At the end of migration, the tip returns to a vertical position and the
rest of the cell mass rounds up. This is the stage at which fruiting body
construction begins (Fig. 1, culmination). Cells which were at the tip of
the grex become incorporated sequentially into the stalk and, as the stalk
increases in length, the cells which were at the rear of the grex ascend it,
finally differentiating into spore cells. Thus the cellular pattern of the
grex is expressed in the fruiting body. An important and significant
observation is that the pattern is size invariant, the fruiting body being

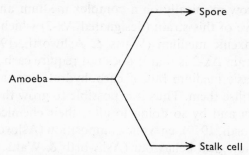

Fig. 2. Cell differentiation in the cellular slime mould.

made up of spore and stalk cells in the same relative proportions despite wide variations in absolute size (Bonner & Slifkin, 1949; Bonner, 1957; Bonner & Dodd, 1962). The spore:stalk cell ratio is approximately 2:1 but can be varied experimentally (see below; Garrod & Ashworth, 1972). Thus the cellular slime mould exhibits pattern formation and morphogenesis at their simplest, the number of cell types involved is small (Fig. 2) and the patterns impressed on the grex and expressed in the fruiting body straightforward.

Although this life cycle represents a very simple form of cell differentiation, the control mechanisms which regulate the pattern formation and morphogenesis appear to have a considerable degree of flexibility and sophistication. Thus Newell, Telser & Sussman (1969), following earlier work by Raper (1940) and Gerisch (1968), have shown that the transition from the grex to the fruiting body stage is very sensitive to environmental conditions, particularly light, pH and ionic strength. Under some conditions it is possible to omit the migrating grex stage altogether from the life cycle and under others to inhibit the progression to the culmination stage completely. Thus there exist at least three types of developmental programme: one that stops at the migrating grex, one that omits the migrating grex stage and one in which the migrating grex makes a transient appearance (Fig. 1 corresponds to this third programme, the 'lightly buffered' condition of Newell et al. 1969). Further, by appropriate changes in pH, ionic strength and illumination during development it is possible to switch a cell population from one programme to another, thus interfering with the normal temporal sequence of events. From an experimental point of view this flexibility is a great advantage.

Until recently it was impossible to grow the vegetative amoebae on anything other than bacteria. Sussman & Sussman (1967) first reported the isolation of a strain of *Dictyostelium discoideum* (designated Ax-1) which would grow axenically on a complex medium and we have isolated a derivative of this strain (designated Ax-2) which will grow in a much simpler axenic medium (Watts & Ashworth, 1970). The great advantage of strain Ax-2 is that it does not require carbohydrates to be added to the basic medium but, if carbohydrates are added, then the amoebae can utilise them. Thus it is possible to grow the amoebae in a variety of media and by so doing to alter their chemical composition (Watts & Ashworth, 1970), enzymic composition (Ashworth & Quance, 1970) and physiological behaviour (Ashworth & Watts, 1970). Despite these differences amoebae grown under different conditions will still differentiate in the same fashion and in the same time as is shown in Fig. 1. Thus the mutually exclusive nature of the two phases of the cellular slime mould life cycle and the microbial nature of the vegetative phase mean that we can induce chemical, enzymic and physiological changes in amoebae during the growth phase and examine the consequences of such changes during their subsequent differentiation. This is equivalent to examining the development of populations of cells which contain identical genomes operating in different metabolic environments. Thus we can investigate, on the one hand, the way in which metabolism and macromolecular syntheses are coupled and, on the other, the way in which metabolism and cellular behaviour interact during the various possible developmental programmes.

DIFFERENTIATION

Two recent reviews have covered this topic (Ashworth, 1971; Newell, 1971) and the reader is referred to these for a comprehensive summary of the burgeoning literature on the biochemical and molecular biological aspects of cell differentiation in this organism. We wish to discuss here what seems to be emerging as a central, if not the central, problem posed by such studies, namely how do the changes in macromolecular synthesis demonstrated during differentiation interact with, on the one hand, the changing metabolism of the cell and, on the other, with the changing pattern of cellular behaviour?

Differential gene activity is invariably associated with cell differentiation, in the slime mould (Loomis, 1969), as in other organisms. The teleological explanation for this seems clear; the differentiated cell needs novel proteins so that it may carry out novel functions and so must, during its differentiation, activate the genes regulating the synthesis of these novel proteins. This implies that gene activation must precede the acquisition of the novel function and thus raises the question of how the cell 'knows' how much activity to allow the gene in question. In the case of repressible bacterial operons the answer to this question is clear; there is a negative feed-back control by the terminal metabolite of a pathway on the activity of the structural gene controlling the synthesis of the first unique enzyme of that pathway. Similarly the activity of inducible operons is a function of the concentration of their inducers. We have studied the synthesis of a number of carbohydrates synthesised during the differentiation of *Dictyostelium discoideum* strain Ax-2 to see if such controls could operate during cell differentiation.

Trehalose [1-(α-D-glucopyranosyl-α-D-glucopyranoside)] accumulates in the spores during the latter stages of culmination (Clegg & Filosa, 1965). Preceding the appearance of trehalose the differentiating cells synthesise trehalose 6-phosphate synthase (Roth & Sussman, 1968) which catalyses the reaction:

UDP-glucose + glucose 6-phosphate = trehalose 6-phosphate + UDP. Trehalose 6-phosphate is converted by the action of a phosphatase(s) to trehalose and, since trehalose 6-phosphate does not accumulate at any stage of cell differentiation, the synthesis of trehalose 6-phosphate presumably represents the rate limiting step in the overall rate of synthesis of trehalose. The synthesis of trehalose 6-phosphate synthase is, in turn, dependent on prior synthesis of a specific RNA molecule (Roth, Ashworth & Sussman, 1968) and thus the synthesis of this enzyme reflects a specific gene activation event which is itself part of the programme of such events which occurs during the cell differentiation.

Amoebae of *Dictyostelium discoideum* strain Ax-2 grown in axenic medium containing glucose and harvested when in the stationary phase of growth have a high glycogen content (Hames, Weeks & Ashworth, 1972). When such amoebae differentiate they form four times as much trehalose as do amoebae grown in media lacking added carbohydrate and thus having initially a low glycogen content (Fig. 3). There is no difference in the time at which the two cell populations make trehalose 6-phosphate synthase, neither is there any difference between them in the amount of this enzyme that they contain (Fig. 4). Amoebae are known to contain a trehalase (Ceccarini, 1967) but this disappears during differentiation in the same time and to the same extent in both populations of cells. What does appear to change, and what thus probably controls the rate and extent of trehalose synthesis, is the level of the two substrates, UDP-glucose and glucose 6-phosphate (Fig. 5) (Hames & Ashworth, 1972). It would seem, therefore, that the amount of trehalose 6-phosphate synthase that is synthesised by the cell is not

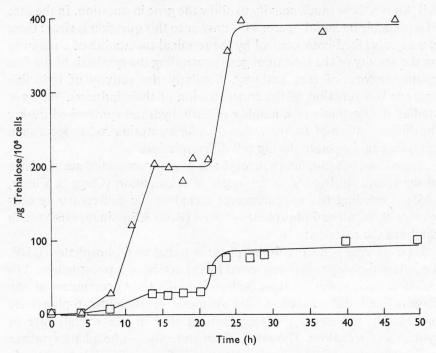

Fig. 3. Trehalose synthesis during differentiation of amoebae of *D. discoideum* Ax-2 containing initially 6.0 mg glycogen/10^8 cells ($\triangle$) or 0.3 mg glycogen/10^8 cells ($\square$).

simply related to the amount of trehalose that the cell synthesises. Comparison of Fig. 3 with Fig. 4 shows that in this case there is a much greater efficiency of utilisation of trehalose 6-phosphate synthase by cells containing a high glycogen content than there is by cells containing a low glycogen content. Similar studies that we have also done on the synthesis of phosphorylase during differentiation have revealed a situation which is apparently the converse of this.

During the differentiation of amoebae having a high glycogen content there is a dramatic fall in the cellular glycogen content whereas in cells containing little glycogen initially a period of glycogen synthesis precedes the final glycogenolysis (Hames *et al.* 1972). The rate of glycogen disappearance from cells having a high glycogen content initially seems constant and quantitatively most of the glycogen has been broken down by 20 h (Fig. 6). However, it is just at this time that there is the appearance of a novel enzyme activity – phosphorylase – whose function is glycogenolysis. Cells which had a high glycogen content initially accumulate, in what seems a curiously belated fashion, twice as much phosphorylase activity as do those that had a low glycogen

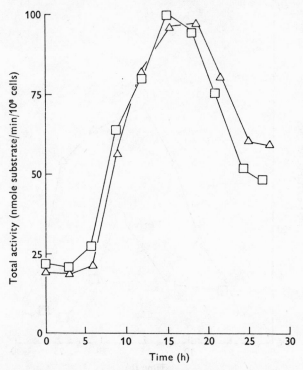

Fig. 4. Changes in trehalose 6-phosphate synthase activity during differentiation of amoebae of *D. discoideum* Ax-2 containing initially 5.8 mg glycogen/10^8 cells (△) or 0.3 mg glycogen/10^8 cells (□).

content initially (Fig. 7). It is possible to rationalise these observations in terms of two pools of glycogen, for which there is other evidence (Hames *et al*. 1972), and we are far from understanding exactly what is going on here, but a comparison of Fig. 6 with Fig. 7 shows a curious lack of correlation between amount of phosphorylase synthesised and total cellular glycogen. During the early stages of differentiation the high rate of glycogen disappearance is presumably catalysed by amylase(s) which are present in high activity in the amoebae (Weiner & Ashworth, 1970). However, despite the considerable differences in rate of glyco-genolysis (Fig. 6) observed there is no difference in amylase activity between the two cell populations – in both cases there is a decrease in total amylase activity during the first few hours of differentiation (Hames & Ashworth, 1972). This reinforces the conclusion that in these cases there appears to be no simple correlation between the amount of an enzyme that is synthesised (or degraded) during differentiation and the extent to which the cell uses (or does not use) that enzyme activity.

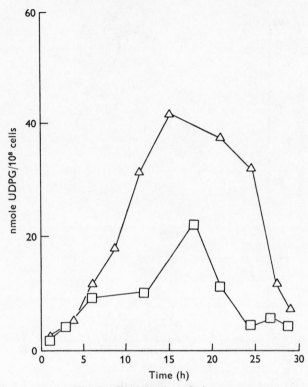

Fig. 5a. For legend see facing page.

There is no evidence, therefore, for a simple feed-back control by intermediary metabolites.

Using a totally different approach Newell & Sussman (1970) have come to essentially similar conclusions. They found that when amoebae grown on bacteria were forced to differentiate under environmental conditions which precluded culmination they synthesised UDP-glucose pyrophosphorylase. This enzyme had been previously shown to be developmentally programmed in the same sense, and using the same criteria, as trehalose 6-phosphate synthase (Ashworth & Sussman, 1967; Roth et al. 1968). However, the kinetics of accumulation of this enzyme in the migrating grex differ from that seen when differentiation proceeds according to Fig. 1. If migrating grex cells are allowed to finish accumulating UDP-glucose pyrophosphorylase and are then allowed (by alteration of the environmental conditions) to resume development they accumulate additional quantities of enzyme (Fig. 8; Newell & Sussman, 1970). We have repeated these experiments and shown that as a consequence of this shift in developmental programme there is if anything

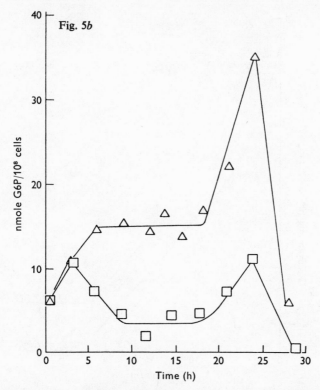

Fig. 5. (a) Changes in the cellular UDP-glucose content during the differentiation of amoebae of *D. discoideum* Ax-2 containing initially 7.8 mg glycogen/10^8 cells ($\triangle$) or 0.26 mg glycogen/10^8 cells ($\square$). (b) Changes in the cellular glucose 6-phosphate content during the differentiation of amoebae of *D. discoideum* Ax-2 containing initially 7.8 mg glycogen/10^8 cells ($\triangle$) or 0.26 mg glycogen/10^8 cells ($\square$).

a decrease in the amount of end-product saccharides formed, so that far from the enhanced enzyme levels of Fig. 8 leading to an increased flux through the UDP-glucose pool it would appear that there is either no change or a diminution in flux.

Another way of interfering with the normal development of the cells is to disaggregate the various morphological stages by trituration. If cells are disaggregated and then redeposited on Millipore filters they rapidly recapitulate the normal morphological sequence and within 2–3 h re-attain the stage at which they were disaggregated and then go on to form fruiting bodies as they would have done if left undisturbed. Newell, Longlands & Sussman (1971) have shown that as a consequence of this disaggregation a further quantity of enzyme is synthesised (Fig. 9). We have repeated these experiments and shown that as a consequence of the disaggregation (using cell aggregates formed from amoebae containing

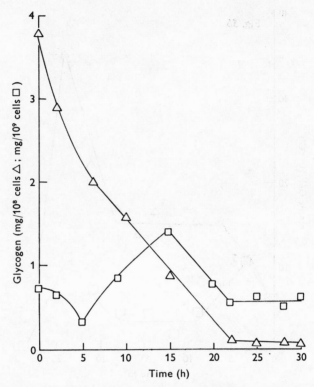

Fig. 6. Changes in the cellular content of glycogen during the differentiation of amoebae of *D. discoideum* Ax-2 containing initially 3.8 mg glycogen/10^8 cells ($\triangle$) or 0.09 mg glycogen/10^9 cells ($\square$). (Ashworth & Weiner, 1972.)

either a high or a low initial glycogen content) there is very little change in the amount of total anthrone-positive material formed. Thus, far from the increased activity of Fig. 9 involving an enhanced flux of UDP-glucose, it seems to involve either no change or a decreased flux. Newell, Franke & Sussman (1972) have extended these disaggregation studies to other enzymes and have shown that as a consequence of disaggregation enhanced quantities of trehalose 6-phosphate synthase, UDP-galactose 4-epimerase and UDP-galactose: polysaccharide transferase are synthesised as well as UDP-glucose pyrophosphorylase.

One notable feature of these experiments is that when the effects of changes in the amoebal growth pattern on development are examined quantitatively, or when the effects of changes in the developmental programme are similarly studied, there appears to be a simple numerical relationship between the changed enzyme or metabolite levels. Thus, in Fig. 3 it can be seen that trehalose synthesis occurs in phases or waves,

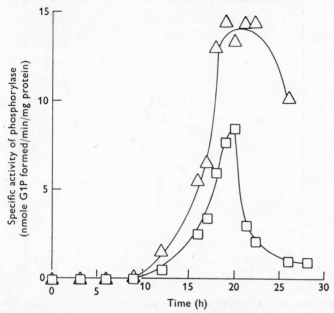

Fig. 7. Changes in specific activity of glycogen phosphorylase during the differentiation of amoebae of *D. discoideum* Ax-2 containing initially 4.9 mg glycogen/10^8 cells (△) or 0.3 mg glycogen/10^8 cells (□).

each of which is a multiple of 50 units (in fact the steps in Fig. 3 are in the ratio 1:2:4:8). Similar observations hold when enzyme-specific activities are considered. The enhanced synthesis of phosphorylase (Fig. 7) by cells derived from high-glycogen-containing amoebae involves a doubling of the specific activity. The alterations of activity reported by Sussman & Newell and their co-workers as a consequence of alterations in the developmental programme also seem to involve unit quantities or 'quanta' of activity (Sussman & Newell, 1972). This situation is very different from that in bacteria where the amount of, say, an enzyme formed is a linear function (within limits) of some parameter such as repressor or inducer concentration. In the slime mould such events seem to be discontinuous functions of, at present, unknown controlling agents. Similar simple numerical relationships have been reported before for other enzymes of the slime mould (Ashworth, 1971), for the amounts of enzymes formed by a number of tissues in inbred strains of mice (K. Paigen & J. Felton, personal communication) and Hess, Boiteux & Kruger (1969) have reported that in yeast there is a simple numerical relationship between the quantities of glycolytic enzymes formed. One does not have to be a Pythagorean to believe that such observations are

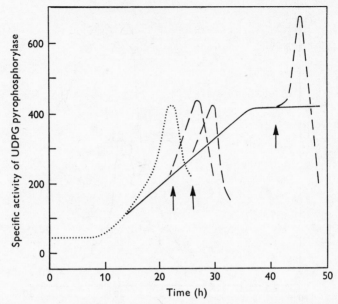

Fig. 8. Changes in specific activity of UDP-glucose pyrophosphorylase during differentiation of amoebae of *D. discoideum* NC-4 grown on *A. aerogenes*. Activity in cells developing without a migration phase (....), activity in migrating grexes (——), activity in migrating grexes transferred at the times indicated by the arrows to conditions which stop migration and initiate resumption of fruiting body formation (– – – –). (Newell, 1971.)

more than coincidence and imply some regularity in the way in which the genetic information is packaged in the chromosome, or in the way in which it is transcribed, or both. Certainly the naive, teleological argument that a differentiating cell makes a novel protein because it needs a novel function is shown to be at best only qualitatively true and is quantitatively quite unreliable.

PATTERN FORMATION

Although manifestation of the cellular pattern – the actual appearance of stalk and spore cells – occurs during fruiting body formation, there are a number of pieces of evidence which, taken together, strongly suggest that the pattern arises during the migrating grex stage of the full life cycle. The cells at the front of the grex are generally thought to become pre-stalk cells and those at the back, pre-spore cells. Some of the evidence for this view is set out in Table 1. It is quite clear, however, that the migrating grex stage is not essential for pattern formation and nor, for that matter, is the chemotactic aggregation stage. This is demonstrated by the observation that pre-aggregation cells can be

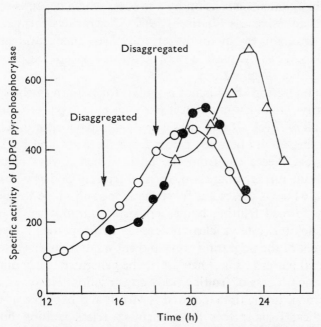

Fig. 9. Changes in specific activity of UDP-glucose pyrophosphorylase during differentiation of amoebae of *D. discoideum* NC-4 grown on *A. aerogenes*. Undisturbed controls (○), cells disaggregated at 15 h (●) or 18 h (△) and redeposited on fresh filters at the same cell density. (Newell, Franke & Sussman, 1972.)

Table 1. *Brief summary of some of the differences which have been found between cells at the front ('pre-stalk' cells) and back ('pre-spore' cells) of the migrating slime mould grex*

Difference	Reference
Behavioural: Pattern-forming ability (see text)	Raper, 1940
Histochemical: PAS staining after amylase digestion	Bonner, Chiquoine & Kolderie, 1955
Ultrastructural:	
Pre-spore vacuoles at back	Gregg & Badman, 1970; Hohl & Hamamoto, 1969
Long membrane-bounded vesicles at front	Miller, Quance & Ashworth, 1969
Narrower separation of membranes at front	Maeda & Takeuchi, 1969
Adhesiveness: To substratum of disaggregated cells	Yabuno, 1971
Biochemical: Differences in specific activity of enzymes	Newell, Ellingson & Sussman, 1969; Miller, Quance & Ashworth, 1969
Immunological: Back cells stain with fluorescent anti-spore serum	Takeuchi, 1963; Gregg, 1965

aggregated in agitated suspension by a process which depends solely on their mutual adhesiveness (Gerisch, 1960). Aggregates formed in this way, when brought to an air–water interface, change directly into fruiting bodies. Thus the pattern may be formed without either chemotactic aggregation or migration.

While the migration stage is not essential for pattern formation, it is not necessarily safe to assume that it plays no part in pattern formation during the full life cycle. Raper (1940) demonstrated a clear difference in pattern-forming ability between pieces from the front and back of the migrating grex. Pieces cut from the back, which always culminated directly without further migration, formed fruiting bodies with normal proportions, whereas pieces cut from the front and made to culminate immediately formed fruiting bodies consisting entirely of stalk. This experiment demonstrates a clear difference in properties between the front and back of the migrating grex, and shows that, in this instance at least, a normal pattern cannot necessarily be generated solely during the morphogenetic process of fruiting body construction. However, there is a clear difference in the time required for front and back pieces to form fruiting bodies; front pieces form their sporeless fruiting bodies in roughly half the time taken by back pieces to form normal fruiting bodies (D. R. Garrod, unpublished results). In order for front pieces to generate a normal pattern, a period of migration of 24 h duration before the onset of fruiting body construction was required (Raper, 1940).

At present the mechanism of pattern formation in the slime mould is a matter for considerable speculation, and several different experimental approaches seem possible. A surprising discovery made recently in our laboratory gives rise to one of these. When *Dictyostelium discoideum* is grown in the presence of glucose, the cells (G cells) differentiate to form a smaller number of larger fruiting bodies than those grown in medium lacking added glucose (N.S. cells). We have found that G fruiting bodies have a larger spore:stalk ratio (3.95:1) than N.S. fruiting bodies (2.70:1), judged by measurements of spore head and stalk volumes (Garrod & Ashworth, 1972). This difference is not related to the size difference between the fruiting bodies because the spore:stalk ratio is size invariant for both fruiting body populations. We feel that some difference in the physiological mechanism which determines proportions is involved but none of the many differences between these two populations of cells that have been discovered indicates what the crucial difference might be.

One suggestion with regard to the mechanism of pattern formation is that even during the vegetative stage there may be cells in the population with tendencies to become either spore or stalk. At a later stage in the life cycle, during aggregation or migration, these may sort out to give rise to the cellular pattern (Takeuchi, 1963, 1969; Bonner, Sieja & Hall, 1971). Sorting out between light and dense cells separated on dextrose gradients (Takeuchi, 1969) and between strains with different spore sizes (Bonner *et al.* 1971) has been obtained and it is clear that the sorting out must take place prior to, or during, the migration stage.

We have been able to demonstrate sorting out between glucose and N.S. cells, mixed during the pre-aggregation stage, by using a temperature-sensitive mutant (D. R. Garrod, C. K. Leach & J. M. Ashworth, unpublished results). The technique is to grow the wild type and the mutant in different media, to mix the cells and assay the results by assessing the ability of the resultant spores to grow at 22° and 27°. We find that G cells tend to become spores whereas N.S. cells tend to become stalks. (Wild type and mutant cells grown under the same conditions do not sort out from each other.) By cutting the tips and backs from migrating grexes, we have shown that sorting out has occurred by this stage, the glucose cells being at the back of the grex (so-called pre-spore region) and the N.S. cells at the front (so-called pre-stalk region).

So far two further experiments have been performed with this system. First, migrating slugs were allowed to form from wild-type N.S. cells and mutant G cells. By the migration stage, pre-stalk and pre-spore tendencies would be expected to have developed and a grex of one type was therefore placed next to a grex of the other type, and the two thoroughly stirred together to mix the cells (Fig. 10). The question was, when the cells re-aggregated and formed fruiting bodies, would they sort out according to their pre-stalk–pre-spore characteristics or their G–N.S. characteristics? Of the spores recovered from such mixes, 85 % were formed from glucose-grown cells, demonstrating that the cells had sorted out predominantly according to their conditions of growth rather than their (presumed) pre-stalk and pre-spore tendencies. Further, the re-aggregates formed from the mixes almost invariably changed into fruiting bodies without the intervention of a migration stage, suggesting that the latter is not *essential* for sorting out.

Secondly, the tips of migrating wild-type N.S. grexes were removed and replaced with tips of mutant G grexes. The cells in these composite grexes were thus (presumably) in the appropriate position with regard to their pre-stalk–pre-spore tendencies but not with regard to their growth conditions. Spores recovered from the resulting fruiting bodies were of the G type (Fig. 11). Thus again, sorting out took place according to growth conditions and not according to pre-stalk–pre-spore tendencies.

14

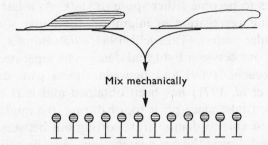

Fig. 10. Similar sized migrating grexes formed from amoebae of *D. discoideum* Ax-2 grown in media lacking added carbohydrate or from amoebae of a growth-temperature-sensitive mutant of *D. discoideum* Ax-2 grown in glucose-containing media (shaded) were dis-aggregated and numerous small fruiting bodies formed from the cell mixture; 85 % of the spores were of mutant genotype.

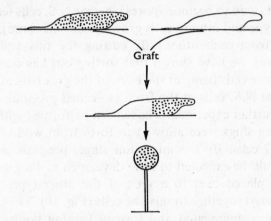

Fig. 11. The front half of a migrating grex formed from amoebae of a growth-temperature-sensitive mutant of *D. discoideum* Ax-2 grown in glucose-containing media (stippled) was grafted onto the back of a grex formed from amoebae of *D. discoideum* Ax-2 grown in media lacking added carbohydrate; 100 % of the spores in the resulting fruiting body had the mutant genotype although the cells from the front of a migrating grex are known to give rise, in undisturbed grexes, to stalk cells. (Raper, 1940.)

Further, development of composite grexes involved no aggregation stage, so it would seem that aggregation is not *essential* for sorting out.

We feel that sorting out of different cell types undoubtedly can take place during the slime mould life cycle, but are not yet convinced that it has any importance with regard to the mechanism of pattern formation. There is no evidence at present which shows that the cells sort out according to spore–stalk tendencies, but only according to density (Takeuchi, 1969), strain difference (Bonner *et al.* 1971) or growth conditions. Sorting out according to these characteristics undoubtedly gives rise to a pattern of sorts, but the stalk–spore pattern may be imposed on this sorting-out pattern by some mechanism which is different from that involved in sorting out but might equally, of course, be connected with it.

Even if sorting out according to spore–stalk tendency does occur, it seems unlikely to represent the whole story for two reasons. Firstly, as we have mentioned above, the cellular pattern is size invariant. In order to achieve this, the ratio of cells with stalk and spore tendencies would have to be fairly precisely controlled during the vegetative phase when the cells are uncommunicative, and the properties of each type entering each aggregate would also have to be controlled. Secondly, regulation of the pattern can take place at later stages in the life cycle, after sorting out according to the characteristics mentioned has taken place. Thus the cells cannot be irreversibly determined as stalk or spore even in the migrating grex. They must presumably be able to develop new tendencies even after they have sorted out according to some characteristic and after they have, as it were, exhausted their sorting-out potential. Reformation of a complete pattern from cells with only one tendency cannot involve sorting out of cells with different tendencies. It is possible that, rather than two distinct tendencies in the vegetative population, there is a range between complete spore tendency and complete stalk tendency, but in this case it is difficult to imagine how a size-invariant pattern can be achieved without some internal mechanism in the grex which at some stage determines the proportions and spatial organisation of the pattern.

The possibility that such a mechanism exists has been suggested by Bonner (1957) and Farnsworth & Wolpert (1971), but there is little indication of what this mechanism might be. It is possible that the following experiment provides a clue. Bonner (1970) showed that if vegetative cells were kept at low density on agar containing 10^{-3} M 3'5-cyclic-AMP, they could change into stalk cells within 48 h. Sometimes only a very small percentage of the cells changed, however. If inhibitors of protein synthesis (actinomycin D, puromycin or ethionine) were incorporated into the agar with cyclic-AMP, a much higher percentage of stalk cell formation could be achieved within a shorter time (24 h) and at a lower cyclic-AMP concentration (10^{-4} M). The inhibitors of protein synthesis were present at fairly low concentration, so it is impossible to be sure to what extent protein synthesis was inhibited. However, the results imply the possibility that vegetative cells do not need to synthesise new proteins in order to form stalk cells. Inhibition of protein synthesis aids stalk formation rather than hindering it. On the other hand, spore formation probably does require protein synthesis (Sussman & Sussman, 1969). It has been shown that pre-spore cells acquire specific vacuoles, recognisable under the electron micro-scope, which are not present in either vegetative cells or pre-stalk cells (Hohl & Hamamoto, 1969; Gregg & Badman, 1970). Formation of these vacuoles was inhibited by actinomycin D, in both normal aggregates and pre-stalk isolates which would normally have regulated (Gregg & Badman, 1970). Also, in very general terms, Raper's experiment (Raper, 1940) shows that pre-spore cells can change into pre-stalk cells quicker than vice versa. We may speculate that the role of cyclic-AMP in this context may be to provide a signal for stalk formation, the signal being necessary to determine the nature of differentiation.

An old observation suggests that there may be more of the chemo-tactic agent acrasin (which is, at least partially, cyclic-AMP) at the front of the migrating grex than at the back (Bonner, 1949). In other words there may be a gradient of cyclic-AMP along the grex, its high point at the tip. Since it is the front cells which form the stalk, the presence of such a gradient coupled with the observation that cyclic-AMP causes cells to become stalk would be extremely interesting in relation to pattern formation. It seems important to test for the presence or absence of a cyclic-AMP gradient in the grex, using modern techniques. Prelimi-nary experiments (A. Malkinson, D. R. Garrod & J. M. Ashworth, un-published observations) using the Gilman (1970) procedure for assaying cyclic-AMP have, however, failed to detect any differences between the front and the back of the grex. Of course, since stalk formation in *Dictyostelium discoideum* occurs after the migration phase has finished, this might not be very significant and it would be interesting to repeat these observations on a related species, *D. mucoroides*, where stalk formation actually occurs during the migration phase (Bonner, 1967).

MORPHOGENESIS

The morphogenetic changes involved in the slime mould life cycle from aggregation onwards have been described briefly in the introduction. A detailed review of the subject has been written by Shaffer (1962, 1964). The stage which has been most studied is that of aggregation itself, the main reason being that this is the stage which lends itself most readily to microscopic observation. Up to and during the aggregation process, the cells remain adherent to the substratum where they are easily and clearly observable (there is some multilayering in aggregation streams), but the grex is a three-dimensional mass in which it is difficult or impossible to discern details of the behaviour of individual cells.

There are two important aspects of the aggregation process, chemotaxis and contact interaction of cells. With the discovery that cyclic-AMP is a chemotactic agent (Konijn, Barkley, Chang & Bonner, 1968) chemotaxis in the slime mould has received a great deal of attention (for review see Bonner, 1971). In general, chemotaxis does not appear to be particularly important in the development of organisms other than the slime mould, and for this reason we propose to concentrate on the other important aspect, cell movement and contact interaction.

Firstly, there appear to be important changes in cellular adhesiveness during the pre-aggregation stage of the life cycle, i.e. the stage between cessation of feeding and the onset of chemotactic aggregation. These changes in adhesiveness were suggested by Shaffer (1957), from observations of the interaction of cells on agar surfaces. Shaffer's suggestions may be summarised as follows: (i) feeding cells have very low mutual adhesiveness (it seems preferable to use the term 'cohesiveness' in relation to cell–cell adhesions, and 'adhesiveness' in relation to cell–substratum adhesions); (ii) pre-aggregation cells are sufficiently cohesive to form clusters; (iii) aggregating cells are strongly cohesive. We have pointed out that it is not necessarily justifiable to interpret the behaviour of slime mould cells on a substratum in terms of cellular cohesiveness (Garrod & Gingell, 1970; Garrod, 1972). Because the feeding cells have been shown to exhibit mutual negative chemotaxis (Sammuel, 1961), whereas positive chemotaxis is a feature of the aggregation stage, the fact that cells either do or do not stick to each other at particular stages may be because of their chemotactic responses rather than their cohesiveness. However, experiments in which the cohesiveness of slime mould cells has been studied in agitated suspension where no chemotactic gradients exist broadly confirm Shaffer's deductions.

The first work of this nature was that of Gerisch (1960) who showed that aggregates formed in suspension by cells which had just finished feeding were smaller than those formed by aggregation-competent cells, nine hours after feeding. More recent work, in which the rate of cohesion of cells in simple salt solutions was followed, suggested that the cells increased in cohesiveness soon after the cessation of feeding (Born & Garrod, 1968). A more detailed study (Garrod, 1972) has been carried out as follows. Cells which had been grown in association with bacteria on agar plates were removed from the food supply and placed on Millipore filters. Under these conditions, chemotactic aggregation begins 8 h after cessation of feeding and measurement of the rate of adhesion of cells in phosphate buffer, at two-hour intervals throughout the pre-aggregation phase, showed: (i) that feeding cells were of very low cohesiveness, (ii) their cohesiveness increased dramatically during the first two hours after feeding, and (iii) from two hours onward up to aggregation no further increase in cohesiveness could be detected. There did not appear to be an increase in cohesiveness associated with aggregation, over and above that which occurred during the first two hours after feeding and six hours before aggregation. However, this type of experiment may not reveal the whole story. Gerisch (1960, 1965, 1968) found that cohesion of cells shortly after feeding could be inhibited with EDTA, whereas that of aggregation-competent cells was not similarly inhibited. He suggested that the EDTA-insensitive cohesion was specific for aggregation and the EDTA-sensitive cohesion was non-specific. Since Gerisch's cells were grown and maintained under different conditions from those used by Born & Garrod (1968) and Garrod (1972), it is difficult to compare the timing of the onset of these different types of cohesiveness. However, Gingell & Garrod (1969) found that the cohesion of cells which had been kept on Millipore filters for two hours after feeding was inhibited by EDTA, so that the onset of EDTA-insensitive cohesiveness presumably occurs between two hours and eight hours of the time of aggregation in cells kept on Millipore filters. It seems probable that the sequence of cohesiveness changes from the feeding stage through to aggregation is as follows: (a) feeding cells are very slightly cohesive; (b) there is a dramatic increase in cohesiveness, probably of a non-specific nature, during the first two hours after cessation of feeding; (c) between two hours and the onset of aggregation there appears an EDTA-insensitive cohesiveness which is specific for aggregation. The non-specific increase in cohesiveness has the same time course as a reduction in electrophoretic mobility of the cells, which also takes place soon after the cessation of feeding (Garrod & Gingell, 1970; Yabuno, 1970). The aggregation-specific increase is probably associated with the appearance of a new surface antigen during the pre-aggregation stage (Gerisch, 1968).

An important finding is that cohesion of aggregation competent cells can be inhibited by univalent antibody fragments produced from an anti-serum obtained with a cell homogenate (Beug, Gerisch, Kempf, Riedel & Cremer, 1970). It is possible that the antibody attaches to a cell surface site which is involved in cohesion. Although the antibody inhibited cell cohesion and adhesion to the substratum, it did not inhibit cell motility or chemotactic response.

Shaffer (1962) has suggested that the cohesion of cells, in addition to chemotaxis, is important in guiding them as they move in aggregation streams towards an aggregation centre. Cells in streams are elongated in the direction of movement and their front and back surfaces remain in contact as they move. This behaviour has been aptly called 'contact following' (Shaffer, 1962). It seems to be important in relation to the movement of cells within the grex at later stages in the life cycle, as well as during aggregation.

Essentially three possible mechanisms for contact following have been suggested. The first (Shaffer, 1964) depends upon the suggestion that the surface of slime mould cells turns over during movement. The idea is that, as the cell moves forward, new surface is produced at a surface source at the front while old surface is resorbed at a surface sink at the rear. The surface in between (i.e. on the top, bottom and lateral surfaces) remains stationary relative to the substratum. During contact following, the anterior and posterior cell surfaces adhering to other cells are stabilised and the surface sources and sinks are transferred from the extreme ends of the cells to annular regions around the terminal contacts. By determining the region of surface production and resorption, terminal contacts could control the direction of cell movement.

Shaffer's suggestion regarding surface turnover was based on the observation of particles attached to the cell surface, which remained stationary relative to the substratum (Shaffer, 1963). Particle behaviour might be an unreliable guide to surface behaviour, however, and use of fluorescent antibody to mark the surface of moving cells (Garrod & Wolpert, 1968) has given results which do not support Shaffer's conclusion. Cellular surfaces remained labelled as the cells moved through several times their own length. Shaffer's surface turnover theory predicts that the whole surface should be renewed when a cell moves *once* through its own length. Further, Beug *et al.* (1970) report that cells treated with cohesion-inhibiting antibody remain unable to cohere for about two hours, even though they are motile. From the data of Sammuel (1961) cells would be able to move through a distance of at least ten times their own length in two hours, so that the persistence of the cohesion-inhibiting antibody at the surface for this length of time is further evidence against rapid surface turnover.

How could 'contact following' be compatible with surface permanence? A possibility is that the adhesions between cell surfaces are resistant to forces which tend to break them in a direction perpendicular to the plane of contact, but not to forces parallel to the plane of contact (Garrod, 1969). In this case, end-to-end adhesions between cells in aggregation streams would be stable, resisting the pull of the cell in front which is in a direction at right angles to the plane of contact. However, the lateral surfaces of cells moving next to each other in the stream would be able to slide past each other, as they are observed to do.

Gingell (1971) has shown that, if cells adhere by a colloidal mechanism (Curtis, 1962), there is no resistance, apart from viscous drag, to prevent relative sliding of their surfaces, but the adhesion resists the surface being pulled apart (see review by Wolpert, 1971). Another possibility is that whatever holds the surfaces together is thixotropic in nature (Curtis, 1961). This again would allow lateral sliding but not perpendicular separation, but might result in lateral cell surfaces being effectively stationary because of the shearing forces generated between them. The thixotropic property might not reside in the unit membrane of the cell but in some intercellular material, so that permanence of the unit membrane might be combined with turnover of a surface coat. The mucopolysaccharide coat of *Amoeba proteus* does not turn over during movement on a substratum (Wolpert & O'Neill, 1962), but is renewed at the front end when the cell is made to crawl through a capillary tube (Jeon & Bell, 1964), possibly because shear is generated within the coat between the wall of the capillary and the advancing unit membrane (Garrod, 1969). This second idea is not incompatible with results from fluorescent antibody labelling, because surface turnover would not be expected in isolated moving cells (except possibly at the side in contact with the substratum) and the unit membrane would not be expected to turn over at any time.

A third possibility (Gerisch, 1968) is that the EDTA-insensitive contacts are localised to certain areas on the cell surfaces and it is these which are involved in end-to-end cohesion. It was found that aggregation-competent cells, immobilised with dinitrophenol and therefore spherical, formed flat aggregates in roller culture. It was suggested that this might indicate an equatorial zone of contact sites. In fact, if an equatorial zone of sites is postulated, it would seem necessary to make the further assumption to account for the formation of flat aggregates that there is some anisotropy about the sites which permits adhesion only when the equatorial planes through the contact sites are aligned. Otherwise it seems that an equatorial zone of sites could not account for the formation of flat aggregates, because the planes through the sites could be oriented at any angle to each other, resulting in the formation of a more spherical aggregate. An alternative possibility is that flat aggregates formed as a consequence of the flow pattern of liquid in the aggregation vessel, and the usual spherical form could not be achieved because the cells were immobilised. It has been suggested that cell motility may be important in increasing the area of mutual adhesion betwen slime mould cells after their initial formation (Garrod & Born, 1971).

Cellular adhesiveness to the substratum, as well as cellular cohesiveness, must be considered in relation to aggregation, even though to do so at present is almost pure speculation. As has been said, it is during the aggregation stage that the cells lose adhesion with the substratum and begin to pile up on top of each other. Aggregation streams may be multi-layers of cells, while in the aggregation centre the cells are piled up into a mound. A possible explanation of this behaviour may be sought in terms of cellular adhesiveness.

Before considering slime mould cells in particular, it is useful to point out a general similarity between the behaviour of cell populations and liquid systems (see Steinberg, 1962). A drop of liquid in a uniform environment adopts a spherical shape because of the cohesion of its molecules. The configuration of minimum surface area, a sphere, is that in which molecular cohesion is minimised. The same configuration is generally adopted by cell aggregates when they are suspended in liquid medium. The behaviour of a liquid drop on a surface depends upon whether or not its constituent molecules adhere to the surface more strongly than they cohere to each other. If they do, as with water molecules on a clean glass surface, the drop will spread over the surface. If they do not, as with water molecules on a 'non-stick' surface, the cohesion between the drop molecules will maintain its spherical shape. Similarly, if a spherical aggregate of chick embryo cells is placed on a surface to which the cells adhere, such as a plasma clot or the surface of a plastic tissue-culture dish, the spherical shape is disrupted as the aggregate adheres and the cells spread out over the substratum. However, if a similar aggregate is placed on agar, it generally maintains its spherical shape. It is also the case that if populations of single cells are placed on these surfaces, at sufficient density for them to make contact with each other, they will remain in monolayer on the adhesive surface, but round up into aggregates on agar.

With slime mould cells, Gerisch (1960) found that aggregates formed from post-feeding cells dispersed when placed on an agar substratum, whereas aggregates formed from aggregation-competent cells did not. Garrod (1972) made aggregates of cells at two-hour intervals during the pre-aggregation stage of cells kept on Millipore filters, i.e. under the same conditions used when it was found that cellular cohesiveness increased during the two hours following feeding. Only the aggregates formed from cells which had been incubated on filters for eight hours, and which had begun, or were about to begin, chemotactic aggregation, remained intact on an agar surface. Thus, the increase in cohesiveness demonstrated in these experiments was not, by itself, able to maintain the aggregates intact, nor, seemingly, to account for the piling up of cells during the aggregation process.

In principle, in order for aggregates to remain intact on a substratum, the cells must cohere to each other more strongly than they adhere to the substratum. Since the demonstrated increase in cohesiveness could not maintain the aggregates of slime mould cells, it is possible that there is either a further increase in cohesiveness during aggregation (for example, the appearance of aggregation-specific cohesions) or a decrease in cellular adhesiveness to the substratum. The experiments of Yabuno (1971) suggest, however, that there may be an increase in adhesiveness to the substratum during the pre-aggregation stage, though other interpretations of these experiments are possible.

The piling up of slime mould cells at aggregation is a clear example of loss of 'contact inhibition of overlapping' (Martz & Steinberg, 1972), a term which is preferable in certain circumstances to 'contact inhibition of movement' (see Abercrombie, 1961), because it refers strictly to the observed monolayering of cells and carries no mechanistic implications. There seem to be essentially two possible explanations of the observed monolayering of fibroblasts in tissue culture (Abercrombie, 1961). Either, contact between cells results in paralysis of their motile apparatus making it impossible for them to continue movement over the substratum in the direction of mutual contact and thus for them to climb on top of each other, or, the cells are more adhesive to the substratum than cohesive to each other. In the case of slime mould cells, contact does not result in paralysis of pseudopodal activity (Shaffer, 1962; Garrod, 1969), making it likely that an explanation of their behaviour should be sought in terms of cohesiveness and adhesiveness.

Recently we have investigated morphogenetic differences between populations of the axenic strain of *Dictyostelium discoideum* grown in the presence and absence of added glucose. Cells grown in the absence of glucose (N.S. cells) formed 2.1 times more fruiting bodies than cells grown in the presence of glucose (G cells) and allowed to differentiate under the same conditions and at the same cell density. The explanation of this difference in fruiting body number appears to be as follows. During aggregation, the N.S. aggregation streams generally break up into numerous secondary aggregation centres which originate as nodular swellings on the aggregation streams. Each secondary centre gives rise to one small grex and hence one small fruiting body. With G cells, the aggregation streams in general do not fragment into secondary centres, but the cells move into the initial aggregation centre. The initial centre then breaks up into a number of grexes which are, however, larger and fewer in number than those formed by the secondary centres of N.S. cells. A number of explanations of this difference in stream behaviour are possible, including one in terms of cellular cohesiveness and adhesiveness. From what has been said above, it can be seen that either a lower cohesiveness or a higher adhesiveness of G cells compared with N.S. cells might account for the lesser tendency of G cell streams to form swellings, since both effects would tend to make the cells less likely to round up into aggregates. However, direct experiments to test this hypothesis remain to be carried out.

CONCLUSION

The cellular slime mould *Dictyostelium discoideum* has a pattern of cell development, and thus differentiation, which appears to be sufficiently straightforward to be accessible to analysis by currently available techniques and yet sufficiently complex to be a useful 'model' for more complicated organisms. Of course Professor Bonner is right to say in his Introduction to this Symposium that 'The main point is not what one system tells you about another but what it tells you about itself' but we cannot help feeling that what *D. discoideum* has to tell us is going to be peculiarly interesting, for it bridges the gap between the unicellular and the multicellular eukaryote organisms. The microbial mode of life adopted by the amoebae allows us to use all the tricks of the biochemist/microbiologist and the multicellular nature of the developmental phase allows us to use, if not all, then at any rate some, of the tricks of the embryologist/cell biologist. Hybrid vigour is a well known biological phenomenon and if the vigorous growth of Molecular Biology is any precedent the coming together of these disciplines is going to lead to a very busy life for the slime moulds.

We thank our colleagues for permission to quote from their unpublished results, Drs Newell, Sussman and Longlands for permission to reproduce Figs. 8 and 9, and the Science Research Council for financial support of our own research.

REFERENCES

ABERCROMBIE, M. (1961). The basis of the locomotory behaviour of fibroblasts. *Experimental Cell Research*, supplement 8, 188–98.

ASHWORTH, J. M. (1971). Cell development in the cellular slime mould *Dictyostelium discoideum*. In *Control Mechanisms of Growth and Differentiation. Symposia of the Society for Experimental Biology*, 25, 27–49.

ASHWORTH, J. M. & QUANCE, J. (1970). Enzyme synthesis in myxamoebae of the cellular slime mould *Dictyostelium discoideum* during growth in axenic culture. *Biochemical Journal*, 126, 601–8.

ASHWORTH, J. M. & SUSSMAN, M. (1967). The appearance and disappearance of UDP-glucose pyrophosphorylase activity during differentiation of the cellular slime mould *Dictyostelium discoideum*. *Journal of Biological Chemistry*, **242**, 1696–700.

ASHWORTH, J. M. & WATTS, D. J. (1970). Metabolism of the cellular slime mould *Dictyostelium discoideum* grown in axenic culture. *Biochemical Journal*, **119**, 175–82.

ASHWORTH, J. M. & WEINER, E. (1972). The lysosomes of the cellular slime mould *Dictyostelium discoideum*. In *Lysosomes in Biology and Pathology*, vol. 3, ed. J. T. Dingle, pp. 36–46. Amsterdam: North-Holland.

BEUG, H., GERISCH, G., KEMPF, S., RIEDEL, V. & CREMER, G. (1970). Specific inhibition of cell contact formation in *Dictyostelium* by univalent antibodies. *Experimental Cell Research*, **63**, 147–58.

BONNER, J. T. (1949). The demonstration of acrasin in the later stages of the development of the slime mold *Dictyostelium discoideum*. *Journal of Experimental Zoology*, **110**, 259–71.

BONNER, J. T. (1957). A theory of the control of differentiation in the cellular slime molds. *Quarterly Review of Biology*, **32**, 232–46.

BONNER, J. T. (1967). *The Cellular Slime Molds*, 2nd ed. pp. 36–42. Princeton University Press.

BONNER, J. T. (1970). Induction of stalk cell differentiation by cyclic AMP in the cellular slime mold *Dictyostelium discoideum*. *Proceedings of the National Academy of Sciences, U.S.A.* **65**, 110–13.

BONNER, J. T. (1971). Aggregation and differentiation in the cellular slime molds. *Annual Review of Microbiology*, **25**, 75–92.

BONNER, J. T., CHIQUOINE, A. D. & KOLDERIE, M. Q. (1955). A histochemical study of differentiation in the cellular slime molds. *Journal of Experimental Zoology*, **130**, 133–57.

BONNER, J. T. & DODD, M. R. (1962). Evidence for gas-induced orientation in the cellular slime moulds. *Developmental Biology*, **5**, 344–61.

BONNER, J. T., SIEJA, T. W. & HALL, E. M. (1971). Further evidence for the sorting out of cells in the differentiation of the cellular slime mould *Dictyostelium discoideum*. *Journal of Embryology and Experimental Morphology*, **25**, 457–65.

BONNER, J. T. & SLIFKIN, M. K. (1949). A study of the control of differentiation: the proportions of stalk and spore cells in the slime mold *Dictyostelium discoideum*. *American Journal of Botany*, **36**, 727–34.

BORN, G. V. R. & GARROD, D. (1968). Photometric demonstration of aggregation of slime mould cells showing effects of temperature and ionic strength. *Nature, London*, **220**, 616–18.

CECCARINI, C. (1967). The biochemical relationship between trehalase and trehalose during growth and differentiation in the cellular slime mould *Dictyostelium discoideum*. *Biochimica et Biophysica Acta*, **148**, 114–24.

CLEGG, J. S. & FILOSA, M. F. (1965). Trehalose in the cellular slime mould *Dictyostelium mucoroides*. *Nature, London*, **192**, 1077–8.

CURTIS, A. S. G. (1961). Timing mechanisms in the specific adhesion of cells. *Experimental Cell Research*, supplement **8**, 107–22.

CURTIS, A. S. G. (1962). Cell contact and adhesion. *Biological Reviews*, **37**, 82–129.

FARNSWORTH, P. A. & WOLPERT, L. (1971). Absence of cell sorting out in the grex of the slime mould *Dictyostelium discoideum*. *Nature, London*, **231**, 329–30.

GARROD, D. R. (1969). The cellular basis of movement of the migrating grex of the slime mould *Dictyostelium discoideum*. *Journal of Cell Science*, **4**, 681–98.

GARROD, D. R. (1972). Acquisition of cohesiveness by slime mould cells prior to morphogenesis. *Experimental Cell Research*, **72**, 588–91.

GARROD, D. R. & ASHWORTH, J. M. (1972). Effect of growth conditions on development of the cellular slime mould, *Dictyostelium discoideum. Journal of Embryology and Experimental Morphology*, **28**, 463–79.

GARROD, D. R. & BORN, G. V. R. (1971). Effect of temperature on the mutual adhesion of pre-aggregation cells of the slime mould, *Dictyostelium discoideum. Journal of Cell Science*, **8**, 751–65.

GARROD, D. R. & GINGELL, G. (1970). A progressive change in the electrophoretic mobility of pre-aggregation cells of the slime mould *Dictyostelium discoideum. Journal of Cell Science*, **6**, 277–84.

GARROD, D. R. & WOLPERT, L. (1968). Behaviour of the cell surface during movement of pre-aggregation cells of the slime mould *Dictyostelium discoideum* studied with fluorescent antibody. *Journal of Cell Science*, **3**, 365–72.

GERISH, G. (1960). Zellfunktionen und Zellfunctionswechsel in der Entwicklung von *Dictyostelium discoideum*. I. Zellagglutination und Induction der Fruchtkörperpolarität. *Wilhelm Roux' Archiv für Entwicklungsmechanik*, **152**, 632–54.

GERISCH, G. (1965). Stadienspezifische aggregationsmuster bei *Dicyostelium discoideum. Wilhelm Roux' Archiv für Entwicklungsmechanik*, **156**, 127–44.

GERISCH, G. (1968). Cell aggregation and differentiation in *Dictyostelium*. In *Current Topics in Developmental Biology*, vol. 3, ed. A. A. Moscona & A. Monroy, pp. 157–97. New York and London: Academic Press.

GILMAN, A. G. (1970). A protein binding assay for adenine 3′:5′-cyclic monophosphate. *Proceedings of the National Academy of Sciences, U.S.A.* **67**, 305–12.

GINGELL, D. (1971). Computed force and energy of membrane interaction. *Journal of Theoretical Biology*, **30**, 121–49.

GINGELL, G. & GARROD, D. R. (1969). Effect of EDTA on the electrophoretic mobility of slime mould cells and its relationship to current theories of cell adhesion. *Nature, London*, **221**, 192–3.

GREGG, J. H. (1965). Regulation in the cellular slime moulds. *Developmental Biology*, **12**, 377–93.

GREGG, J. H. & BADMAN, W. S. (1970). Morphogenesis and ultrastructure in *Dictyostelium. Developmental Biology*, **22**, 96–111.

HAMES, B. D. & ASHWORTH, J. M. (1972). The control of trehalose synthesis during the differentiation of myxamoebae of the cellular slime mould *Dictyostelium discoideum* grown axenically. *Biochemical Journal*, in press.

HAMES, B. D., WEEKS, G. & ASHWORTH, J. M. (1972). Glycogen synthetase and the control of glycogen synthesis in the cellular slime mould *Dictyostelium discoideum* during cell aggregation. *Biochemical Journal*, **126**, 627–33.

HESS, B., BOITEUX, A. & KRUGER, J. (1969). Co-operation of glycolytic enzymes. In *Advances in Enzyme Regulation*, vol. 7, ed. G. Weber, pp. 149–67. Oxford: Pergamon Press.

HOHL, H. R. & HAMAMOTO, S. T. (1969). Ultrastructure of spore differentiation in *Dictyostelium*: the prespore vacuole. *Journal of Ultrastructure Research*, **26**, 442–53.

JEON, K. W. & BELL, L. G. E. (1964). Behaviour of cell membrane in relation to locomotion in *Amoeba proteus. Experimental Cell Research*, **33**, 531–9.

KONIJN, T. M., BARKLEY, D. S., CHANG, Y. Y. & BONNER, J. T. (1968). Cyclic AMP: a naturally occurring acrasin in the cellular slime molds. *American Naturalist*, **102**, 225–33.

LOOMIS, W. F. JR (1969). Temperature sensitive mutants of *Dictyostelium discoideum. Journal of Bacteriology*, **99**, 65–9.

MAEDA, Y. & TAKEUCHI, I. (1969). Cell differentiation and fine structures in the development of the cellular slime molds. *Development, Growth and Differentiation*, **11**, 232–45.

MARTZ, E. & STEINBERG, M. S. (1972). The role of cell–cell contact in 'contact' inhibition of cell division: a review and new evidence. *Journal of Cellular Physiology*, **79**, 189–210.

MILLER, Z. I., QUANCE, J. & ASHWORTH, J. M. (1969). Biochemical and cytological heterogeneity of the differentiating cells of the cellular slime mould *Dictyostelium discoideum*. *Biochemical Journal*, **114**, 815–18.

NEWELL, P. C. (1971). The development of the cellular slime mould *Dictyostelium discoideum*: a model system for the study of cellular differentiation. In *Essays in Biochemistry*, vol. 7, ed. P. N. Campbell & F. Dickens, pp. 87–126. New York and London: Academic Press.

NEWELL, P. C., ELLINGSON, J. S. & SUSSMAN, M. (1969). Synchrony of enzyme accumulation in a population of differentiating slime mold cells. *Biochimica et Biophysica Acta*, **177**, 610–14.

NEWELL, P. C., FRANKE, J. & SUSSMAN, M. (1972). Regulation of four functionally related enzymes during shifts in the developmental program of *Dictyostelium discoideum*. *Journal of Molecular Biology*, **63**, 373–82.

NEWELL, P. C., LONGLANDS, M. & SUSSMAN, M. (1971). The control of enzyme synthesis by cellular interaction during development of the cellular slime mold *Dictyostelium discoideum*. *Journal of Molecular Biology*, **58**, 541–4.

NEWELL, P. C. & SUSSMAN, M. (1970). Regulation of enzyme synthesis by slime mold cell assemblies embarked upon alternative developmental programs. *Journal of Molecular Biology*, **49**, 627–37.

NEWELL, P. C., TELSER, A. & SUSSMAN, M. (1969). Alternative developmental pathways, determined by environmental conditions in the cellular slime mould *Dictyostelium discoideum*. *Journal of Bacteriology*, **100**, 763–8.

RAPER, K. B. (1940). Pseudoplasmodium formation and organisation in *Dictyostelium discoideum*. *Journal of the Elisha Mitchell Scientific Society*, **56**, 241–82.

ROTH, R., ASHWORTH, J. M. & SUSSMAN, M. (1968). Periods of genetic transcription required for the synthesis of three enzymes during cellular slime mold development. *Proceedings of the National Academy of Sciences, U.S.A.* **59**, 1235–42.

ROTH, R. & SUSSMAN, M. (1968). Trehalose-6-phosphate synthetase and its regulation during slime mold development. *Journal of Biological Chemistry*, **243**, 5081–7.

SAMMUEL, E. W. (1961). Orientation and rate of locomotion of individual amoebae in the life cycle of the cellular slime mold *Dictyostelium mucoroides*. *Developmental Biology*, **3**, 317–35.

SHAFFER, B. M. (1957). Properties of slime mould amoebae of significance for aggregation. *Quarterly Journal of Microscopical Science*, **98**, 377–92.

SHAFFER, B. M. (1962). The Acrasina. *Advances in Morphogenesis*, **2**, 109–82.

SHAFFER, B. M. (1963). Behaviour of particles adhering to amoebae of the slime mould *Polysphondylium violaceum* and the fate of the cell surface during locomotion. *Experimental Cell Research*, **32**, 603–6.

SHAFFER, B. M. (1964). Intracellular movement and locomotion of cellular slime mold amoebae. In *Primitive Motile Systems in Cell Biology*, ed. R. D. Allen & N. Kaniya, pp. 387–405. New York and London: Academic Press.

STEINBERG, M. (1962). Reconstruction of tissues by dissociated cells. *Science, New York*, **141**, 401–8.

SUSSMAN, M. (1966). Biochemical and genetic methods in the study of cellular slime mold development. In *Methods in Cell Physiology*, vol. 2, ed. D. M. Prescott, pp. 397–410. New York and London: Academic Press.

SUSSMAN, M. & NEWELL, P. C. (1972). Quantal control. In *Molecular Genetics and Developmental Biology*, ed. M. Sussman. Prentice-Hall. (In press.)

Sussman, M. & Sussman, R. R. (1967). Cultivation of *Dictyostelium discoideum* in axenic medium. *Biochemical and Biophysical Research Communications*, **29**, 53–6.

Sussman, M. & Sussman, R. R. (1969). Patterns of RNA synthesis and of enzyme accumulation and disappearance during cellular slime mould cytodifferentiation. In *Microbial Growth. Symposium of the Society for General Microbiology*, **19**, 403–35.

Takeuchi, I. (1963). Immunochemical and immunohistochemical studies on the development of the cellular slime mould *Dictyostelium mucoroides*. *Developmental Biology*, **8**, 1–26.

Takeuchi, I. (1969). Establishment of polar organisation during slime mold development. In *Nucleic Acid Metabolism, Cell Differentiation and Cancer Growth*, ed. E. V. Cowdry & S. Seno, pp. 297–304. Oxford: Pergamon Press.

Watts, D. J. & Ashworth, J. M. (1970). Growth of myxamoebae of the cellular slime mould *Dictyostelium discoideum* in axenic culture. *Biochemical Journal*, **119**, 171–4.

Weiner, E. & Ashworth, J. M. (1970). The isolation and characterisation of lysosomal particles from myxamoebae of the cellular slime mould *Dictyostelium discoideum*. *Biochemical Journal*, **118**, 505–12.

Wolpert, L. (1971). Cell movement and cell contact. In *The Scientific Basis of Medicine Annual Reviews*, pp. 81–98.

Wolpert, L. & O'Neill, C. H. (1962). Dynamics of the membrane of *Amoeba proteus* studied with labelled specific antibody. *Nature, London*, **196**, 1261–6.

Yabuno, Y. Y. (1970). Changes in electronegativity of the cell surface during the development of the cellular slime mold, *Dictyostelium discoideum*. *Development, Growth and Differentiation*, **12**, 229–39.

Yabuno, Y. Y. (1971). Changes in cellular adhesiveness during the development of the slime mould *Dictyostelium discoideum*. *Development, Growth and Differentiation*, **13**, 181.

INDEX